J. E. Lagnese
Günter Leugering
E. J. P. G. Schmidt

Modeling, Analysis and Control of Dynamic Elastic Multi-Link Structures

Springer Science+Business Media, LLC

J. E. Lagnese
Department of Mathematics
Georgetown University
Washington, DC 20057, USA

Günter Leugering
Fakultät für Mathematik und Physik
University of Bayreuth
D-W 8580 Bayreuth, Germany

E. J. P. G. Schmidt
Dept. of Mathematics and Statistics
McGill University
Montreal, Canada H3A 2K6

Library of Congress Cataloging-in-Publication Data

Lagnese, J.
Modelling, analysis, and control of dynamic elastic multi-link structures / J. E. Lagnese, Günter Leugering, E. J. P. G. Schmidt.
p. cm. -- (Systems & control)
Includes bibliographical references and index.

ISBN 978-0-8176-3705-7 ISBN 978-1-4612-0273-8 (eBook)

DOI 10.1007/978-1-4612-0273-8

1. Flexible structures--Mathematical models. 2. Distributed parameter systems--Mathematical models. 3. Elastic analysis (Engineering) I. Leugering, Günter, 1953- . II. Schmidt, E. J. P. G., 1940- . III. Title. IV. Series.
TA660.F53L34 1994 94-3096
624.1'7--dc20 CIP

Printed on acid-free paper

Originally published by Birkhäuser Boston in 1994

ISBN 978-0-8176-3705-7

Typeset by the authors.

9 8 7 6 5 4 3 2 1

Contents

List of Figures

Preface

The purpose of this monograph is threefold. First, mathematical models of the transient behavior of some or all of the state variables describing the motion of multiple-link *flexible structures* will be developed. The structures which we have in mind consist of finitely many interconnected flexible elements such as strings, beams, plates and shells or combinations thereof and are representative of trusses, frames, robot arms, solar panels, antennae, deformable mirrors, etc., currently in use. For example, a typical subsystem found in almost all aircraft and space vehicles consists of beam, plate and/or shell elements attached to each other in a rigid or flexible manner. Due to limitations on their weights, the elements themselves must be highly flexible, and due to limitations on their initial configuration (i.e., before deployment), those aggregates often have to contain several links so that the substructure may be unfolded or telescoped once it is deployed. The point of view we wish to adopt is that in order to understand completely the dynamic response of a complex elastic structure it is not sufficient to consider only its global motion but also necessary to take into account the flexibility of individual elements and the interaction and transmission of elastic effects such as bending, torsion and axial deformations at junctions where members are connected to each other. The second object of this book is to provide rigorous mathematical analyses of the resulting models. Its third purpose is to develop control-theoretic properties of multiple-link (multi-link) flexible structures based on the control-theoretic properties of the models.

To be more specific, *distributed parameter models* of multi-link flexible structures will be derived. While there is already a vast literature concerning various aspects of modeling, control and stabilization of multi-body systems, it deals almost exclusively with rigid rather than flexible elements, with finite-dimensional modal (hence global) modeling rather than adopting a spatially dependent (local) approach through distributed parameter modeling. As a result, the transmission and conversion of bending, torsion, and axial deformation at the joints is usually not taken into account. Recent engineering research strongly indicates, however, that it is important to account for such phenomena in order to understand completely transmission effects in trusses and frames and in other aggregates of this kind. The starting points in our derivations of models of the structures considered in this work are the *principles of continuum mechanics.* From these are obtained, under reasonable *constraints on the geometry of the admissible deformations*, a set of *coupled partial differential equations* governing the evolution of the displacements and other *internal variables* of the structure. The models will account for full three dimensional transient behavior, various possible joint interactions, couplings between flexible motions and the dynamics of joints and/or actuators, superimposed rigid rotations, viscous and thermal effects, initial curvature and other effects.

Once the models are developed, the next objective is to establish that they comprise well-posed mathematical problems. It at this point that the various possible choices of controls will find their mathematical formulations. There are normally three different locations within a multi-link body where controls can reasonably be implemented: (a) base points, where the substructure (as a whole) is attached to a massive main body, (b) joints (nodes) within the structure, and (c) endpoints, or tips of the structure. The types of control instruments that can be employed depend very much on where the controllers are to be located. For example, at base points one can usually only control the strains, whereas at endpoints one can influence the stresses, shear forces and moments. The controllers employed at internal joints usually perform a rigid rotation, or amount to dry friction. The object of the control action is to change the system response in some specified way. Our emphasis is on the controllability of the *full infinite dimensional system.* To this end we shall employ methods from *control theory of distributed parameter systems* to formulate and solve some specific control problems which are motivated by typical control objectives for elastic structures, such as stabilizabilizing or exactly controlling the transient motion of the structure. The first control objective is related to he problem of maintaining the stability of a structure under various disturbances. The second control objective is concerned with the configuration of the structure and amounts to describing its complete *configuration space.* In any consideration of controllability or stabilizability it is necessary to distinguish between controls at base points, at interior joints and at tips, as indicated above. In particular, the control of structures through application of control mechanisms at joints may be viewed as a particular issue within the general question of appropriate joint conditions as such, and represents a major research topic in itself. It is clear that a distinction between various applications has to be made: in contrast to a robot arm (or a satellite arm unfolding a solar panel), in a truss one does not consider any large rotations in the joints. However, a global rigid rotation of the overall structure may very well be applied in both cases. There are many different possibilities which, because of space limitations, cannot all be considered here. When dealing with the problem of stabilization and, in particular, with robustness one has to specify classes of disturbances which need to be counterbalanced within the structure. The classical viewpoint is that regardless of the nature of the disturbance, the result of any disturbance is a new set of initial conditions distinct from the equilibrium state. Therefore, a primary goal is to guarantee uniform energy decay of all possible trajectories starting near the equilibrium.

We shall also provide, in Chapter V, a number of *numerical simulations* of structures composed of one dimensional elements. The purpose of the simulations is to validate the models and to illustrate the control concepts and their effectiveness. We believe that these simulations strongly support the efficacy of the models developed in this monograph, although much work along these lines, especially with regard to structures containing two dimensional elements, remains to be done. In particular, a rigorous numerical analysis of the transient behavior of elastic structures and their associated control systems based on the models presented in this book is beyond the scope of this work.

It is our hope that this book will prove to be useful not only to engineers and mathematicians interested in design, modeling and control of flexible structures, but also to those wishing to develop some understanding of the sorts of complicated interactions between elements that can and do occur in complex elastic systems. While a substantial knowledge of partial differential equations and functional analysis is necessary for a complete understanding of the details of the mathematical analyses, we believe that this book will also be useful to those who simply wish to understand the dynamics of elastic networks at a descriptive level. To this end we have tried, in particular, to present a self-contained treatment of the mechanical modeling aspects of the research described in this book.

The authors wish to convey their appreciation to those organizations whose financial support made possible the research which led to this work. The research of the first two authors was supported by the Air Force of Scientific Research. The research of the second author was also supported by the Deutsche Forschungsgemeinschaft (DFG), Heisenbergreferat. The research of the third author was supported by grants from the Natural Sciences and Engineering Research Council of Canada and from the Ministère de l'Education du Quebec.

CHAPTER I

Introduction

1. General Overview

The need to control multi-body structures is evident and does not require further justification. What may be necessary to justify however, is the relevance of *distributed parameter models* and their control properties in such a context. Indeed, ultimately every *practically implemented* control law has to be computed in one sense or another from measurements and data taken from the structure via a *finite dimensional* model of the structure. It is, however, well known that controls which are computed from a reduced order model may lead - and in many cases do lead - to the *spillover* of energy into unmodeled higher modes of the system. This is one reason for studying control problems for flexible structures in the context of distributed systems: control results for highly flexible structures obtained without taking into account the infinite dimensional nature of the system are, in principle, questionable.

Another, even more important reason for distributed parameter system (DPS) modeling and control of flexible structures stems from the fact that such structures are often deployed in hostile environments in the sense that the location may be difficult to access (space-structures), that some of the states are hard to observe and to difficult to control, that the physical parameters are subject to drastic changes (e.g., the temperature), and so forth. Often the system cannot be adequately tested under its expected working conditions. All of these considerations make *precise mechanical and mathematical modeling* of such structures mandatory. There are numerous examples of situations where poor modeling has led to catastrophic results, even in the presence of superb control techniques. We only need mention EXPLORER 1, launched in 1958, which began to tumble after a few orbits because of unmodeled internal damping stemming from flexible antennae attached to the main body, an effect which was not accounted for in the control device. Another famous example is provided by MARINER

10 (1973). In that case the control regime activated the 7th frequency of oscillation, which was not considered in the model. See Bremer [9]. A third example is provided by the HUBBLE telescope (1989) which has not operated as anticipated, partly because the abrupt and large changes in temperature when the spacecraft passes from the earth's shadow into sunlight cause tremendous thermal stresses in the beams carrying the solar panels. Those thermal stresses are converted into long bending waves, which ultimately cause some panels to flip. Examples of this kind are by no means restricted to ambitious projects in outer space. Consider, for instance, a simple system consisting of a beam with a mass attached to it, rotating around its centerline. The combined objects can be viewed as a rigid body as long as the angular velocity is small. However for large angular velocities the system behaves like a deformable elastic body. This is even more apparent if the axis of rotation is perpendicular to the centerline, that is, if one considers robot arms, crane mechanisms, etc. In all of these examples it is clear that a thorough mechanical and mathematical modeling of the system *and controls* is indispensable. This monograph is intended precisely to contribute to this aspect of flexible structure control.

We do not claim that the approach to modeling which we take here is *conceptually* new from a mechanical viewpoint. (However, we do not know of any similar work on the mechanical modeling of interconnected membranes and plates). Nor do we suggest that our approach is superior to others as far as practical implementations of control stragegies are concerned. Rather, in this work we wish to emphasis the *transient behavior of solutions*, the *interaction phenomena at mechanical junctions*, the *occurrence and propagation of singularities through junctions regions* and to study *questions of well-posedness and stability*, *spectral properties* and *exact, spectral or approximate controllability* of the transients. In these respects this monograph is almost disjoint from the vast literature on flexible structure control. Indeed, the sparsity of *mathematical literature* on interaction problems of the kind considered here is very surprising. In most papers on the subject, even after some mechanical justification of balance laws the system is represented by a *finite element approximation* or a *modal reduction* in terms of generalized coordinates, which necessarily have finitely many degrees of freedom only. We may cite Hughes [39], Kane et al. [41], Shabana [99] as exceptional examples, where the modeling is based on principles of continuum mechanics and where the relation between the distributed model and its finite dimensional approximation is kept transparent. In most engineering papers, however, multi-link structures are *introduced* through their finite dimensional representations rather than through a system of partial differential equations or integrodifferential equations with appropriate junction and boundary conditions, as it is done in this monograph.

It is well known that many phenomena and distinctions which are present in (linear) infinite dimensional systems are lost in its finite dimensional approximations. Examples are the distinctions which exist in infinite dimensional systems between various notions of stability (weak, strong, uniform) and different concepts of controllability (spectral, approximate, exact). In the context of an elastic frame containing closed circuits, the role of *comeasurability of the optical lengths* of the structural elements in a circuit, which is crucial to understanding the lack of (even weak) stabilizability and (even approximate) controllability of such a frame, disappears in finite dimensional approximations of the frame. In fact, it is the fine structure of the *frequency spectrum and its asymptotic behavior* which can be viewed as being responsible for these differences. Because of the tradition and tendency in the engineering literature to analyze finite dimensional approximations only, the fine structure of the spectrum of multi-link structures is generally not considered (see, however, Wittrick and Williams [110], [111]). In the mathematical literature, exceptions are to found in the work of von Below [106], [105] and Nicaise [75], [77], who consider the spectrum of second order differential operators acting on scalar functions on 1-d networks.

We do not have sufficient space to give appropriate credit to the huge *engineering* literature related to control of flexible structures. We simply cite [39], [99] and [41] as general references and refer the reader to their excellent bibliographies. Rather, we concentrate on the *mathematical* literature concerned with the problems to be addressed in this monograph. One may identify two distinct groups of papers within that literature. The first group considers multi-link flexible structures as being obtained from 3-d elasticity. In this philosophy a frame (with rigid joints) is obtained from a cube of dense material by "carving" out appropriate parts. The approach is made mathematically rigorous by a singular perturbation (asymptotic) analysis with respect to a certain "thickness" parameter using ideas due to Ph. Ciarlet and Dstuynder [18]. It has subsequently been used by many authors to obtain models of elastic junctions for a variety of multi-link structures. (See in particular the monographs of Ciarlet [19] and LeDret [57] and references therein.) This method is very powerful and universal. It has, however, some shortcoming when it comes to dynamical problems. In particular, as a consequence of the scaling of the Lamé moduli with respect to the asymptotic parameter, it occurs that the in-plane (or longitudinal) stiffness of the material becomes infinite in the limit of the asymptotic parameter . This being the case, the interaction conditions at the junctions, in particular, the continuity condition on the *vector displacements* of all of the elements meeting at the junction, can be violated. Therefore, from our perspective, the transmission of waves at the joints is not as transparent as in the models developed here. In some important cases, however, the junction conditions obtained here can be specialized to the ones obtained within the framework of asymptotic analysis.

The second group of papers is concerned with constructive approaches to modeling interconnected flexible bodies. This is the approach which will be followed in this work. One begins by modeling single elements (see Wempner [108], Danielson and Hodges [22], Kane et. al. [41] and many others) and then considers various possible connections between such elements such as rigid, revolutive, prismatic, spherical or pinned (in 1-d networks). Such considerations lead to various *continuity conditions*, sometimes called *compatibility conditions* or *geometric junction conditions*, between elements sharing a joint and to *dynamic transmission conditions*, also called *dynamic junction conditions*. The latter conditions result from *energy principles of mechanics*. The physical interpretations of the dynamic conditions are as *balance laws for forces and moments at the joints*. The energy principles are to be used within the context of appropriate *energy spaces*, also known as *ramification spaces* (see Lumer [66]). This approach has been used by Schmidt [93], Schmidt and Leugering [58], Lagnese, Leugering and Schmidt [49], [50], [40], Lagnese and Leugering [48] and Lagnese [46], Schmidt and Ming [95], Puel and Zuazua [85] for networks of strings, beams, membranes, plates and combinations thereof. In the context of diffusion problems, reaction-diffusion problems, nerve-pulse propagation problems, and scalar wave propagation on networks (out-of-plane displacement), this approach has been used by Nicaise [75], [77], Ali-Mehmeti [67], [69], Ali-Mehmeti and Nicaise [70], von Below [106], [105] and others.

As for numerical simulations of structures based on the types of models discussed above, we may refer to [19] for simulations utilizing models obtained from 3-d elasticity via asymptotic analysis and to [39] for a discussion of simulations of models obtained via energy principles. In [39], however, only the general outline is given and no particular network is simulated. Here we shall utilize *finite difference*, *finite element* and *implicit Runge-Kutta techniques* in simulations of both uncontrolled and controlled structures. Finite differences are used for simple one-dimensional models involving absorbing boundary conditions; specifically, we use the Lax-Friedrichs scheme, see Le Roux [91]. For more complex structures, in particular for networks of Timoshenko beams, a finite element implementation along the lines of [39] is employed. Finally, for problems involving nonsmooth damping (such as dry friction), a fully implicit Runge-Kutta method such as Radau IIa in its radau5-implementation of Hairer and Wanner [33] is utilized; see also [8]. Other current research in this area is to be found in the work of Lubich [64], [65]. The latter is in the spirit of [99], where the geometric joint conditions are treated as kinematic constraints. This approach leads to the occurrence of generalized forces and Lagrange multipliers in the state equations and has the advantage of leading to simply structured stiffness matrices even in the case of structures containing closed circuits of elements.

To resolve the question of well-posedness of the mathematical models, which is implicit in the Ciarlet-Dystunder approach, we note that we consider here mainly models obtained by linearization around an equilibrium of the structure. Then well-posedness is easily settled by using *hyperbolic estimates* (for 1-d networks of strings and beams) as in [40], or *energy estimates* obtained from standard *energy multipliers*, followed by generic semi-group arguments. While in the 1-d networks no singularities develop at the junctions, the situation is much more complicated in 2-d and higher dimensional cases. Indeed, the regularity issues in those cases are for the most part unsettled at this point. There is substantial work in this direction by Nicaise [78], [81] for *scalar* functions on 2-d networks. We note that Ali-Mehmeti [67], [69], von Below [106], [105] and others have obtained well-posedness of some quasilinear scalar multi-domain models. Again, well-posedness of the quasilinear problems describing the three dimensional, or even planar, motion of (even one-dimensional) networks appear to be open problems.

Once appropriate models are developed and analyzed, control problems for the models can be formulated in a mathematically meaningful way. As mentioned above, we will concentrate on basic control concepts such as exact, spectral or approximate controllability, or on stabilizability. In the case of exact (null) controllability, one looks for a finite time T and a set of admissible controls, which are to be applied at junctions or "boundaries" of the structure, such that for each set of initial data (with certain regularity) there is a control which steers the corresponding solution of the system to rest. This turns out to be equivalent to the problem of describing the entire space of reachable configurations of the structure, if the structure is initially in an equilibrium state . The notion of spectral controllability involves the possibility of exactly controlling the span of any set of *finitely* many of the eigenmodes of the underlying system operator. Approximate controllability, finally, is the most natural weakening of the concept of exact controllability. In this case the goal is to reach an arbitrarily small neighborhood of a desired target without requiring that the target be hit exactly. Questions of uniform or nonuniform stabilizability are known to be intimately related to the controllability concepts just mentioned. Control theory of distributed parameter systems has become a rich field of research in the last three decades. Readers who are not familiar with the concepts of control of distributed parameter systems are referred to Curtain and Pritchard [21] for an elementary introduction, to Russell [92] for and excellent review of methods until 1978, and to Lions [61] for a description of the *Hilbert Uniqueness Method* and its application to exact controllability problems for distributed parameter systems. As for the vast literature treating controllability/stabilizability problems for single elastic elements such as strings, beams, membranes and plates, we refer only to the monographs by Lagnese and Lions [51], Lagnese [52], Lions [60] and the bibliographies include therein.

Controllability problems for nonlinear dynamic networks are not investigated in this monograph. Let us note, however, that through an appropriate application of the implicit function theorem, generically there is local controllability in the nonlinear (semi-linear) situation as soon as the corresponding global results for the linear approximation have been established. We do, on the other hand, present some stabilizability results for linear dynamic models of 1-d networks to which are applied nonlinear feedback controls at some of the nodes of the network.

Although optimal control problems, other than minimum norm control reachability problems, are not considered in this monograph, it is our hope that the availability of distributed parameter models of multi-link flexible structures will stimulate future research in this direction. There are important (sometimes combinatorial) optimization problems which we do consider numerically but for which a rigorous mathematical analyses is lacking. One such example involves the optimal location of a given finite number of (directing or absorbing) controllers at junctions and boundaries of a large truss or frame. Another optimization problems of considerable importance related to structures is that of optimal design. In addition, there are obviously many problems of parameter identification which can be addressed within the framework presented below but which we shall not consider.

2. On the Contents of the Book

In this section the organization of the book will be described. Chapter II is devoted to dynamic networks of strings and is intended to serve as a paradigm for the philosophy and approach used throughout the entire book. First, the nonlinear equations of motion for a single element are derived from first principles of continuum mechanics. Next, our basic approach to obtaining the network models by the use of energy principles is developed. The resulting nonlinear model is then linearized about an equilibrium and well-posedness of the linear model is investigated by means of energy estimates. Some of the energy estimates developed in Chapters II and IV are particular to 1-d networks and are based on the possibility of recasting the dynamic equations as a system of symmetric hyperbolic equations of first order. The energy estimates are then used to establish a result on exact controllability of tree-like networks of elastic strings. In particular, given a tree structure as the underlying reference configuration and given that the root node is clamped while the nodes at the tips of all the branches are controlled, then generically one obtains exact controllability. The optimal control time is related the longest path in the graph. It is shown that even approximate controllability fails to hold for networks containing circuits. It is also shown that exact controllability is sensitive to boundary conditions at uncontrolled boundaries.

Modeling of thin, thermoelastic nonlinear 3-d beams and networks of such elements is considered in Chapter III. Even though there are many papers dealing with one or another class of beams models, we were unable to find in the literature a treatment in which all of the features of interest are incorporated. Indeed, most papers do not go beyond the stress-strain constitutive relations and, therefore, do not contain a *practical set of partial differential equations in terms of the displacements and internal variables.* We therefore took some pains to provide a concise development of beam theory. The approach used has been outlined by Wempner [108] and also by Danielson and Hodges [22] and Kane et. al. [41]. We provide a list of beam models which includes all of the well known models but also some models which appear not to have been described in the literature before, such as a 3-d Timoshenko beam, a 2-d thermoelastic nonlinear shearable beam, etc. The point is that all these models can be derived immediately and without difficulty from our general set of equations. Once the single element models are obtained we proceed to formulate the network models following the general procedure outlined in Chapter II. As we have to account for shearing and twisting at the joints, the continuity conditions at the joints generally combine local displacements and shear/twist angles. In the plane, however, there is no twist and since in this case all shear angles have their axes normal to the plane of motion, the displacements decouple from the shear angles in the continuity conditions. As in Chapter II, balance laws of linear and angular momenta at the junctions are obtained from energy principles.

In Chapter IV we introduce a general model for networks of one-dimensional elements whose dynamics are described by hyperbolic systems. This model includes not only the string network of Chapter II but a variety of planar and three dimensional beam models of the Bresse-Timoshenko type which are derived in Chapter III. It also includes "mixed networks" which we illustrate in a rudimentary way. The existence theory for the rather complicated system of equations and joint conditions is obtained by the same standard variational and semigroup methods which were developed in detail in Chapter II. The state space and the spaces of controls are so chosen that one can simultaneously use Dirichlet and Neumann controls. The a priori energy estimates which are essential to results on controllability and stabilizability are then proved in a rather general form. The results on controllability which were given in Chapter II generalize readily as do the results on stabilizability. In fact the treatment of stabilizability is considerably extended from what was done for string networks in Chapter II, incorporating recent treatments of maximal monotone velocity feedback systems.

Chapter V contains a discussion of the eigenvalue problems for networks of strings and of Timoshenko beams and also contains numerical simulations of a variety 1-d network problems. In the first part of Chapter V, an appropriate notation for eigenvalue problems on networks, due to von Below [106] for scalar problems, is introduced. The existence and completeness of an orthonormal eigensystem for the string operator follows from results of Chapter II. Our principle goal is to give an explicit description of the spectrum and of the eigenfunctions. This is accomplished along the lines of [106]. The most explicit insight into the spectrum of a network of strings is achieved for uniform networks with Neumann conditions at all nodes. It turns out – exactly as in the scalar case discussed by von Below [106], [105] and Nicaise [75] – that the spectrum is determined by the underlying graph. Indeed, the eigenvalues are computed from the generalized eigenvalues of the adjacency matrix and the diagonal matrix carrying the edge degrees at the diagonal entries. To some extent these results also carried over to planar networks of Timoshenko beams. It is found that the asymptotic behavior of the spectrum of such a network is similar to that of a string network on the same graph, whose asymptotic behavior is governed by a Weyl's formula. These spectral properties are extremely important in order to fully understand the degree of controllability of the network. For example, the lack of approximate controllability can easily be checked by looking at the eigenmodes. If the control is exerted at a vertex where some eigenmode has a zero displacement, then approximate controllability fails to hold if the control is in the Neumann condition at that vertex. On the other hand, if the spectrum is simple and a Weyl formula holds, we may deduce spectral controllability on a time interval which can be computed by the Beuerling Malliavin density. At this point, however, the picture is far from complete and it remains a challenging problem to achieve a more detailed knowledge of the spectrum and the corresponding spectral family. For instance, there does not seem to be a mathematical proof for the fact that the eigenvalues, which have multiplicities proportional to the number of vertices in the case of uniform networks, split into a sequence of simple eigenvalues in the case of rationally independent optical lengths. This, however, is strongly suggested by a numerical computation of the spectrum, which is based on methods developed in this chapter. Another important open problem is to completely characterize approximately or spectrally controllable networks.

In the second part of Chapter V a variety of numerical simulations of 1-d networks are presented. The controlled and uncontrolled motion of a single string is treated first. Attention is restricted to closed-loop controls, in particular, to absorbing feedback controls. A unit square network of strings is used to demonstrate how the energy is transmitted through the joints. We consider also the problem of stabilizing networks containing circuits. It is shown that dissipative feedbacks at the joints are not capable of extracting all of the total energy out of the circuit. Rather, there remains a residual motion, which has zero displacement at the vertices of the circuit, in the network. Also presented is numerical support to techniques of "channeling the energy" through the network. These techniques employ either the direction of the flux of energy or dry friction at the joints. In this connection there are many open problems of both numerical and theoretical natures.

In particular, any generalization to 2-d elements would be of considerable interest. It is striking how sparse the literature is on the numerical work related to such networks.

Our consideration of interconnected 2-d elements is initiated in Chapter VI with a study of interconnected membranes. We first give a rigorous derivation of a model of a single nonlinear elastic membrane in 3-d space and then turn to the problem of modeling interconnected membranes. The notion of a joint between two membranes, which is necessarily more involved than in the 1-d case, is defined and the appropriate compatibility conditions at the joint are described. The corresponding dynamical joint conditions are derived from Hamilton's principle. The overall system is then linearized by replacing the nonlinear strain components by their linear approximation. Existence and uniqueness of weak solutions of the linear system is settled along lines of Chapter II. However, as mentioned above, in this situation singularities may be expected to develop at corners and possibly along the joints and it is an open problem to describe those membrane networks for which the weak solution is actually a classical solutions in the case of regular data. As discussed in Chapter II, exact controllability (exact reachability) is obtained by investigating the homogeneous adjoint problem, in particular, appropriate *observability estimates* on solutions of the adjoint problem are sought. This is a very difficult problem since inplane motions are typically coupled to the transverse displacement at the joints as may be seen by considering, for example, a 3-d configurations of interconnected membranes such as a thin-walled rectangular tube. Nevertheless, under various geometrical assumptions on the equilibrium shape of the network and under the standing assumption of sufficient regularity of solutions, the crucial observability inequality and corresponding exact controllability result are obtained. A particular consequence of our analysis is an exact controllability result for serially connected membranes and, as a further consequence, the same controllability result for so-called transmission problems.

Chapter VII is devoted to the modeling, analysis and control of linked thin, elastic plates. According to the general program outlined in the previous chapters, the equations of motions for a single nonlinear plate in three dimensions are first derived. The model is again an infinitesimal strain -moderate rotation model along the lines of Wempner [108] (as in the beam case) and accounts for transverse shearing of cross-sections. Upon introduction of the Kirchhoff hypothesis (the analog of the Euler-Bernoulli hypothesis for beams) one may obtain the von Kármán system, and, by linearization, the Kirchhoff plate system. If the general nonlinear equations are properly linearized, the classical equations of Reissner-Mindlin plate theory result. Modeling of interconnected plates is considered next. It turns out to be a very subtle task to properly formulate the problem of interconnected plates. We first study linked Reissner-Mindlin plates.

As in the case of beams, there are a number of distinct types of connections between plates which can be imagined. For example, a joint between plates can be considered as very stiff and rigid in the sense that it does not bend and twist, or it might bend in one direction but not in another. It might rotate if it is straight, and so on. Various possible types of joints between membranes are defined in terms of appropriate restrictions on the deformation of the joint. These involve compatibility relations among displacements, tangential and transversal shearing angles as well as in-plane twisting angles at the joint and possible restrictions on the geometry of the deformation. The compatibility conditions are nonlinear, in general, but are linearized in order to obtain a model whose the analysis is tractable. Upon the introduction of appropriate ramification spaces, the corresponding dynamic joint conditions are derived and the well-posedness of the linear model is established. Only the existence of weak solutions is ascertained. As in the case of linked membranes, the problem of when weak solutions are classical solutions, if the data are smooth enough, is completely open. Exact controllability results for linked Reissner-Mindlin plates may be obtained in a way similar to that in the linked membrane case. The computations, however, are much more involved and the results which could be obtained would in the end be only formal because of the unresolved regularity issues. For these reasons attention is restricted to transmission problems in which a smooth domain is embedded into a larger (not necessarily smooth) master domain. Observability estimates are derived and the configuration space is described. The final section of this Chapter is devoted to linked Kirchhoff plates. Rather than give a separate derivation of the model, which is possible, we obtain the system of equations from the Reissner-Mindlin model by letting the shear stiffness increase to infinity. In this manner, models of interconnected Kirchhoff plates corresponding to various types of joints are obtained and convergence of the solutions of the parameterized Reissner-Mindlin model to those of the Kirchhoff model is established.

The final Chapter studies the coupling of plates and beams. Two specific configurations are considered. In the first one a beam is attached to the edge of a plate in such a manner that the centerline of the beam is orthogonal to the tangent along the edge of the plate at its reference surface. The second configuration is one in which a beam is orthogonally attached to one of the faces of the plate. In each case the compatibility condition at the interface is the requirement that the deformation of the cross-section of the beam match that of the (3-d) plate throughout the junction region and thus this condition describes a rigid connection between the plate and beam. The dynamic conditions are then obtained from Hamilton's principle and lead (upon linearization) to a system which couples the Reissner-Mindlin plate system to the Timoshenko beam system and the equation for torsion in a rod. The dynamic coupling equations are *nonlocal* since they balance forces and moments at the centerline of the beam with *averages* of forces and moments in the plate taken over the junction region. We also investigate the exact controllability problem in the case of the first configuration and show that the plate-beam system may be exactly controlled through the application of controls in the forces and moments along a portion of the edge of the plate which excludes the junction region. Further, in the case of the first configuration, a simpler model is obtained by the process of passing to the limit as the shear stiffnesses on the plate and beam increase beyond bound, in analogy with the process carried out in Chapter VII. The result is a combined Kirchhoff plate – Raleigh beam system coupled to an ordinary differential equation which governs a time dependent (rigid) rotation of the beam around its centerline.

CHAPTER II

Modeling of Networks of Elastic Strings

This chapter deals with three dimensional networks of elastic strings. Such networks, originally introduced in [94], provide a paradigm for the much more complicated networks of beams, membranes and plates which are the central subject matter of this monograph. Many of the main ideas of this monograph can be illustrated in the relatively simple context of string networks and certain questions, such as those concerning controllability and stabilizability, can be answered more fully than is possible for the more complex structures to be considered later.

1. Modeling of Nonlinear Elastic Strings

We begin by considering vibrations in $\mathbb{R}^3$ of a single elastic string segment. If the "natural" length of that string is l it is reasonable to consider the string as parameterized by the rest arc length x with $x \in [0, l]$. The position at time t of the point corresponding to the parameter x is to be denoted by the vector $\mathbf{R}(x, t)$. If ρ is the density of the string and if the string is assumed to satisfy Hooke's law with constant h the kinetic and potential energies are given by

$$\mathcal{K}(t) = \frac{1}{2} \int_0^l \rho |\dot{\mathbf{R}}(x,t)|^2 \, dx. \tag{1.1}$$

and

$$\mathcal{U}(t) = \frac{1}{2} \int_0^l h[|\mathbf{R}_{,x}(x,t)| - 1]^2 \, dx, \tag{1.2}$$

respectively, where $|\cdot|$ denotes the Euclidean norm, the dot "$\dot{}$" indicates partial differentiation with respect to t and the subscript "$_{,x}$" partial differentiation with respect to x. The total energy is then

$$\mathcal{E}(t) = 1/2 \int_0^l \left[\rho|\dot{\mathbf{R}}(x,t)|^2 + h[|\mathbf{R}_{,x}(x,t)| - 1]^2\right] dx.$$

We apply Hamilton's principle to the Lagrangian functional

$$\mathcal{L}(\mathbf{R}) = 1/2 \int_0^T \int_0^l \left[\rho|\dot{\mathbf{R}}(x,t)|^2 - h[|\mathbf{R}_{,x}(x,t)| - 1]^2\right] dx\, dt.$$

This requires $\mathcal{L}$ to be stationary at $\mathbf{R}$. Of course the domain of $\mathcal{L}$ needs to be specified. On the one hand this involves requirements on the regularity of $\mathbf{R}$ about which we do not wish to be precise at this point. On the other hand $\mathbf{R}$ may also be required to satisfy some *geometric condition* such as a Dirichlet boundary condition. Suppose, for example, that we require

$$\mathbf{R}(0,t) = \mathbf{U}_0(t) \tag{1.3}$$

and that no condition is placed on $\mathbf{R}(l,t)$, in which case we say that the right end point is *free*. In that case one can consider a perturbation $\mathbf{R}(x,t) + \lambda\delta\mathbf{R}(x,t)$ where $\delta\mathbf{R}(0,t) = \mathbf{0}$ and there is again no condition at the other end. In fact we also assume that $\delta\mathbf{R}$ and $\delta\dot{\mathbf{R}}$ vanish for $t = 0$ and $t = T$. Now the stationary condition becomes

$$(1.4)\quad 0 = \frac{d}{d\lambda}\mathcal{L}(\mathbf{R} + \lambda\delta\mathbf{R})|_{\lambda=0} =$$

$$\int_0^T \int_0^l \left[\rho\dot{\mathbf{R}}(x,t)\cdot\delta\dot{\mathbf{R}}(x,t) - h[|\mathbf{R}_{,x}| - 1]\frac{\mathbf{R}_{,x}}{|\mathbf{R}_{,x}|}(x,t)\cdot\delta\mathbf{R}_{,x}(x,t)\right] dxdt.$$

for all admissible $\delta\mathbf{R}$. Assuming that $\mathbf{R}(x,t)$ is twice differentiable (and that the integrals are indeed convergent) one integrates by parts to obtain

$$\begin{aligned}(1.5)\quad \int_0^T \int_0^l \left[-\rho\ddot{\mathbf{R}} + h\left[\mathbf{R}_{,x} - \frac{\mathbf{R}_{,x}}{|\mathbf{R}_{,x}|}\right]_{,x}\right](x,t)\cdot\delta\mathbf{R}(x,t)\,dx\,dt \\ - h\left[\mathbf{R}_{,x} - \frac{\mathbf{R}_{,x}}{|\mathbf{R}_{,x}|}\right](l,t)\cdot\delta\mathbf{R}(l,t) = 0.\end{aligned}$$

Considering first arbitrary perturbations $\delta\mathbf{R}(x,t)$ satisfying $\delta\mathbf{R}(l,t) = \mathbf{0}$, one obtains the non-linear system of partial differential equations

$$\rho\ddot{\mathbf{R}} = h\left[\mathbf{R}_{,x} - \frac{\mathbf{R}_{,x}}{|\mathbf{R}_{,x}|}\right]_{,x}. \tag{1.6}$$

If one now takes variations which do not vanish at l one is led to the *free* (or *natural*) boundary condition

$$h\left[\mathbf{R}_{,x} - \frac{\mathbf{R}_{,x}}{|\mathbf{R}_{,x}|}\right](l,t) = \mathbf{0}. \tag{1.7}$$

The boundary conditions (1.3) and (1.7) need to be complemented by initial conditions prescribing $\mathbf{R}(\cdot,0)$ and $\dot{\mathbf{R}}(\cdot,0)$

It is easy to verify by differentiation with respect to t that the total energy $\mathcal{E}(t)$ is conserved for solutions of the equation (1.6) with boundary conditions (1.3) (with data which is independent of t) and (1.7).

We can also suppose that a distributed force $\mathbf{F}(x,t)$ is acting along the string and that a force $\mathbf{U}_l(t)$ acts on the endpoint at $x = l$. Then we introduce the following work functional:

$$\mathcal{W}(t) = \int_0^l \mathbf{F}(x,t)\cdot\mathbf{R}(x,t)\,dx + \mathbf{U}_l(t)\cdot\mathbf{R}(l,t). \tag{1.8}$$

Applying Hamilton's principle to the modified Lagrangian functional

$$\mathcal{L}(\mathbf{R}) = \int_0^T [\mathcal{K}(t) + \mathcal{W}(t) - \mathcal{U}(t)]\,dt$$

we get the inhomogeneous equation

$$\rho\ddot{\mathbf{R}} = h\left[\mathbf{R}_{,x} - \frac{\mathbf{R}_{,x}}{|\mathbf{R}_{,x}|}\right]_{,x} + \mathbf{F}, \tag{1.9}$$

as well as the inhomogeneous free boundary condition

$$h\left[\mathbf{R}_{,x} - \frac{\mathbf{R}_{,x}}{|\mathbf{R}_{,x}|}\right](l,t) = \mathbf{U}_l(t). \tag{1.10}$$

More complicated boundary conditions are possible. Suppose for example that the endpoint of the string corresponding to $x = 0$ is constrained to slide along a frictionless rod located along the line given parametrically by $\mathbf{c} + \sigma\mathbf{b}$, a geometric boundary condition which can be expressed as

$$[\mathbf{R}(0,t) - \mathbf{c}] \times \mathbf{b} = \mathbf{0}, \tag{1.11}$$

where $\times$ denotes the cross product. Then the perturbations $\delta\mathbf{R}(x,t)$ must be such that $\delta\mathbf{R}(0,t)$ is a multiple of $\mathbf{b}$; in that case $\delta\mathbf{R}(0,t) = \phi(t)\mathbf{b}$ and, since $\phi(t)$ can be an arbitrary smooth function, one concludes from Hamilton's Principle that (1.11) must be complemented by the dynamic boundary condition

$$h\left[\mathbf{R}_{,x} - \frac{\mathbf{R}_{,x}}{|\mathbf{R}_{,x}|}\right](l,t)\cdot\mathbf{b} = \mathbf{0}. \tag{1.12}$$

REMARK 1.1. We have no idea how existence theory for the highly nonlinear equation (1.6) can be approached, except in the stationary case.

REMARK 1.2. The assumption that Hooke's law holds is certainly only reasonable for a limited range of extension (corresponding to $|\mathbf{R}_{,x}| > 1$) or contraction (corresponding to $|\mathbf{R}_{,x}| < 1$). Hooke's law can be replaced by replacing the quadratic potential energy function $h/2[|\mathbf{R}_{,x}| - 1]^2$ by more general functions of the form $u(|\mathbf{R}_{,x}| - 1)$; this leads to a variety of intriguing nonlinear systems.

REMARK 1.3. For related nonlinear string models derived in other ways see Carrier [14] and Antman [2].

2. Networks of Nonlinear Elastic Strings

A network consists of more than one string segment each of which is modeled as above. We suppose that the strings are indexed by $i \in \mathcal{I} = \{1, \ldots, n\}$. These strings may have different physical characteristics associated with constants l_i, ρ_i and h_i. Let the spatial location of the i-th string be described by $\mathbf{R}^i(x,t)$. We then let $\mathbf{R}$ denote the $n-$tuple of all these location functions. We suppose that strings are joined together at certain endpoints in such a way that the network is connected.

The model involves conditions at the *nodes* by which we mean the locations of the endpoints of the strings. We suppose that the positions of these nodes are given by $\mathbf{N}^j(t)$ with $j \in \mathcal{J} = \{j = 1 \ldots m\}$. We distinguish between *simple* nodes, indexed by $j \in \mathcal{J}_S$ which are the endpoints of only one string, and *multiple* nodes, indexed by $j \in \mathcal{J}_M$ at which several strings meet. Corresponding to $j \in \mathcal{J}_M$ we introduce the notation

$$\mathcal{I}_j = \{i \in \mathcal{I} : \text{ the } i\text{-th string has an end at } \mathbf{N}^j\}.$$

For $i \in \mathcal{I}_j$ we also introduce the parameter values x_{ij} to be equal to l_i, if $\mathbf{N}^j(t) = \mathbf{R}^i(l_i, t)$ or 0 if $\mathbf{N}^j(t) = \mathbf{R}^i(0,t)$. For the purposes of integration by parts we introduce the sign function ϵ_{ij} to equal 1 if $x_{ij} = l_i$ and -1 if $x_{ij} = 0$. Then $\epsilon_{ij}\mathbf{R}^i_{,x}(x_{ij}, t)$ denotes the "outward pointing" derivative at x_{ij}.

Implicit in the above notation is the following *geometric multiple node condition* which requires the continuity of the location functions at each multiple node:

$$\text{the } \mathbf{R}^i(x_{ij}, t) \text{ are equal for all } i \in \mathcal{I}_j,\ j \in \mathcal{J}_M, \tag{2.1}$$

the common value of course being $\mathbf{N}^j(t)$.

At both simple and multiple nodes various conditions can be imposed. We shall say that a node is *clamped* if $\mathbf{N}^j(t)$ is prescribed in a fixed and time independent way. Clamped nodes will be indexed by $j \in \mathcal{J}^C$ so that the explicit condition is

$$\mathbf{N}^j(t) = \check{\mathbf{U}}^j \quad \text{for } j \in \mathcal{J}^C. \tag{2.2}$$

Some nodes, indexed by $j \in \mathcal{J}^D$, may be subject to a *Dirichlet* condition which takes the form

$$\mathbf{N}^j(t) = \mathbf{U}^j(t) \quad \text{for } j \in \mathcal{J}^D, \tag{2.3}$$

where the control functions $\mathbf{U}^j(t)$ belongs to a specified set of admissible controls. Other nodes, indexed by $j \in \mathcal{J}^N$, may be subject to a *Neumann* control which physically corresponds to the application of a force $\mathbf{U}^j(t)$ at the j-th node. The precise form of such a condition (namely (2.12) below) will emerge from the variational argument. The remaining nodes will be referred to as *free* nodes and correspond to indices $j \in \mathcal{J}^F$; at these nodes the *free* (or *natural*) node conditions (2.13) are a consequence of Hamilton's principle. We refine the index notation by allowing both subscripts and superscripts with the obvious interpretation: for example $\mathcal{J}_M^N = \mathcal{J}^N \cap \mathcal{J}_M$. In most of our applications, and certainly for the string networks we consider in this chapter, these index sets will be disjoint. This point will be remarked upon again in Section 2.6. To complete our terminology, we say that the clamped and Dirichlet conditions are *geometric* conditions, while the free and Neumann conditions are *dynamic* conditions. The former act directly as constraints on the underlying geometric structure, while the latter are embodied in the formulation of the Lagrangian functional and hence a consequence of Hamilton's principle.

The kinetic and potential energies of the system are respectively given by

$$\mathcal{K}(t) = \sum_{i \in \mathcal{I}} \frac{1}{2} \int_0^{l_i} \rho_i |\dot{\mathbf{R}}^i(x,t)|^2 \, dx \tag{2.4}$$

and

$$\mathcal{U}(t) = \sum_{i \in \mathcal{I}} \int_0^{l_i} \frac{h_i}{2} [|\mathbf{R}^i_{,x}(x,t)| - 1]^2 \, dx. \tag{2.5}$$

The work functional corresponding to distributed forces $\mathbf{F}^i(x,t)$ acting on the i-th string as well as forces $\mathbf{U}^j(t)$ acting on the nodes indexed by $j \in \mathcal{J}^N$ is

$$\mathcal{W}(t) = \sum_{i \in \mathcal{I}} \int_0^{l_i} \mathbf{F}^i(x,t) \cdot \mathbf{R}^i(x,t) \, dx + \sum_{j \in \mathcal{J}^N} \mathbf{U}^j(t) \cdot \mathbf{N}^j(t). \tag{2.6}$$

Then the Lagrangian functional takes the form

$$\begin{aligned} \mathcal{L}(\mathbf{R}) = \int_0^T \Bigg[\sum_{i \in \mathcal{I}} \int_0^{l_i} \Big[\frac{\rho_i}{2} |\dot{\mathbf{R}}^i(x,t)|^2 + \mathbf{F}^i(x,t) \cdot \mathbf{R}^i(x,t) \\ - \frac{h_i}{2} [|\mathbf{R}^i_{,x}(x,t)| - 1]^2 \Big] \, dx + \sum_{j \in \mathcal{J}^N} \mathbf{U}^j(t) \cdot \mathbf{N}^j(t) \Bigg] \, dt. \end{aligned} \tag{2.7}$$

We apply Hamilton's principle to the Lagrangian (2.7) with variations $\delta\mathbf{R}$ which respect the geometric node conditions (2.1), (2.2) and (2.3) so that

$$\begin{cases} \delta\mathbf{R}^i(x_{ij},t) = \delta\mathbf{N}^j(t) & \text{for } i \in \mathcal{I}_j, \\ \delta\mathbf{N}^j(t) = \mathbf{0} & \text{for } j \in \mathcal{J}^C \cup \mathcal{J}^D. \end{cases} \tag{2.8}$$

One obtains

$$\frac{d}{d\lambda}\mathcal{L}(\mathbf{R}+\lambda\delta\mathbf{R})|_{\lambda=0} = \int_0^T \Bigg[\sum_{i\in\mathcal{I}} \int_0^{l_i} [\rho_i \dot{\mathbf{R}}^i \cdot \delta\dot{\mathbf{R}}^i + \mathbf{F}^i \cdot \delta\mathbf{R}^i \\ - h[\mathbf{R}^i_{,x} - 1]\frac{\mathbf{R}^i_{,x}}{|\mathbf{R}^i_{,x}|} \cdot \delta\mathbf{R}^i_{,x}](x,t)\,dx + \sum_{j\in\mathcal{J}^N} \mathbf{U}^j(t)\cdot\delta\mathbf{N}^j(t)\Bigg]\,dt = 0. \tag{2.9}$$

Integrating by parts with respect to both t and x, taking into account (2.8) and assuming that $\delta\mathbf{R}(\cdot,t)$ vanishes for $t=0$ and $t=T$ one gets

$$\int_0^T \Bigg[\sum_{i\in\mathcal{I}} \int_0^{l_i} [-\rho_i \ddot{\mathbf{R}}^i + h_i \left[\mathbf{R}^i_{,x} - \frac{\mathbf{R}^i_{,x}}{|\mathbf{R}^i_{,x}|}\right]_{,x} + \mathbf{F}^i] \cdot \delta\mathbf{R}^i(x,t)\,dx \\ + \sum_{j\in\mathcal{J}^N} [-\sum_{i\in\mathcal{I}_j} \epsilon_{ij} h_i \left[\mathbf{R}^i_{,x} - \frac{\mathbf{R}^i_{,x}}{|\mathbf{R}^i_{,x}|}\right](x_{ij},t) + \mathbf{U}^j(t)] \cdot \delta\mathbf{N}^j(t) \\ - \sum_{j\in\mathcal{J}^F} [\sum_{i\in\mathcal{I}_j} \epsilon_{ij} h_i \left[\mathbf{R}^i_{,x} - \frac{\mathbf{R}^i_{,x}}{|\mathbf{R}^i_{,x}|}\right](x_{ij},t)] \cdot \delta\mathbf{N}^j(t)\Bigg]\,dt = 0.$$

By using variations $\delta\mathbf{R}$ which vanish near the nodes one obtains the partial differential equations

$$\rho_i \ddot{\mathbf{R}}^i = h_i \left[\mathbf{R}^i_{,x} - \frac{\mathbf{R}^i_{,x}}{|\mathbf{R}^i_{,x}|}\right]_{,x} + \mathbf{F}^i,\ i \in \mathcal{I}. \tag{2.10}$$

If one now considers $\delta\mathbf{R}$ with support concentrated about a particular Neumann controlled or free node one is led to the *dynamic node conditions*

$$\sum_{i\in\mathcal{I}_j} \epsilon_{ij} h_i \left[\mathbf{R}^i_{,x} - \frac{\mathbf{R}^i_{,x}}{|\mathbf{R}^i_{,x}|}\right](x_{ij},t) = \begin{cases} \mathbf{U}^j & \text{for } j \in \mathcal{J}^N, \\ \mathbf{0} & \text{for } j \in \mathcal{J}^F. \end{cases} \tag{2.11}$$

The above equations together with the geometric conditions (2.1), (2.2) and (2.3) have to be supplemented by initial conditions which prescribe $\mathbf{R}(0)$ and $\dot{\mathbf{R}}(0)$.

If the Dirichlet data are time independent and the forces $\mathbf{F}^i$ and $\mathbf{U}^j$ all vanish it is easy to verify, at least formally, that the total energy

$$(2.12)\qquad \mathcal{E}(t) = \frac{1}{2}\sum_{i\in\mathcal{I}}\int_0^{l_i}\left[\rho_i|\dot{\mathbf{R}}^i(x,t)|^2 + h_i[|\mathbf{R}^i_{,x}(x,t)|-1]^2\right]dx$$

of the network is conserved. If one differentiates with respect to t and integrates by parts with respect to x, taking into account the equation (2.10), one finds

$$\begin{aligned}\frac{d\mathcal{E}}{dt} &= \sum_{i\in\mathcal{I}}\int_0^{l_i}\left[\dot{\mathbf{R}}^i(x,t)\cdot\rho_i\ddot{\mathbf{R}}^i(x,t) + \dot{\mathbf{R}}^i_{,x}(x,t)\cdot h_i\big[\mathbf{R}^i_{,x} - \frac{\mathbf{R}^i_{,x}}{|\mathbf{R}^i_{,x}|}\big](x,t)\right]dx\\ &= \sum_{i=1}^{n}\left[\dot{\mathbf{R}}^i(x,t)\cdot h_i\big[\mathbf{R}^i_{,x} - \frac{\mathbf{R}^i_{,x}}{|\mathbf{R}^i_{,x}|}\big](x,t)\right]_{x=0}^{x=l_i}\\ &= \sum_{j\in\mathcal{J}}\dot{\mathbf{N}}^j\cdot\sum_{i\in\mathcal{I}_j}\epsilon_{ij}h_i\big[\mathbf{R}^i_{,x} - \frac{\mathbf{R}^i_{,x}}{|\mathbf{R}^i_{,x}|}\big](x_{ij},t) = 0,\end{aligned}$$

where we have used the node conditions (2.1), (2.2), (2,3), and (2.11) together with the time independence of the Dirichlet data and the vanishing of the Neumann data.

3. Linearization

In this section we linearize the network equations about equilibrium configurations which we shall denote by $\check{\mathbf{R}}$. The latter are obtained by solving

$$(3.1)\qquad h_i\left[\mathbf{R}^i_{,x} - \frac{\mathbf{R}^i_{,x}}{|\mathbf{R}^i_{,x}|}\right]_{,x}(x) = \check{\mathbf{F}}^i(x),\ \text{ for } i\in\mathcal{I},$$

subject to (2.1), (2.2), (2,3) and (2.11) with all data $\check{\mathbf{F}}^i$ and $\check{\mathbf{U}}^j$ independent of time. In fact there are generally many possibilities. In this regard see [95] (and Section 2.7) for networks in which gravitational forces are taken into account. Here we shall *suppose that at equilibrium there are no distributed forces acting along the strings* so that $\check{\mathbf{F}}^i(x) = \mathbf{0}$. The individual equations in (3.1) then have two evident classes of solutions. The first, very large, class of solutions consists of all functions $\check{\mathbf{R}}^i(x)$ for which $|\check{\mathbf{R}}^i_{,x}(x)| = 1$ and describe *limp* strings. The second class of *stretched* or *compressed* solutions have the linear form

$$(3.2)\qquad \check{\mathbf{R}}^i(x) = \mathbf{R}^i_0 + xs_i\mathbf{e}^i,$$

where $\mathbf{e}^i$ is the unit direction pointing along the i-th string segment and s_i is a constant measuring the extent to which the string is stretched (if $s_i > 1$) or compressed (if $s_i < 1$). For networks one must also satisfy the geometric node conditions (2.1), (2.2) and (2.3) as well as the dynamic conditions (2.11) which take the form

$$\sum_{i\in\mathcal{I}_j} \epsilon_{ij} h_i[s_i - 1]\mathbf{e}^i = \begin{cases} \check{\mathbf{U}}^j & \text{for } j \in \mathcal{J}^N, \\ \mathbf{0} & \text{for } j \in \mathcal{J}^F. \end{cases} \tag{3.3}$$

If $\check{\mathbf{U}}^j \neq \mathbf{0}$ the strings adjoining the j-th node cannot all be limp. If all $\check{\mathbf{U}}^j = \mathbf{0}$ networks of limp strings automatically satisfy (3.3). It is easy to give examples of networks and boundary conditions with equilibria in which different elements may be limp, stretched or compressed. We shall avoid such situations and suppose that the boundary conditions are such that in the equilibrium configuration *all strings are stretched.* We then linearize about such a stretched equilibrium configuration.

To carry out the linearization, we let

$$\begin{aligned} \mathbf{R}^i(x,t) &= \check{\mathbf{R}}^i(x) + \mathbf{r}^i(x,t) = \check{\mathbf{R}}^i_0 + x s_i \mathbf{e}^i + \mathbf{r}^i(x,t), \\ \mathbf{N}^j(t) &= \check{\mathbf{N}}^j + \mathbf{n}^j(t) \quad \text{for } j \in \mathcal{J}, \\ \mathbf{U}^j(t) &= \check{\mathbf{U}}^j + \mathbf{u}^j(t) \quad \text{for } j \in \mathcal{J}^D, \\ \mathbf{F}^i(x,t) &= \mathbf{f}^i(x,t) \quad \text{for } i \in \mathcal{I}, \end{aligned}$$

where $\mathbf{r}_i(x,t)$ and $\mathbf{n}^j(t)$ are "small" displacements corresponding to "small" perturbations $\mathbf{u}^j(t)$ of the nodal data, "small" forces $\mathbf{f}^i(x,t)$ distributed along the strings and "small" perturbations of the initial data. Consequently, the geometric conditions (2.1), (2.2) and (2.3) require the displacement functions to be subject to the node conditions

$$\mathbf{r}^i(x_{ij},t) = \mathbf{n}^j(t) \quad \text{for } i \in \mathcal{I}_j, \tag{3.4}$$

$$\mathbf{n}^j(t) = \mathbf{0} \quad \text{for } j \in \mathcal{J}^C, \mathbf{n}^j(t) = \mathbf{u}^j(t) \quad \text{for } j \in \mathcal{J}^D. \tag{3.5}$$

Obviously $\ddot{\mathbf{R}}^i(x,t) = \ddot{\mathbf{r}}^i(x,t)$ so that the left hand side of equation (2.10) can be immediately rewritten and remains linear. One possibility is now to linearize the right hand side of (2.10) as well as the left hand sides of the node conditions (2.11) directly, using the following quadratic Taylor expansion in $\mathbf{r}^i$ about $\mathbf{0}$:

$$\begin{aligned} |\mathbf{R}^i_{,x}| = [\mathbf{R}^i_{,x} \cdot \mathbf{R}^i_{,x}]^{1/2} &= [s_i^2 + 2s_i v^i \cdot \mathbf{r}^i_{,x} + |\mathbf{r}^i_{,x}|^2]^{1/2} \\ &= s_i\big[1 + \frac{2}{s_i}\mathbf{e}^i \cdot \mathbf{r}^i_{,x} + \frac{1}{s_i}|\mathbf{r}^i_{,x}|^2\big]^{1/2} \\ &= s_i[1 + \frac{1}{s_i}\mathbf{e}^i \cdot \mathbf{r}^i_{,x} + \frac{1}{2s_i^2}|\mathbf{r}^i_{,x}|^2 - \frac{1}{2s_i^2}(\mathbf{e}^i \cdot \mathbf{r}^i_{,x})^2 + \cdots]. \end{aligned}$$

Instead we return to the variational formulation expanding up to quadratic terms in $\mathbf{r}^i_{,x}$. The kinetic energy and the work functionals are simply

$$\mathcal{K}(t) = \sum_{i\in\mathcal{I}} \int_0^{l_i} \rho_i |\dot{\mathbf{r}}^i|^2 \, dx, \tag{3.6}$$

and

$$\mathcal{W}(t) = \sum_{i\in\mathcal{I}} \int_0^{l_i} \mathbf{f}^i(x,t)\cdot[\check{\mathbf{R}}^i + \mathbf{r}^i]\,dx + \sum_{j\in\mathcal{J}^N} \mathbf{U}^j\cdot[\check{\mathbf{N}}^j + \mathbf{n}^j]. \tag{3.7}$$

Both of these expressions contain no terms of order higher than quadratic. As far as the potential energy is concerned, note first that

$$h_i[|\mathbf{R}^i_{,x}| - 1]^2 =$$
$$h_i\left[(s_i-1)^2 + 2(s_i-1)\mathbf{e}^i\cdot\mathbf{r}^i_{,x} + (1-\frac{1}{s_i})|\mathbf{r}^i_{,x}|^2 + \frac{1}{s_i}[\mathbf{e}^i\cdot\mathbf{r}^i_{,x}]^2\right] + \cdots.$$

The quadratic terms in this expression can be written as $\mathbf{Q}_i\mathbf{r}^i_{,x}\cdot\mathbf{r}^i_{,x}$ with

$$\mathbf{Q}_i = h_i\left[(1-\frac{1}{s_i})\mathbf{I} + \frac{1}{s_i}{\mathbf{e}^i}^T\mathbf{e}^i\right], \tag{3.8}$$

where $\mathbf{I}$ is the 3×3 identity matrix, $\mathbf{e}_i$ is written as a column and ${\mathbf{e}_i}^T$ denotes the transpose. We then obtain from (2.5)

$$\mathcal{U}(t) = \sum_{i\in\mathcal{I}} \int_0^{l_i} \frac{1}{2}\left[h_i[(s_i-1)^2 + 2(s_i-1)\mathbf{e}^i\cdot\mathbf{r}^i_{,x}] + \mathbf{Q}_i\mathbf{r}^i_{,x}\cdot\mathbf{r}^i_{,x}\right]dx + \cdots$$

If in the above expression for $\mathcal{U}$ we discard terms of order higher than quadratic in $\mathbf{r}$ and if we then apply Hamilton's principle to the corresponding Lagrangian functional of $\mathbf{r} = (\mathbf{r}^i)_{i\in\mathcal{I}}$ we obtain

$$\int_0^T \Bigg[\sum_{i\in\mathcal{I}} \int_0^{l_i} [\rho_i\dot{\mathbf{r}}^i\cdot\delta\dot{\mathbf{r}}^i - h_i(s_i-1)\mathbf{e}^i\cdot\delta\mathbf{r}^i_{,x} - \mathbf{Q}_i\mathbf{r}^i_{,x}\cdot\delta\mathbf{r}^i_{,x}]\,dx$$
$$+ \sum_{j\in\mathcal{J}^N} \mathbf{U}^j\cdot\delta\mathbf{n}^j\Bigg]\,dt = 0.$$

Here the variations $\delta\mathbf{r}^i$ and the corresponding $\delta\mathbf{n}^j$ must respect the conditions (3.4), (3.5) and (3.6) and can also be required to vanish for $t=0$ and

$t = T$. Integrating by parts and grouping terms we then get

$$\int_0^T \Big[\sum_{i\in\mathcal{I}} \int_0^{l_i} \left[-\rho_i \ddot{\mathbf{r}}^i + \mathbf{Q}_i \mathbf{r}^i_{,xx}\right] \cdot \delta\mathbf{r}^i \, dx$$
$$- \sum_{j\in\mathcal{J}^F} \big[\sum_{i\in\mathcal{I}_j} [\epsilon_{ij} h_i (s_i - 1)\mathbf{e}^i + \epsilon_{ij} \mathbf{Q}_i \mathbf{r}^i_{,x}] - \mathbf{U}^j\big] \cdot \delta\mathbf{n}^j$$
$$- \sum_{j\in\mathcal{J}^F} \sum_{i\in\mathcal{I}_j} \left[\epsilon_{ij} h_i (s_i - 1)\mathbf{e}^i + \epsilon_{ij} \mathbf{Q}_i \mathbf{r}^i_{,x}\right] \cdot \delta\mathbf{n}^j \Big] \, dt = 0.$$

Hence one is led to the linear system

$$\rho_i \ddot{\mathbf{r}}^i = \mathbf{Q}_i \mathbf{r}^i_{,xx} \quad \text{for } i \in \mathcal{I}, \tag{3.9}$$

as well as the conditions

$$\sum_{i\in\mathcal{I}_j} [\epsilon_{ij} h_i (s_i - 1)\mathbf{e}^i + \epsilon_{ij} \mathbf{Q}_i \mathbf{r}^i_{,x}] = \begin{cases} \mathbf{U}^j & \text{for } j \in \mathcal{J}^N, \\ \mathbf{0} & \text{for } j \in \mathcal{J}^F. \end{cases}$$

In view of (3.3) the latter simplify to

$$\sum_{i\in\mathcal{I}_j} \epsilon_{ij} \mathbf{Q}_i \mathbf{r}^i_{,x} = \begin{cases} \mathbf{u}^j & \text{for } j \in \mathcal{J}^N, \\ \mathbf{0} & \text{for } j \in \mathcal{J}^F. \end{cases} \tag{3.10}$$

Now we can streamline the formulation of the linearized problem by associating it with the following kinetic energy, potential energy and work functionals:

$$\mathcal{K}(t) = \sum_{i\in\mathcal{I}} \int_0^{l_i} \rho_i |\dot{\mathbf{r}}^i|^2 \, dx,$$
$$\mathcal{W}(t) = \sum_{i\in\mathcal{I}} \int_0^{l_i} \mathbf{r}^i \cdot \mathbf{f}^i \, dx + \sum_{j\in\mathcal{J}^F} \mathbf{u}^j \cdot \mathbf{n}^j,$$
$$\mathcal{U}(t) = \sum_{i\in\mathcal{I}} \frac{1}{2} \int_0^{l_i} \mathbf{Q}_i \mathbf{r}^i_{,x} \cdot \mathbf{r}^i_{,x} \, dx.$$

One easily verifies that Hamilton's principle applied to the corresponding Lagrangian directly yields (3.10)and (3.11).

We note that $\mathbf{Q}_i$ as defined by (3.9) is a symmetric matrix with eigenvalues h_i and $h_i(1 - s_i^{-1})$ corresponding to the eigenspace spanned by $\mathbf{e}^i$ and the orthogonal complement of $\mathbf{e}^i$, respectively. Both eigenvalues are positive because $s_i > 1$ so that the equations (3.10) are hyperbolic. For each string the vibration can be uncoupled into transverse and longitudinal vibrations. The node conditions scramble these modes.

The total energy of the network in the linearized model is

$$(3.11)\qquad \mathcal{E}(t) = \frac{1}{2}\sum_{i\in\mathcal{I}}\int_0^{l_i} \left[\rho_i|\dot{\mathbf{r}}^i(x,t)|^2 + \mathbf{Q}_i\mathbf{r}^i_{,x}(x,t)\cdot\mathbf{r}^i_{,x}(x,t)\right] dx.$$

It is easy to check that if $\mathbf{u}^j(t) = \mathbf{0}$ for $j \in \mathcal{J}^F$ and $\mathbf{u}^j(t)$ is independent of t for $j \in \mathcal{J}^D_S$ then energy is conserved, as before.

4. Well-posedness of the Network Equations

In this section we consider the well-posedness of the linear problem obtained in the last section. In its entirety this system is as follows:

$$(4.1)\qquad \begin{cases} \rho_i\ddot{\mathbf{r}}^i = \mathbf{Q}_i\mathbf{r}^i_{,xx} + \mathbf{f}^i \ \text{ for } i \in \mathcal{I}, \\ \mathbf{r}^i(x_{ij},t) \ \text{ coincide for } j \in \mathcal{J}_M,\ i \in \mathcal{I}_j, \\ \mathcal{D}^j\mathbf{r}(t) = \mathbf{u}^j(t) \text{ for } j \in \mathcal{J}^D,\ \mathcal{D}^j\mathbf{r}(t) = \mathbf{0} \text{ for } j \in \mathcal{J}^C, \\ \mathcal{N}^j\mathbf{r}(t) = \mathbf{u}^j(t) \text{ for } j \in \mathcal{J}^N,\ \mathcal{N}^j\mathbf{r}(t) = \mathbf{0} \text{ for } j \in \mathcal{J}^F, \\ \mathbf{r}(0) = \mathbf{r}_0,\ \ \dot{\mathbf{r}}(0) = \dot{\mathbf{r}}_0, \end{cases}$$

where $\mathbf{r} = (\mathbf{r}_i)_{i\in\mathcal{I}}$, $\dot{\mathbf{r}} = (\dot{\mathbf{r}}_i)_{i\in\mathcal{I}}$ and the Dirichlet and Neumann boundary operators $\mathcal{D}^j$ and $\mathcal{N}^j$ are defined at the nodes by

$$(4.2)\qquad \begin{cases} \mathcal{D}^j\mathbf{r}(t) = \mathbf{n}^j(t) = \mathbf{r}^i(x_{ij},t) \ \text{ for } i \in \mathcal{I}_j, \\ \mathcal{N}^j\mathbf{r}(t) = \sum_{i\in\mathcal{I}_j}\epsilon_{ij}\mathbf{Q}_i\mathbf{r}^i_{,x}(x_{ij},t). \end{cases}$$

The governing equations can be rewritten in the form

$$(4.3)\qquad \frac{\partial}{\partial t}\begin{pmatrix}\mathbf{r}\\ \dot{\mathbf{r}}\end{pmatrix} = \begin{pmatrix}\mathbf{0} & \mathbf{I}\\ \mathbf{A} & \mathbf{0}\end{pmatrix}\begin{pmatrix}\mathbf{r}\\ \dot{\mathbf{r}}\end{pmatrix} + \begin{pmatrix}\mathbf{0}\\ \mathbf{f}/\rho\end{pmatrix}$$

with $\mathbf{f}/\rho = (\mathbf{f}^i/\rho_i)_{i\in\mathcal{I}}$ and

$$(4.4)\qquad \mathbf{A}\mathbf{r} = ({\rho_i}^{-1}\mathbf{Q}_i\mathbf{r}^i_{,xx})_{i\in\mathcal{I}}$$

defined in a formal fashion for the moment. Two variants of this operator will be defined below on domains which incorporate several of the side conditions occurring in the system (4.1).

The precise formulation of the problem depends on the definition of appropriate spaces. We shall usually require $\dot{\mathbf{r}}(\cdot,t)$ and the initial data $\dot{\mathbf{r}}_0$ to belong to

$$\mathbf{H} = \prod_{i\in\mathcal{I}} L_2(0,l_i;\mathbb{R}^3),$$

where $L_2(0, l_i; I\!R^3)$ is the space of square integrable functions mapping $(0, l_i)$ into $I\!R^3$. We provide $\mathbf{H}$ with the inner product

$$(4.5) \qquad (\mathbf{r}, \mathbf{s})_{\mathbf{H}} = \sum_{i \in \mathcal{I}} \frac{1}{2} \int_0^{l_i} \rho_i \mathbf{r}^i \cdot \mathbf{s}^i \, dx,$$

which corresponds to the kinetic energy. We shall require $\mathbf{r}(\cdot, t)$ and the initial data $\mathbf{r}_0$ to belong to the space

$$\mathbf{V} = \left\{ \mathbf{r} \in \prod_{i \in \mathcal{I}} H^1(0, l_i; I\!R^3) \middle| \begin{cases} \mathbf{r}^i(x_{ij}) \text{ coincide for } j \in \mathcal{J}_M \text{ and } i \in \mathcal{I}_j, \\ \mathcal{D}^j \mathbf{r} = 0 \text{ for } j \in \mathcal{J}^C \end{cases} \right\},$$

where $H^k(0, l_i; I\!R^3)$ denotes the space of functions which have square integrable distributional derivatives up to order k. On $\mathbf{V}$ we we define

$$(4.6) \qquad (\mathbf{r}, \mathbf{s})_{\mathbf{V}} = \sum_{i \in \mathcal{I}} \frac{1}{2} \int_0^{l_i} \mathbf{Q}_i \mathbf{r}^i_{,x} \cdot \mathbf{s}^i_{,x} \, dx,$$

which corresponds to the potential energy, as well as

$$(4.7) \qquad (\mathbf{r}, \mathbf{s})_{\mathbf{V},\lambda} = (\mathbf{r}, \mathbf{s})_{\mathbf{V}} + \lambda (\mathbf{r}, \mathbf{s})_{\mathbf{H}}.$$

For $\lambda > 0$, $\| \mathbf{r} \|_{\mathbf{V},\lambda} = (\mathbf{r}, \mathbf{r})^{1/2}_{\mathbf{V},\lambda}$ is a norm for $\mathbf{V}$ equivalent to the standard product norm. In case $\mathcal{J}^C \neq \emptyset$ it follows from the connectedness of the network that $\| \mathbf{r} \|_{\mathbf{V}} = \| \mathbf{r} \|_{\mathbf{V},0}$ is also an equivalent norm. We shall need the following subspace of $\mathbf{V}$:

$$\mathbf{V}_0 = \left\{ \mathbf{r} \in \mathbf{V} : \mathcal{D}^j \mathbf{r} = 0 \text{ for } j \in \mathcal{J}^D \right\}.$$

This space is densely imbedded in $\mathbf{H}$. Only in the case of homogeneous Dirichlet data can the solution $\mathbf{r}(\cdot, t)$ be expected to remain in $\mathbf{V}_0$. Note that $\| \cdot \|_{\mathbf{V}}$ is an inner product on $\mathbf{V}_0$ if and only if $\mathcal{J}^C \cup \mathcal{J}^D \neq \emptyset$. Finally we introduce the energy spaces

$$\mathcal{H} = \mathbf{V} \times \mathbf{H}, \qquad \mathcal{H}_0 = \mathbf{V}_0 \times \mathbf{H}.$$

In view of (4.3) it is natural to try to obtain the solution of (4.1) with the help of semigroups (which will turn out to be unitary groups) acting on $\mathcal{H}$ or $\mathcal{H}_0$ generated by operators of the form

$$(4.8) \qquad \mathcal{A} = \begin{pmatrix} 0 & \mathbf{I} \\ \mathbf{A} & 0 \end{pmatrix}$$

acting on appropriate domains.

In order to specify suitable domains for $\mathbf{A}$ we consider first the following stationary problem which if $\lambda = 0$ and $\mathbf{f}$ is replaced by $\mathbf{f}/\rho$ is just the stationary version of (4.1):

$$
(4.9) \qquad \begin{cases} \lambda \mathbf{r} - \mathbf{A}\mathbf{r} = \mathbf{f}, \\ \mathbf{r}^i(x_{ij}) \text{ coincide for } j \in \mathcal{J}_M,\ i \in \mathcal{I}_j, \\ \mathcal{D}^j\mathbf{r} = \mathbf{u}^j \text{ for } j \in \mathcal{J}^D,\ \mathcal{D}^j\mathbf{r} = \mathbf{0} \text{ for } j \in \mathcal{J}^C, \\ \mathcal{N}^j\mathbf{r} = \mathbf{u}^j \text{ for } j \in \mathcal{J}^N,\ \mathcal{N}^j\mathbf{r} = \mathbf{0} \text{ for } j \in \mathcal{J}^F. \end{cases}
$$

Given $\mathbf{f} \in \mathbf{H}$ and vectors $\mathbf{u}_j \in \mathbb{R}^3$ for $j \in \mathcal{J}^D \cup \mathcal{J}^N$ we say that $\mathbf{r}$ is a *weak solution* of (4.9) if and only if $\mathbf{r}$ belongs to $\mathbf{V}$ and satisfies $\mathcal{D}^j\mathbf{r} = \mathbf{u}^j$ for $j \in \mathcal{J}^D$ as well as the weak identity

$$
(4.10) \qquad (\mathbf{r}, \boldsymbol{\phi})_{\mathbf{V},\lambda} = (\mathbf{f}, \boldsymbol{\phi})_{\mathbf{H}} + \sum_{j \in \mathcal{J}^N} \mathbf{u}^j \cdot \mathcal{D}^j \boldsymbol{\phi}
$$

for all $\boldsymbol{\phi} \in \mathbf{V}_0$. It is easy to convince oneself that any sufficiently smooth weak solution indeed satisfies the equation as well as the latter two dynamic conditions in (4.9).

The uniqueness of solution to (4.9) is easily verified by subtracting the weak identities for two given solutions, then setting $\boldsymbol{\phi}$ equal to the difference of solutions to get $(\boldsymbol{\phi}, \boldsymbol{\phi})_{\mathbf{V},\lambda} = 0$ and hence $\boldsymbol{\phi} = \mathbf{0}$. Existence of weak solutions is obtained by a standard application of the Riesz representation theorem in case $\lambda > 0$ and also, if $\mathcal{J}^C \cup \mathcal{J}^D \neq \emptyset$, when $\lambda = 0$. One introduces $\tilde{\mathbf{r}} \in \mathbf{V}$ satisfying $\tilde{\mathcal{N}}^j = \mathbf{u}^j$ for $j \in \mathcal{J}^D$, where $\tilde{\mathcal{N}}^j$ denote the nodal positions corresponding to $\tilde{\mathbf{r}}$. One then seeks a solution to (4.9) in the form $\mathbf{r} = \tilde{\mathbf{r}} + \mathbf{s}$. Then $\mathbf{s}$ must belong to $\mathbf{V}_0$ satisfy

$$
(4.11) \qquad (\mathbf{s}, \boldsymbol{\phi})_{\mathbf{V},\lambda} = (\tilde{\mathbf{r}}, \boldsymbol{\phi})_{\mathbf{V},\lambda} - (\mathbf{f}, \boldsymbol{\phi})_{\mathbf{H}} + \sum_{j \in \mathcal{J}^N} \mathbf{u}^j \cdot \mathcal{D}^j \boldsymbol{\phi}
$$

for all $\boldsymbol{\phi} \in \mathbf{V}_0$. The right hand side of the latter identity defines a linear functional on $\mathbf{V}_0$, which is then representable by some $\mathbf{s}$ in terms of the inner product $(\cdot,\cdot)_{\mathbf{V},\lambda}$.

In particular we can solve the weak stationary problem with homogeneous Dirichlet and Neumann data, in which case the solution lies in $\mathbf{V}_0$. One can for $\lambda > 0$ (and, in case $\mathcal{J}^C \cup \mathcal{J}^D \neq \emptyset$ also for $\lambda = 0$) define the solution operator $\mathbf{S}_\lambda : \mathbf{H} \mapsto \mathbf{V}_0 \subset \mathbf{H}$ by $\mathbf{S}_\lambda \mathbf{f} = \mathbf{r}$ where $\mathbf{r}$ is the unique solution of (4.10) with each $\mathbf{u}^j = \mathbf{0}$. The operator is injective, since if $\mathbf{S}_\lambda \mathbf{f} = \mathbf{0}$ (4.10) implies that $(\mathbf{f}, \boldsymbol{\phi})_{\mathbf{H}} = 0$ for all $\boldsymbol{\phi} \in \mathbf{V}_0$; thus $\mathbf{f} = \mathbf{0}$ since $\mathbf{V}_0$ is dense in $\mathbf{H}$. Setting $\boldsymbol{\phi} = \mathbf{S}_\lambda \mathbf{g}$ in (4.10) one gets the identity

$$
(4.12) \qquad (\mathbf{S}_\lambda \mathbf{f}, \mathbf{S}_\lambda \mathbf{g})_{\mathbf{V},\lambda} = (\mathbf{f}, \mathbf{S}_\lambda \mathbf{g})_{\mathbf{H}}.
$$

From this it follows that $\mathbf{S}_\lambda : \mathbf{H} \mapsto \mathbf{H}$ is bounded, symmetric and, taking into account the injectivity, positive definite. If we set $\mathbf{f} = \mathbf{g}$ and then apply Schwarz's inequality to the right hand side we also get the estimate $\| \mathbf{S}_\lambda \mathbf{f} \|_{\mathbf{V},\lambda} \leq \| \mathbf{f} \|_{\mathbf{H}}$. It then follows from Rellich's theorem, which asserts the compactness of the imbedding of $H^1(0, l_i; \mathbb{R}^3)$ into $L_2(0, l_i; \mathbb{R}^3)$, that $\mathbf{S}_\lambda : \mathbf{H} \mapsto \mathbf{H}$ is a compact operator. Consequently the spectrum consists of a sequence of eigenvalues of finite multiplicity accumulating at 0 and corresponding to a complete orthonormal system of eigenfunctions. The eigenvalues are positive.

The image of $\mathbf{S}_\lambda$ is dense in $\mathcal{H}$. We define the self adjoint operator $\mathbf{A}_\lambda = -\mathbf{S}_\lambda^{-1}$ on $\mathrm{dom}(\mathbf{A}_\lambda) = \mathrm{im}(\mathbf{S}_\lambda)$. The domain does not depend on λ because

$$(\mathbf{r}, \phi)_{\mathbf{V},\lambda} = (\mathbf{f}, \phi)_{\mathbf{H}} \tag{4.13}$$

implies

$$(\mathbf{r}, \phi)_{\mathbf{V},\mu} = (\mathbf{f} + [\mu - \lambda]\mathbf{r}, \phi)_{\mathbf{H}} \tag{4.14}$$

and therefore if $\mathbf{r} = \mathbf{S}_\lambda \mathbf{f}$ one has $\mathbf{r} = \mathbf{S}_\mu(\mathbf{f} + [\mu - \lambda]\mathbf{r})$. One can in fact verify that $\mathrm{dom}(\mathbf{A}_\lambda)$ is the subset of $\prod_{i \in \mathcal{I}} H^2(0, l_i; \mathbb{R}^3)$ consisting of elements satisfying all the homogeneous geometric and dynamic side conditions associated with (4.10). Finally, in case $(\cdot, \cdot)_{\mathbf{V}}$ is not an inner product on $\mathbf{V}_0$ so that the above arguments did not apply for $\lambda = 0$, we define $\mathbf{A}_0 = \mathbf{A}_\lambda + \lambda \mathbf{I}$ with $\mathrm{dom}(\mathbf{A}_0) = \mathrm{dom}(\mathbf{A}_\lambda)$. This definition is indeed independent of λ because it follows from (4.13) and (4.14) that $\mathbf{A}_\mu \mathbf{r} = \mathbf{A}_\lambda \mathbf{r} - [\mu - \lambda]\mathbf{r}$ so that $[\mathbf{A}_\lambda + \lambda \mathbf{I}]\mathbf{r} = [\mathbf{A}_\mu + \mu \mathbf{I}]\mathbf{r}$. In this way we obtain a self-adjoint operator $\mathbf{A}_0$ in $\mathbf{H}$ for which it is easy to verify that

$$(\mathbf{A}_0 \mathbf{r}, \mathbf{\Phi})_H = -(\mathbf{r}, \mathbf{\Phi})_V \quad \text{for } \mathbf{r} \in \mathrm{dom}(\mathbf{A}_0), \ \phi \in \mathbf{V}_0.$$

Therefore $\mathbf{A}_0$ is negative semi-definite (and definite if $(\cdot, \cdot)_{\mathbf{V}}$ is an inner product on $\mathbf{V}_0$) and the spectrum consists of an decreasing sequence of non-positive real eigenvalues of finite multiplicity accumulating at ∞ and corresponding to a complete orthonormal system of eigenfunctions. The first eigenvalue is zero if and only if there are neither clamped nor Dirichlet controlled nodes and the corresponding eigenspace then consists of $\mathbf{r}$'s which are identically constant across the network.

Returning to the non-stationary problem we introduce the operator $\boldsymbol{\mathcal{A}}_0$ on $\boldsymbol{\mathcal{H}}_0$ by

$$\begin{cases} \boldsymbol{\mathcal{A}}_0[\mathbf{r} \times \dot{\mathbf{r}}] = \dot{\mathbf{r}} \times \mathbf{A}_0 \mathbf{r}, \\ \text{with } \mathbf{r} \times \dot{\mathbf{r}} \in \mathrm{dom}(\boldsymbol{\mathcal{A}}_0) = \mathrm{dom}(\mathbf{A}_0) \times \mathbf{V}_0. \end{cases} \tag{4.15}$$

In case $\mathcal{J}^C \cup \mathcal{J}^D \neq \emptyset$ (so that $(\cdot, \cdot)_{\mathbf{V}}$ is an inner product on V_0) it is a simple exercise in definitions to show that this is a skew adjoint operator on $\boldsymbol{\mathcal{H}}_0$ with respect to the energy inner product

$$(\mathbf{r} \times \dot{\mathbf{r}}, \mathbf{s} \times \dot{\mathbf{s}})_E = (\mathbf{r}, \mathbf{s})_{\mathbf{V}} + (\dot{\mathbf{r}}, \dot{\mathbf{s}})_{\mathbf{H}}$$

and hence generates a group of unitary operators $\mathcal{U}_0(t)_{t\in\mathbb{R}}$ on $\mathcal{H}_0$. If, on the other hand, $\mathcal{J}^C \cup \mathcal{J}^D = \emptyset$, this is not quite true. We then use the inner product $(\cdot,\cdot)_{\mathbf{V},1}$ on $\mathbf{V}_0 = \mathbf{V}$ and set $\mathbf{V}_{00} = \ker(\mathbf{A}_0)$ and $\mathbf{V}_{01}$ equal to the orthogonal complement of $\mathbf{V}_{00}$ in $\mathbf{V}$. In the present situation V_{00} consists of $\mathbf{r}$'s which are constant across the network. On $\mathbf{V}_{01}$ $(\cdot,\cdot)_V$ is again an inner product. We also set $\mathbf{H}_{00} = \mathbf{V}_{00}$ and $\mathbf{H}_{01}$ equal to the orthogonal complement of $\mathbf{H}_{00}$ in $\mathbf{H}$; the latter is simply the direct sum of all the eigenspaces corresponding to non-zero eigenvalues. One easily verifies that V_{01} is dense in H_{01} and that the restriction $\mathbf{A}_{01}$ of $\mathbf{A}_0$ to $\text{dom}(\mathbf{A}_{01}) = \text{dom}(\mathbf{A}_0) \cap \mathbf{H}_{01}$ is a self-adjoint operator in H_{01}. Then

$$(4.16)\qquad \begin{cases} \mathcal{A}_{01}[\mathbf{r}\times\dot{\mathbf{r}}] = \dot{\mathbf{r}} \times \mathbf{A}_{01}\mathbf{r}, \\ \text{with } \mathbf{r}\times\dot{\mathbf{r}} \in \text{dom}(\mathcal{A}_{01}) = \text{dom}(\mathbf{A}_{01}) \times \mathbf{V}_{01} \end{cases}$$

is a skew adjoint operator on $\mathcal{H}_{01} = \mathbf{V}_{01} \times \mathbf{H}_{01}$ with respect to the energy inner product, and as before generates a unitary group $\mathcal{U}_{01}(t)_{t\in\mathbb{R}}$ on $\mathcal{H}_{01}$. We define $\mathcal{H}_{00} = \mathbf{V}_{00} \times \mathbf{H}_{00}$ and define $\mathcal{U}_{00}(t)_{t\in\mathbb{R}}$ on $\mathcal{H}_{00}$ by

$$\mathcal{U}_{00}(t)[\mathbf{r}_{00} \times \dot{\mathbf{r}}_{00}] = (\mathbf{r}_{00} + t\dot{\mathbf{r}}_{00}) \times \dot{\mathbf{r}}_{00}.$$

This is a group of operators which is generated by the bounded operator $\mathcal{A}_{00}[\mathbf{r}_{00}\times\dot{\mathbf{r}}_{00}] = \dot{\mathbf{r}}_{00} \times \mathbf{0}$. This group preserves energy but cannot be called unitary since energy is not a norm on $\mathcal{H}_{00}$! Given $\mathbf{r}_0 \times \dot{\mathbf{r}}_0 \in \mathcal{H}_0$ we can decompose it into the sum of elements of $\mathcal{H}_{00}$ and $\mathcal{H}_{01}$:

$$\mathbf{r}_0 \times \dot{\mathbf{r}}_0 = \mathbf{r}_{00} \times \dot{\mathbf{r}}_{00} + \mathbf{r}_{01} \times \dot{\mathbf{r}}_{01}.$$

One can then define

$$(4.17)\qquad \mathcal{U}_0(t)[\mathbf{r}_0 \times \dot{\mathbf{r}}_0] = \mathcal{U}_{00}(t)[\mathbf{r}_{00} \times \dot{\mathbf{r}}_{00}] + \mathcal{U}_{01}(t)[\mathbf{r}_{01} \times \dot{\mathbf{r}}_{01}].$$

$\mathbf{V}_0$ is a proper subspace of $\mathbf{V}$ if and only if $\mathcal{J}^D \neq \emptyset$. In that case we would like to extend the semigroup from $\mathbf{V}_0$ to $\mathbf{V}$. In order to do so let $\mathbf{V}_1$ denote the orthogonal complement of $\mathbf{V}_0$ with respect to the inner product $(\cdot,\cdot)_{\mathbf{V}}$. Then the condition that $\mathbf{r}_1$ belongs to $\mathbf{V}_1$ is equivalent to it being a weak solution of

$$\begin{cases} \mathbf{Ar} = \mathbf{0}, \\ \mathbf{r}^i(x_{ij}) \text{ coincide for } j \in \mathcal{J}_M,\, i \in \mathcal{I}_j, \\ \mathcal{D}^j\mathbf{r} = \mathcal{D}^j\mathbf{r}_1 \text{ for } j \in \mathcal{J}^D,\ \mathcal{D}^j\mathbf{r} = \mathbf{0} \text{ for } j \in \mathcal{J}^C, \\ \mathcal{N}^j\mathbf{r} = \mathbf{0} \text{ for } j \in \mathcal{J}^N \cup \mathcal{J}^F. \end{cases}$$

Then $\mathbf{r}_1$ can be regarded as a stationary solution to the hyperbolic system with Dirichlet data determined by itself. Hence given $\mathbf{r}_0 \times \dot{\mathbf{r}}_0 \in \mathcal{H}$ we can first decompose $\mathbf{r}_0$ in $\mathbf{V}$ as $\mathbf{r}_0 = \mathbf{r}_{00} + \mathbf{r}_1$ with $\mathbf{r}_{00} \in \mathbf{V}_0$ and $\mathbf{r}_1 \in \mathbf{V}_1$. One has $\mathcal{D}^j\mathbf{r}_1 = \mathcal{D}^j\mathbf{r}_0$ for each $j \in \mathcal{J}^D$ and the orthogonality condition is the weak formulation of $\mathbf{A}\mathbf{r}_1 = 0$ and $\mathcal{N}^j\mathbf{r} = \mathbf{0}$ for $j \in \mathcal{J}^N \cup \mathcal{J}^F$. The initial data can also be decomposed into

$$\mathbf{r}_0 \times \dot{\mathbf{r}}_0 = [\mathbf{r}_{00} \times \dot{\mathbf{r}}_0] + [\mathbf{r}_1 \times \mathbf{0}],$$

and one then defines

$$\mathcal{U}(t)[\mathbf{r}_0 \times \dot{\mathbf{r}}_0] = \mathcal{U}_0(t)[\mathbf{r}_0 \times \dot{\mathbf{r}}_0] + [\mathbf{r}_1 \times \mathbf{0}].$$

This gives a unitary semigroup on $\mathcal{H}$ which solves the initial value problem for initial data in $\mathcal{H}$, with vanishing Neumann data and distributed forces and time independent Dirichlet data $\mathbf{u}^j = \mathcal{D}^j\mathbf{r}_0$. The generator of this group is the operator defined by

$$\begin{cases} \mathcal{A}[\mathbf{r} \times \dot{\mathbf{r}}] = \dot{\mathbf{r}} \times \mathbf{A}\mathbf{r}, \\ \text{with } \mathbf{r} \times \dot{\mathbf{r}} \in [\mathrm{dom}(\mathbf{A}_0) + \mathbf{V}_1] \times \mathbf{V}. \end{cases}$$

In case $\mathbf{V}_0 = \mathbf{V}$ (or $\mathbf{V}_1 = \{\mathbf{0}\}$) we shall from now on simply write $\mathcal{U}(t) = \mathcal{U}_0(t)$.

Given initial data $\mathbf{r}_0 \times \dot{\mathbf{r}}_0 \in \mathcal{H}$ and $\mathbf{f}(t) \in L_1(0,T;\mathbf{H})$ one obtains a solution to the system (4.1) with time independent Dirichlet data $\mathbf{u}^j = \mathcal{D}^j\mathbf{r}_0$ and vanishing Neumann data by setting

$$\mathbf{r}(t) \times \dot{\mathbf{r}}(t) = \mathcal{U}(t)[\mathbf{r}_0 \times \dot{\mathbf{r}}] + \int_0^t \mathcal{U}(t-s)[\mathbf{0} \times \mathbf{f}(s)/\rho]\, ds. \tag{4.18}$$

Solutions obtained in this way are called "mild solutions" and are extensively discussed in Chapter 4 of Pazy [83]. In particular the following regularity result can be inferred from Theorem 2.4 and Corollary 2.5 of that Chapter.

LEMMA 4.1. *Let $\mathcal{B}$ be the generator of a strongly continuous semigroup $\mathcal{T}(t)$ on a Banach space $\mathbf{X}$. Suppose that $\mathbf{f}(t) \in C_0^k((0,T];\mathbf{X})$ and that $\mathbf{x}_0 \in dom(\mathcal{B}^k)$. Then the mild solution of the equation*

$$\frac{d}{dt}\mathbf{x}(t) = \mathcal{B}\mathbf{x}(t) + \mathbf{f}(t), \quad x(0) = x^0$$

belongs to $C^k([0,T];\mathbf{X}) \cap C^{k-1}([0,T];\mathrm{dom}(\mathcal{B}))$, where $\mathrm{dom}(\mathcal{B})$ is normed by the graph norm $\| \mathbf{x} \|_{\mathcal{B}}^2 = \| \mathcal{B}\mathbf{x} \|_{\mathbf{X}}^2 + \| \mathbf{x} \|_{\mathbf{X}}^2$. Moreover, for $i = 1, \ldots, k$, the i-th order time derivatives $\mathbf{x}^{(i)}(t)$ satisfy the equation

$$\frac{d}{dt}\mathbf{x}^{(i)}(t) = \mathcal{B}\mathbf{x}^{(i)}(t) + \mathbf{f}^{(i)}(t), \quad \mathbf{x}^{(i)}(0) = \mathcal{B}^i\mathbf{x}_0.$$

In particular one can conclude that if each $\mathbf{f}^i \in C^\infty([0,l_i] \times [0,T]; \mathbb{R}^3)$ and if $\mathbf{r}_0 \times \dot{\mathbf{r}} \in \operatorname{dom}(\mathcal{A}^\infty)$ the solution $\mathbf{r}(t) \times \dot{\mathbf{r}}(t)$ lies in $C^\infty([0,T]; \mathbf{V} \times \mathbf{H})$ and consequently that each $\mathbf{r}^i(x,t) \in C^\infty([0,l_i] \times [0,T]; \mathbb{R}^3)$.

We next consider time dependent Dirichlet data and inhomogeneous Neumann data in (4.1). The Dirichlet data has to be compatible with the initial data. We start with data satisfying

$$
\begin{cases} \mathbf{u}_j(t) - \mathcal{D}^j \mathbf{r}_0 \in \mathbf{C}_0^\infty((0,T]; \mathbb{R}^3) & \text{for } j \in \mathcal{J}^D, \\ \mathbf{u}_j(t) \in \mathbf{C}_0^\infty((0,T]; \mathbb{R}^3) & \text{for } j \in \mathcal{J}^N. \end{cases} \tag{4.19}
$$

Since we already know how to obtain mild solutions corresponding to initial data and distributed forces we can, for the moment, let these vanish. In that case compatibility requires that $\mathbf{u}_j(t) \in \mathbf{C}_0^\infty((0,T]; \mathbb{R}^3)$ for $j \in \mathcal{J}^D$. Note that functions in $\mathbf{C}_0^\infty((0,T]; \mathbb{R}^3)$ vanish near $t = 0$. Let $\tilde{\mathbf{r}}(t)$ denote the weak solution of (4.9) for arbitrary, but fixed, t in $[0,T]$ with $\mathbf{f} = \mathbf{0}$, $\mathbf{u}^j = \mathbf{u}^j(t)$ for $j \in \mathcal{J}^D \cup \mathcal{J}^N$. Then $\tilde{\mathbf{r}}(t) \in \mathbf{C}^\infty((0,T]; \mathbf{V})$ and it is also easy to check that each $\tilde{\mathbf{r}}^i(x,t) \in C^\infty([0,l_i] \times [0,T]; \mathbb{R}^3)$. Now one seeks a solution to (4.1) having the form $\mathbf{r} = \tilde{\mathbf{r}} + \mathbf{s}$ where $\mathbf{s}$ has to satisfy

$$
\begin{cases} \rho_i \ddot{\mathbf{s}}^i = \mathbf{Q}_i \mathbf{s}^i_{,xx} + \tilde{\mathbf{f}}^i \ \text{ for } i \in \mathcal{I}, \\ \mathbf{s}^i(x_{ij}, t) \text{ coincide for } j \in \mathcal{J}_M,\ i \in \mathcal{I}_j, \\ \mathcal{D}^j \mathbf{r}(t) = \mathbf{0} \ \text{ for } j \in \mathcal{J}^D \cup \mathcal{J}^C, \\ \mathcal{N}^j \mathbf{r}(t) = \mathbf{0} \ \text{ for } j \in \mathcal{J}^F \cup \mathcal{J}^N, \\ \mathbf{s}(0) = \mathbf{0}, \ \ \dot{\mathbf{s}}(0) = \mathbf{0}, \end{cases} \tag{4.20}
$$

with $\tilde{\mathbf{f}}^i = -\rho_i \ddot{\tilde{\mathbf{r}}}^i + \lambda \tilde{\mathbf{r}}^i$. This system has the mild solution

$$
\mathbf{s}(t) \times \dot{\mathbf{s}}(t) = \int_0^t \mathcal{U}(t-s)[\mathbf{0} \times \tilde{\mathbf{f}}(s)/\rho]\, ds.
$$

It again follows from Lemma (4.1) that each $\mathbf{r}^i(x,t) \in C^\infty([0,l_i] \times [0,T]; \mathbb{R}^3)$ and hence the same is true for each component $\mathbf{r}^i(x,t)$ of the solution $\mathbf{r} = \tilde{\mathbf{r}} + \mathbf{s}$.

We have therefore proved the existence of smooth solutions to (4.1) corresponding to given smooth data. This existence result can be extended to less regular data by first proving an a priori estimate for solutions corresponding to smooth data.

Here and later the following notation will be useful:

$$
e_i(x,t) = \frac{1}{2}\left[\rho_i |\dot{\mathbf{r}}^i(x,t)|^2 + \mathbf{Q}_i \mathbf{r}^i_{,x}(x,t) \cdot \mathbf{r}^i_{,x}(x,t)\right],
$$

and

$$
\mathcal{E}_i(t; x_1, x_2) = \int_{x_1}^{x_2} e_i(x,t)\, dx, \qquad \mathcal{E}_i^\dagger(x; t_1, t_2) = \int_{t_1}^{t_2} e_i(x,t)\, dt.
$$

The next lemma states a multiplier identity together with a resulting estimate; such identities go back at least to Rellich and have played an important role in the control of hyperbolic equations (see Lop Fat Ho [37] and Lions [60]).

LEMMA 4.2. *Let* $\mathbf{r}$ *be a smooth solution of*

$$\rho\ddot{\mathbf{r}}(x,t) = [\mathbf{Q}\mathbf{r}_{,x}]_{,x} \quad \text{for } (x,t) \in [0,l] \times [0,T], \tag{4.21}$$

where we allow $\mathbf{Q}$ *and* ρ *to depend on* x *and* $m(x)$ *is a given smooth* multiplier function. *Then*

$$\begin{aligned}\int_0^T \Big[e(x,t)m(x)\Big]_{x=0}^{x=l} dt = \int_0^T \int_0^l e(x,t)m_{,x}\,dxdt \\ + \frac{1}{2}\int_0^T \int_0^l \big[\rho_x|\dot{\mathbf{r}}|^2 - \mathbf{Q}_{,x}\mathbf{r}_{,x}\cdot\mathbf{r}_{,x}\big](x,t)m(x)\,dxdt \\ + \int_0^l \Big[m\rho\dot{\mathbf{r}}\cdot\mathbf{r}_{,x}\Big]_{t=0}^{t=T} dx.\end{aligned} \tag{4.22}$$

As a consequence

$$\mathcal{E}^\dagger(0;0,T) + \mathcal{E}^\dagger(l;0,T) \le C\left[\int_0^T \mathcal{E}(t)\,dt + \mathcal{E}(0) + \mathcal{E}(T)\right]. \tag{4.23}$$

PROOF. The idea is to multiply the equation by $m(x)\mathbf{r}_{,x}$ and then integrate by parts to get the desired identity. We first treat the two terms separately. Integration by parts with respect to t gives

$$\begin{aligned}\int_0^T \int_0^l \rho\ddot{\mathbf{r}}\cdot m\mathbf{r}_{,x}\,dxdt &= -\int_0^T \int_0^l m\rho\dot{\mathbf{r}}\cdot\dot{\mathbf{r}}_{,x}\,dxdt \\ &\quad + \int_0^l \Big[m\rho\dot{\mathbf{r}}\cdot\mathbf{r}_{,x}\Big]_{t=0}^{t=T} dx \\ &= \frac{1}{2}\int_0^T \int_0^l \big[m_{,x}\rho|\dot{\mathbf{r}}|^2 + m\rho_{,x}|\dot{\mathbf{r}}|^2\big]\,dxdt \\ &\quad - \int_0^T \Big[\frac{m}{2}\rho|\dot{\mathbf{r}}|^2\Big]_{x=0}^{x=l} dt + \int_0^l \Big[m\rho\dot{\mathbf{r}}\cdot\mathbf{r}_{,x}\Big]_{t=0}^{t=T} dx,\end{aligned} \tag{4.24}$$

where we have used the identity

$$m\rho\dot{\mathbf{r}}\cdot\dot{\mathbf{r}}_{_x} = \frac{1}{2}m\rho[|\dot{\mathbf{r}}|^2]_{_x} = \frac{1}{2}[m\rho|\dot{\mathbf{r}}|^2]_{,x} - \frac{1}{2}m_{,x}\rho|\dot{\mathbf{r}}|^2 - \frac{1}{2}m\rho_{,x}|\dot{\mathbf{r}}|^2,$$

and integrated $[m\rho|\dot{\mathbf{r}}|^2]_{,x}$ with respect to x. To obtain the next identity we note that

$$[\mathbf{Q}\mathbf{r}_{,x}]_{,x}\cdot\mathbf{r}_{,x} = \frac{1}{2}[\mathbf{Q}\mathbf{r}_{,x}\cdot\mathbf{r}_{,x}]_{,x} + \frac{1}{2}\mathbf{Q}_{,x}\mathbf{r}_{,x}\cdot\mathbf{r}_{,x}$$

and then integrate by parts with respect to x to get

$$
\begin{aligned}
(4.25)\qquad -\int_0^T\int_0^l [\mathbf{Q}\mathbf{r}_{,x}]_{,x}\cdot m\mathbf{r}_{,x}\,dxdt &= -\int_0^T\Big[\frac{m}{2}\mathbf{Q}\mathbf{r}_{,x}\cdot\mathbf{r}_{,x}\Big]_{x=0}^{x=l}dt \\
&+\int_0^T\int_0^l\big[\frac{m_{,x}}{2}\mathbf{Q}\mathbf{r}_{,x}\cdot\mathbf{r}_{,x}-\frac{m}{2}\mathbf{Q}_{,x}\mathbf{r}_{,x}\cdot\mathbf{r}_{,x}\big]\,dxdt
\end{aligned}
$$

If we add together the two identities (4.24) and (4.25) and regroup terms we get (4.22).

Finally we obtain (4.23) by setting $m(x) = -1 + 2x/l$ in (4.22) and by estimating all the integrands on the right hand side of that identity by suitable multiples of $e(x,t)$. □

In the remainder of this section we shall be making many estimates involving a constant C, which will simply be redefined from estimate to estimate with no change of notation.

LEMMA 4.3. $\mathbf{r}$ *be a smooth solution of* (4.1). *Then*

$$
(4.26)\qquad \mathcal{E}(t)\le C\left[\mathcal{E}(0)+\int_0^t\Big[\sum_{j\in\mathcal{J}^D}|\dot{\mathbf{u}}^j(\tau)|^2+\sum_{j\in\mathcal{J}^N}|\mathbf{u}^j(\tau)|^2\Big]\,d\tau\right].
$$

PROOF. We differentiate $\mathcal{E}(t)$, integrate by parts with respect to x and regroup the terms taking into account the system (4.1). This gives

$$
\begin{aligned}
\frac{d\mathcal{E}}{dt}(\tau) &= \sum_{i\in\mathcal{I}}\int_0^{l_i}\big[\rho_i\dot{\mathbf{r}}^i\cdot\ddot{\mathbf{r}}^i+\mathbf{Q}_i\mathbf{r}^i_{,x}\cdot\dot{\mathbf{r}}^i_{,x}\big](x,\tau)\,dx \\
&= \sum_{i\in\mathcal{I}}\Big[\mathbf{Q}_i\mathbf{r}^i_{,x}\cdot\dot{\mathbf{r}}^i\Big]_{x=0}^{x=l_i} \\
&= \sum_{j\in\mathcal{J}}\sum_{i\in\mathcal{I}_j}\epsilon_{ij}[\mathbf{Q}_i\mathbf{r}^i_{,x}\cdot\dot{\mathbf{r}}^i](x_{ij},\tau) \\
&= \sum_{j\in\mathcal{J}^D}\boldsymbol{\mathcal{N}}^j\mathbf{r}\cdot\dot{\mathbf{u}}^j(\tau)+\sum_{j\in\mathcal{J}^N}\boldsymbol{\mathcal{D}}^j\dot{\mathbf{r}}\cdot\mathbf{u}^j(\tau).
\end{aligned}
$$

By using the estimate $ab\le\epsilon a^2+b^2/4\epsilon$, this gives

$$
\begin{aligned}
(4.27)\qquad \frac{d\mathcal{E}}{dt}(\tau) &\le \epsilon\Big[\sum_{j\in\mathcal{J}^D}|\boldsymbol{\mathcal{N}}^j\mathbf{r}(\tau)|^2+\sum_{j\in\mathcal{J}^N}|\boldsymbol{\mathcal{D}}^j\dot{\mathbf{r}}(\tau)|^2\Big] \\
&\quad+\frac{1}{4\epsilon}\Big[\sum_{j\in\mathcal{J}^D}|\dot{\mathbf{u}}^j(\tau)|^2+\sum_{j\in\mathcal{J}^N}|\mathbf{u}^j(\tau)|^2\Big].
\end{aligned}
$$

Note that

$$\text{for any } j \in \mathcal{J} \quad \begin{cases} |\mathcal{D}^j \dot{\mathbf{r}}(\tau)|^2 \le C e_i(x_{ij}, t) \ \text{ for any } i \in \mathcal{I}_j, \\ |\mathcal{N}^j \mathbf{r}(\tau)|^2 \le C \sum_{i \in \mathcal{I}_j} e_i(x_{ij}, t). \end{cases}$$

Hence, integrating (4.27) from 0 to t and using (4.23) we obtain

$$\begin{aligned} \mathcal{E}(t) &\le \mathcal{E}(0) + \epsilon \int_0^t \Big[\sum_{j \in \mathcal{J}^D} |\mathcal{N}^j \mathbf{r}|^2 + \sum_{j \in \mathcal{J}^N} |\mathcal{D}^j \dot{\mathbf{r}}(\tau)|^2 \Big] \, d\tau \\ &\qquad + \frac{1}{4\epsilon} \int_0^t \Big[\sum_{j \in \mathcal{J}^D} |\dot{\mathbf{u}}^j(\tau)|^2 + \sum_{j \in \mathcal{J}^N} |\mathbf{u}^j(\tau)|^2 \Big] \, d\tau \\ &\le \mathcal{E}(0) + \epsilon C \Big[\int_0^t \mathcal{E}(\tau) \, d\tau + \mathcal{E}(0) + \mathcal{E}(t) \Big] \, d\tau \\ &\qquad + \frac{1}{4\epsilon} \int_0^t \Big[\sum_{j \in \mathcal{J}^D} |\dot{\mathbf{u}}^j(\tau)|^2 + \sum_{j \in \mathcal{J}^N} |\mathbf{u}^j(\tau)|^2 \Big] \, d\tau, \end{aligned}$$

where C is a positive constant. We can choose ϵ so that $\epsilon C < 1$ and then bring the term $\mathcal{E}(t)$ to the left and divide by $1 - \epsilon C$ to obtain the estimate

$$\begin{aligned} \mathcal{E}(t) \le C \Big[\mathcal{E}(0) &+ \int_0^t \mathcal{E}(\tau) \, d\tau \\ &+ \int_0^t \Big[\sum_{j \in \mathcal{J}^D} |\dot{\mathbf{u}}^j(\tau)|^2 + \sum_{j \in \mathcal{J}^N} |\mathbf{u}^j(\tau)|^2 \Big] d\tau \Big]. \end{aligned} \tag{4.28}$$

An easy application of Gronwall's inequality yields (4.26). □

THEOREM 4.1. *Suppose that*
(i) $\mathbf{r}_0 \in \mathbf{V}$, $\dot{\mathbf{r}}_0 \in \mathbf{H}$, $\mathbf{f} \in L_2(0, T; \mathbf{H})$;
(ii) $\mathbf{u}^j \in H^1(0, T; \mathbb{R}^3)$ *and satisfies* $\mathbf{u}^j(0) = \mathcal{D}^j \mathbf{r}_0$ *for each* $j \in \mathcal{J}^D$;
(iii) $\mathbf{u}^j(t) \in L_2(0, T; \mathbb{R}^3)$ *for each* $j \in \mathcal{J}^N$.
Then there exists a unique $\mathbf{r} \in C([0, T] : \mathbf{V}) \cap C^1([0, T] : \mathbf{H})$ *such that*

$$\mathcal{D}^j \mathbf{r}(t) = \mathbf{u}^j(t) \quad \text{for } j \in \mathcal{J}^D$$

and which satisfies the remaining requirements of (4.1) *in the following weak sense: for any* $\phi \in \mathbf{V}_0$ *the function* $(\mathbf{r}(t), \phi)_{\mathbf{H}}$ *has an absolutely continuous first derivative in* t *and almost everywhere satisfies*

$$\frac{d^2}{dt^2} (\mathbf{r}(t), \phi)_{\mathbf{H}} + (\mathbf{r}(t), \phi)_{\mathbf{V}} = (\mathbf{f}(t)/\rho, \phi)_{\mathbf{H}} + \sum_{j \in \mathcal{J}^N} \mathbf{u}^j(t) \cdot \mathcal{D}^j \phi \tag{4.29}$$

as well as the initial conditions

$$(\mathbf{r}(0), \phi)_{\mathbf{H}} = (\mathbf{r}_0, \phi)_{\mathbf{H}}, \ \frac{d}{dt}(\mathbf{r}(t), \phi)_{\mathbf{H}}\Big|_{t=0} = (\dot{\mathbf{r}}_0, \phi)_{\mathbf{H}}. \tag{4.30}$$

PROOF. Note first that the requirements on the regularity of $(\mathbf{r}(t), \phi)_{\mathbf{H}}$ as well as (4.29) and (4.30) are all neatly contained in the identity

$$(\mathbf{r}(t), \phi)_{\mathbf{H}} = (\mathbf{r}_0, \phi)_{\mathbf{H}} + t(\dot{\mathbf{r}}_0, \phi)_{\mathbf{H}} + \int_0^t \int_0^\tau \Big[-(\mathbf{r}(\sigma), \phi)_{\mathbf{V}} + (\mathbf{f}(\sigma)/\rho, \phi)_{\mathbf{H}} + \sum_{j \in \mathcal{J}^N} \mathbf{u}^j(\sigma) \cdot \mathcal{D}^j \phi\Big] \, d\sigma \, d\tau. \tag{4.31}$$

Next we prove the uniqueness of solutions. Given two solutions corresponding to the same data we let $\mathbf{s}$ denote the difference. Then $\mathbf{s}(t) \in \mathbf{V}_0$ and

$$(\mathbf{s}(t), \phi)_{\mathbf{H}} = \int_0^t \int_0^\tau \Big[-(\mathbf{s}(\sigma), \phi)_{\mathbf{V}}\Big] \, d\sigma \, d\tau \tag{4.32}$$

for all $\phi \in \mathbf{V}_0$. Now we can take as test functions the eigenfunctions of $\mathbf{A}_0$. If such an eigenfunction corresponds to the eigenvalue $-\mu$ we have $(\mathbf{s}(t), \phi)_{\mathbf{V}} = \mu(\mathbf{s}(t), \phi)_{\mathbf{H}}$. Consequently (4.31) becomes

$$(\mathbf{s}(t), \phi)_{\mathbf{H}} = \int_0^t \int_0^\tau \Big[-\mu(\mathbf{s}(\sigma), \phi)_{\mathbf{H}}\Big] \, d\sigma \, d\tau.$$

This implies that $(\mathbf{s}(t), \phi)_{\mathbf{H}} = y(t)$ satisfies

$$\ddot{y}(t) = -\mu y(t), \quad y(0) = 0, \ \dot{y}(0) = 0,$$

and hence vanishes identically. From the completeness of the system of eigenfunctions one concludes that $\mathbf{s}(t) = \mathbf{0}$ for all t, which completes the proof of uniqueness.

For the proof of existence we note that this has already been established for smooth data. It is easy to verify that the corresponding smooth solutions satisfy the identity (4.31). Smooth data is dense in the class of data specified by the hypotheses. We then obtain solutions corresponding to arbitrary data by "continuity", making use of the a priori inequality (4.26). □

5. Controllability of Networks of Elastic Strings

We introduce the spaces of controls. First let $\mathbf{u}^D(t)$ and $\mathbf{u}^N(t)$ respectively denote $(\mathbf{u}^j(t))_{j \in \mathcal{J}^D}$ and $(\mathbf{u}^j(t))_{j \in \mathcal{J}^N}$. These are required to belong to the spaces

$$\mathbf{U}^D = [H^1((0,T]; I\!R^3)]^{m^D} \quad \text{and} \quad \mathbf{U}^N = [L_2(0,T; I\!R^3)]^{m^N},$$

where m^D and m^N are the respective cardinalities of $\mathcal{J}^D$ and $\mathcal{J}^N$ and where $H^1((0,T];\mathbb{R}^3)$ is the subspace of $H^1(0,T;\mathbb{R}^3)$ consisting of functions vanishing at $t=0$. On these spaces we can use the inner products

$$(\mathbf{u}^D,\mathbf{v}^D)_{\mathbf{U}^D} = \sum_{j\in\mathcal{J}^D}\int_0^T \dot{\mathbf{u}}^j(t)\cdot\dot{\mathbf{v}}^j(t)\,dt$$

and

$$(\mathbf{u}^N,\mathbf{v}^N)_{\mathbf{U}^N} = \sum_{j\in\mathcal{J}^N}\int_0^T \mathbf{u}^j(t)\cdot\mathbf{v}^j(t)\,dt.$$

It is also convenient to let $\mathbf{U}=\mathbf{U}^D\times\mathbf{U}^N$ with elements $\mathbf{u}=\mathbf{u}^D\times\mathbf{u}^N$.

By Theorem 4.1 we know that for $\mathbf{u}\in\mathbf{U}$ there exists a unique solution to the following system, in which the distributed forces and the initial data have been set equal to zero:

$$(5.1)\quad \begin{cases} \rho_i\ddot{\mathbf{r}}^i = \mathbf{Q}_i\mathbf{r}^i_{,xx} \text{ for } i\in\mathcal{I}, \\ \mathbf{r}^i(x_{ij},t) \text{ coincide for } j\in\mathcal{J}_M,\ i\in\mathcal{I}_j, \\ \mathcal{D}^j\mathbf{r}(t) = \mathbf{u}^j(t) \text{ for } j\in\mathcal{J}^D,\ \mathcal{D}^j\mathbf{r}(t)=\mathbf{0} \text{ for } j\in\mathcal{J}^C, \\ \mathcal{N}^j\mathbf{r}(t) = \mathbf{u}^j(t) \text{ for } j\in\mathcal{J}^N,\ \mathcal{N}^j\mathbf{r}(t) = \mathbf{0} \text{ for } j\in\mathcal{J}^F, \\ \mathbf{r}(0) = \mathbf{0},\ \dot{\mathbf{r}}(0)=\mathbf{0}. \end{cases}$$

The solution defines a continuous trajectory $\mathbf{r}(t)\times\dot{\mathbf{r}}(t)$ in $\mathcal{H}$. We define the control to state map $\mathcal{C}_T:\mathbf{U}\mapsto\mathcal{H}$ by $\mathcal{C}_T\mathbf{u}=\mathbf{r}(T)\times\dot{\mathbf{r}}(T)$. This is a bounded linear operator. We would like to prove that this operator is surjective for T sufficiently large, which is equivalent to the assertion that the network is exactly controllable, and also closely related to the stabilizability of the network which will be discussed later.

It is well known that the surjectivity of the bounded linear operator $\mathcal{C}_T$ is equivalent to the following estimate on the adjoint operator $\mathcal{C}_T^\star:\mathcal{H}\mapsto\mathbf{U}$:

$$(5.2)\quad \|\boldsymbol{\lambda}_0\times\dot{\boldsymbol{\lambda}}_0\|_{\mathcal{H}}\le C\,\|\mathcal{C}_T^\star[\boldsymbol{\lambda}_0\times\dot{\boldsymbol{\lambda}}_0]\|_{\mathbf{U}},\quad \text{for all } \boldsymbol{\lambda}_0\times\dot{\boldsymbol{\lambda}}_0\in\mathcal{H}.$$

Here the spaces $\mathcal{H}$ and $\mathbf{U}$ have been identified with their dual spaces. First we need to identify $\mathcal{C}_T^\star[\boldsymbol{\lambda}_0\times\dot{\boldsymbol{\lambda}}_0]$. The form of this operator does of course depend on the choice of inner products.

We introduce the following adjoint system:

$$(5.3)\quad \begin{cases} \rho_i\ddot{\boldsymbol{\lambda}}^i = \mathbf{Q}_i\boldsymbol{\lambda}^i_{,xx} \text{ for } i\in\mathcal{I}, \\ \boldsymbol{\lambda}^i(x_{ij},t) \text{ coincide for } j\in\mathcal{J}_M,\ i\in\mathcal{I}_j, \\ \mathcal{D}^j\boldsymbol{\lambda}(t) = \mathcal{D}^j\boldsymbol{\lambda}_0 \text{ for } j\in\mathcal{J}^D,\ \mathcal{D}^j\boldsymbol{\lambda}(t)=\mathbf{0} \text{ for } j\in\mathcal{J}^C, \\ \mathcal{N}^j\boldsymbol{\lambda}(t) = \mathbf{0} \text{ for } j\in\mathcal{J}^N\cup\mathcal{J}^F, \\ \boldsymbol{\lambda}(T) = \boldsymbol{\lambda}_0,\ \dot{\boldsymbol{\lambda}}(T)=\dot{\boldsymbol{\lambda}}_0. \end{cases}$$

We then have the following characterization of $\mathcal{C}_T^\star$.

PROPOSITION 5.1. *Suppose that* $\mathcal{J}^C \neq \emptyset$. *Let* $\boldsymbol{\lambda}(t)$ *be the solution of* (4.3) *corresponding to terminal data* $\boldsymbol{\lambda}_0 \times \dot{\boldsymbol{\lambda}}_0 \in \boldsymbol{\mathcal{H}}$. *Then*

$$\mathcal{C}_T^\star[\boldsymbol{\lambda}_0 \times \dot{\boldsymbol{\lambda}}_0] = \left\{\int_0^t \boldsymbol{\mathcal{N}}^j \boldsymbol{\lambda}(\tau)\,d\tau\right\}_{j\in\mathcal{J}^D} \times \left\{\boldsymbol{\mathcal{D}}^j \dot{\boldsymbol{\lambda}}(t)\right\}_{j\in\mathcal{J}^N}, \tag{5.4}$$

where $\boldsymbol{\lambda}$ *is the solution of the adjoint system* (5.3).

PROOF. The hypothesis allows us to use $(\cdot,\cdot)_{\mathbf{V}}$ as inner product on $\mathbf{V}$-components of elements of $\boldsymbol{\mathcal{H}}$. We evaluate

$$\begin{aligned}
&\frac{d}{dt}\Big[(\mathbf{r},\boldsymbol{\lambda})_{\mathbf{V}} + (\dot{\mathbf{r}},\dot{\boldsymbol{\lambda}})_{\mathbf{H}}\Big] \\
&\quad = \frac{d}{dt}\sum_{i\in\mathcal{I}}\int_0^{l_i}\frac{1}{2}\Big[\mathbf{Q}_i\mathbf{r}^i_{,x}\cdot\boldsymbol{\lambda}^i_{,x} + \rho_i\dot{\mathbf{r}}^i\cdot\dot{\boldsymbol{\lambda}}^i\Big]\,dx \\
&\quad = \sum_{i\in\mathcal{I}}\int_0^{l_i}\frac{1}{2}\Big[\mathbf{Q}_i\mathbf{r}^i_{,x}\cdot\dot{\boldsymbol{\lambda}}^i_{,x} + \mathbf{Q}_i\dot{\mathbf{r}}^i_{,x}\cdot\boldsymbol{\lambda}^i_{,x} + \rho_i\dot{\mathbf{r}}^i\cdot\ddot{\boldsymbol{\lambda}}^i + \rho_i\ddot{\mathbf{r}}^i\cdot\dot{\boldsymbol{\lambda}}^i\Big]\,dx \\
&\quad = \sum_{i\in\mathcal{I}}\int_0^{l_i}\frac{1}{2}\Big[\big[-\mathbf{Q}_i\mathbf{r}^i_{,xx} + \rho_i\ddot{\mathbf{r}}^i\big]\cdot\dot{\boldsymbol{\lambda}}^i + \dot{\mathbf{r}}^i\cdot\big[-\mathbf{Q}_i\boldsymbol{\lambda}^i_{,xx} + \rho_i\ddot{\boldsymbol{\lambda}}^i\big]\Big]\,dx \\
&\qquad + \sum_{j\in\mathcal{J}}\sum_{i\in\mathcal{I}_j}\epsilon_{ij}\big[\mathbf{Q}_i\mathbf{r}^i_{,x}\cdot\dot{\boldsymbol{\lambda}}^i + \mathbf{Q}_i\dot{\mathbf{r}}^i\cdot\boldsymbol{\lambda}^i_{,x}\big](x_{ij},t),
\end{aligned}$$

where we have integrated by parts with respect to x. Thus we obtain

$$\begin{aligned}
\frac{d}{dt}(\mathbf{r}\times\dot{\mathbf{r}},\boldsymbol{\lambda}\times\dot{\boldsymbol{\lambda}})_{\boldsymbol{\mathcal{H}}} &= \sum_{j\in\mathcal{J}}\Big[\boldsymbol{\mathcal{D}}^j\dot{\boldsymbol{\lambda}}\cdot\boldsymbol{\mathcal{N}}^j\mathbf{r} + \boldsymbol{\mathcal{N}}^j\boldsymbol{\lambda}\cdot\boldsymbol{\mathcal{D}}^j\dot{\mathbf{r}}\Big] \\
&= \sum_{j\in\mathcal{J}^D}\boldsymbol{\mathcal{N}}^j\boldsymbol{\lambda}\cdot\dot{\mathbf{u}}^j + \sum_{j\in\mathcal{J}^N}\boldsymbol{\mathcal{D}}^j\dot{\boldsymbol{\lambda}}\cdot\mathbf{u}^j.
\end{aligned} \tag{5.5}$$

Integrating with respect to t from 0 to T one finds

$$(\mathcal{C}_T\mathbf{u},\boldsymbol{\lambda}_0\times\dot{\boldsymbol{\lambda}}_0)_{\boldsymbol{\mathcal{H}}} = \int_0^T\Big[\sum_{j\in\mathcal{J}^D}\boldsymbol{\mathcal{N}}^j\boldsymbol{\lambda}\cdot\dot{\mathbf{u}}^j + \sum_{j\in\mathcal{J}^N}D^j\dot{\boldsymbol{\lambda}}\cdot\mathbf{u}^j\Big]\,dt.$$

Hence (5.4) follows by Riesz identification, taking into account the inner products we have chosen for $\mathbf{U}^D$ and $\mathbf{U}^N$ and noting that

$$\int_0^T\boldsymbol{\mathcal{N}}^j\boldsymbol{\lambda}\cdot\dot{\mathbf{u}}^j\,dt = \int_0^T\dot{\boldsymbol{\mu}}^j\cdot\dot{\mathbf{u}}^j\,dt \quad \text{with } \boldsymbol{\mu}^j(t) = \int_0^t\boldsymbol{\mathcal{N}}^j\boldsymbol{\lambda}(\tau)\,d\tau.$$

□

Because of Proposition 5.1 the estimate (5.2), characterising exact controllability in time T, becomes

$$\text{(5.6)} \quad \|\boldsymbol{\lambda}_0 \times \dot{\boldsymbol{\lambda}}_0\|^2_{\mathcal{H}} \leq C \int_0^T \Big[\sum_{j \in \mathcal{J}^D} |\mathcal{N}^j \boldsymbol{\lambda}(t)|^2 + \sum_{j \in \mathcal{J}^N} |\mathcal{D}^j \dot{\boldsymbol{\lambda}}(t)|^2 \Big] \, dt.$$

This is a nontrivial estimate which will be established below under the strong hypothesis that the network forms a "tree" with controls acting at all simple nodes except for one, which is clamped. The estimate will in fact hold when T is larger than a certain critical time $T_\star$, for which we give an explicit expression. Examples will then be given of networks containing "closed circuits" for which the kernel of the adjoint operator $\boldsymbol{C}_T^\star$ is nontrivial for all $T > 0$, so that the network is not exactly controllable over any finite time interval.

5.1. Exact Controllability of Tree Networks. The proof of the estimate (5.6) for tree networks depends on the energy estimate stated in the following Lemma. This is a special case of the more general Theorem 4.3 of Chapter IV and no proof will be given for the moment. However the estimate is intuitively plausible if one remembers that in hyperbolic equations the energy flows along and between characteristics.

LEMMA 5.1. *Let* $\mathbf{r} \in W^2([0,l] \times [t_1, t_2]; \mathbb{R}^3$ *be a solution of*

$$\rho \ddot{\mathbf{r}} = \mathbf{Q} \mathbf{r}_{,xx} \quad on\ [0,l] \times [t_1, t_2],$$

where $\mathbf{Q}$ *has eigenvalues* h *and* $h(1 - s^{-1})$*, and the corresponding families of characteristics are given by*

$$x \pm \mu_k t = c, \ for\ k = 1,2 \ with\ \mu_1^2 = \frac{h(1-s^{-1})}{\rho},\ \mu_2^2 = \frac{h}{\rho}.$$

Then for $t_2 - t_1 \geq T_\star = 2S_\star = 2l/\mu_1$ *one has the estimate*

$$\text{(5.7)} \qquad \mathcal{E}^\uparrow(0; t_1 + S_\star, t_2 - S_\star) + \mathcal{E}(t; 0, l) \leq C \mathcal{E}^\uparrow(l; t_1, t_2),$$

which is valid for all $t \in [t_1 + S_\star, t_2 - S_\star]$. *The same estimate holds with* 0 *and* l *interchanged.*

This lemma can be applied to each element of the network; in the process we need to introduce the times $T_\star^i = 2S_\star^i$ associated with the i-th element. These times represent the maximal time it takes for a wave to move back and forth along the i-th element.

A second lemma is an immediate consequence of the geometric and dynamic conditions at multiple nodes.

LEMMA 5.2. *Suppose that* $\mathbf{r}$ *is a solution of the network equations* (5.1) *with the homogeneous initial condition replaced by the requirement* $\mathbf{r}(0) \times \dot{\mathbf{r}}(0) \in \boldsymbol{\mathcal{H}}$. *Let* $i_0 \in \mathcal{I}_j$ *for* $j \in \mathcal{J}_M$. *Then*

$$\mathcal{E}_{i_0}^{\uparrow}(x_{i_0 j}, t_1, t_2) \leq C \sum_{i \in \mathcal{I}_j \setminus \{i_0\}} \mathcal{E}_i^{\uparrow}(x_{ij}, t_1, t_2) \tag{5.8}$$

We are now in a position to state and prove our principal result on exact controllability. We let $[i_1, i_2, \ldots, i_r]$ denote the path which leads along the network elements labelled successively by $i_1, i_2, \ldots, i_r$, with the understanding that successive indices differ. A closed circuit in a network is then a path for which $i_r = i_1$.

THEOREM 5.1. *Suppose that the network contains no closed circuit, that all multiple nodes and no simple nodes are free, that exactly one simple node is clamped and that the remaining simple nodes are controlled by Dirichlet or Neumann controls. Then the network is exactly controllable for each time* $T \geq T_\star$, *where*

$$T_\star = \max\Big\{T_\star^{i_1} + T_\star^{i_1} + \cdots + T_\star^{i_r} : [i_1, i_2, \ldots, i_r] \text{ leads from the clamped node to a controlled node}\Big\}. \tag{5.9}$$

REMARK 5.1. The time $T_\star$ is the best possible for the general statement of the theorem as can easily be verified by considering the degenerate network consisting of a single string clamped at one end and controlled from the other, for which $T_\star$ coincides with the minimum time for controllability established, for example, with the help of the d'Alembert solution.

REMARK 5.2. the fact that one simple node is clamped ensures that the energy in fact is a norm on $\boldsymbol{\mathcal{H}}$. One can in fact allow a clamped multiple (rather than simple) node with controls operative at all simple nodes. In case no node is clamped the energy is no longer a norm, but our methods would easily give controllability in the quotient space of $\boldsymbol{\mathcal{H}}$ with the subspace consisting of states of zero energy. We are not sure whether exact controllability also holds in the whole space!

We introduce some notational conventions involving simple nodes. For $j \in \mathcal{J}_S$ the set $\mathcal{I}_j$ consists of a single index i_j. We then write $x_j = x_{i_j j}$ and $\epsilon_j = \epsilon_{i_j j}$.

To prove Theorem 1.1 one needs to establish the estimate (5.6) under the given hypotheses. This estimate is an immediate consequence of the more general estimate contained in the following theorem as it applies to the solution $\boldsymbol{\lambda}$ of (5.3), since

- The boundary conditions satisfied by $\boldsymbol{\lambda}$ imply that,

$$\begin{cases} ce_{i_j}(x_j,t) \le |\mathcal{N}^j \boldsymbol{\lambda}(t)|^2 \le Ce_{i_j}(x_j,t) & \text{for } j \in \mathcal{J}_S^D, \\ ce_{i_j}(x_j,t) \le |\mathcal{D}^j \dot{\boldsymbol{\lambda}}(t)|^2 \le Ce_{i_j}(x_j,t) & \text{for } j \in \mathcal{J}_S^N. \end{cases}$$

- The boundary conditions also imply energy conservation so that $\mathcal{E}(\frac{T}{2}) = \mathcal{E}(T)$.
- By Remark 5.2, $\mathcal{E}(T)$ is equivalent to $\| \boldsymbol{\lambda}_0 \times \dot{\boldsymbol{\lambda}}_0 \|^2_{\mathcal{H}}$.

THEOREM 5.2. *Assume that the hypotheses on the network made in the previous theorem are satisfied. Let* $\mathbf{r}$ *be a solution of the system* (5.1) *with the homogeneous initial condition replaced by the requirement* $\mathbf{r}(0) \times \dot{\mathbf{r}}(0) \in \mathcal{H}$. *Then*

$$\mathcal{E}(\frac{T_\star}{2}) = \sum_{i=1}^{n} \mathcal{E}_i(\frac{T_\star}{2}) \le C \sum_{j \in \mathcal{J}_S \setminus \{m\}} \mathcal{E}^\dagger_{i_j}(x_j; 0, T_\star), \tag{5.10}$$

where m *is the index of the clamped node.*

PROOF. We note first that because of the estimate in Lemma 4.3 it is enough to establish (5.10) for smooth solutions corresponding to smooth data, thus avoiding problems with regularity. The proof is by induction on the number n of elements. For $n = 1$ the result follows from Lemma 5.1. Suppose now that the estimate (5.10) holds for networks with fewer than n elements. Given a network with n elements, we can locate a multiple node, indexed by j_0 say, such that all "neighbouring" nodes except at most one are simple. To find this node one chooses a "longest" path leading from the single clamped node to another simple node. Here "length" is to be understood in terms of the number of elements traversed, with the understanding that no element is to be traversed more than once. Then we choose the penultimate node along that path. That node cannot have more than two neighbouring multiple nodes. If it did one would have to lie along the longest path from the clamped node and the path could be continued beyond the other to yield a longer path and hence a contradiction. Generally, unless the longest path traverses only two elements, the j_0-th node has exactly one multiple node as neighbour. Note that the step we have just described depends on the assumptions that the network is connected and contains no closed circuits.

We now consider the reduced network obtained by discarding the elements leading from our selected node to a neighbouring simple node other than, possibly, the clamped node. The reduced network has fewer than n elements and the l-th node is now simple. It is convenient to relable the original network in such a way that the discarded nodes have indices $1 \le j < j_0$, the discarded elements leading to the discarded nodes have indices $1 \leq i < j_0$ and the element leading to the j_0-th node has index j_0. By the induction hypothesis we have for the reduced network

$$\sum_{i=j_0}^{n} \mathcal{E}_i(\frac{\tilde{T}_\star}{2}) \leq C \sum_{j \in \tilde{\mathcal{J}}_S \setminus \{m\}} \mathcal{E}_{i_j}^{\dagger}(x_j; 0, \tilde{T}_\star), \tag{5.11}$$

where the tilde identifies quantities pertaining to the reduced network. We note that

$$T_\star = \tilde{T}_\star + \max_{1 \leq i < j_0} T_\star^i,$$

and that

$$\tilde{\mathcal{J}}_S = \{j_0\} \cup [\mathcal{J}_S \setminus \{j : 1 \leq j < j_0\}]. \tag{5.12}$$

By a shift of the time variable (5.11) implies

$$\begin{aligned} \sum_{i=j_0}^{n} \mathcal{E}_i(\frac{T_\star}{2}) &\leq C \sum_{j \in \tilde{\mathcal{J}}_S \setminus \{m\}} \mathcal{E}_{i_j}^{\dagger}(x_j; \frac{T_\star}{2} - \frac{\tilde{T}_\star}{2}, \frac{T_\star}{2} + \frac{\tilde{T}_\star}{2}) \\ &\leq C \left[\sum_{j \in \tilde{\mathcal{J}}_S \setminus \{m, j_0\}} \mathcal{E}_{i_j}^{\dagger}(x_j; 0, T_\star) + \mathcal{E}_{j_0}^{\dagger}(x_{j_0}; \frac{T_\star}{2} - \frac{\tilde{T}_\star}{2}, \frac{T_\star}{2} + \frac{\tilde{T}_\star}{2}) \right]. \end{aligned} \tag{5.13}$$

Here there is a small notational inconsistency arising from the fact that the node indexed by j_0 is multiple for the original network but simple for the reduced network: we have written x_{j_0} in place of $x_{j_0 j_0}$. Now by Lemma 5.2

$$\mathcal{E}_{j_0}^{\dagger}(x_{j_0}; \frac{T_\star}{2} - \frac{\tilde{T}_\star}{2}, \frac{T_\star}{2} + \frac{\tilde{T}_\star}{2}) \leq C \sum_{i=1}^{j_0 - 1} \mathcal{E}_i^{\dagger}(x_{i j_0}; \frac{T_\star}{2} - \frac{\tilde{T}_\star}{2}, \frac{T_\star}{2} + \frac{\tilde{T}_\star}{2}). \tag{5.14}$$

Using (5.13) and (5.14) we now get

$$\begin{aligned} \sum_{i=1}^{n} \mathcal{E}_i(\frac{T_\star}{2}) \leq C \Bigg[&\sum_{j \in \tilde{\mathcal{J}}_S - \{m, j_0\}} \mathcal{E}_{i_j}^{\dagger}(x_j; 0, T_\star) \\ &+ \sum_{i=1}^{j_0 - 1} \left[\mathcal{E}_i(\frac{T_\star}{2}) + \mathcal{E}_i^{\dagger}(x_{i j_0}; \frac{T_\star}{2} - \frac{\tilde{T}_\star}{2}, \frac{T_\star}{2} + \frac{\tilde{T}_\star}{2}) \right] \Bigg]. \end{aligned} \tag{5.15}$$

The proof of Theorem 5.2 is completed by noting that for $i = 1, \ldots, j_0 - 1$, Lemma 5.1 gives

$$\begin{aligned} \mathcal{E}_i(\frac{T_\star}{2}) + \mathcal{E}^{\dagger}(x_{i j_0}; \frac{T_\star}{2} - \frac{\tilde{T}_\star}{2}, \frac{T_\star}{2} + \frac{\tilde{T}_\star}{2}) &\leq C \mathcal{E}_i^{\dagger}(x_i; \frac{T_\star}{2} - \frac{\tilde{T}_\star}{2} - S_\star^i, \frac{T_\star}{2} + \frac{\tilde{T}_\star}{2} + S_\star^i) \\ &\leq C \mathcal{E}_i^{\dagger}(x_i; 0, T_\star). \end{aligned}$$

□

5.2. Lack of Controllability for Networks with Closed Circuits. In this section we shall show that if certain "rationality" conditions on the physical constants are satisfied a system with closed loops will not be exactly controllable. These conditions allow one to show that the adjoint operator $\boldsymbol{C}_T^\star$ has a non-trivial kernel for all positive T, in which case $\boldsymbol{C}_T$ can never be surjective. This raises a number of questions concerning the image of $\boldsymbol{C}_T$ which we have not been able to resolve.

We try to construct an element of the kernel of $\boldsymbol{C}_T^\star$ with the help of an "aberrant" eigenfunction $\boldsymbol{\phi} = (\boldsymbol{\phi}^i)_{i\in\mathcal{I}}$ corresponding to an eigenvalue ω^2 which satisfies not only the eigenvalue equations

$$(5.16)\qquad \begin{cases} \rho_i\omega^2\boldsymbol{\phi}^i &= \mathbf{Q}_i\boldsymbol{\phi}^i_{,xx} \ \text{ for } i\in\mathcal{I},\\ \boldsymbol{\phi}^i(x_{ij}) & \text{coincide for } j\in\mathcal{J}_M,\ i\in\mathcal{I}_j,\\ \boldsymbol{\mathcal{D}}^j\boldsymbol{\phi} &= \mathbf{0} \ \text{ for } j\in\mathcal{J}^D\cup\mathcal{J}^C,\\ \boldsymbol{\mathcal{N}}^j\boldsymbol{\phi}(t) &= \mathbf{0} \ \text{ for } j\in\mathcal{J}^N\cup\mathcal{J}^F. \end{cases}$$

but also the supplementary conditions

$$(5.17)\qquad \begin{cases} \boldsymbol{\mathcal{D}}^j\boldsymbol{\phi} &= \mathbf{0} \ \text{ for } j\in\mathcal{J}^N,\\ \boldsymbol{\mathcal{N}}^j\boldsymbol{\phi} &= \mathbf{0} \ \text{ for } j\in\mathcal{J}^D. \end{cases}$$

If we then define $\boldsymbol{\lambda} = \cos\omega(t-T)\boldsymbol{\phi}$ we obtain a solution of the adjoint system (5.3) corresponding to terminal data $\boldsymbol{\lambda}_0 = \boldsymbol{\phi}$ and $\dot{\boldsymbol{\lambda}}_0 = \mathbf{0}$ with $\boldsymbol{C}_T^\star[\boldsymbol{\lambda}_0\times\dot{\boldsymbol{\lambda}}_0] = 0$ by (5.4).

The system (5.16) determines the eigenvalues associated with the stationary network equations and hence, as discussed in Section 1.4, gives rise to a sequence of eigenvalues of finite multiplicity. The supplemental conditions (5.17) lead to an overdetermined system which can only be expected to have a solution under special circumstances. We shall illustrate a situation where these arise.

We consider a single closed circuit of r strings indexed by $i = 1,\ldots,r$, where such a circuit might be thought of as standing alone or might in fact be part of a larger network. We suppose that the indexing and parametrisation are such that the circuit is described by the path $[1,2,\ldots,r]$ with parameter x increasing from 0 to l_i as the path traverses the i-th string. It is also convenient to identify the index $r+1$ with 0 in accordance with the cyclic nature of the circuit. We rewrite the systems (5.16) together with

(5.17) explicitly for $(\phi)_{i=1}^{r}$ as

$$(5.18)\qquad \begin{cases} \rho_i\omega^2\phi^i &= \mathbf{Q}_i\phi^i_{,xx} \quad \text{for } i=1,\ldots,r,\\ \phi^i(l_i) &= \phi^{i+1}(0) = \mathbf{0} \quad \text{for } i=1,\ldots,r,\\ \mathbf{Q}_i\phi^i_{,x}(l_i) &= \mathbf{Q}_{i+1}\phi^{i+1}_{,x}(0) \quad \text{for } i=1,\ldots,r. \end{cases}$$

This corresponds to a situation where in the primal system controls are active at all the nodes.

In an attempt to solve (5.18) we consider first a single equation, dropping the index i,

$$(5.19)\qquad \begin{cases} \rho\omega^2\phi &= \mathbf{Q}\phi_{,xx},\\ \phi(0) &= \phi(l) \;=\; \mathbf{0}. \end{cases}$$

One can associate with $\mathbf{Q}$ an orthonormal basis $\{\mathbf{e}_1,\mathbf{e}_2,\mathbf{e}_3\}$ of $\mathbb{R}^3$ consisting of eigenvectors of $\mathbf{Q}$. We can take $\mathbf{e}_1$ to be the longitudinal direction of the string at equilibrium. Then

$$(5.20)\qquad \mathbf{Q}\mathbf{e}_k = \omega_k^2\mathbf{e}_k, \quad \text{with } \omega_1^2 = h,\ \omega_2^2=\omega_3^2 = h\big(1-\frac{1}{s}\big).$$

One now easily verifies that if for some k

$$(5.21)\qquad \omega\frac{\sqrt{\rho}l}{\omega_k\pi} \quad \text{is an integer,}$$

then (5.19) has the solution

$$(5.22)\qquad \phi_k(x) = \xi_k(x)\mathbf{e}_k \quad \text{with } \xi_k(x) = \sin\big(\frac{\sqrt{\rho}}{\omega_k}\omega x\big).$$

For a single element such a choice of ω is of course possible and in fact one then obtains a sequence of suitable values for ω.

Returning to the system (5.18) we again index all quantities by i and make the very strong, yet easily satisfied, assumption that

$$(5.23)\qquad \frac{\sqrt{\rho^i}l^i\omega_k^1}{\sqrt{\rho^1}l^1\omega_k^i} \quad \text{is rational for each } i=1,\ldots,r \text{ and } k=1,2,3.$$

From this it follows readily that one can find ω so that

$$(5.24)\qquad \omega\frac{\sqrt{\rho_i}l_i}{\omega_k^i\pi} \;=\; n_k^i \text{ for each } i=1,\ldots,r \text{ and } k=1,2,3,$$

where n_k^i is an integer which we can, without loss of generality, assume to be even. If we then let

$$\phi^i(x) = \sum_{k=1}^{3} c_k^i\xi_k^i(x)\mathbf{e}_k^i \quad \text{with } \xi_k^i(x) = \sin\big(\frac{\sqrt{\rho_i}}{\omega_k^i}\omega x\big),$$

where the c_i^k are arbitrary coefficients we obtain a solution to the first two conditions of (5.18). We now show that the coefficients can be chosen in such a way that the third condition is also satisfied. That condition translates into

$$\mathbf{Q}_i \sum_{k=1}^{3} c_k^i \xi_{k,x}^i(l_i)\mathbf{e}_k^i = \mathbf{Q}_{i+1} \sum_{k=1}^{3} c_k^{i+1} \xi_{k,x}^{i+1}(0)\mathbf{e}_k^{i+1} \quad \text{for } i = 1, \ldots, r,$$

or, taking into account (5.20) and (5.24) and cancelling ω from both sides,

$$\sum_{k=1}^{3} c_k^i \omega_k^i \sqrt{\rho_i}\mathbf{e}_k^i = \sum_{k=1}^{3} c_k^{i+1} \omega_k^i \sqrt{\rho_{i+1}}\mathbf{e}_k^{i+1} \quad \text{for } i = 1, \ldots, r.$$

To see that this can always be solved we introduce the orthogonal matrix $\mathbf{S}_i$ which has a_1^i, a_2^i and a_3^i as its columns and we incorporate the coefficients $c_k^i \omega_k^i \sqrt{\rho_i}$ with $k = 1, \ldots, 3$ into a column vector $\mathbf{c}^i$. The latter equations then take the form

$$\mathbf{S}_i \mathbf{c}^i = \mathbf{S}_{i+1}\mathbf{c}^{i+1} \quad \text{for } i = 1, \ldots, r,$$

and it is a trivial matter to see that this has a non-trivial solution given by $\mathbf{c}^i = \mathbf{S}_i^T \mathbf{S}_1 \mathbf{c}^1$ where $\mathbf{c}^1$ can be chosen arbitrarily.

If we construct such an eigenfunction on a circuit within a larger network it can of course be extended to the other elements by simply setting $\boldsymbol{\phi}^i = \mathbf{0}$ for $i = r+1, \ldots, n$. One is then led to the following theorem:

THEOREM 5.3. *Suppose that a string network contains a circuit along which the various parameters satisfy condition* (5.24). *Then* $\ker(\mathcal{C}_T^*) \neq \emptyset$ *so that the system is not approximately controllable for any positive T.*

REMARK 5.3. The result of this section raises a variety of mathematically intrigueing questions which are, perhaps, of limited interest from the point of view of applications where a distinction between rational and irrational parameters does not seen "physically stable." Regardless, it may be of interest to pose some of these questions:

- Is the existence of an eigenfunction satisfying the overdetermined system (5.16) and (5.17) *equivalent* to the conditions $\ker(\mathcal{C}_T^*) \neq \emptyset$?
- In general, what is the structure of the image of $\mathcal{C}_T$?
- Is there any case of a network containing closed loops which is exactly controllable?

6. Stabilizability of String Networks

In this section we shall prove only a rather rudimentary result, leaving some of the details and much of the elaboration to Chapter IV, where we shall be dealing with a more general class of networks including beam networks. We proceed along the usual lines of introducing a velocity feedback into the system in such a way that energy is obviously dissipated; one then verifies that the decay is actually exponential or polynomial, depending on the nature of the feedback.

To motivate the selection of a suitable feedback we once again (as, for example, in the proof of Lemma 4.3) differentiate the energy with respect to t to get, in case all multiple nodes are free and one simple node, indexed by m, is clamped,

$$\frac{d\mathcal{E}}{dt}(t) = \sum_{j\in\mathcal{J}_S\setminus\{m\}} \mathcal{N}^j\mathbf{r}\cdot\mathcal{D}^j\dot{\mathbf{r}}(t). \tag{6.1}$$

Consequently the "feedback law" $\mathcal{N}^j\mathbf{r}(t) = -\alpha_j\mathcal{D}^j\dot{\mathbf{r}}(t)$ for $j\in\mathcal{J}_S\setminus\{m\}$, where each α_j is a positive scalar, ensures energy dissipation of the "feedback system"

$$\begin{cases}\rho_i\ddot{\mathbf{r}}^i = \mathbf{Q}_i\mathbf{r}^i_{,xx} \text{ for } i\in\mathcal{I},\\ \mathbf{r}^i(x_{ij},t) \text{ coincide for } j\in\mathcal{J}_M,\ i\in\mathcal{I}_j,\\ \mathcal{N}^j\mathbf{r}(t) = \mathbf{0} \text{ for } j\in\mathcal{J}_M,\\ \mathcal{D}^m\mathbf{r} = \mathbf{0},\\ \mathcal{N}^j\mathbf{r}(t) = -\alpha_j\mathcal{D}^j\dot{\mathbf{r}}(t) \text{ for } j\in\mathcal{J}_S\setminus\{m\},\\ \mathbf{r}(0) = \mathbf{r}_0,\ \dot{\mathbf{r}}(0)=\dot{\mathbf{r}}_0.\end{cases} \tag{6.2}$$

In the case of tree networks we shall see that for this linear feedback the decay is in fact exponential. This depends on the same energy estimate, namely that given by Theorem 5.2, which was previously used to prove exact controllability. Before showing this we point out that the solution to (6.2) can be given in terms of a contraction semigroup $\mathcal{U}_s(t)$ on the space $\mathcal{H}$. We do not prove this here but refer to the more general result Theorem IV.6.1. Assuming the existence of this semi-group we now use an argument due to Rauch and Taylor [87].

It follows from (6.1) and the feedback law that

$$\|\mathcal{U}_s(T_\star)[\mathbf{r}_0\times\dot{\mathbf{r}}_0]\|^2_{\mathcal{H}} = \|\mathbf{r}_0\times\dot{\mathbf{r}}_0\|^2_{\mathcal{H}} - \sum_{j\in\mathcal{J}_S\setminus\{m\}}\alpha_j\int_0^{T_\star}|\mathcal{D}^j\dot{\mathbf{r}}(t)|^2\,dt. \tag{6.3}$$

Since $\mathcal{D}^j\dot{\mathbf{r}}(t) = \dot{\mathbf{r}}^{i_j}(x_j,t)$ and $\mathcal{N}^j\mathbf{r}(t) = \epsilon_j\mathbf{Q}_{i_j}\mathbf{r}^{i_j}_{,x}(x_j,t)$ the feedback law implies that

$$e_{i_j}(x_j,t)\le C|\mathcal{D}^j\dot{\mathbf{r}}(t)|^2.$$

Applying Theorem 5.2 to $\mathbf{r}$ and reversing the inequality given there one has after readjustment of the constant C

$$- \sum_{j \in \mathcal{J}_S \setminus \{m\}} \alpha_j \int_0^{T_\star} |\mathcal{D}^j \dot{\mathbf{r}}(t)|^2 \, dt \leq -C \, \| \mathcal{U}_s(\frac{T_\star}{2})[\mathbf{r}_0 \times \dot{\mathbf{r}}_0] \|_{\mathcal{H}}^2$$
$$\leq -C \, \| \mathcal{U}_s(T_\star)[\mathbf{r}_0 \times \dot{\mathbf{r}}_0] \|_{\mathcal{H}}^2,$$

where in the last step we have used the contraction property of the semigroup. From (6.3) it then follows that

$$\| \mathcal{U}_s(T_\star)[\mathbf{r}_0 \times \dot{\mathbf{r}}_0] \|_{\mathcal{H}} \leq \beta \, \| \mathbf{r}_0 \times \dot{\mathbf{r}}_0 \|_{\mathcal{H}}, \text{ with } \beta = (1+C)^{-\frac{1}{2}} < 1.$$

Therefore $\| \mathcal{U}_s(T_\star) \| < 1$ and the group $\mathcal{U}_s(t)$ decays exponentially.

The argument used above also applies, with no significant changes, to the non-linear feedback law

$$\mathcal{N}^j \mathbf{r}(t) = -\mathbf{g}^j(\mathcal{D}^j \dot{\mathbf{r}}(t))$$

under the assumption of continuity and

$$\mathbf{g}_j(\mathbf{u}) \cdot \mathbf{u} \geq \delta[|\mathbf{u}|^2 + |\mathbf{g}(\mathbf{u})|^2].$$

REMARK 6.1. It follows from the construction in Section 6.2 of the aberrant eigenfunction ϕ corresponding to a closed loop under the condition (5.23) that the feedback system will not be stabilizable. One simply considers $\mathbf{r} = (\cos \omega t)\phi$ which is a solution of (6.2) but does not decay.

In Chapter 4 we treat these matters for a wide class of networks. There we shall also introduce nonlinear feedbacks yielding only polynomial decay rates, as well as setvalued feedbacks for which we can prove only that the energy does decay to zero.

7. String Networks with Masses at the Nodes

In complex flexible structures it often happens that the various elements are linked together by joints which are rigid bodies. Since these are usually small in relation to the flexible elements it seems reasonable, at least as a first approximation, to think of these as point masses. Gravitation will of course act on these nodal masses as well as on the flexible elements. If the latter are very flexible the nature of the equilibria, and hence of the equations of motion linearized about an equilibrium, will be different from what we have described in the earlier parts of this chapter. These modelling issues, as they relate to networks of elastic strings, have been treated in the technical report [95]. Many aspects of the controllability of such a system are discussed in the doctoral thesis of Wei Ming [72]. Hanson and Zuazua have treated the exact controllability of a "two string with one point mass" system in [34]. A generalization of their results and of those of Wei Ming are the subject of [96].

Here we shall merely sketch some of the features of a model which incorporates point masses and the action of gravity.

Let m_j be the mass located at the multiple node indexed by j. Let $\mathbf{v}$ denote the vertical direction and g be the gravitational constant The kinetic and potential energies of the system are respectively

$$\mathcal{K}(t) = \sum_{i \in \mathcal{I}} \frac{1}{2} \int_0^{l_i} \rho_i |\dot{\mathbf{R}}^i(x,t)|^2 \, dx + \sum_{j \in \mathcal{J}_M} \frac{1}{2} m_j |\dot{\mathbf{N}}^j(t)|^2 \tag{7.1}$$

and

$$\mathcal{U}(t) = \sum_{i \in \mathcal{I}} \int_0^{l_i} \frac{h_i}{2} \Big[[|\mathbf{R}^i_{,x}(x,t)| - 1]^2 + g\rho_i \mathbf{R}^i(x,t) \cdot \mathbf{v} \Big] \, dx + \sum_{j \in \mathcal{J}_M} g m_j \mathbf{N}^j \cdot \mathbf{v}. \tag{7.2}$$

For simplicity, suppose that no additional forces are acting on the strings or on the masses and that we apply Dirichlet controls (2.2) and (2.3) at all the simple nodes, so that $\mathcal{J}_S = \mathcal{J}^C \cup \mathcal{J}^D$. Hamilton's principle then gives

$$\rho_i \ddot{\mathbf{R}}^i = h_i \left[\mathbf{R}^i_{,x} - \frac{\mathbf{R}^i_{,x}}{|\mathbf{R}^i_{,x}|} \right]_{,x} - \rho_i g \mathbf{v}, \quad i \in \mathcal{I}, \tag{7.3}$$

as well as the multiple node condition

$$m_j \ddot{\mathbf{N}}^j(t) + \sum_{i \in \mathcal{I}_j} \epsilon_{ij} h_i \left[\mathbf{R}^i_{,x} - \frac{\mathbf{R}^i_{,x}}{|\mathbf{R}^i_{,x}|} \right] (x_{ij}, t) = -m_j g \mathbf{v} \quad \text{for } j \in \mathcal{J}_M. \tag{7.4}$$

One can begin by studying the equilibria subject to time independendent Dirichlet data. In [95] the existence of unique "stretched equilibria", described by functions $\check{\mathbf{R}}^i$ satisfying $|\check{\mathbf{R}}^i_{,x}| > 1$, is proved by showing that the potential energy function achieves its minimum. What is interesting about this argument is that the potential energy function is not convex and that there are sometimes also other equilibria which involve "compressed strings" for which $|\check{\mathbf{R}}^i_{,x}| < 1$; this latter class of solutions is presumably unstable. In a stretched equilibrium each string occupies a convex curve in a vertical plane, which is of course consistent with ones intuition. If one uses the procedure of Section 3 to linearize about the equilibrium one obtains a linear system

$$\rho_i \ddot{\mathbf{r}}^i(x,t) = [\mathbf{Q}_i(x) \mathbf{r}^i_{,x}(x,t)]_{,x} \quad \text{for } i \in \mathcal{I}, \tag{7.5}$$

along with the conditions

$$m_j \ddot{\mathbf{n}}^j(t) + \sum_{i \in \mathcal{I}_j} \epsilon_{ij} \mathbf{Q}_i(x_{ij}) \mathbf{r}^i_{,x}(x_{ij}, t) = \mathbf{0} \text{ for } j \in \mathcal{J}_M. \tag{7.6}$$

Here $\mathbf{Q}(x)$ is the symmetric 3×3 matrix whose eigenvalues are h and $h\left(1 - |\check{\mathbf{R}}_{,x}(x)|^{-1}\right)$, with the corresponding eigenspaces consisting of the span of $\check{\mathbf{R}}_{,x}(x)$ and the orthogonal complement of $\check{\mathbf{R}}_{,x}(x)$. Since $|\check{\mathbf{R}}_{,x}| > 1$, $\mathbf{Q}$ is positive definite.

Well-posedness of the system can be established using variational and semigroup arguments; we refer to [95] for details. The question of exact controllability has an interesting feature which was first noticed by Hanson and Zuazua. They consider two strings which are both joined to a point mass, with the other end of one string clamped and the remaining end of the second string controlled. There is then an increase of regularity as one crosses the mass from the controlled to the clamped string. One way of seeing this is to note that the multiple node condition (7.6) together with energy estimates on the terms appearing on the left hand side of that condition ensure that at the mass $\ddot{\mathbf{n}}(t)$ is square integrable. Therefore the equation for the string with a clamped end now has Dirichlet data with an extra order of smoothness which raises the order of regularity of the solution. We can therefore no longer expect exact controllability into the energy space which was previously used for string networksl, although approximate controllability is easily proved within that space. In [34] and in [72] (which also treats in detail serial networks with arbitrarily many elements) exact controllability results are obtained using spaces which reflect the increase in regularity. In [96] these ideas are developed for general networks.

CHAPTER III

Modeling of Networks of Thermoelastic Beams

1. Modeling of a Thin Thermoelastic Curved Beam

We consider the deformation of a thin beam of length ℓ with a given initial curvature and torsion, and with variable doubly symmetric cross section. To be precise, the undeformed beam, in its initial reference configuration, occupies the region

$$\Omega := \{\mathbf{r} := \mathbf{r}_0(x_1) + x_2\mathbf{e}_2(x_1) + x_3(x_1)\mathbf{e}_3 | \quad x_1 \in [0,\ell], \\ (x_2,x_3) := x_2\mathbf{e}_2(x_1) + x_3\mathbf{e}_3(x_1) \in A(x_1)\},$$

where $\mathbf{r}_0 : [0,\ell] \mapsto I\!R^3$ is a smooth function representing the centerline of the beam at rest, and the orthonormal triads $\mathbf{e}_1(\cdot)$, $\mathbf{e}_2(\cdot)$, $\mathbf{e}_3(\cdot)$ are chosen to be smooth functions of x_1 such that $\mathbf{e}_1$ is the direction of the tangent of the centerline with respect to the variable x_1, i.e. $\frac{d}{dx_1}\mathbf{r}_0(x_1) = \mathbf{e}_1(x_1)$, and such that $\mathbf{e}_2(x_1)$, $\mathbf{e}_3(x_1)$ span the orthogonal cross section at x_1. The meaning of the variables x_i are as follows: x_1 denotes the length along the undeformed centerline, x_2 and x_3 denote the lengths along lines orthogonal to the reference line. The set Ω can then be viewed as obtained by translating the reference curve $\mathbf{r}_0(x_1)$ to the position $x_2\mathbf{e}_2 + x_3\mathbf{e}_3$ within the cross section vertical to the tangent of $\mathbf{r}_0$. The *cross section* at x_1 is defined as

$$A(x_1) = \{x_2\mathbf{e}_2 + x_3\mathbf{e}_3 | \, x_1\mathbf{e}_1 + x_2\mathbf{e}_2 + x_3\mathbf{e}_3 \in \Omega\}.$$

$A(x_1)$ is assumed to vary smoothly with respect to x_1 in a sense that will be made clear later on. In order that certain moments vanish, we will restrict our attention to cross sections which are *doubly symmetric* in the sense that $(x_2,x_3) = x_2\mathbf{e}_2 + x_3\mathbf{e}_3 \in A(x_1)$ implies that $(-x_2,x_3)$, $(x_2,-x_3) \in A(x_1)$, and hence $-(x_2,x_3) \in A(x_1)$.

The triad $\mathbf{e}_i(x_1)$, the initial body frame, can be written in terms of the standard base $\mathbf{e}_i^0$ in $\mathbf{R}^3$ by the way of a rotation $\mathbf{\Gamma}_0$ as

$$\mathbf{e}_i = \mathbf{\Gamma}_0 \mathbf{e}_i^0,$$

where $\mathbf{\Gamma}_0\mathbf{\Gamma}_0^t = \mathbf{I}$. We have

$$\frac{\partial}{\partial x_1}\mathbf{e}_i = \frac{\partial}{\partial x_1}\mathbf{\Gamma}\mathbf{e}_i^0 = (\frac{\partial}{\partial x_1}\mathbf{\Gamma}_0)\mathbf{\Gamma}_0^t\mathbf{e}_i =: \mathbf{\Omega}_0\mathbf{e}_i$$

Obviously, $\mathbf{\Omega}_0$ is skew-symmetric, and hence $\mathbf{\Omega}_{0ii} = 0, \mathbf{\Omega}_{0ij} = -\mathbf{\Omega}_{0ji}$. With this notation the initial curvatures and twist of the beam or, more precisely, of its centerline in the reference configuration, can be described using Frenet–type formulae as

$$\begin{aligned}\kappa_2 &=, \mathbf{e}_2 \cdot \mathbf{e}_{1,1} = -\mathbf{e}_1 \cdot \mathbf{e}_{2,1} = -\mathbf{\Omega}_{012},\\ \kappa_3 &= \mathbf{e}_3 \cdot \mathbf{e}_{1,1} = -\ e_1 \cdot \mathbf{e}_{3,1} = -\mathbf{\Omega}_{013},\\ \tau &= \mathbf{e}_3 \cdot \mathbf{e}_{2,1} = -\mathbf{e}_2 \cdot \mathbf{e}_{3,1} = -\mathbf{\Omega}_{023}.\end{aligned}$$

The matrix $\mathbf{\Omega}_0$, in the basis $\mathbf{e}_i$, is given by

$$\mathbf{\Omega}_0 =: \begin{pmatrix} 0 & -\kappa_2 & -\kappa_3 \\ \kappa_2 & 0 & -\tau \\ \kappa_3 & \tau & 0 \end{pmatrix}.$$

Let us also introduce the axial vector

$$\boldsymbol{\omega}_0 := \tau\mathbf{e}_1 - \kappa_3\mathbf{e}_2 + \kappa_2\mathbf{e}_3$$

which satisfies $\mathbf{\Omega}_0\boldsymbol{\omega}_0 = 0$, hence

$$\frac{\partial}{\partial x_1}\mathbf{e}_i = \boldsymbol{\omega}_0 \times \mathbf{e}_i.$$

After *deformation*, the particle has the position

$$\mathbf{R}(x_1, x_2, x_3, t).$$

We introduce the *displacement* vector by

$$\mathbf{V} := \mathbf{R} - \mathbf{r}.$$

A position on the reference line $(x_2 = x_3 = 0)$, denoted by

$$\mathbf{r}_0 = \mathbf{r}(x_1, 0, 0),$$

is deformed to

$$\mathbf{R}_0 = \mathbf{R}(x_1, 0, 0).$$

Accordingly, the displacement vector of a particle on the reference line is

$$\mathbf{W} := \mathbf{V}(x_1, 0, 0) = \mathbf{R}_0 - \mathbf{r}_0.$$

The tangent vector at any particle in the undeformed configuration is denoted by

$$\mathbf{r}_{,i} = \mathbf{g}_i,$$

where the subscript separated by a comma means spatial derivation with respect to variable x_i. Accordingly, the tangent at this particle in the deformed configuration is

$$\mathbf{G}_i = \mathbf{R}_{,i} = \mathbf{g}_i + \mathbf{V}_i.$$

The triads $\mathbf{g}_i, \mathbf{G}_i$ represent the tangents along the coordinate lines in the undeformed and deformed configuration, respectively. These are right–handed systems as the determinant of the deformation matrix $(\mathbf{G}_i)_i$ given with respect to the base $\mathbf{g}_i$ is positive definite. In particular, at the reference line we have the triads

$$\mathbf{e}_i := \mathbf{g}_i(x_1, 0, 0), \quad \mathbf{E}_i := \mathbf{G}_i(x_1, 0, 0), \quad i = 1, 2, 3.$$

Note, however, that the second triad, $\mathbf{E}_i$ *is not necessarily orthogonal, due to shearing.* We introduce the vector $\hat{\mathbf{e}}_1 = \mathbf{R}_{,1}/\|\mathbf{R}_{,1}\|$ in the direction of the tangent $\mathbf{R}_{,1}(x_1, 0, 0, t)$ reference curve. *We postulate that the location of the deformed point can be written as*

$$\mathbf{R} = \mathbf{R}_0 + x_2\mathbf{E}_2 + x_3\mathbf{E}_3 + \gamma(x_1)\phi(x_2, x_3)\hat{\mathbf{e}}_1. \tag{1.1}$$

In (1.1) the function ϕ is related to the solution χ (called the torsion function) of the St.Venant torsion problem in the following way: χ satisfies the equation $\Delta\chi + 2 = 0$ (Δ signifies the Laplace operator) over the cross section and vanishes at the boundary of the cross section. This function χ can be written as $\phi_1 - \frac{1}{2}(x_2^2 + x_3^2)$ with a harmonic function ϕ_1, the Prandtl stress function. With this notation, ϕ above is the conjugate harmonic function to ϕ_1. Hence, ϕ is itself harmonic. By the assumed double symmetry and the application of Schwartz' reflection principle, ϕ has the following symmetries: $\phi(x_2, x_3) = \phi(-x_2, -x_3), \phi(-x_2, x_3) = \phi(x_2, -x_3) = -\phi(-x_2, -x_3)$.

For a canonical example, we may think of the case of an elliptical cross section $\frac{x_2^2}{a^2} + \frac{x_3^2}{b^2} = 1$ where the stress function ϕ_1 is given by

$$\phi_1 = \frac{1}{2}\frac{a^2 - b^2}{a^2 + b^2}(x_2^2 - x_3^2) + \frac{a^2 b^2}{a^2 + b^2},$$

with the conjugate function

$$\phi = -\frac{a^2 - b^2}{a^2 + b^2}x_2 x_3.$$

Another canonical example is a rectangular cross section with length a, b along the x_2, x_3 directions, respectively. We only reproduce from Sokolnikoff [104] the formula for ϕ:

$$\phi = x_2 x_3 - \frac{8a^2}{\pi^3} \sum_{n=0}^{\infty} \frac{(-1)^n}{(2n+1)^3} \frac{\sinh k_n x_3}{\cosh(k_n b/2)} \sin k_n x_2$$

with $k_n = (2n+1)\pi/a$. The above mentioned properties of ϕ are easily verified in these examples. The function $\gamma(x_1)$ measures the amount of warping of the cross-section at x_1. Warping is considered small if the beam is thin, and hence $\gamma\phi$ is small compared to the breadth of the beam. We note that (1.1) restricts the geometry of the deformed beam, and that it would be more accurate to account for higher order terms in the variables x_2, x_3.

Some remarks are in order.

REMARK 1.1. (1) First of all, with the assumption above we are looking for a *first order theory*, that is a theory with Taylor series of the deformed point in the variables x_2, x_3 truncated after the linear terms. There are various papers in the literature in which theories of arbitrary order are developed, but we shall not dwell on these here; see for instance Antman and Warner [4].

(2) There are various ways to introduce moving triads in order to represent the deformed position $\mathbf{R}$; see Green [29], Antman and Warner [4], Antman [1], Reissner [89], Hodges [22], Kane et al [41], Simo [100], Marsden, Krishnaprasad, Simo [101] etc. We will essentially follow the presentation of Wempner [108], which is based on the same philosophy as in [29], [4]. This will be explained in due course and the differences between the possible choices will be discussed. At this point it is sufficient to say that we do not, at the onset, assume that the there is no distortion within the plane of cross section. As a result, the vectors $\mathbf{E}_2, \mathbf{E}_3$, which span the deformed cross section, are not necessarily perpendicular; neither are they necessarily perpendicular to $\mathbf{E}_1$, due to shearing. In [22], [100], [101] and in many other publications, where a director approach (as in the Cosserat continuum) is taken, the moving triads representing the deformed body are taken to be orthogonal. Orthogonal triads are more appealing from the point of view of *geometrically exact theories of rods*, as the configuration space can entirely be represented by translations and rotations. However, whenever it comes down to explicit relations between strains and displacements, additional restrictions on the admissible configurations have to be introduced anyway and then there is no apparent advantage of one method over the other. For this reason we want to choose a representation in which the role of the rotation of infinitesimal elements *and* the shear angles are most transparent *and* one which leads to a *practical* set of equations of motions under a sufficiently general set of assumptions.

(3) In order to define the strains at the reference curve, there are also at least two possibilities. One can either take the triad $\mathbf{E}_i$ itself, as we will do here, and measure the angles between the plane of cross section spanned by $\mathbf{E}_2, \mathbf{E}_3$ and the tangent to the deformed reference curve, $\mathbf{E}_1$, or one can work with the triad $\mathbf{E}_2, \mathbf{E}_3, \mathbf{E}_2 \times \mathbf{E}_3 = \mathbf{N}$ and measure the angle between $\mathbf{N}$ and the tangent. We will see that in the case where there is no distortion within the plane of cross sections, these two concepts coincide. The choice made here seems to have the advantage of revealing a more transparent relation between the Green-St.Venant- (Lagrange) strain tensor to be introduced below and the reference strains just mentioned.

(4) As we are also interested in superimposed rigid body dynamics, as in Ballieul and Levi [5], [6], Kane [41] and others, we also want to present a *frame indifferent* setup of strain and stretch tensors. The equations of motion will first be derived within the context of *spatial* variables, that is with respect to a moving triad, and then, in order to obtain the differential equations in terms of displacements and (local and global) rotations, those equations are either to be transformed to *material* coordinates by means of a *pull back*, see [101], or to *convected* coordinates using the *Piola transform*, see [101], [18]. □

Let us now attempt to make the notion of strain explicit. We have,

$$\mathbf{r}_{,1} = (1 - x_2\kappa_2 - x_3\kappa_3)\mathbf{e}_1 + x_2\tau\mathbf{e}_3 - x_3\tau\mathbf{e}_2$$
$$\mathbf{r}_{,2} = \mathbf{e}_2, \quad \mathbf{r}_{,3} = \mathbf{e}_3.$$

Note that in the presence of an initial twist the triad $\mathbf{g}_i$ is not orthogonal. Introducing the length s_i of an undeformed x_1–coordinate line passing through (x_2, x_3), we have the increments

$$ds_1 = |\mathbf{r}_{,1}|dx_1, \quad ds_2 = dx_2, \quad ds_3 = dx_3.$$

With this notation we are able to write down the strain components as follows:

$$\varepsilon_{ij} := \frac{1}{2}\left(\frac{\partial \mathbf{R}}{\partial s_i} \cdot \frac{\partial \mathbf{R}}{\partial s_j} - \frac{\partial \mathbf{r}}{\partial s_i} \cdot \frac{\partial \mathbf{r}}{\partial s_j}\right).$$

Note that the reference configuration already includes curvatures and twist. The matrices $(\frac{\partial \mathbf{r}}{\partial s_i} \cdot \frac{\partial \mathbf{r}}{\partial s_j})$ and $(\frac{\partial \mathbf{R}}{\partial s_i} \cdot \frac{\partial \mathbf{R}}{\partial s_j})$ give rise to Riemannian structures on the reference body and the deformed body, respectively, and the strain ε just defined is then the *Green–St. Venant strain tensor*; see Ciarlet [18] or Antman and Warner [4]. This definition of a strain has an obvious physical interpretation, as it compares the length of an infinitesimal line in the deformed configuration with the respective length in the undeformed configuration. It is also important to note that this strain measure is obviously invariant under rigid body motions.

Now

$$\frac{\partial \mathbf{R}}{\partial s_i} = \mathbf{G}_i \frac{dx_i}{ds_i}, \quad \frac{\partial \mathbf{r}}{\partial s_i} = \mathbf{g}_i \frac{dx_i}{ds_i},$$

where

$$\frac{dx_2}{ds_2} = \frac{dx_3}{ds_3} = 1, \quad \frac{ds_1}{dx_1} = |\mathbf{r}_{,1}| = (\mathbf{g}_i \cdot \mathbf{g}_i)^{\frac{1}{2}}.$$

Hence the components of the strain tensor can be written as

$$(1.2) \qquad \begin{cases} \varepsilon_{11} = \frac{1}{2}(\frac{\mathbf{G}_1 \cdot \mathbf{G}_1}{|\mathbf{r}_{,1}|^2} - 1), \\ \varepsilon_{12} = \frac{1}{2}\frac{1}{|\mathbf{r}_{,1}|}(\mathbf{G}_1 \cdot \mathbf{G}_2 - \mathbf{g}_1 \cdot \mathbf{g}_2), \\ \varepsilon_{13} = \frac{1}{2}\frac{1}{|\mathbf{r}_{,1}|}(\mathbf{G}_1 \cdot \mathbf{G}_3 - \mathbf{g}_1 \cdot \mathbf{g}_3), \\ \varepsilon_{22} = \frac{1}{2}(\mathbf{G}_2 \cdot \mathbf{G}_2 - 1), \\ \varepsilon_{33} = \frac{1}{2}(\mathbf{G}_3 \cdot \mathbf{G}_3 - 1), \\ \varepsilon_{23} = \frac{1}{2}(\mathbf{G}_2 \cdot \mathbf{G}_3). \end{cases}$$

The last three strains are strains within the cross section. *Those are usually neglected in theories of thin beams.*

The deformation of $\mathbf{r}$ *into* $\mathbf{R}(\cdot, t)$ *will be considered as a succession of two motions:* (1) *a rotation* $\mathbf{\Gamma}_1$ *carrying the triad* $\mathbf{e}_i(x_1)$ *to an intermediate triad* $\hat{\mathbf{e}}_i(x_1, t)$, *followed by* (2) *a deformation into the non-orthogonal triad* $\mathbf{E}_i(x_1, t)$. The two triads $\hat{\mathbf{e}}_i$ and $\mathbf{E}_i$ then differ on account of a strain $\bar{\varepsilon}$ to be specified below. We choose to orient the intermediate (right–handed) triad $\hat{\mathbf{e}}_i$, which serves as a moving orthonormal reference frame, such that

$$(1.3) \qquad \mathbf{E}_1 = (\mathbf{E}_1 \cdot \hat{\mathbf{e}}_1)\hat{\mathbf{e}}_1 = |\mathbf{E}_1|\hat{\mathbf{e}}_1, \quad \hat{\mathbf{e}}_2 \cdot \mathbf{E}_3 = \hat{\mathbf{e}}_3 \cdot \mathbf{E}_2$$

holds. Note that the triad $\hat{\mathbf{e}}$ is neither the canonical Frenet-system nor the one employed in, e.g., [100]. Nevertheless, it is uniquely determined by property (1.3).

A strain $\bar{\varepsilon}$, related to the deformation carrying the triad $\hat{\mathbf{e}}_i$ to the triad $\mathbf{E}_i$, is defined by

$$(1.4) \qquad \begin{cases} \hat{\mathbf{e}}_1 \cdot \mathbf{E}_1 =: 1 + \bar{\varepsilon}_{11}, \\ \hat{\mathbf{e}}_2 \cdot \mathbf{E}_2 =: 1 + \bar{\varepsilon}_{22}, \\ \hat{\mathbf{e}}_3 \cdot \mathbf{E}_3 =: 1 + \bar{\varepsilon}_{33}, \\ \hat{\mathbf{e}}_1 \cdot \mathbf{E}_2 =: 2\bar{\varepsilon}_{12}, \\ \hat{\mathbf{e}}_1 \cdot \mathbf{E}_3 =: 2\bar{\varepsilon}_{13}, \\ \hat{\mathbf{e}}_2 \cdot \mathbf{E}_3 =: \bar{\varepsilon}_{23}. \end{cases}$$

The remaining strains are defined by requiring symmetry $\bar{\varepsilon}_{ij} = \bar{\varepsilon}_{ji}$. We may now use the strains $\bar{\varepsilon}_{ij}$ in order to express the triad $\mathbf{E}_i$ in terms of the moving orthonormal reference triad $\hat{\mathbf{e}}_i$ as follows:

$$
(1.5)\qquad \begin{cases} \mathbf{E}_1 = \hat{\mathbf{e}}_1 + \bar{\varepsilon}_{11}\hat{\mathbf{e}}_1, \\ \mathbf{E}_2 = \hat{\mathbf{e}}_2 + \bar{\varepsilon}_{22}\hat{\mathbf{e}}_2 + \bar{\varepsilon}_{23}\hat{\mathbf{e}}_3 + 2\bar{\varepsilon}_{21}\hat{\mathbf{e}}_1, \\ \mathbf{E}_3 = \hat{\mathbf{e}}_3 + \bar{\varepsilon}_{33}\hat{\mathbf{e}}_3 + \bar{\varepsilon}_{23}\hat{\mathbf{e}}_2 + 2\bar{\varepsilon}_{31}\hat{\mathbf{e}}_1. \end{cases}
$$

REMARK 1.2. If the distortion of the plane of cross section is neglected, that is if we neglect the strains $\bar{\varepsilon}_{22}, \bar{\varepsilon}_{33}, \bar{\varepsilon}_{23}$, then we can express the normal $\mathbf{N}$ to the cross section as

$$
\mathbf{N} = \mathbf{E}_2 \times \mathbf{E}_3 = \hat{\mathbf{e}}_1 - 2\bar{\varepsilon}_{21}\hat{\mathbf{e}}_2 - 2\bar{\varepsilon}_{31}\hat{\mathbf{e}}_3,
$$

and then the strain vector γ introduced by

$$
\gamma := \mathbf{E}_1 - \mathbf{N} = \bar{\varepsilon}_{11}\hat{\mathbf{e}}_1 + 2\bar{\varepsilon}_{21}\hat{\mathbf{e}}_2 + 2\bar{\varepsilon}_{31}\hat{\mathbf{e}}_3
$$

is precisely the one introduced in [22], [89], [100, 101] etc., where the assumption of rigidly moved cross sections is made at the onset. In addition, in the case that the strains $\bar{\varepsilon}_{ij}$ are small in the sense that the products $\bar{\varepsilon}_{ij}\bar{\varepsilon}_{kl}$ may be ignored regardless of the magnitude of the rotation that carries the triad $\mathbf{e}_i$ into $\hat{\mathbf{e}}_i$, we obtain

$$
(1.6)\qquad \bar{\varepsilon}_{ij} \approx \frac{1}{2}(\mathbf{E}_i \cdot \mathbf{E}_j - \delta_{ij}).
$$

However, the latter is exactly the strain $\varepsilon_{ij}(x_1, 0, 0)$ at the deformed reference curve. □

We proceed to express the motion carrying the triad $\mathbf{e}_i$ into the triad $\mathbf{E}_i$. We have, for a certain orthogonal matrix $\mathbf{\Gamma}_1(x)$,

$$
\hat{\mathbf{e}}_i(x_1) = \mathbf{\Gamma}_1(x_1)\mathbf{e}_i(x_1) = \mathbf{\Gamma}_1(x_1)\mathbf{\Gamma}_0(x_1)\mathbf{e}_i^0
$$

and

$$
\begin{aligned} \frac{\partial}{\partial x_1}\hat{\mathbf{e}}_i(x_1) &= (\frac{\partial}{\partial x_1}\mathbf{\Gamma}_1(x_1))\mathbf{\Gamma}_1(x_1)^t\hat{\mathbf{e}}_i(x_1) \\ &\quad + (\mathbf{\Gamma}_1(x_1)\frac{\partial}{\partial x_1}\mathbf{\Gamma}_0(x_1)\mathbf{\Gamma}_0(x_1)^t)\mathbf{\Gamma}_1(x_1)^t\hat{\mathbf{e}}(x_1). \end{aligned}
$$

Therefore, upon defing the skew-symmetric matrices

$$
\begin{aligned} \mathbf{\Omega}_1(x_1) &:= (\frac{\partial}{\partial x_1}\mathbf{\Gamma}_1(x_1))\mathbf{\Gamma}_1(x_1)^t, \\ \mathbf{\Omega}(x_1) &:= \mathbf{\Omega}_1(x_1) + \mathbf{\Gamma}_1(x_1)\mathbf{\Omega}_0(x_1)\mathbf{\Gamma}_1(x_1)^t, \end{aligned}
$$

we can express the derivatives of the triads with respect to x_1 as

$$
\begin{aligned} \frac{\partial}{\partial x_1}\hat{\mathbf{e}}_i(x_1) &= \mathbf{\Omega}(x_1)\hat{\mathbf{e}}_i(x_1), \\ \frac{\partial}{\partial x_1}\mathbf{e}_i(x_1) &= \mathbf{\Omega}_0(x_1)\hat{\mathbf{e}}_i(x_1). \end{aligned}
$$

We may now introduce "curvatures" K_i and "twist" T in the same way as with the undeformed reference curve, but now based on the moving triad $\hat{\mathbf{e}}_i$, as follows

$$\begin{cases} K_2 := \hat{\mathbf{e}}_2 \cdot \hat{\mathbf{e}}_{1,1} = -\hat{\mathbf{e}}_1 \cdot \hat{\mathbf{e}}_{2,1}, \\ K_3 := \hat{\mathbf{e}}_3 \cdot \hat{\mathbf{e}}_{1,1} = -\hat{\mathbf{e}}_1 \cdot \hat{\mathbf{e}}_{3,1}, \\ T := \hat{\mathbf{e}}_3 \cdot \hat{\mathbf{e}}_{2,1} = -\hat{\mathbf{e}}_2 \cdot \hat{\mathbf{e}}_{3,1}. \end{cases} \tag{1.7}$$

With this notation $\boldsymbol{\Omega}(x_1)$ can be represented as

$$\boldsymbol{\Omega} = \begin{pmatrix} 0 & -K_2 & -K_3 \\ K_2 & 0 & -T \\ K_3 & T & 0 \end{pmatrix}, \tag{1.8}$$

where we have omitted the arguments. We also define the axial vector

$$\boldsymbol{\omega} := T\hat{\mathbf{e}}_1 - K_3\hat{\mathbf{e}}_2 + K_2\hat{\mathbf{e}}_3$$

satisfying $\boldsymbol{\Omega}\boldsymbol{\omega} = 0$ and

$$\frac{\partial}{\partial x_1}\hat{\mathbf{e}}_i = \boldsymbol{\omega} \times \hat{\mathbf{e}}_i.$$

It is possible to express the rotations above by means of the Rodrigues formula, see [22], [101] (in fact, one might want to use Euler or Rodrigues parameters). However, it turns out that the use of *dextral rotations* provides a more transparent interpretation of the angles within the reference frame $\mathbf{e}_i$. The latter approach has been used by Wempner [108] and Kane et al [41]. It is our goal to express the strains ε_{ij} in terms of the reference strains $\bar{\varepsilon}_{ij}$ and the strains related to the rotation. That rotation is obtained by successive dextral rotations about the $\mathbf{e}_1$, $\mathbf{e}_2$ and $\mathbf{e}_3$ axes, respectively, through respective angles Θ_1, Θ_2 and Θ_3. The result is

$$\begin{cases} \hat{\mathbf{e}}_1 = (\cos\Theta_2\cos\Theta_3)\mathbf{e}_1 + (\cos\Theta_2\sin\Theta_3)\mathbf{e}_2 - \sin\Theta_2\mathbf{e}_3, \\ \hat{\mathbf{e}}_2 = (\sin\Theta_1\sin\Theta_2\cos\Theta_3 - \cos\Theta_1\sin\Theta_3)\mathbf{e}_1 \\ \qquad + (\cos\Theta_1\cos\Theta_3 + \sin\Theta_1\sin\Theta_2\sin\Theta_3)\mathbf{e}_2 + (\sin\Theta_1\cos\Theta_2)\mathbf{e}_3, \\ \hat{\mathbf{e}}_3 = (\cos\Theta_1\sin\Theta_2\cos\Theta_3 + \sin\Theta_1\sin\Theta_3)\mathbf{e}_1 \\ \qquad - (\sin\Theta_1\cos\Theta_3 - \cos\Theta_1\sin\Theta_2\sin\Theta_3)\mathbf{e}_2 + (\cos\Theta_1\cos\Theta_2)\mathbf{e}_3. \end{cases} \tag{1.9}$$

This makes it possible to express the curvature and twist, with respect to the frame $\mathbf{e}_i$, explicitly as

$$(1.10)\quad \begin{cases} K_2 = \Theta_3' - \Theta_1 \sin\Theta_2 + \kappa_2 \cos\Theta_1 \cos\Theta_2 \\ \qquad + \kappa_3(\sin\Theta_1 \cos\Theta_3 - \sin\Theta_2 \sin\Theta_3 \cos\Theta_1) \\ \qquad\qquad + \tau(\sin\Theta_1 \sin\Theta_3 + \sin\Theta_2 \cos\Theta_1 \cos\Theta_3, \\ K_3 = \Theta_2' \cos\Theta_3 - \Theta_1 \sin\Theta_3 \cos\Theta_2 - \kappa_2 \sin\Theta_1 \cos\Theta_2 \\ \qquad + \kappa_3(\cos\Theta_1 \cos\Theta_3 \sin\Theta_1 \sin\Theta_2 \sin\Theta_3) \\ \qquad\qquad + \tau(\sin\Theta_3 \cos\Theta_1 - \sin\Theta_3 \sin\Theta_2 \cos\Theta_3), \\ T = \Theta_1' \cos\Theta_2 \cos\Theta_3 - \Theta_2' \sin\Theta_3 \\ \qquad - \kappa_2 \sin\Theta_2 - \kappa_3 \sin\Theta_3 \cos\Theta_2 + \tau \cos\Theta_2 \cos\Theta_3. \end{cases}$$

In addition to that, we can write down the *exact* relations between the reference frame $\mathbf{e}_i$ and the tangent frame $\mathbf{E}_i$.

(1.11)

$$\begin{cases} \mathbf{E}_1 = (1+\bar{\varepsilon}_{11})[(\cos\Theta_2 \cos\Theta_3)\mathbf{e}_1 + (\sin\Theta_3 \cos\Theta_2)\mathbf{e}_2 - (\sin\Theta_2)\mathbf{e}_3], \\ \mathbf{E}_2 = [-\sin\Theta_3 \cos\Theta_1 + \sin\Theta_1 \sin\Theta_2 \sin\Theta_3 \\ \qquad + 2\bar{\varepsilon}_{21} \cos\Theta_2 \cos\Theta_3 \\ \qquad - \bar{\varepsilon}_{22}(\sin\Theta_3 \cos\Theta_1 - \sin\Theta_1 \sin\Theta_2 \cos\Theta_3) \\ \qquad + \bar{\varepsilon}_{23}(\sin\Theta_1 \sin\Theta_3 + \sin\Theta_2 \cos\Theta_1 \cos\Theta_3)]\mathbf{e}_1 \\ \qquad + [\cos\Theta_1 \cos\Theta_3 + \sin\Theta_1 \sin\Theta_2 \sin\Theta_3 \\ \qquad + 2\bar{\varepsilon}_{21} \sin\Theta_3 \cos\Theta_2 \\ \qquad + \bar{\varepsilon}_{22}(\cos\Theta_1 \cos\Theta_2 + \sin\Theta_1 \sin\Theta_2 \sin\Theta_3) \\ \qquad - \bar{\varepsilon}_{23}(\sin\Theta_1 \cos\Theta_2 - \sin\Theta_2 \sin\Theta_3 \cos\Theta_1)]\mathbf{e}_2 \\ \qquad + [\sin\Theta_1 \cos\Theta_2 - 2\bar{\varepsilon}_{12} \sin\Theta_2 \\ \qquad + \bar{\varepsilon}_{22} \sin\Theta_1 \cos\Theta_2 + \bar{\varepsilon}_{23} \cos\Theta_1 \cos\Theta_2]\mathbf{e}_3, \\ \mathbf{E}_3 = [\sin\Theta_1 \sin\Theta_3 + \sin\Theta_2 \cos\Theta_1 \cos\Theta_3 \\ \qquad + 2\bar{\varepsilon}_{31} \cos\Theta_2 \cos\Theta_3 \\ \qquad + \bar{\varepsilon}_{33}(\sin\Theta_1 \sin\Theta_3 + \sin\Theta_2 \cos\Theta_1 \cos\Theta_3) \\ \qquad - \bar{\varepsilon}_{23}(\sin\Theta_3 \cos\Theta_1 - \sin\Theta_1 \sin\Theta_2 \cos\Theta_3)]\mathbf{e}_1, \\ \qquad + [-\sin\Theta_1 \cos\Theta_3 + \sin\Theta_2 \sin\Theta_3 \cos\Theta_1 \\ \qquad + 2\bar{\varepsilon}_{31} \sin\Theta_3 \cos\Theta_2 \\ \qquad - \bar{\varepsilon}_{33}(\sin\Theta_1 \cos\Theta_3 - \sin\Theta_2 \sin\Theta_3 \cos\Theta_1) \\ \qquad + \bar{\varepsilon}_{23}(\cos\Theta_1 \cos\Theta_3 + \sin\Theta_1 \sin\Theta_2 \sin\Theta_3]\mathbf{e}_2 \\ \qquad + [\cos\Theta_1 \cos\Theta_2 - 2\bar{\varepsilon}_{31} \sin\Theta_2 \\ \qquad + \bar{\varepsilon}_{33} \cos\Theta_1 \cos\Theta_2 + \bar{\varepsilon}_{23} \sin\Theta_1 \cos\Theta_2]\mathbf{e}_3. \end{cases}$$

These calculations amply demonstrate that in order to obtain any practical set of equations of motions we have to approximate the rotations. There are two stages of approximation, which have been termed *finite rotation*, and *moderate rotation*, based on power series expansions, to be found in the literature. A finite rotation approximation to any given rotation described in terms of the three angles above amounts to keeping all quadratic terms in the angles, while a moderate rotation, which has also come to be known as *infinitesimal rigid displacement* (see Ciarlet[18, Section 6.3]), reflects the linear terms only. Another way of thinking of moderate and finite rotations is to consider a one parameter group of rotations in $\mathbf{R}^3$ generated by a skew symmetric matrix Ω. The linear approximation is precisely the moderate rotation, while the quadratic approximation coincides with the finite rotation. The finite rotation approximation to (1.9) is as follows. We have

$$(1.12)\quad \begin{cases} \hat{\mathbf{e}}_1 \doteq (1-\frac{1}{2}(\Theta_2^2+\Theta_3^2))\mathbf{e}_1 + \Theta_3\mathbf{e}_2 - \Theta_2\mathbf{e}_3, \\ \hat{\mathbf{e}}_2 \doteq (\Theta_1\Theta_2-\Theta_3)\mathbf{e}_1 + (1-\frac{1}{2}(\Theta_1^2+\Theta_3^2))\mathbf{e}_2 + \Theta_1\mathbf{e}_3, \\ \hat{\mathbf{e}}_3 \doteq (\Theta_2+\Theta_1\Theta_3)\mathbf{e}_1 - (\Theta_1-\Theta_2\Theta_3)\mathbf{e}_2 + (1-\frac{1}{2}(\Theta_1^2+\Theta_2^2))\mathbf{e}_3. \end{cases}$$

The dot above the equality sign indicates an approximate identity. The finite rotation approximation of (1.10) is

$$\begin{aligned} K_2-\kappa_2 &\doteq \Theta_{3,1}+\kappa_3\Theta_1+\tau\Theta_2-\Theta_2\Theta_{1,1} \\ &\qquad +\tau\Theta_1\Theta_3-\kappa_3\Theta_2\Theta_3-\frac{1}{2}\kappa_2(\Theta_1^2+\Theta_3^2), \\ K_3-\kappa_3 &\doteq -\Theta_{2,1}-\kappa_2\Theta_1+\tau\Theta_3-\Theta_3\Theta_{1,1}-\tau\Theta_1\Theta_2-\frac{1}{2}\kappa_3(\Theta_1^2+\Theta_3^2), \\ T-\tau &\doteq \Theta_{1,1}-\kappa_2\Theta_2-\kappa_3\Theta_3-\Theta_{2,1}\Theta_3-\frac{1}{2}\tau(\Theta_2^2+\Theta_3^2). \end{aligned}$$

These expressions are analogous to the corresponding terms in Reissner [89], where a slightly different triad is used. The quadratic approximation to (1.11) is given by

$$\begin{aligned} \mathbf{E}_1 &\doteq (1+\bar{\varepsilon}_{11}-\frac{1}{2}(\Theta_2^2+\Theta_3^2))\mathbf{e}_1+(1+\bar{\varepsilon}_{11})\Theta_3\mathbf{e}_2-(1+\bar{\varepsilon}_{11})\Theta_2\mathbf{e}_3, \\ \mathbf{E}_2 &\doteq (-\Theta_3+\Theta_1\Theta_2+2\bar{\varepsilon}_{21}+\bar{\varepsilon}_{23}\Theta_2-\bar{\varepsilon}_{22}\Theta_3)\mathbf{e}_1 \\ &\qquad +(1+\bar{\varepsilon}_{22}-\frac{1}{2}(\Theta_1^2+\Theta_3^2)+2\bar{\varepsilon}_{21}\Theta_3-\bar{\varepsilon}_{23}\Theta_1)\mathbf{e}_2 \\ &\qquad +(\Theta_1+\bar{\varepsilon}_{23}-2\bar{\varepsilon}_{21}\Theta_2+\bar{\varepsilon}_{22}\Theta_1)\mathbf{e}_3, \\ \mathbf{E}_3 &\doteq (\Theta_2+\Theta_1\Theta_3+2\bar{\varepsilon}_{31}-\bar{\varepsilon}_{23}\Theta_3+\bar{\varepsilon}_{33}\Theta_2)\mathbf{e}_1 \\ &\qquad +(-\Theta_1+\Theta_2\Theta_3+\bar{\varepsilon}_{23}+2\bar{\varepsilon}_{31}\Theta_3-\bar{\varepsilon}_{33}\Theta_1)\mathbf{e}_2 \\ &\qquad +(1+\bar{\varepsilon}_{33}-\frac{1}{2}(\Theta_1^2+\Theta_2^2)-2\bar{\varepsilon}_{31}\Theta_2+\bar{\varepsilon}_{23}\Theta_1)\mathbf{e}_3. \end{aligned}$$

It is possible to derive the equations of motion including *all* of the strains above, in particular including the strains which are responsible for the distortion of the plane of cross section. In theories of thin beams, however, the strains $\bar{\varepsilon}_{22}, \bar{\varepsilon}_{23}, \bar{\varepsilon}_{33}$ are usually neglected, and theories based on this assumption are still called geometrically exact within the literature. Under this assumption, we can express the triad $\mathbf{E}_i$ as follows:

$$\begin{aligned}
\mathbf{E}_1 \doteq & [1+\bar{\varepsilon}_{11} - \frac{1}{2}(\Theta_2^2+\Theta_3^2)]\mathbf{e}_1 + (1+\bar{\varepsilon}_{11})\Theta_3\mathbf{e}_2 - (1+\bar{\varepsilon}_{11})\Theta_2\mathbf{e}_3,\\
\mathbf{E}_2 \doteq & (-\Theta_3+\Theta_1\Theta_2+2\bar{\varepsilon}_{21})\mathbf{e}_1 + [1-\frac{1}{2}(\Theta_1^2+\Theta_3^2)+2\bar{\varepsilon}_{21}\Theta_3]\mathbf{e}_2\\
& + (\Theta_1 - 2\bar{\varepsilon}_{21}\Theta_2)\mathbf{e}_3,\\
\mathbf{E}_3 \doteq & (\Theta_2+\Theta_1\Theta_3+2\bar{\varepsilon}_{31})\mathbf{e}_1 + (-\Theta_1+\Theta_2\Theta_3+2\bar{\varepsilon}_{31}\Theta_3)\mathbf{e}_2\\
& + [1-\frac{1}{2}(\Theta_1^2+\Theta_2^2) - 2\bar{\varepsilon}_{31}\Theta_2]\mathbf{e}_3.
\end{aligned}$$

We calculate the normal $\mathbf{N} = \mathbf{E}_2 \times \mathbf{E}_3$ to the cross section with respect to the reference triad $\mathbf{e}_i$ and take the quadratic approximation

$$\begin{aligned}
\mathbf{N} \doteq & [1+2\bar{\varepsilon}_{21}\Theta_3 - 2\bar{\varepsilon}_{31}\Theta_2 - \frac{1}{2}(\Theta_2^2+\Theta_3^2)]\mathbf{e}_1\\
& + (\Theta_3 - 2\bar{\varepsilon}_{21} + 2\Theta_1\bar{\varepsilon}_{31})\mathbf{e}_2 - (\Theta_2 + 2\bar{\varepsilon}_{31} + 2\Theta_1\bar{\varepsilon}_{21})\mathbf{e}_3.
\end{aligned}$$

Let us now introduce yet another rotation, namely, the one taking the reference triad $\mathbf{e}_i$ into the triad $\mathbf{N}, \mathbf{E}_2, \mathbf{E}_3$. Once again, we describe this rotation by the way of three dextral rotations $\vartheta_1, \vartheta_2, \vartheta_3$ about $\mathbf{e}_1, \mathbf{e}_2, \mathbf{e}_3$, respectively. If we take the quadratic approximation (the finite rotation approximation) of the latter, we can make the following identifications:

$$(1.13) \qquad \begin{cases}
\vartheta_1 \doteq \Theta_1,\\
\vartheta_2 \doteq \Theta_2 + 2\bar{\varepsilon}_{31} + 2\Theta_1\bar{\varepsilon}_{21},\\
\vartheta_3 \doteq \Theta_3 - 2\bar{\varepsilon}_{21} + 2\Theta_1\bar{\varepsilon}_{31},\\
\boldsymbol{\vartheta} := \vartheta_1\mathbf{e}_1 + \vartheta_2\mathbf{e}_2 + \vartheta_3\mathbf{e}_3.
\end{cases}$$

These angles are now being interpreted as the *global rotations*. The shear strains can then be expressed as

$$(1.14) \qquad \begin{cases}
2\bar{\varepsilon}_{21} = -\vartheta_3 + \Theta_3 + \Theta_1(\vartheta_2 - \Theta_2),\\
2\bar{\varepsilon}_{31} = \vartheta_2 - \Theta_2 + \Theta_1(\vartheta_3 - \Theta_3).
\end{cases}$$

It is obvious from these relations that the shear strains vanish if and only if the angles Θ_i, ϑ_i coincide, respectively, or, what is the same, if the normal $\mathbf{N}$ to the cross section coincides with $\mathbf{E}_1$.

In order to complete the representation of the reference strains in terms of the angles above and the displacements $W_i := \mathbf{W} \cdot \mathbf{e}_i$, we compute

$\bar{\varepsilon}_{11} = \mathbf{E}_1 \cdot \hat{\mathbf{e}}_1 - 1$ as

$$\begin{aligned}\bar{\varepsilon}_{11} &= (\mathbf{W}_{,1} + \mathbf{e}_1) \cdot \hat{\mathbf{e}}_1 - 1 \\ &= W_{1,1} - \kappa_2 W_2 - \kappa_3 W_3 + \frac{1}{2}(\Theta_2^2 + \Theta_3^2).\end{aligned}$$

Now $\mathbf{W}_{,1} = \mathbf{E}_1 - \mathbf{e}_1 \doteq \bar{\varepsilon}_{11}\mathbf{e}_1 + \Theta_3\mathbf{e}_2 - \Theta_2\mathbf{e}_3$ and hence we have

$$\begin{aligned}\Theta_2 &= -\mathbf{W}_{,1} \cdot \mathbf{e}_3 = -(W_{3,1} + \kappa_3 W_1 + \tau W_2), \\ \Theta_3 &= \mathbf{W}_{,1} \cdot \mathbf{e}_3 = W_{2,1} + \kappa_2 W_1 - \tau W_3.\end{aligned}$$

Therefore, we can express the reference strains as

$$\begin{aligned}\bar{\varepsilon}_{11} =& W_{1,1} - \kappa_2 W_2 - \kappa_3 W_3 + \frac{1}{2}[(W_{3,1} + \kappa_3 W_1 + \tau W_2)^2 \\ &+ (W_{2,1} + \kappa_2 W_1 - \tau W_3)^2], \\ \bar{\varepsilon}_{21} =& \frac{1}{2}(W_{2,1} - \vartheta_3 + \kappa_2 W_1 - \tau W_3 + \vartheta_1(\vartheta_2 + W_{3,1} + \kappa_3 W_1 + \tau W_2)), \\ \bar{\varepsilon}_{31} =& \frac{1}{2}(W_{3,1} + \vartheta_2 + \kappa_3 W_1 + \tau W_2 + \vartheta_1(\vartheta_3 - W_{2,1} - \kappa_2 W_1 + \tau W_3)).\end{aligned}$$

The columns of the deformation gradient $(\mathbf{G}_i)_i$ with respect to the moving triad $\hat{\mathbf{e}}_i$ can be expressed as follows:

$$\begin{aligned}\mathbf{G}_1 =& [1 + \bar{\varepsilon}_{11} + x_2(-K_2 + 2\bar{\varepsilon}_{21,1}) + x_3(-K_3 + 2\bar{\varepsilon}_{31}) + \gamma_{,1}\phi]\hat{\mathbf{e}}_1 \\ &+ [x_2 2\bar{\varepsilon}_{21}K_1 + x_3(2\bar{\varepsilon}_{31}K_1 - T) + \gamma\phi K_2]\hat{\mathbf{e}}_2 \\ &+ [x_2(2\bar{\varepsilon}_{21} + T) + x_3 2\bar{\varepsilon}_{31}K_3 + \gamma\phi K_3]\hat{\mathbf{e}}_3, \\ \mathbf{G}_2 =& \hat{\mathbf{e}}_2 + 2\bar{\varepsilon}_{21}\hat{\mathbf{e}}_1 + \gamma\phi_{,2}\hat{\mathbf{e}}_1, \\ \mathbf{G}_3 =& \hat{\mathbf{e}}_3 + 2\bar{\varepsilon}_{31}\hat{\mathbf{e}}_1 + \gamma\phi_{,3}\hat{\mathbf{e}}_1,\end{aligned}$$

We are going to evaluate the scalar products $\mathbf{G}_i \cdot \mathbf{G}_j$ required for the Green -St.Venant strain tensor introduced above. Again, we take the *quadratic approximations in all variables*, i.e., in the strains $\bar{\varepsilon}_{11}$, $\bar{\varepsilon}_{21}$, $\bar{\varepsilon}_{31}$, K_1, K_2, T, $\bar{\varepsilon}_{21,1}$, $\bar{\varepsilon}_{31,1}$, $\gamma, \gamma_{,1}$ and the coordinates x_2, x_3. Thus we write

$$(1.15) \qquad \left\{\begin{aligned}\mathbf{G}_1 \cdot \mathbf{G}_1 =& 1 + 2\bar{\varepsilon}_{11} + 2\gamma_{,1}\phi - 2x_2(K_2 - 2\bar{\varepsilon}_{21,1}) \\ &- 2x_3(K_3 - 2\bar{\varepsilon}_{31,1}) + \bar{\varepsilon}_{11}^2, \\ \mathbf{G}_1 \cdot \mathbf{G}_2 =& 2\bar{\varepsilon}_{21} + \gamma\phi_{,2} + 2\bar{\varepsilon}_{11}\bar{\varepsilon}_{21} - x_3 T, \\ \mathbf{G}_1 \cdot \mathbf{G}_3 =& 2\bar{\varepsilon}_{31} + \gamma\phi_{,3} + 2\bar{\varepsilon}_{11}\bar{\varepsilon}_{31} + x_2 T.\end{aligned}\right.$$

As for the norm of the tangent to the reference line we take the approximation

$$|\mathbf{r}_{,1}| \doteq 1 - x_2\kappa_2 - x_3\kappa_3.$$

We rewrite $\mathbf{G}_1 \cdot \mathbf{G}_1$ in (1.15) as

$$\mathbf{G}_1 \cdot \mathbf{G}_1 = 1 - 2\kappa_2 - 2\kappa_3 + 2\bar{\varepsilon}_{11} + 2\gamma_{,1}\,\phi - 2x_2(K_2 - \kappa_2 - 2\bar{\varepsilon}_{21,1}) \\ - 2x_3(K_3 - \kappa_3 - 2\bar{\varepsilon}_{31,1}).$$

Now let us introduce our final set of bending strains by setting

$$(1.16) \qquad \begin{cases} \tilde{\kappa}_2 := K_2 - \kappa_2 - 2\bar{\varepsilon}_{12,1}, \\ \tilde{\kappa}_3 := K_3 - \kappa_3 - 2\bar{\varepsilon}_{13,1}, \\ \tilde{\tau} := T - \tau. \end{cases}$$

In order to evaluate ε_{ij} we recall the identities $\mathbf{g}_1 \cdot \mathbf{g}_2 = -x_3\tau$ and $\mathbf{g}_1 \cdot \mathbf{g}_3 = x_2\tau$. Then the *quadratic approximation* of ε_{ij} is given by

$$(1.17) \qquad \begin{cases} \varepsilon_{11} = (1 + \frac{1}{2}\bar{\varepsilon}_{11})\bar{\varepsilon}_{11} - x_2\tilde{\kappa}_2 - x_3\tilde{\kappa}_3 + \gamma_{,1}\,\phi, \\ \varepsilon_{12} = (1 + \bar{\varepsilon}_{11})\bar{\varepsilon}_{12} - \frac{1}{2}(x_3\tilde{\tau} - \gamma\phi_{,2}), \\ \varepsilon_{13} = (1 + \bar{\varepsilon}_{11})\bar{\varepsilon}_{13} + \frac{1}{2}(x_2\tilde{\tau} + \gamma\phi_{,3}). \end{cases}$$

It is important to note that the introduction of the new strains (1.16) is not just a matter of notational convenience; in fact it has a physical interpretation in that the quantities $\bar{\varepsilon}_{21,1}, \bar{\varepsilon}_{31,1}$ can be viewed as the entries of the axial vector $(0, \bar{\varepsilon}_{21,1}, \bar{\varepsilon}_{31,1})$ describing the rotation of the cross section into the plane spanned by $\hat{\mathbf{e}}_2, \hat{\mathbf{e}}_3$. Let us also note once again that the only smallness assumption we have adopted up to this point is the consistent use of quadratic approximations. It is customary within the engineering literature, however, to use two different degrees of approximation: a quadratic approximation in all rotations, and a linear approximation in all strains $\bar{\varepsilon}_{ij}$. This is what has come to be known as the theory of rods with *infinitesimal strains and moderate rotations.* With our approximations, we obtain

$$(1.18) \qquad \begin{cases} \bar{\varepsilon}_{11} = W_{1,1} - \kappa_2 W_2 - \kappa_3 W_3 + \frac{1}{2}(\vartheta_2^2 + \vartheta_3^2), \\ \bar{\varepsilon}_{21} = \frac{1}{2}(W_{2,1} - \vartheta_3 + \kappa_2 W_1 - \tau W_3 \\ \qquad\qquad + \vartheta_1(\vartheta_2 + W_{3,1} + \kappa_3 W_1 + \tau W_2)), \\ \bar{\varepsilon}_{31} = \frac{1}{2}(W_{3,1} + \vartheta_2 + \kappa_3 W_1 + \tau W_2 \\ \qquad\qquad + \vartheta_1(\vartheta_3 - W_{2,1} - \kappa_2 W_1 + \tau W_3)), \end{cases}$$

and

$$(1.19) \qquad \begin{cases} \tilde{\kappa}_2 = \vartheta_{3,1} + \kappa_3\vartheta_1 + \tau\vartheta_2, \\ \tilde{\kappa}_3 = -\vartheta_{2,1} - \kappa_2\vartheta_1 + \tau\vartheta_3, \\ \tilde{\tau} = \vartheta_{1,1} - \kappa_2\vartheta_2 - \kappa_3\vartheta_3. \end{cases}$$

2. The Equations of Motion

Let $\Delta T(x_1, x_2, x_3)$ denote the difference of the temperature $T(x_1, x_2, x_3)$ of the deformed state and the *uniform* temperature T_0 of the reference state. As the body is supposed to be very thin, we may take the temperature difference ΔT to satisfy

$$\Delta T = \theta_1(x_1) + x_2\theta_2(x_1) + x_3\theta_3(x_1).$$

In addition, let E denote Young's modulus and G the shear modulus of the beam at the point x_1 on the reference curve. We have assumed from the beginning that the cross section $A(x_1)$ is doubly symmetric with respect to the lines $x_2 = 0, x_3 = 0$.

We introduce the moments of the cross section as follows:

$$I_{22} := \iint_A x_2^2 dA, \quad I_{33} := \iint_A x_3^2 dA, \quad I := I_{22} + I_{33},$$

$$\begin{aligned} J :&= \iint_A \{(\phi_{,2} - x_3)^2 + (\phi_{,3} + x_2)^2\} dA \\ &= I + \iint_A \{-\phi_{,2}x_3 + \phi_{,3}x_2\} dA, \end{aligned}$$

$$\Gamma := \iint_A \phi^2 dA.$$

The second identity for J follows from the double symmetry of the cross section. Note that these quantities are still functions of x_1 since $A(x_1)$ varies with x_1. From now on we assume in addition that

- the cross section dimensions are negligible compared to the initial radii of curvature (i.e. $1/\kappa_i, 1/\tau$);
- the material is linearly elastic and transversely isotropic.

The free energy per unit volume $\Phi(x_1, x_2, x_3)$ may then be taken to be of the form

$$\Phi(x_1, x_2, x_3) = \frac{E}{2}\varepsilon_{11}^2 + 2G(\varepsilon_{12}^2 + \varepsilon_{13}^2) - \alpha\varepsilon_{11}\Delta T. \tag{2.1}$$

Note that the trace of the strain tensor, which is the coefficient of the temperature difference in (2.1), is just ε_{11}. The total strain energy then is

$$W_I := \iiint_\Omega \{\frac{E}{2}\varepsilon_{11}^2 + 2G(\varepsilon_{21}^2 + \varepsilon_{31}^2) - \alpha\varepsilon_{11}\Delta T\} dx_1 dx_2 dx_3, \tag{2.2}$$

This is similar to the corresponding expression typical for 3-d St.Venant -Kirchhoff materials, namely

$$W_I = \iiint_\Omega \{\frac{\lambda}{2}(\mathrm{tr}(\varepsilon))^2 + \mu\mathrm{tr}(\varepsilon^2) - (3\lambda + 2\mu)\tilde{\alpha}\mathrm{tr}(\varepsilon)\Delta T\} dx_1 dx_2 dx_3,$$

where λ, μ are the Lamé moduli and $\mathrm{tr}(\cdot)$ indicated the trace of a matrix.

We may express (2.2) in terms of our reference strains $\bar{\varepsilon}_{ij}$ as follows. We have

$$\begin{aligned}\varepsilon_{11}^2 =& \bar{\varepsilon}_{11}^2 - 2x_2\tilde{\kappa}_2\bar{\varepsilon}_{11} - 2x_3\tilde{\kappa}_3\bar{\varepsilon}_{11} + 2\gamma_{,1}\phi\bar{\varepsilon}_{11} + x_2^2\tilde{\kappa}_2^2 + 2x_3x_2\tilde{\kappa}_3\tilde{\kappa}_2 \\ &- 2x_2\gamma_{,1}\phi\tilde{\kappa}_2 + \gamma_{,1}^2\phi^2 2x_3\gamma_{,1}\phi\tilde{\kappa}_3 + x_3^2\tilde{\kappa}_3^2, \\ \varepsilon_{21}^2 =& \bar{\varepsilon}_{21}^2 + \frac{1}{4}\gamma^2\phi_{,2}^2 + \gamma\phi_{,2}\bar{\varepsilon}_{21} - \frac{1}{2}x_3\gamma\phi_{,2}\tilde{\tau} - x_3\tilde{\tau}\bar{\varepsilon}_{21} + \frac{1}{4}x_3^2\tilde{\tau}^2, \\ \varepsilon_{31}^2 =& \bar{\varepsilon}_{31}^2 + \frac{1}{4}\gamma^3\phi_{,2}^2 - \gamma\phi_{,3}\bar{\varepsilon}_{31} + \frac{1}{2}x_2\gamma\phi_{,3}\tilde{\tau} - x_2\tilde{\tau}\bar{\varepsilon}_{31} + \frac{1}{4}x_2^2\tilde{\tau}^2.\end{aligned}$$

In order to evaluate (2.2), we use the fact that due to the assumption of double symmetry of the cross section and due to the properties of the harmonic function ϕ related to the Prandtl-stress function as indicated above, the following integrals vanish: $\iint_A \phi dx_2dx_3$, $\iint_A \phi x_2dx_2dx_3$, $\iint_A \phi x_3dx_2dx_3$, $\iint_A \phi_{,2}dx_2dx_3$, $\iint_A \phi_{,3}dx_2dx_3$. The vanishing of the first three integrals is easily seen to follow from the symmetry properties of ϕ and the double symmetry of the cross section, which implies that the domain $A(x_1)$ is a succession of four reflections of a domain given in the first quadrant. As for the other two integrals, we apply Green's theorem and use the symmetry properties of the normal. The total strain (or potential) energy is then calculated as

$$\begin{aligned}(2.3)\quad W_I = \int_0^\ell U dx_1 = \int_0^\ell \{&\frac{EA}{2}\bar{\varepsilon}_{11}^2 + 2GA(\bar{\varepsilon}_{12}^2 + \bar{\varepsilon}_{13}^2) + \frac{EI_{22}}{2}\tilde{\kappa}_2^2 \\ &+ \frac{EI_{33}}{2}\tilde{\kappa}_3^2 + \frac{GI}{2}\tilde{\tau}^2 + \frac{E\Gamma}{2}\gamma_{,1}^2 - G(I-J)\gamma\tilde{\tau} + \frac{G}{2}(I-J)\gamma^2 \\ &- \alpha A\bar{\varepsilon}_{11}\theta_1 + \alpha I_{22}\tilde{\kappa}_2\theta_2 + \alpha I_{33}\tilde{\kappa}_3\theta_3\}dx.\end{aligned}$$

Let us now focus on the stress–strain relations. Let s_{ij} denote the entries of the *second Piola–Kirchhoff stress tensor* $\mathbf{S}$. Under the standing assumption that the body is very thin, we have $s_{22} = s_{33} = s_{23} = 0$. By the symmetry of this stress tensor, the only three remaining entries are to be given in terms of the strains and the temperature. As for the stress–strain constitutive relation we require

$$(2.4)\quad \begin{cases} s_{11} = E\varepsilon_{11} - \alpha\Delta T = E(\bar{\varepsilon}_{11} - x_2\tilde{\kappa}_2 - x_3\tilde{\kappa}_3 + \gamma_{,1}\phi) - \alpha\Delta T \\ s_{12} = 2G\varepsilon_{12} = G(2\bar{\varepsilon}_{12} - x_3\tilde{\tau} + \gamma\phi_{,2}) \\ s_{13} = 2G\varepsilon_{13} = G(2\bar{\varepsilon}_{13} + x_2\tilde{\tau} + \gamma\phi_{,3}). \end{cases}$$

This is the assumption of a St. Venant-Kirchhoff material, see Ciarlet [18]. Note that in taking the St.Venant strain tensor $\boldsymbol{\varepsilon}$ we keep quadratic terms in the *reference* strains. Dropping those reduces (2.4) to the classical Hooke's law. By definition, the deformation matrix is $\mathbf{G} := (\mathbf{G}_1, \mathbf{G}_2, \mathbf{G}_2)^T$, where the $\mathbf{G}_i$'s are taken to be columns. Hence,

$$s_{11} = \frac{\partial U}{\partial \bar{\varepsilon}_{11}}, \quad s_{12} = \frac{1}{2}\frac{\partial U}{\partial \bar{\varepsilon}_{21}}, \quad s_{13} = \frac{1}{2}\frac{\partial U}{\partial \bar{\varepsilon}_{31}},$$

as they should be. If we apply the stress matrix $\mathbf{S} := (s_{ij})$ to $\mathbf{G}^T$ we obtain the transpose of the *first Piola–Kirchhoff stress tensor* $\mathbf{T}$. Upon introducing the *Piola–Kirchhoff stress vector* by

$$\mathbf{t} = (s_{11}\mathbf{G}_1 + s_{12}\mathbf{G}_2 + s_{13}\mathbf{G}_3),$$

we can express $\mathbf{T}$ as

$$\mathbf{T} = (\mathbf{t},\ s_{12}\mathbf{G}_1,\ s_{13}\mathbf{G}_1).$$

We can also introduce the second Piola-Kirchhoff stress vector $\mathbf{s} = \mathbf{S}_{1,(\cdot)}$, the first row of the second Piola-Kirchhoff matrix. As we have $\mathbf{T} = \mathbf{GS}$, we can write

$$\mathbf{s} = \mathbf{G}^{-1}\mathbf{t} = s_{11}\mathbf{e}_1 + s_{12}\mathbf{e}_2 + s_{13}\mathbf{e}_3 \tag{2.5}$$

The resultant force, $\mathbf{F}$, and couple, $\mathbf{M}$, on a cross section are given by

$$\mathbf{F} := \iint_A \mathbf{t}dA, \tag{2.6}$$

$$\mathbf{M} := \iint_A (\mathbf{R} - \mathbf{R}_0) \times \mathbf{t}dA. \tag{2.7}$$

Let $\mathbf{f}_0, \mathbf{c}_0, \mathbf{t}_0$ denote the external body force, body couple, and surface traction, respectively, per unit undeformed volume and surface, respectively. Also, denote by ρ_0 the density per unit reference volume and by $m_0(x_1) := \rho_0 A(x_1)$ the density per unit reference length ($m := \int_0^\ell m_0 dx_1$ is then the total mass). Then the body force $\mathbf{P}$ and the body couple $\mathbf{C}$ are given by

$$\begin{cases} \mathbf{P} := \iint_A \rho_0(\mathbf{f}_0 - \mathbf{R}_{,tt})dA + \int_S \mathbf{t}_0 dS, \\ \mathbf{C} := \iint_A [\mathbf{c}_0 + (\mathbf{R} - \mathbf{R}_0)] \times \rho_0(\mathbf{f}_0 - \mathbf{R}_{,tt})dA \\ \qquad\qquad + \int_S (\mathbf{R} - \mathbf{R}_0) \times \mathbf{t}_0 dS, \end{cases} \tag{2.8}$$

where S is the curve circumscribing the cross section. As we are mainly interested in forces and couples applied to the end points of the beam we set $\mathbf{f}_0, \mathbf{c}_0, \mathbf{t}_0$ equal to zero. We consider a small segment (slice) Ω_Q of the beam located at a point Q on the undeformed centerline with length $2\Delta x_1$ in the direction of $\mathbf{e}_1$. The traction on the cross section at $Q - \Delta x_1$ is $-\mathbf{t}(Q - \Delta x_1)$ and the corresponding traction at $Q + \Delta x_1$ is $\mathbf{t}(Q + \Delta x_1)$. Let us denote the end surfaces of the slice Ω_Q by Σ_Q, and the surface tractions by $\mathbf{g}$. The body force density on that volume element is provided by the inertial force density $-\mathbf{R}_{,tt}$. Let us finally denote by $\mathbf{A} : \mathbf{B}$ the inner product of matrices, i.e. $\mathbf{A} : \mathbf{B} = \mathrm{tr}(\mathbf{A}^T\mathbf{B})$, and by $\nabla\delta\mathbf{R}$ the Jacobian of the virtual displacement

$$\delta\mathbf{R} = \delta\mathbf{W} + x_2\delta\mathbf{E}_2 + x_3\delta\mathbf{E}_3 + \delta\gamma\phi\hat{\mathbf{e}}_1 := \delta\mathbf{W} + \delta\tilde{\mathbf{R}},$$

where $\tilde{\mathbf{R}} := \mathbf{R} - \mathbf{R}_0$ and $\delta\mathbf{R}_0 = \delta\mathbf{W}$. Then the *principle of virtual work* can be expressed as

$$\iiint_{\Omega_Q} \mathbf{T} : \nabla\delta\mathbf{R}dx = \iiint_{\Omega_Q} \mathbf{f}\cdot\delta\mathbf{R}dx + \iint_{\Sigma_Q} \mathbf{g}\cdot\delta\mathbf{R}d\sigma \tag{2.9}$$

for all sufficiently smooth vector fields $\delta\mathbf{R} : \bar{\Omega} \to I\!R^3$ (see Ciarlet [18]), in fact, for all variations $\delta\mathbf{W}$, $\delta\vartheta$, $\delta\gamma$ satisfying some specified geometrical boundary conditions. It is also possible to express the principle of virtual work in terms of the second Piola Kirchhoff tensor $\mathbf{S}$ and the St.Venant-strain tensor $\boldsymbol{\varepsilon}$, since $\mathbf{T} : \nabla\delta\mathbf{R} = \mathbf{T} : \delta\mathbf{G} = \mathbf{S} : \mathbf{G}^T\delta\mathbf{G} = \mathbf{S} : \delta\boldsymbol{\varepsilon}$. Using the notation introduced above we find that (2.9) is equivalent to

$$\begin{aligned}\int_{Q-\Delta}^{Q+\Delta} [\iint_A \{\mathbf{t}\cdot\delta\mathbf{R}_{,1} + (s_{12}\mathbf{G}_1\cdot\delta\mathbf{R}_{,2} + s_{13}\mathbf{G}_1\cdot\delta\mathbf{R}_{,3})\}dA]dx_1 \\ = \int_{Q-\Delta}^{Q+\Delta} \{\iint_A (-\rho_0\mathbf{R}_{,tt})\cdot\delta\mathbf{R}dA\}dx_1 \\ + \iint_{A(Q+\Delta)} \mathbf{t}\cdot\delta\mathbf{R}dA - \iint_{A(Q-\Delta)} \mathbf{t}\cdot\delta\mathbf{R}dA.\end{aligned}$$

Following Wempner [108] (see also Hodges [22]), we introduce an *average external force* f_1, an *average resultant force* $\mathbf{D}$ and a *surface action* ξ reflecting the warping of the cross sections, as follows:

$$\begin{aligned} f_1 :&= \hat{\mathbf{e}}_1 \cdot \iint_A (-\rho_0\mathbf{R}_{,tt})\phi dA \\ &= \hat{\mathbf{e}}_1 \cdot \iint_A \mathbf{T}\nabla\phi dA, \end{aligned}$$

$$\mathbf{D} := \iint_A \mathbf{t}\phi dA,$$

$$\xi := \hat{\mathbf{e}}_1 \cdot \iint_A (s_{12}\phi_{,2} + s_{13}\phi_{,3})\mathbf{G}_1 dA.$$

We obtain the following approximations for the variations of $\mathbf{R}$:

$$\begin{gathered}\delta\mathbf{R} = \delta\mathbf{W} + \delta\vartheta\times\tilde{\mathbf{R}} + \delta\gamma\phi\hat{\mathbf{e}}_1, \\ \delta\mathbf{R}_{,1} = \delta\mathbf{W}_{,1} + \delta\vartheta_{,1}\times\tilde{\mathbf{R}} + \delta\vartheta\times\tilde{\mathbf{R}}_{,1} + \delta\gamma_{,1}\phi\hat{\mathbf{e}}_1 + \delta\gamma\phi(K_2\hat{\mathbf{e}}_2 + K_3\hat{\mathbf{e}}_3), \\ \delta\mathbf{R}_{,2} = \delta\vartheta\times\hat{\mathbf{e}}_2 + \delta\gamma_{,2}\hat{\mathbf{e}}_1, \quad \delta\mathbf{R}_{,3} = \delta\vartheta\times\hat{\mathbf{e}}_3 + \delta\gamma_{,3}\hat{\mathbf{e}}_1, \\ \tilde{\mathbf{R}}_{,1} = -x_2(K_2\hat{\mathbf{e}}_1 - T\hat{\mathbf{e}}_3) - x_3(K_3\hat{\mathbf{e}}_1 + T\hat{\mathbf{e}}_2) + \gamma_{,1}\phi\hat{\mathbf{e}}_1 + \gamma\phi(K_2\hat{\mathbf{e}}_2 + K_3\hat{\mathbf{e}}_3).\end{gathered}$$

After some standard calculations, using these approximations, we obtain from the principle of virtual work the following balance laws, formulated for the whole beam:

$$(2.10)\qquad \begin{cases} \mathbf{F}_{,1} + \mathbf{P} = 0, \\ \mathbf{M}_{,1} + \hat{\mathbf{e}}_1 \times \mathbf{F} + \mathbf{C} = 0, \\ D_{1,1} - \xi + f_1 = 0. \end{cases}$$

Associated boundary conditions are

$$(2.11)\qquad \mathbf{F}|_0^\ell = \tilde{\mathbf{F}}|_0^\ell, \quad \mathbf{M}|_0^\ell = \tilde{\mathbf{M}}|_0^\ell, \quad D_{,1}|_0^\ell = 0,$$

with given boundary forces $\tilde{\mathbf{F}}$ and couples $\tilde{\mathbf{M}}$ (see [108]). In this formulation, equations (2.10) are the classical equations of motion. Boundary forces are not, of course, to be prescribed where geometric boundary conditions are imposed.

To equations (2.10) we have to add an equation governing the flow of heat. We have assumed that the body has, at its reference configuration, a homogeneous reference temperature T_0. Introducing the heat capacity c_ν and relying on Newton's law for the heat flux, we obtain

$$(2.12)\qquad \rho_0 c_\nu \Delta T_{,t} = \sum_i \Delta T_{,ii} - T_0 \alpha \sum_i \varepsilon_{ii,t},$$

together with the boundary conditions, e.g.

$$(2.13)\qquad \Delta T|_0^\ell = \tilde{\Delta T}|_0^\ell.$$

In order to explicate the system of balance and heat equations we could use the moving reference frame $\hat{\mathbf{e}}_i$, as it is done in Wempner's book. However, as we have expressed the deformation of the initial triad $\mathbf{e}_i$ into the triad $\mathbf{E}_i$ in terms of the initial triad, it is most natural to express all the quantities in terms of the triad $\mathbf{e}_i$, e.g., $\mathbf{F} = \sum_i F_i \mathbf{e}_i$, $\mathbf{M} = \sum_i M_i \mathbf{e}_i$,

$$\mathbf{F}_{,1} = (F_{1,1} - \kappa_2 F_2 - \kappa_3 F_3)\mathbf{e}_1 + (F_{2,1} + \kappa_2 F_1 - \tau F_3)\mathbf{e}_2 + (F_{3,1} + \kappa_3 F_1 + \tau F_2)\mathbf{e}_2,$$

$$\hat{\mathbf{e}}_1 \times \mathbf{F} = (\Theta_3 F_3 + \Theta_2 F_2)\mathbf{e}_1 - (F_3 + \Theta_2 F_1)\mathbf{e}_2 + (F_2 - \Theta_3 F_1)\mathbf{e}_3.$$

Ultimately we want to express rotations in terms of the angles ϑ_i rather than through the auxiliary angles Θ_i. We can accomplish this by our assumption that the shear is small compared to the rotation, thus expressions like $\varepsilon_{ab}F_a, \Theta_a F_a$, where $a, b = 2, 3$, are to be neglected. Therefore, we use the approximation

$$\hat{\mathbf{e}}_1 \times \mathbf{F} = -(F_3 + \vartheta_2 F_1)\mathbf{e}_2 + (F_2 - \vartheta_3 F_1)\mathbf{e}_3.$$

This helps us to express the balance equations more explicitly as

$$(2.14)\qquad \begin{cases} F_{1,1} - \kappa_2 F_2 - \kappa_3 F_3 + P_1 = 0, \\ F_{2,1} + \kappa_2 F_1 - \tau F_3 + P_2 = 0, \\ F_{3,1} + \kappa_3 F_1 + \tau F_2 + P_3 = 0, \\ M_{1,1} - \kappa_2 M_2 - \kappa_3 M_3 + C_1 = 0, \\ M_{2,1} + \kappa_2 M_1 - \tau M_3 - \vartheta_2 F_1 - F_3 + C_2 = 0, \\ M_{3,1} + \kappa_3 M_1 + \tau M_2 - \vartheta_3 F_1 + F_2 + C_3 = 0, \\ D_{1,1} - \xi + f_1 = 0. \end{cases}$$

According to our assumptions, we obtain

$$(2.15)\qquad \begin{cases} F_i = \iint_A s_{1i} dA, \\ M_1 = \iint_A (s_{13}x_2 - s_{12}x_3) dA, \\ M_2 = \iint_A s_{11}x_3 dA, \quad M_3 = -\iint_A s_{11}x_2 dA \\ D_1 = \iint_A s_{11}\phi dA, \quad \xi = \iint_A (s_{12}\phi_{,2} + s_{13}\phi_{,3}) dA. \end{cases}$$

We may then use the stress–strain constitutive equations (2.4) in (2.15) to obtain

$$(2.16)\qquad \begin{cases} F_1 = EA\bar{\varepsilon}_{11} - \alpha\theta_1, \\ F_2 = 2GA\bar{\varepsilon}_{12}, \\ F_3 = 2GA\bar{\varepsilon}_{13}, \\ M_1 = GI\tilde{\tau} - G(I-J)\gamma, \\ M_2 = -EI_{33}\tilde{\kappa}_3 - \alpha\theta_3, \\ M_3 = EI_{22}\tilde{\kappa}_2 + \alpha\theta_2, \\ D_1 = E\Gamma\gamma_{,1}, \\ \xi = G(I-J)(\gamma - \tilde{\tau}). \end{cases}$$

Relations (2.16) constitute a set of eight functions of the eight strains $\bar{\varepsilon}_{11}, \bar{\varepsilon}_{12}, \bar{\varepsilon}_{13}, \tilde{\kappa}_2, \tilde{\kappa}_3, \tilde{\tau}, \gamma, \gamma_{,1}$. The components of **P** and **C** may be approximated as

$$(2.17)\qquad P_i = -m_0 W_{i,tt}, \quad C_i = -m_0 \vartheta_{i,tt},$$

whereas the average force f_1 is

$$(2.18)\qquad f_1 = -\rho_0 \Gamma \gamma_{,tt}.$$

REMARK 2.1. The equations (2.14) together with the constitutive relations (2.16) and the expression of forces and moments (2.17), (2.17) constitute a complete set of partial differential equations governing the motion of a general beam in terms of displacements and rotations, if complemented by the relations (1.18), (1.19), the heat equation (2.12) and appropriate boundary conditions (2.11), (2.13). □

We remark that the first three equations of (2.14) are determined by the axial and in–plane forces F_i, and the second three equations describe twist (M_1) about the $\mathbf{e}_1$–axis and bending M_2, M_3 in the $\mathbf{e}_1\mathbf{e}_3$ and $\mathbf{e}_1\mathbf{e}_2$ planes, respectively. The last equation of (2.14) accounts for warping of the cross section. It is apparent that the initial curvatures κ_i and the twist τ introduce a linear coupling between stretching, bending, twisting and the temperature. It is also apparent that bending and twisting are *nonlinearly* coupled to shear. Furthermore, under the *Euler–Bernoulli–Kirchhoff hypothesis*, namely, that the cross section remains perpendicular to the centerline, the shear strains $\bar{\varepsilon}_{12}, \bar{\varepsilon}_{13}$ are then equal to zero, implying that F_2, F_3 are identically zero.

2.1. Some Remarks on Warping and Torsion. In the absence of the average body force f_1 and in view of (2.16), the last equation of (2.14) reduces to the equation

$$(E\Gamma\gamma_{,1})_{,1} - G(I-J)(\gamma-\tilde{\tau}) = 0$$

or, equivalently, to

$$G(I-J)\tilde{\tau} = G(I-J)\gamma - (E\Gamma\gamma_{,1})_{,1}\,. \tag{2.19}$$

If the warping strain $\gamma_{,1}$ is neglected and if the beam is initially straight and untwisted, then (2.19) implies $\gamma = \vartheta_{1,1}$, and hence $M_1 = GJ\vartheta_{1,1}$, or $M_1 = GI\vartheta_{1,1}$ if also $\phi \equiv 0$. If the torsion is considered to be nonuniform to the effect that the warping strain $\gamma_{,1}$ is present, the right hand side of (2.19) can be used as an approximation of the torsion if γ is still taken to be equal to $\vartheta_{1,1}$. Note that in this case the torsion is given by a third order differential expression in the variable ϑ_1 (see Kane et al [41]).

Before writing down the differential equations in some examples of particular interest, let us reflect on the potential energy (2.3). It is interesting to observe that in the case where no warping is present (in particular if $\Gamma = 0,\ J = I,\ \gamma = \tilde{\tau}$), the only term representing torsion is $\frac{1}{2}GI\tilde{\tau}^2$, whereas if warping is present *and γ is set equal to $\vartheta_{1,1}$*, torsion is represented by $\frac{1}{2}GJ\vartheta_{1,1}^2 + \frac{1}{2}E\Gamma\vartheta_{1,11}^2$ (if the beam is initially straight and untwisted). As has been indicated above, it is in this case also possible to take (2.19) as an approximation for the torsion and to account for torsion through

$$\frac{1}{2G(I-J)}\left(G(I-J)\vartheta_{1,1} - (E\Gamma\vartheta_{1,11})_{,1}\right)^2. \tag{2.20}$$

This is done in Kane et al [41]. It is obvious that in this case the first approximation, based on the potential energy (2.3), results in a fourth order differential equation governing the torsion angle ϑ_1, whereas the second approximation, based on (2.3) with (2.20) as the only term to represent torsion, results in a sixth order differential equation in ϑ_1. In this paper we either take warping into account while keeping the body force f_1, to the extent that the approximations just mentioned do not hold, or we neglect the warping entirely.

3. Rotating Beams

In this section we consider the motion of a flexible beam, as described in the previous section, which is supposed to be attached to a rigid body. The rigid body together with its attachment is assumed to undergo a rigid body transformation: a translation and a rotation. We do not take into account warping of the cross section. There is a considerable literature dealing with precisely this problem. See Antman and Nachman [3], Kane et al [41], Ballieul and Levi [5], [6], Delfour et al [26] and many others. In most papers, however, attention is devoted to either special kinematical assumptions or special material laws. The equations which are derived here are most easily compared with those obtained in [5].

Let m_b denote the mass of the rigid body and I_b its moment of inertia. For simplicity we assume that the center of mass of the rigid body is in the origin of the body frame. We consider planar rotations about the $\mathbf{e}_2$ -axis, only. That is, we consider a rotation

$$\mathbf{\Pi}(t) = \begin{pmatrix} \cos(\theta(t)) & 0 & -\sin(\theta(t)) \\ 0 & 1 & 0 \\ \sin(\theta(t)) & 0 & \cos(\theta(t)) \end{pmatrix}.$$

We also consider a translation in that plane

$$\mathbf{y}(t) = (y_1(t), 0, y_2(t))^T.$$

In the global inertia frame the position vector $\mathcal{R}$ is given by

$$\mathcal{R} = \mathbf{\Pi}(t)\mathbf{R} + \mathbf{y}(t).$$

The velocity $\mathcal{R}_{,t}$ in the inertia frame is given by

$$\mathcal{R}_{,t} = \mathbf{\Pi}(t)(\mathbf{\Pi}(t)^T\dot{\mathbf{\Pi}}(t)\mathbf{R} + \mathbf{\Pi}(t)\mathbf{R}_{,t}) + \dot{\mathbf{y}},$$

where the dot denotes the derivative of functions depending only on the time parameter t. In the given case of planar rotations the matrix $\mathbf{\Pi}^T\dot{\mathbf{\Pi}}$ has a simple representation,

$$\mathbf{\Pi}^T\dot{\mathbf{\Pi}} = \dot{\theta}\mathbf{J},$$

where

$$\mathbf{J} := \begin{pmatrix} 0 & 0 & -1 \\ 0 & 0 & 0 \\ 1 & 0 & 0 \end{pmatrix}.$$

It is convenient to introduce the notation, adopted from [6],

$$D\mathbf{R} := \mathbf{R}_{,t} + \dot{\theta}\mathbf{J}\mathbf{R},$$

so that the velocity in the inertia frame is given by

$$\mathcal{R}_{,t} = \mathbf{\Pi} D\mathbf{R} + \dot{\mathbf{y}}.$$

We can now express the body forces and couples (2.8) as follows:

$$\mathbf{P} = -m_0\mathbf{\Pi}(\mathbf{R}_{0,tt} + 2\dot{\theta}\mathbf{J}\dot{\mathbf{R}}_0 - \dot{\theta}^2\tilde{\mathbf{I}}\mathbf{R}_0 + \ddot{\theta}\mathbf{J}\mathbf{R}_0) + m_0\ddot{\mathbf{y}}, \tag{3.1}$$

where we have used $\mathcal{R}_{,tt} = \mathbf{\Pi} D^2\mathbf{R}$, $\mathbf{R} = \mathbf{R}_0 + x_2\mathbf{E}_2 + x_3\mathbf{E}_3$ and the assumption of double symmetry of the cross section. Also, we have used the notation $\tilde{\mathbf{I}} = \mathbf{diag}(1, 0, 1)$. As for the couples we obtain

(3.2)
$$\begin{aligned} \mathbf{C} &= \iint_A \rho_0(\boldsymbol{\mathcal{R}} - \boldsymbol{\mathcal{R}}_0) \times (-\boldsymbol{\mathcal{R}}_{,tt})dA \\ &= -\rho_0 \iint_A (\boldsymbol{\mathcal{R}} - \boldsymbol{\mathcal{R}}_0) \times (\boldsymbol{\mathcal{R}}_{,tt} - \boldsymbol{\mathcal{R}}_{0,tt})dA \\ &= -\rho_0 \iint_A x_3\mathbf{\Pi}\mathbf{E}_3 \times (x_3\mathbf{\Pi}(\mathbf{E}_{3,tt} + 2\dot{\theta}\mathbf{J}\mathbf{E}_{3,t} - \dot{\theta}^2\tilde{\mathbf{I}}\mathbf{E}_3 + \ddot{\theta}\mathbf{J}\mathbf{E}_3)dA \\ &= -\rho_0 I_{33}\mathbf{\Pi}(\mathbf{E}_3 \times \mathbf{E}_{3,tt} + 2\dot{\theta}\mathbf{E}_3 \times \mathbf{J}\mathbf{E}_{3,t} + \ddot{\theta}\mathbf{E}_3 \times \mathbf{J}\mathbf{E}_3) \end{aligned}$$

Again, we base our calculations on the quadratic approximation of rotations. As we are working in the $\mathbf{e}_1\mathbf{e}_3$ plane, we have

$$\begin{aligned} \mathbf{E}_3 &= \vartheta_2\mathbf{e}_1 + (1 - \frac{1}{2}\vartheta_2^2)\mathbf{e}_3, \\ \mathbf{E}_{3,t} &= \vartheta_{2,t}\mathbf{e}_1 - \vartheta_2\vartheta_{2,t}\mathbf{e}_3, \\ \mathbf{E}_{3,tt} &= \vartheta_{2,tt}\mathbf{e}_1 - (\vartheta_{2,t}^2 + \vartheta_2\vartheta_{2,tt})\mathbf{e}_3. \end{aligned}$$

Taking the cross products in (3.2) and retaining only quadratic terms, we obtain

$$\mathbf{C} = C_2\mathbf{e}_2 = -m_0 I_{33}\mathbf{\Pi}\{\vartheta_{2,tt} - \ddot{\theta}\}\mathbf{e}_2. \tag{3.3}$$

We note that the deformation gradient in the rotated frame is $\mathbf{\Pi}\mathbf{G}$, and therefore the corresponding stress vector is $\mathbf{\Pi}\mathbf{t}$. Hence the forces and moments in the rotated frame are $\mathbf{\Pi}\mathbf{F}, \mathbf{\Pi}\mathbf{M}$, respectively. We also note that $\mathbf{\Pi}\hat{\mathbf{e}}_1 \times \mathbf{\Pi}\mathbf{F} = \mathbf{\Pi}(\hat{\mathbf{e}}_1 \times \mathbf{F})$. Therefore, the balance equations (2.10) with respect to the rotated frame can multiplied through by the transpose of $\mathbf{\Pi}$. We obtain

$$\text{(3.4)} \qquad \begin{cases} \mathbf{F}_{,1} = m_0 D^2\mathbf{R} - m_0 \mathbf{\Pi}^T \ddot{\mathbf{y}}, \\ \mathbf{M}_{,1} + \hat{\mathbf{e}}_1 \times \mathbf{F} = m_0 I_{33}\{\vartheta_{2,tt} - \ddot{\theta}\}\mathbf{e}_2. \end{cases}$$

This set of equations has to be complemented by the balance of linear and total angular momentum

$$\text{(3.5)} \qquad \begin{cases} m_b \ddot{\mathbf{y}} + \iiint_\Omega \rho_0 \boldsymbol{\mathcal{R}}_{,tt} dx_1 dx_2 dx_3 = 0, \\ I_b \ddot{\theta} - \iiint_\Omega \rho_0 \boldsymbol{\mathcal{R}} \times \boldsymbol{\mathcal{R}}_{,tt} dx_1 dx_2 dx_3 = 0. \end{cases}$$

We can express the integral in the second equation of (3.5) as

$$\begin{aligned} \text{(3.6)} \quad & \iiint_\Omega \rho_0 \boldsymbol{\mathcal{R}} \times \boldsymbol{\mathcal{R}}_{,tt} dx_1 dx_2 dx_3 \\ &= \iiint_\Omega \rho_0 \mathbf{\Pi}(\mathbf{R}_0 + x_3\mathbf{E}_3) \times (\mathbf{\Pi}(D^2\mathbf{R}_0 + x_3 D^2\mathbf{E}_3) + \ddot{\mathbf{y}}) dx \\ &\doteq \mathbf{\Pi}(\iiint_\Omega \rho_0 \mathbf{R}_0 \times D^2\mathbf{R}_0 + \rho_0 \mathbf{R}_0 \times \mathbf{\Pi}^T \ddot{\mathbf{y}} dx_1 dx_2 dx_3 \\ &\qquad\qquad - \int_0^\ell m_0 I_{33} dx_1 \ddot{\theta} + \int_0^\ell I_{33} \vartheta_{2,tt} dx_1. \end{aligned}$$

As $\mathbf{R}_0 = x\mathbf{e}_1 + \mathbf{W}$, we find

$$\begin{aligned} D^2\mathbf{R}_0 = (W_{1,tt} - \ddot{\theta}W_3 - 2\dot{\theta}W_{3,t} - \dot{\theta}^2(W_1 + x))\mathbf{e}_1 \\ + (W_{3,tt} + \ddot{\theta}W_1 + 2\dot{\theta}W_{1,t} - \dot{\theta}^2 W_3 + x\ddot{\theta})\mathbf{e}_3. \end{aligned}$$

The first equation of (3.5) can be rewritten as

$$\text{(3.7)} \qquad (m_b + m_0\ell)\ddot{\mathbf{y}} + m_0 \mathbf{\Pi} \int_0^\ell D^2\mathbf{R}_0 dx = 0.$$

Therefore, (3.4) and (3.6) can be written as

$$\text{(3.8)} \qquad \begin{cases} F_{1,1} - \kappa_3 F_3 = m_0(W_{1,tt} - \ddot{\theta}W_3 - 2\dot{\theta}W_{3,t} - \dot{\theta}^2(W_1 + x)) \\ \qquad\qquad -m_0(\cos(\theta)\ddot{y}_1 + \sin(\theta)\ddot{y}_3), \\ F_{3,1} + \kappa_3 F_1 = m_0(W_{3,tt} + \ddot{\theta}W_1 + 2\dot{\theta}W_{1,t} - \dot{\theta}^2 W_3 + x\ddot{\theta}) \\ \qquad\qquad -m_0(-\sin(\theta)\ddot{y}_1 + \cos(\theta)\ddot{y}_3), \\ M_{2,1} - \vartheta_2 F_1 - F_3 = \rho_0 I_{33}\{-\vartheta_{2,tt} + \ddot{\theta}\}, \end{cases}$$

and

(3.9)

$$\begin{cases}(m_b+m)\ddot{y}_1 = -\int_0^\ell\{\cos(\theta)(W_{1,tt}-\ddot{\theta}W_3-2\dot{\theta}W_{3,t}-\dot{\theta}^2(W_1+x)) \\ \qquad\qquad -\sin(\theta)(W_{3,tt}+\ddot{\theta}W_1+2\dot{\theta}W_{1,t}-\dot{\theta}^2W_3+x\ddot{\theta})\}dx_1, \\ (m_b+m)\ddot{y}_2 = -\int_0^\ell\{\sin(\theta)(W_{1,tt}-\ddot{\theta}W_3-2\dot{\theta}W_{3,t}-\dot{\theta}^2(W_1+x)) \\ \qquad\qquad +\cos(\theta)(W_{3,tt}+\ddot{\theta}W_1+2\dot{\theta}W_{1,t}-\dot{\theta}^2W_3+x\ddot{\theta})\}dx_1, \\ (I_b+\hat{I})\ddot{\theta} = -\int_0^\ell m_0\{-W_3(W_{1,tt}-\ddot{\theta}W_3-2\dot{\theta}W_{3,t}) \\ \qquad +(W_1+x)(W_{3,tt}+\ddot{\theta}(W_1+x)+2\dot{\theta}W_{1,t})+I_{33}\vartheta_{2,tt} \\ \qquad +W_3(\cos(\theta)\ddot{y}_1+\sin(\theta)\ddot{y}_2) \\ \qquad -(x+W_1)(-\sin(\theta)\ddot{y}_1+\cos(\theta)\ddot{y}_2)\}dx_1,\end{cases}$$

where $\hat{I}=\int_0^\ell I_{33}m_0dx_1$. These equations are precisely the ones obtained in [5] for a uniform rod. We can also easily write down the kinetic energy which is also found to be exactly as in Ballieul and Levi [6]:

$$T=\frac{1}{2}I_b\dot{\theta}^2+\frac{1}{2}\int_0^\ell I(\vartheta_{2,t}^2+\dot{\theta}^2)dx_1 \\ +\frac{1}{2}m_b\|\dot{\mathbf{y}}\|^2+\frac{1}{2}\int_0^\ell \rho_0\|\mathbf{\Pi}\mathbf{D}\mathbf{R}_0+\dot{\mathbf{y}}\|^2dx_1 \tag{3.10}$$

3.1. Dynamic Stiffening. In this subsection we consider an initially straight beam ($\kappa_3=0$) rotating about the origin in the $\mathbf{e}_1\mathbf{e}_3$ plane (as above). We neglect the inertia forces with respect to stretching and retain as the only nonlinearity in the kinematics the terms with factor $\dot{\theta}^2$ (centrifugal forces). We can then integrate the first equation of (3.8) with respect to x and obtain

$$F_1(x)=\frac{m_0}{2}\dot{\theta}^2(\ell^2-x^2),$$

where we take into account that $F_1(\ell)=0$. As the origin is assumed fixed, we neglect the balance of total linear momentum. Then the set of equations (3.8), (3.9) reduces to

$$\begin{cases}F_{3,1}=m_0W_{3,tt}-\dot{\theta}^2W_3+x\ddot{\theta}, \\ M_{2,1}-\frac{m_0}{2}\dot{\theta}^2(\ell^2-x^2)\vartheta_2-F_3=m_0I_{33}(\vartheta_{2,tt}-\ddot{\theta}), \\ (I_b+\hat{I}\ell)\ddot{\theta}=-m_0\int_0^\ell(xW_{3,tt}+I_{33}\vartheta_{2,tt})dx.\end{cases} \tag{3.11}$$

If we further differentiate the second equation with respect to x, we can eliminate F_3 by and obtain instead

$$M_{2,11}-\left(\frac{m_0}{2}\dot{\theta}^2(\ell^2-x^2)\vartheta_2\right)_{,1}=m_0W_{3,tt}+\rho_0I_{33}\vartheta_{2,1tt}-\dot{\theta}^2W_3+x\ddot{\theta}.$$

In the case where there is no shear strain ($\bar{\varepsilon}_{31} = 0$), the Euler- Bernoulli hypothesis, we have $\vartheta_2 = -W_{3,1}$ and M_2 reduces to $M_2 = -EI_{33}W_{3,11}$. Then the equation above is equivalent to

$$m_0 W_{3,tt} - \rho_0 I_{33} W_{3,11tt} + EI_{33} W_{3,1111} - \dot{\theta}^2 W_3 + x\ddot{\theta} - \left(\frac{m_0}{2}\dot{\theta}^2(\ell^2 - x^2)W_{3,1}\right)_{,1} = 0.$$

In this equation it becomes apparent that the term involving $-(m_0\dot{\theta}^2(\ell^2 - x^2)W_{3,1})_{,1}$ introduces a *dynamic stiffening.* The last equation can be found for instance in Ballieul and Levi [5] and other papers. In most of the engineering papers, such a stiffening term is introduced by the way of additional forces, if it is considered at all. The point here is that we obtain this significant feature as a result of the *infinitesimal strain/large displacement* or the *small strain/finite rotation* theory of rods.

4. Straight, Untwisted, Nonshearable Nonlinear 3–d Beams

As we have only one spatial variable x_1 to take into account, we will consequently replace the subscript $_{,1}$ with a prime in order to indicate spatial derivation, and replace $_{,t}$ by a superimposed dot. We set $\kappa_2 = \kappa_3 = \tau = 0$. By the *Euler-Bernoulli-Kirchhoff hypothesis,* $\bar{\varepsilon}_{12} = \bar{\varepsilon}_{13} = 0$. In addition, we *neglect the nonlinearities in the shear strains.* We have

$$\vartheta_3 = W_2', \ \vartheta_2 = -W_3',$$
$$\tilde{\kappa}_2 = W_2'', \ \tilde{\kappa}_3 = W_3'', \ \tilde{\tau} = \vartheta_1'.$$

The reference strains are

$$\bar{\varepsilon}_{11} = W_1' + \frac{1}{2}((W_2')^2 + (W_3')^2), \ \bar{\varepsilon}_{12} = \bar{\varepsilon}_{13} = 0$$

As for the body forces and couples, we find

$$P_i = -m_0 \ddot{W}_i, \quad i = 1, 2, 3,$$
$$C_1 = -\rho_0 I \ddot{\vartheta}_1, \quad C_2 = \rho_0 I_{33} \ddot{W}_3', \quad C_3 = -\rho_0 I_{22} \ddot{W}_2'.$$

We make use of the approximations (2.16) and insert the relations above into the balance laws (2.14). Then we differentiate with respect to x_1 the second and third of the equations representing the balance of angular momentum, and insert the second and third equation of the balance of forces, respectively. The result is a set of five equations governing the longitudinal displacement W_1, the two lateral displacements W_2, W_3, the twist ϑ_1 and the warping γ. This set has to be complemented by the equations governing the temperatures θ_i, and by boundary and initial conditions.

Equations of Motion

[Longitudinal motion]

$$m_0\ddot{W}_1 = [EA(W_1' + \frac{1}{2}(W_2')^2 + \frac{1}{2}(W_3')^2)]' - \alpha\theta_1'. \tag{4.1}$$

[Vertical motion]

$$m_0\ddot{W}_2 - [\rho_0 I_{22}\ddot{W}_2']' + [EI_{22}W_2'']'' + \alpha\theta_2'' = [(EA(W_1' + \frac{1}{2}(W_2')^2 + \frac{1}{2}(W_3')^2) - \alpha\theta_1)W_2']'. \tag{4.2}$$

[Lateral motion]

$$m_0\ddot{W}_3 - [\rho_0 I_{33}\ddot{W}_3']' + [EI_{33}W_3'']'' + \alpha\theta_3'' = [(EA(W_1' + \frac{1}{2}(W_2')^2 + \frac{1}{2}(W_3')^2) - \alpha\theta_1)W_3']'. \tag{4.3}$$

[Torsional motion]

$$\rho_0 I\ddot{\vartheta}_1 = [GI\vartheta_1' - G(I-J)\gamma]'. \tag{4.4}$$

[Warping]

$$\rho_0\Gamma\ddot{\gamma} = [E\Gamma\gamma']' - G(I-J)(\gamma - \vartheta_1'). \tag{4.5}$$

The equation of heat conduction reduces to

$$\rho_0 c_\nu\{\dot{\theta}_1 + \dot{\theta}_2 x_2 + \dot{\theta}_3 x_3\} = \theta_1'' + x_2\theta_2'' + x_3\theta_3'' - T_0\alpha\{\dot{\bar{\varepsilon}}_{11} - x_2\dot{\bar{\kappa}}_2 - x_3\dot{\bar{\kappa}}_3\}.$$

Upon taking the average and the first moments with respect to the x_i axis, we obtain a set of three 1-d heat equations for the variables θ_i as follows:

[Heat conduction]

$$\begin{cases} \rho_0 c_\nu\dot{\theta}_1 = \theta_1'' - \alpha T_0\frac{\partial}{\partial t}\{W_1' + \frac{1}{2}((W_2')^2 + (W_3')^2)\}, \\ \rho_0 c_\nu\dot{\theta}_2 = \theta_2'' + \alpha T_0\dot{W}_2'', \\ \rho_0 c_\nu\dot{\theta}_3 = \theta_3'' + \alpha T_0\dot{W}_3''. \end{cases} \tag{4.6}$$

These equations have to be complemented by appropriate boundary conditions. As a typical example, we choose a beam which is clamped and thermally insulated at the boundary $x = 0$, and to which bending moments and shear forces as well as heat are applied at the end $x = \ell$.
Boundary conditions

[Geometric boundary conditions]

$$\begin{cases} W_i(0) = 0, \quad i = 1,2,3, \\ W_2'(0) = W_3'(0) = 0, \\ \vartheta_1(0) = 0, \quad \gamma(0) = 0. \end{cases} \tag{4.7}$$

[Insulated boundary conditions]

$$\theta_i'(0) = 0, \quad i = 1,2,3. \tag{4.8}$$

[Dynamical boundary conditions]

$$\begin{cases} EA(W_1' + \frac{1}{2}((W_2')^2 + (W_3')^2))(\ell) - \alpha\theta_1(\ell) = f_1, \\ -EI_{33}W_3''(\ell) - \alpha\theta_3(\ell) = m_2, \\ EI_{22}W_2''(\ell) + \alpha\theta_2(\ell) = m_3, \end{cases} \tag{4.9}$$

$$\begin{aligned} &- [EI_{33}W_3'']'(\ell) - \alpha\theta_3'(\ell) \\ &\quad + [EA(W_1' + \frac{1}{2}((W_2')^2 + (W_3')^2)) - \alpha\theta_1]W_3'(\ell) \\ &\qquad + \rho_0 I_{33}\ddot{W}_3'(\ell) = f_2, \end{aligned} \tag{4.10}$$

$$\begin{aligned} &[EI_{22}W_2'']'(\ell) + \alpha\theta_2'(\ell) \\ &\quad - [EA(W_1' + \frac{1}{2}((W_2')^2 + (W_3')^2)) - \alpha\theta_1]W_2'(\ell) \\ &\qquad - \rho_0 I_{22}\ddot{W}_2'(\ell) = f_3, \end{aligned} \tag{4.11}$$

$$GI\vartheta_1'(\ell) = m_1, \quad \gamma'(\ell) = 0. \tag{4.12}$$

[Heat flux boundary conditions]

$$\theta_1'(\ell) = \varphi_1 \; \theta_2'(\ell) = \varphi_2 \; \theta_3'(\ell) = \varphi_3. \tag{4.13}$$

Initial conditions

$$\begin{aligned} W_i(t=0,\cdot) = W_{i0}(\cdot), \quad &\dot{W}_i(t=0,\cdot) = W_{i1}(\cdot), \\ \vartheta_1(t=0,\cdot) = 0, \quad &\dot{\vartheta}_1(t=0,\cdot) = 0, \\ \gamma(t=0,\cdot) = 0, \quad &\dot{\gamma}(t=0,\cdot) = 0, \\ \theta_i(t=0,\cdot) = \theta_{i0}(\cdot), \quad &i = 1,2,3. \end{aligned}$$

As the body is supposed to be initially straight and untwisted, we have set the torsion and warping distributed along the beam at the time $t = 0$ to be equal zero. A twist can be exerted, however, at the free end of beam (i.e. through m_1).

In the foregoing, we have assumed that the material is homogeneous to the effect that the mass, the Young and shear moduli, as well as the thermal coefficient α and the reference temperature T_0 can be treated as constant. The cross section, however, is not constant in x_1, and so neither are the moments I_i. It is clear that the general inhomogeneous material can be included without any extra effort, as long as we have isotropy. It is important to note that the isothermal homogeneous process (4.1)–(4.5) with θ_i set equal to zero conserves the total energy $U+T$, where in this case

$$
(4.14)\quad \begin{cases} U = \displaystyle\int_0^\ell \{\frac{EA}{2}(W_1' + \frac{1}{2}((W_2')^2 + (W_3')^2))^2 \\ \qquad + \dfrac{EI_{22}}{2}(W_2'')^2 + \dfrac{EI_{33}}{2}(W_3'')^2 \\ \qquad + \dfrac{GJ}{2}(\vartheta_1')^2 + \dfrac{1}{2}E\Gamma(\gamma')^2 + \dfrac{1}{2}G(I-J)(\vartheta_1' - \gamma)^2\}dx, \\ T = \displaystyle\int_0^\ell \{\frac{m_0}{2}(\dot W_1^2 + \dot W_2^2 + \dot W_3^2) + \frac{\rho_0}{2}(I_{22}(\dot W_2')^2 + I_{33}(\dot W_3')^2) \\ \qquad + \dfrac{\rho_0 I}{2}\dot\vartheta_1^2 + \dfrac{1}{2}\dot\gamma^2\}dx. \end{cases}
$$

The temperature dependent system is easily seen to be dissipative. In fact, in the temperature dependent case, we may work with T as in (4.14), but with U in (4.14) replaced with

$$
U + \sum_{i=1}^{3} \int_0^\ell \frac{\rho_0 c_\nu}{2T_0}\theta_i^2 dx.
$$

The reason for this is the fact that the energy (2.3) is not purely quadratic as it contains terms mixed in strains and temperatures. That energy, however, is easily seen to be bounded above by some constant times the potential energy just defined. Therefore, dissipativity with respect to the new "energy" is transferred to the physical energy. If only planar isothermal motion is considered, where we may look at the 1–2 or 1–3 plane (with $\vartheta_1 = 0$), the system above reduces to the one we have discussed in Lagnese and Leugering [47] which, in turn, reduces to a one–dimensional analog to the von Karman plate if the inertia of the longitudinal motion is replaced with a constant load. If, finally, we neglect the rotational inertia ($I_{22} = I_{33} = 0$), the model reduces to an Euler-Bernoulli type system.

4.1. Approximation–Generalizations. When the dimensions in the x_2, x_3 directions are very small, the heat conduction with respect to these directions may be represented by

$$(4.15)\qquad \begin{cases} \rho_0 c_\nu \dot{\theta}_1 = \theta_1'' - \alpha T_0 \frac{\partial}{\partial t}\{W_1' + \frac{1}{2}((W_2')^2 + (W_3')^2)\}, \\ \rho_0 c_\nu \dot{\theta}_2 = -\mu_2 \theta_2 + \alpha T_0 \dot{W}_2'', \\ \rho_0 c_\nu \dot{\theta}_3 = -\mu_3 \theta_3 + \alpha T_0 \dot{W}_3''. \end{cases}$$

Then the second and the third equations can be solved by variation of parameters to give

$$\theta_i(t) = \frac{\alpha T_0}{\rho_0 c_\nu} \int_{-\infty}^{t} e^{-\mu_i(t-s)} \dot{W}_i''(s)ds =: \int_{-\infty}^{t} a_i(t-s)\dot{W}_i''(s)ds, \quad i = 2, 3.$$

If we use this in (4.1)–(4.4), and neglect warping we obtain

[Longitudinal motion]

$$(4.16)\qquad m_0 \ddot{W}_1 = [EA(W_1' + \frac{1}{2}(W_2')^2 + \frac{1}{2}(W_3')^2)]' - \alpha\theta_1'.$$

[Vertical motion]

$$(4.17)\quad m_0 \ddot{W}_2 - \rho_0 I \ddot{W}_2'' + [EI_{22} W_2'']'' + \alpha \int_{\infty}^{t} a_2(t-s)\dot{W}_2''''(s)ds$$
$$= [(EA(W_1' + \frac{1}{2}(W_2')^2 + \frac{1}{2}(W_3')^2) - \alpha\theta_1)W_2']'.$$

[Lateral motion]

$$(4.18)\quad m_0 \ddot{W}_3 - \rho_0 I \ddot{W}_3'' + [EI_{33} W_3'']'' + \alpha \int_{\infty}^{t} a_3(t-s)\dot{W}_3''''(s)ds$$
$$= [(EA(W_1' + \frac{1}{2}(W_2')^2 + \frac{1}{2}(W_3')^2) - \alpha\theta_1)W_3']'.$$

[Torsional motion]

$$(4.19)\qquad \rho_0 I \ddot{\vartheta}_1 = [GI\vartheta_1']'.$$

The equation of heat conduction which with we are left is

$$\rho_0 c_\nu \dot{\theta}_1 = \theta_1'' - \alpha T_0 \frac{\partial}{\partial t}\{W_1' + \frac{1}{2}((W_2')^2 + (W_3')^2)\}.$$

Equations (4.16)–(4.19) are now integro–partial differential equations, describing a thin thermoelastic beam in three dimensions. Upon replacing the convolution kernels $e^{-\mu_i t}$ by more general functions, for instance completely monotone kernels, we can refine the assumptions made in order to arrive at (4.15). Also, this could be viewed as a particular kind of a viscoelastic, in fact thermo–viscoelastic beam model. Furthermore, instead of the simplifying assumption that the original heat equations (4.6) are replaced with (4.15), we could replace Fourier's law with a more general heat flux law accounting for a long memory of the material, and end up with a even more general thermo–viscoelastic model. If then kernels are chosen to approximate the Dirac distribution, we arrive at what has come to be known as the slow flow approximation, or the Kelvin–Voigt viscoelastic damping. There are other models for internal damping, in particular the so called structural damping of Chen and Russell [16] or, as an almost equivalent analog to that, the shear diffusion damping model described in [90], which can be introduced here.

5. Straight, Untwisted Shearable Linear 3–d Beams

We also do not consider warping and thermal coupling. We obtain

$$\begin{gathered}\bar\varepsilon_{11} = W_1',\ \bar\varepsilon_{21} = \frac{1}{2}(W_2' - \vartheta_3),\ \bar\varepsilon_{31} = \frac{1}{2}(W_3' + \vartheta_2),\\ F_1 = EAW_1',\ F_2 = GA(W_2' - \vartheta_3),\ F_3 = GA(W_3' + \vartheta_2),\\ M_1 = GI\vartheta_1',\ M_2 = EI_{33}\vartheta_2',\ M_3 = EI_{22}\vartheta_3',\\ P_i = -m_0\ddot W_i,\ C_i = -m_0\ddot\vartheta_i.\end{gathered}$$

This leads to the following system of 3–d Bresse-Timoshenko beam equations.

Equations of motion

$$(5.1)\quad \begin{cases} m_0\ddot W_1 = [EAW_1]' & \text{[Longitudinal motion]}\\ m_0\ddot W_2 = [GA(W_2' - \vartheta_3)]' & \text{[Lateral motion]}\\ m_0\ddot W_3 = [GA(W_3 + \vartheta_2]' & \text{[Vertical motion]}\\ \rho_0 I\ddot\vartheta_1 = [GI\vartheta_1']' & \text{[Torsional motion]}\\ \rho_0 I_{33}\ddot\vartheta_2 = [EI_{33}\vartheta_2']' - GA(W_3' + \vartheta_2) & \text{[Shear around } \hat{\mathbf{e}}_2]\\ \rho_0 I_{22}\ddot\vartheta_3 = [EI_{22}\vartheta_3']' + GA(W_2' - \vartheta_3) & \text{[Shear around } \hat{\mathbf{e}}_3] \end{cases}$$

REMARK 5.1. It is obvious from (5.1) that there is neither a linear coupling between longitudinal and vertical/lateral motion nor between any displacement and twist. □

As for the boundary conditions, we consider a beam clamped at $x = 0$ and free at $x = \ell$ and obtain:

Boundary conditions
[Geometric boundary conditions]

$$W_i(0) = 0,\ \vartheta_i(0) = 0,\ i = 1, 2, 3.$$

[Dynamic boundary conditions]

$$EAW_1'(\ell) = f_1, \quad GA(W_2' - \vartheta_3)(\ell) = f_2, \quad GA(W_3' + \vartheta_2)(\ell) = f_3,$$
$$GI\vartheta_1'(\ell) = m_1, \quad EI_{33}\vartheta_2'(\ell) = m_2, EI_{22}\vartheta_3'(\ell) = m_3.$$

We also have initial conditions as usual:

Initial conditions

$$W_i(t=0,\cdot) = W_{i0}(\cdot), \quad \dot{W}_i(t=0,\cdot) = W_{i1}(\cdot),$$
$$\vartheta_i(t=0,\cdot) = 0, \quad \dot{\vartheta}_i(t=0,\cdot) = 0,$$
$$i = 1,2,3.$$

The approximations and generalizations are as in the previous case.

6. Shearable Nonlinear 2–d Beams with Curvature

Here we set $W_2, \tau, \kappa_2, \vartheta_3, \vartheta_1$ equal to zero. Since there is no torsion, obviously neither is there warping. We have

$$\tilde{\kappa}_2 = 0, \quad \tilde{\kappa}_3 = -\vartheta_2', \quad \tilde{\tau} = 0,$$
$$\bar{\varepsilon}_{11} = W_1' - \kappa_3 W_3 + \frac{1}{2}\vartheta_2^2, \ \bar{\varepsilon}_{12} = 0,$$
$$\bar{\varepsilon}_{13} = \frac{1}{2}(\vartheta_2 + W_3' + \kappa_3 W_1).$$

F, M are given by (2.16), and C, P reduce to

$$P_i = -m_0 \ddot{W}_i,$$
$$C_2 = -\rho_0 I_{33}\ddot{\vartheta}_2, \quad C_1 = C_3 = 0.$$

We obtain the following system of equations.

Equations of motion
[Longitudinal motion]

$$m_0 h\ddot{W}_1 = [Eh(W_1' - \kappa_3 W_3 + \frac{1}{2}\vartheta_2^2) - \alpha\theta_1]' \\ - \kappa_3 Gh(\vartheta_2 + W_3' + \kappa_3 W_1) + \alpha[\kappa_3\theta_3]'$$

[Vertical motion]

$$m_0\ddot{W}_3 = [Gh(\vartheta_2 + W_3' + \kappa_3 W_1)]' + \kappa_3[Eh(W_1' - \kappa_3 W_3 + \frac{1}{2}\vartheta_2^2) - \alpha\theta_1]$$

[Shear motion]

$$\rho_0 I_{33}\ddot{\vartheta}_2 = EI_{33}\vartheta_2'' - \vartheta_2[Eh(W_1' - \kappa_3 W_3 + \frac{1}{2}\vartheta_2^2) - \alpha\theta_1] \\ - Gh(\vartheta_2 + W_3' + \kappa_3 W_1) - \alpha\theta_3'$$

[Heat flux]

$$
\begin{cases} \rho_0 c_\nu \dot{\theta}_1 = \theta_1'' - \alpha T_0 \frac{\partial}{\partial t}\{W_1' - \kappa_3 W_3 + \frac{1}{2}\vartheta_2^2\}, \\ \rho_0 c_\nu \dot{\theta}_3 = \theta_3'' - \alpha T_0 \dot{\vartheta}_2'. \end{cases} \tag{6.1}
$$

Again, this set of equations has to be complemented by appropriate boundary and initial conditions. We choose the situation analogous to the preceding section where, at $x = 0$, the beam is clamped and insulated, and at $x = \ell$ it is driven by forces, moments, and heat supplies.

Boundary conditions

[Geometric boundary conditions]

$$
W_i(0) = 0,\ i = 1,3, \quad \vartheta_2(0) = 0. \tag{6.2}
$$

[Insulated boundary conditions]

$$
\theta_i'(0) = 0,\ i = 1,3. \tag{6.3}
$$

[Dynamical boundary conditions]

$$
\begin{gathered} Eh[W_1' - \kappa_3 W_3 + \frac{1}{2}\vartheta_2^2](\ell) - \alpha\theta_1(\ell) = f_1, \\ Gh[\vartheta_2 + W_3' + \kappa_3 W_1](\ell) = f_2, \\ EI_{33}\vartheta_2'(\ell) - \alpha\theta_3(\ell) - \vartheta_2[Eh(W_1' - \kappa_3 W_3 + \frac{1}{2}\vartheta_2^2) - \alpha\theta_1](\ell) = m_1. \end{gathered}
$$

[Heat flux boundary conditions]

$$
\theta_1'(\ell) = \varphi_1, \ \theta_3'(\ell) = \varphi_2. \tag{6.4}
$$

Initial conditions

$$
\begin{gathered} W_i(t=0,\cdot) = W_{i0}(\cdot), \quad \dot{W}_i(t=0,\cdot) = W_{i1}(\cdot), \ i = 1,3, \\ \vartheta_2(t=0,\cdot) = \vartheta_{20}, \quad \dot{\vartheta}_2(t=0,\cdot) = \vartheta_{21}, \\ \theta_i(t=0,\cdot) = \theta_{i0}(\cdot), \ i = 1,3. \end{gathered}
$$

From this seemingly complicated system very interesting specializations can be obtained. In particular, the isothermal linear system is exactly the system obtained by Bresse [10] (1856), which is, in fact, more general than the Timoshenko system (this has been pointed out by J. Schmidt in [97]); for this reason we should in the future call the latter the Bresse–Timoshenko system. We should also mention that the equations of balance for precurved and pretwisted beams have been discussed in Love [63] (also before Timoshenko), who traced the literature back to Kirchhoff in 1859. This once again shows that it might be more appropriate not to attach names to the various beam models to begin with, as many of these equations seem to have been rediscovered repeatedly. Again, let us note that the homogeneous isothermal model conserves the total energy $U+T$, where now

$$U = \int_0^\ell \{\frac{Eh}{2}(W_1' - \kappa_3 W_3 + \frac{1}{2}\vartheta_2^2)^2 + \frac{Gh}{2}(\vartheta_2 + W_3' + \kappa_3 W_1)^2 + \frac{EI_{33}}{2}\vartheta_{2,1}^2\}dx,$$

$$T = \int_0^\ell \{\frac{m_0}{2}(\dot{W}_1^2 + \dot{W}_3^2) + \frac{\rho_0 I}{2}\dot{\vartheta}_2^2\}dx.$$

The temperature dependent system is then easily seen to be dissipative.

6.1. Approximation–Generalizations. We apply the same consideration as in the preceding section to the *linear* equations, and replace the heat equations (6.1) with

[Heat flux]

$$\begin{cases} \rho_0 c_\nu \dot{\theta}_1 = \theta_1'' - \alpha T_0 \frac{\partial}{\partial t}\{W_1' - \kappa_3 W_3\}, \\ \rho_0 c_\nu \dot{\theta}_3 = -\mu\theta_3 - \alpha T_0 \dot{\vartheta}_2', \end{cases} \tag{6.5}$$

and integrate the second of these equations. Then we obtain the system of integro–partial differential equations

[Longitudinal motion]

$$m_0 \ddot{W}_1 = [Eh(W_1' - \kappa_3 W_3) - \alpha\theta_1]' - \kappa_3 Gh(\vartheta_2 + W_3' + \kappa_3 W_1) - \alpha^2 \int_{-\infty}^t e^{-\mu(t-s)}[\kappa_3 \frac{T_0}{\rho_0 c_\nu}\dot{\vartheta}_2]'(s)ds. \tag{6.6}$$

[Vertical motion]

$$m_0 \ddot{W}_3 = [Gh(\vartheta_2 + W_3' + \kappa_3 W_1)]' + \kappa_3[Eh(W_1' - \kappa_3 W_3) - \alpha\theta_1]. \tag{6.7}$$

[Shear motion]

$$\rho_0 I_{33}\ddot{\vartheta}_2 = [EI_{33}\vartheta_2']' - Gh(\vartheta_2 + W_3' + \kappa_3 W_1) + \alpha^2 \int_{-\infty}^t e^{-\mu(t-s)}[\kappa_3 \frac{T_0}{\rho_0 c_\nu}\dot{\vartheta}_2']'(s)ds. \tag{6.8}$$

The same remarks as in the preceding section apply to this system. Systems of this type do not seem to have appeared in the literature. They offer the opportunity of introducing very subtle damping mechanisms into the model.

7. A List of Beam Models

For easier reference, we represent the beam models obtained so far in form of a compact list. We use the same labelling as in the text. We restrict ourselves to display the equations of motion.

Initially straight and untwisted, nonshearable, nonlinear beams

[Longitudinal motion]

$$m_0\ddot{W}_1 = [EA(W_1' + \frac{1}{2}(W_2')^2 + \frac{1}{2}(W_3')^2)]' - \alpha\theta_1'.$$

[Vertical motion]

$$m_0\ddot{W}_2 - [\rho_0 I_{22}\ddot{W}_2']' + [EI_{22}W_2'']'' + \alpha\theta_2''$$
$$= [(EA(W_1' + \frac{1}{2}(W_2')^2 + \frac{1}{2}(W_3')^2) - \alpha\theta_1)W_2']'.$$

[Lateral motion]

$$m_0\ddot{W}_3 - [\rho_0 I_{33}\ddot{W}_3']' + [EI_{33}W_3'']'' + \alpha\theta_3''$$
$$= [(EA(W_1' + \frac{1}{2}(W_2')^2 + \frac{1}{2}(W_3')^2) - \alpha\theta_1)W_3']'.$$

[Torsional motion]

$$\rho_0 I\ddot{\vartheta}_1 = [GI\vartheta_1' - G(I-J)\gamma]'.$$

[Warping]

$$\rho_0\Gamma\ddot{\gamma} = [E\Gamma\gamma']' - G(I-J)(\gamma - \vartheta_1').$$

[Heat conduction]

$$\rho_0 c_\nu\dot{\theta}_1 = \theta_1'' - \alpha T_0\frac{\partial}{\partial t}\{W_1' + \frac{1}{2}((W_2')^2 + (W_3')^2)\},$$
$$\rho_0 c_\nu\dot{\theta}_2 = \theta_2'' + \alpha T_0\dot{W}_2'',$$
$$\rho_0 c_\nu\dot{\theta}_3 = \theta_3'' + \alpha T_0\dot{W}_3''.$$

Planar isothermal approximation

[Longitudinal motion]

$$m_0\ddot{W}_1 = [EA(W_1' + \frac{1}{2}(W_3')^2)]'.$$

[Lateral motion]

$$m_0\ddot{W}_3 - [\rho_0 I_{33}\ddot{W}_3']' + [EI_{33}W_3'']'' = [(EA(W_1' + \frac{1}{2}(W_3')^2))W_3']'$$

A reduction of model to transverse motion

We replace the inertial force with respect to the longitudinal motion by a load f_1, integrate the first equation and insert the result into the second equation of 3.2. We define $F = -\int_x^\ell f_1(s)ds$. In the two dimensional analog, this procedure leads to the Airy stress function and hence to the von Karman system. In the one–dimensional case we just get an axial force:

$$m_0\ddot{W}_3 - [\rho_0 I_{33}\ddot{W}_3']' + [EI_{33}W_3'']'' = [FW_3']'.$$

The Rayleigh beam model

We neglect the longitudinal motion entirely:

$$m_0\ddot{W}_3 - [\rho_0 I_{33}\ddot{W}_3']' + [EI_{33}W_3'']'' = 0.$$

The Euler–Bernoulli beam model

We neglect the rotatory inertia of the cross sections:

$$m_0\ddot{W}_3 + [EI_{33}W_3'']'' = 0.$$

Initially straight and untwisted linear shearable beam

$$\begin{aligned}
m_0\ddot{W}_1 &= [EAW_1]' && \text{[Longitudinal motion]}\\
m_0\ddot{W}_2 &= [GA(W_2' - \vartheta_3)]' && \text{[Lateral motion]}\\
m_0\ddot{W}_3 &= [GA(W_3 + \vartheta_2)]' && \text{[Vertical motion]}\\
\rho_0 I\ddot{\vartheta}_1 &= [GI\vartheta_1']' && \text{[Torsional motion]}\\
\rho_0 I_{33}\ddot{\vartheta}_2 &= [EI_{33}\vartheta_2']' - GA(W_3' + \vartheta_2), && \text{[Shear around } \hat{e}_2\text{]}\\
\rho_0 I_{22}\ddot{\vartheta}_3 &= [EI_{22}\vartheta_3']' + GA(W_2' - \vartheta_3) && \text{[Shear around } \hat{e}_3\text{]}
\end{aligned}$$

Nonlinear planar shearable beams with initial curvature

[Longitudinal motion]

$$\begin{aligned}
m_0\ddot{W}_1 = [Eh(W_1' - \kappa_3 W_3 + \frac{1}{2}\vartheta_2^2) - \alpha\theta_1]'&\\
- \kappa_3 Gh(\vartheta_2 + W_3' + \kappa_3 W_1) + \alpha[\kappa_3\theta_3]'&
\end{aligned}$$

[Vertical motion]

$$m_0\ddot{W}_3 = [Gh(\vartheta_2 + W_3' + \kappa_3 W_1)]' + \kappa_3[Eh(W_1' - \kappa_3 W_3 + \frac{1}{2}\vartheta_2^2) - \alpha\theta_1]$$

[Shear motion]

$$\begin{aligned}
\rho_0 I_{33}\ddot{\vartheta}_2 = EI_{33}\vartheta_2'' - \vartheta_2[Eh(W_1' - \kappa_3 W_3 + \frac{1}{2}\vartheta_2^2) - \alpha\theta_1]&\\
- Gh(\vartheta_2 + W_3' + \kappa_3 W_1) - \alpha\theta_3'&
\end{aligned}$$

[Heat flux]

$$\rho_0 c_\nu \dot{\theta}_1 = \theta_1'' - \alpha T_0 \frac{\partial}{\partial t}\{W_1' - \kappa_3 W_3 + \frac{1}{2}\vartheta_2^2\},$$
$$\rho_0 c_\nu \dot{\theta}_3 = \theta_3'' - \alpha T_0 \dot{\vartheta}_2'.$$

The Bresse system

[Longitudinal motion]

$$m_0 \ddot{W}_1 = [Eh(W_1' - \kappa_3 W_3)]' - \kappa_3 Gh(\vartheta_2 + W_3' + \kappa_3 W_1).$$

[Vertical motion]

$$m_0 \ddot{W}_3 = [Gh(\vartheta_2 + W_3' + \kappa_3 W_1)]' + \kappa_3 Eh[W_1' - \kappa_3 W_3]$$

[Shear motion]

$$\rho_0 I_{33} \ddot{\vartheta}_2 = EI_{33}\vartheta_2'' - Gh(\vartheta_2 + W_3' + \kappa_3 W_1).$$

The Timoshenko system

[Vertical motion]

$$m_0 \ddot{W}_3 = [Gh(\vartheta_2 + W_3')]'.$$

[Shear motion]

$$\rho_0 I_{33} \ddot{\vartheta}_2 = EI_{33}\vartheta_2'' - Gh(\vartheta_2 + W_3').$$

Rotating beams

See Section 3. For each particular beam model we can easily write down the corresponding rotating beam models: the fully nonlinear and the linearizations (including dynamic stiffening and damping).

Damping. Damping can be introduced in all of the models as outlined in the text. Possible sources of (internal) damping are:

- thermal damping
- viscous damping
- viscoelastic damping
- thermo–viscoelastic damping
- structural or fractional power–type damping.

8. Networks of Beams

Once we have achieved a sufficiently rich theory for a single beam, we may proceed to develop models of networks of such beams. This is the main purpose of this section. As with the reference arc in the case of a single beam, we have now a connected graph, consisting of a set of smooth open arcs which are connected through multiple nodes, labelled with N, and which may or may not end at a simple node. This network of arcs serves then as the reference net. It would be possible to impose a static prestretching on this net. Then additional equilibrium conditions would have to be taken into account. This has been carried out for a network of strings in Schmidt [93], and for Euler–Bernoulli beams in Leugering and Schmidt [58]. It turns out that a prestretching leads to a coupling between axial strains and bending. For simplicity, we assume that no such prestretching is present here. In the case of strings and Euler–Bernoulli beams it is sufficient to consider the centerline only, which makes the theory much simpler. In all the other cases, where we deal with shear, a joint (multiple node) between two or more beams is considerably more complicated to model. However, apart from the multiple nodes each beam satisfies the requirements of the preceding sections, in particular, the balance laws. This implies that each beam, individually, satisfies the equations of motion which we have derived so far. This also is easily seen from the viewpoint of energy principles; the potential and kinetic energy have to be replaced with their sum over all members. Coupling between longitudinal, transversal, lateral and torsional motion occurs at the multiple nodes. It is, for instance, obvious that in the case of a carpenter's square, the longitudinal displacement of one beam translates to a lateral displacement of the other and vice versa. Also, a twist of one beam results in a bending moment at the other beam, and vice versa.

8.1. Geometric Joint Conditions.

8.1.1. *Rigid Joints.* Here we consider two or more beams (of length ℓ_i) of the nature described in the previous sections which are supposed to end at a joint. Let $\mathcal{I}(N)$ denote the set of indices i such that the i-th beam meets at the node N. When talking about a connected structure at all, the first condition which has to be satisfied at any joint, regardless of its particular nature, is the continuity relation

$$\mathbf{W}^i(N) = \mathbf{W}^j(N), \quad \forall i,j \in \mathcal{I}(N). \tag{8.1}$$

In addition, we always assume that the temperatures ΔT^i are all the same at each joint.

$$\theta^j_k(N) = \theta^i_k(N), \quad \forall i,j \in \mathcal{I}(N),\ k = 1,2,3. \tag{8.2}$$

In order to specify other features of the joints, we have to distinguish between geometrical requirements and force/moment balances at the joints.

Two different types of joints will be investigated: rigid joints and pinned joints. Rigid joints are usually encountered in frames, whereas pinned joints occur in trusses. In this section we discuss rigid joints. A joint will be called *rigid* if the cross sections of the beams connected to each other at the node N undergo the same "moderate rotation," characterized by the vector $\boldsymbol{\vartheta}$, i.e.,

$$\boldsymbol{\vartheta}^i(N) = \boldsymbol{\vartheta}^j(N), \quad \forall i, j \in \mathcal{I}(N). \tag{8.3}$$

This condition allows the joint to rotate as a whole and to twist, but the shape of the joint before and after deformation is forced to be same. If there is no shear involved, rigidity of a joint means nothing more than the tangents to the deformed reference curves span the same angle as the tangents to undeformed reference curves.

It would be of great interest to derive a *mathematically precise* model of joints that are essentially rigid, but where the angle between two consecutive beam elements is allowed to switch if the deflection exceeds a certain threshold. Joints of this type are of common occurrence in satellite structures, where arms or solar panels are to be unfolded in space.

It should be noted that flexible joint conditions, such as the ones described here, are usually not reflected in the engineering literature on frames and trusses. The reason for this is that the equations of motions are formulated in a modal representation to begin with. In effect, this means that a joint of two flexible elements is replaced with a joint described by the two first finite elements of the connecting beams, which of course behave like rigid bodies. As has been demonstrated by Howard in his recent thesis [38], the transmission of elastic energy in a truss or a frame can only be fully understood if the flexibility of the links is taken into account at the joints. He shows that (and how) bending energy of a truss structure (or a frame) is converted to axial energy, and vice versa, through the joints.

EXAMPLE 8.1. We confine ourselves to the illustration of a 3–dimensional carpenter's square, a rectangular configuration of two beams. We neglect the warping.

We have two initial triads $\mathbf{e}_1^i, \mathbf{e}_2^i, \mathbf{e}_3^i$, $i = 1, 2$. However, in this special situation we easily see that we may choose

$$\mathbf{e}_1^1 = -\mathbf{e}_3^2, \quad \mathbf{e}_2^1 = \mathbf{e}_2^2, \quad \mathbf{e}_3^1 = \mathbf{e}_1^2.$$

This implies that the first joint condition (8.1) can be written as

$$W_1^1 = -W_3^2, \quad W_2^1 = W_2^2, \quad W_3^1 = W_1^2.$$

The temperature conditions (8.2) are independent of the triads. Accordingly, (8.3) is then

$$\vartheta_1^1 = -\vartheta_3^2, \quad \vartheta_2^1 = \vartheta_2^2, \quad \vartheta_3^1 = \vartheta_1^2.$$

Although the first set of conditions obviously coincides with intuition, the second set deserves further explanation. We focus on the Euler-Bernoulli framework first. Here we find ($\bar{e}_{1j} = 0,\ j = 2,3$)

$$\begin{aligned} \vartheta_2 &= -(W_3' + \kappa_3 W_1 + \tau W_2), \\ \vartheta_3 &= \ (W_2' + \kappa_2 W_2 - \tau W_3). \end{aligned}$$

Hence, (8.3) becomes

$$\begin{aligned} \vartheta_1^1 &= -((W_2^2)' + \kappa_2^2 W_2^2 - \tau^2 W_3^2), \\ (W_3^1)' + \kappa_3^1 W_1^1 + \tau^1 W_2^1 &= (W_3^2)' + \kappa_3^2 W_1^2 + \tau^2 W_2^2, \\ (W_2^1)' + \kappa_2^1 W_2^1 - \tau^1 W_3^1 &= \vartheta_1^2. \end{aligned}$$

The interpretation of these conditions is particularly simple if the beams are initially straight ($\kappa_j^i = \tau^i = 0$). Then it is apparent that a twist ϑ_1^1 of one beam causes a vertical strain $-(W_2^2)'$ on the other, whereas the orthogonality for the tangents is preserved in the 1–3 plane throughout the deformation.

8.1.2. *Pinned Joints.* A joint is called a pinned joint if no bending waves are transmitted, i.e., if the bending moment of each member at the joint is equal to zero, or prescribed by a moment applied from the outside. In addition, the sum of all shear forces has to be zero, or equal to a driving force. This means that the members, individually, may perform a moderate rotation about the joint. Joints of this type are typical for truss structures. The geometrical or compatibility conditions are now (8.1), (8.2), but not (8.3).

8.2. Dynamic Joint Conditions.

8.2.1. *Rigid Joints.* Dynamic joint conditions can obtained by balance of force and momentum, just as in (2.14). Then the geometric node conditions (8.1), (8.2), (8.3) serve as compatibility conditions. The dynamic node conditions can also be obtained via the variational formulation of the boundary value problem associated with the structure. Then the test functions have to satisfy the geometric joint conditions (8.1), (8.2), (8.3), in addition to the boundary conditions at the simple nodes (where only one beams ends). We assume that all multiple nodes (joints) are rigid. We give a discussion only in the case of the linearized dynamics. The nonlinear node conditions are far from being completely understood in the general case. They are, however, fully understood in the context of nonlinear Euler–Bernoulli beam networks, but for the sake of brevity we do not elaborate on this here.

In order to be able to assign to a typical element an outward pointing normal at a multiple node N, we introduce the notation

$$\varepsilon_i(N) := \begin{cases} -1 & \text{if } \mathbf{R}^i(N) = \mathbf{R}^i(0), \\ +1 & \text{if } \mathbf{R}^i(N) = \mathbf{R}^i(\ell_i). \end{cases}$$

In words, this means that if beam $\#i$ ends in the node at $x = 0$, then we assign -1 to that end, and we assign $+1$ otherwise. This is, essentially, the notation used in the paper by Wittenburg [109], where networks of rigid bodies have been discussed.

We will give the dynamic node conditions in the following two exemplaric cases:

- The initially straight untwisted Euler–Bernoulli beam in $I\!R^3$;
- The initially curved Bresse beam.

In addition, we will then reduce the first class to planar motion, and the second to the Timoshenko system.

Dynamic node conditions for a network of initially straight and untwisted linear, thermoelastic Euler-Bernoulli beams. The geometrical node conditions are (8.1), (8.3) with $\vartheta_2^i = -(W_3^i)'$, $\vartheta_3^i = (W_2^i)'$. With the above notation we obtain the following condition on the sum of the shear forces at N:

$$(8.4)\quad \sum_{i\in\mathcal{I}} \varepsilon_i(N)[-((E_iI_{i33}(W_3^i)'')' + \alpha^i\theta_3^i)\mathbf{e}_3^i - ((E_iI_{i22}(W_2^i)'')' + \alpha^i\theta_2^i)\mathbf{e}_2^i + (E_iA_i(W_1^i)' - \alpha^i\theta_1^i)\mathbf{e}_1^i](N) = \mathbf{F}_N + m_N\ddot{\mathbf{W}}^{\bar{i}}(N),$$

where $\mathbf{F}_N$ is a force applied to the joint N, and m_N is the mass of this joint. The bar on top of an index indicates that all values are the same for all i according to a particular node condition, and that one of those values is taken as a representative. The sum of moments at N satisfies

$$(8.5)\quad \sum_{i} \varepsilon_{i\in\mathcal{I}}(N)[G_iI_i(\vartheta_1^i)'\mathbf{e}_1^i - (E_iI_{i33}(W_3^i)'' + \alpha^i\theta_3^i)\mathbf{e}_2^i + (E_iI_{i22}(W_2^i)'' + \alpha^i\theta_2^i)\mathbf{e}_3^i](N) = \mathbf{M}_N + I_N\ddot{\boldsymbol{\vartheta}}^{\bar{i}}(N),$$

where $\mathbf{M}_N$ are couples applied to the joint N, which has a moment of inertia I_N. In the case of planar motion in the 1–3 plane, conditions (8.4) and (8.5) reduce to, respectively,

$$(8.6)\quad \sum_{i\in\mathcal{I}} \varepsilon_i(N)[-((E_iI_{i33}(W_3^i)'')' + \alpha^i\theta_3^i)\mathbf{e}_3^i + (E_iA_i(W_1^i)' - \alpha^i\theta_1^i)\mathbf{e}_1^i](N) = \mathbf{F}_N + m_N\ddot{\mathbf{W}}^{\bar{i}}(N),$$

$$(8.7)\quad -\sum_{i\in\mathcal{I}} \varepsilon_i(N)[E_iI_{i33}(W_3^i)'' + \alpha^i\theta_3^i](N) = M_N - I_N(\ddot{W}_3^{\bar{i}})'(N).$$

One may, however, treat the joints N as mass points, so that I_N is taken to be zero. The inertia part in (8.4) is not to be confused with the rotary inertia of the cross section which is present in the Rayleigh beam model. One may suppose the joint to be massless if the beams are considered as welded to each other.

EXAMPLE 8.2. In the example of the carpenter's square, with no forces applied at the joint, which is also assumed to carry no extra mass, condition (8.4) at N is

$$\begin{cases} (E_1 I_{133}(W_3^1)'')' + \alpha^1\theta_3^1 = E_2 A_2 (W_1^2)' - \alpha^2\theta_1^2, \\ (E_2 I_{233}(W_3^2)'')' + \alpha^2\theta_3^2 = E_1 A_1 (W_1^1)' - \alpha^1\theta_1^1, \\ (E_1 I_{122}(W_2^1)'')' + \alpha_2^1 = -(E_2 I_{222}(W_2^2)'')' + \alpha^2\theta_2^2. \end{cases} \tag{8.8}$$

The moment condition (8.5) at N may be written

$$\begin{cases} E_1 I_{122}(W_2^1)'' + \alpha^1\theta_2^1 = -G_2 I_2 (\vartheta_1^2)', \\ E_2 I_{222}(W_2^2)'' + \alpha^2\theta_2^2 = -G_1 I_1 (\vartheta_1^1)', \\ E_1 I_{133}(W_3^1)'' + \alpha^1\theta_3^1 = -E_2 I_{233}(W_3^2)'' - \alpha^2\theta_3^2. \end{cases} \tag{8.9}$$

The reduction to the planar case is obvious, as is the reduction to the isothermal situation.

REMARK 8.1. We note that the node conditions obtained in this case are in accordance with the results of Le Dret [56] if the inertia term in the equation governing the longitudinal motion in the vertical beam is neglected, and if the resulting longitudinal stress is expressed in terms of the weight of the vertical beam. Also, Le Dret allows the vertical beam to move out of the plane causing a twist of the horizontal beam expressed in terms of the weight and of surface tractions. The corresponding node condition is obtained from our more general ones by neglecting the inertia term in the equation governing the torsional motion and by expressing the twist ϑ_1^1, which appears in the node conditions, in terms of appropriate forces and couples.

Dynamic node conditions for a planar network of linear initially curved Bresse beams. Here we confine ourselves to the 1–3 plane, and proceed as mentioned above. The balance law for the forces at N is

$$\begin{aligned} \sum_{i\in\mathcal{I}} \varepsilon_i(N)[(E_i h_i((W_1^i)' - \kappa_3^i W_3^i) - \alpha^i\theta_1^i + \alpha^i\kappa_3^i\theta_3^i)\mathbf{e}_1^i \\ + G_i h_i(\vartheta_2^i + (W_3^i)' + \kappa_3^i W_1^i)\mathbf{e}_3^i](N) = \mathbf{F}_N + m_N \ddot{\mathbf{W}}^i(N). \end{aligned} \tag{8.10}$$

For the transmission of moments at N we obtain

$$\sum_{i\in\mathcal{I}} \varepsilon_i(N)[E_i I_{i33}(\vartheta_2^i)' - \alpha^i\theta_3^i](N) = M_N + I_N\ddot{\vartheta}_2^{\bar{i}}(N). \tag{8.11}$$

It is obvious that, for zero initial curvature, (8.10) reduces to

$$\begin{aligned}\sum_{i\in\mathcal{I}} \varepsilon_i(N)((E_i h_i(W_1^i)' - \alpha^i\theta_1^i)\mathbf{e}_1^i + G_i A_i(\vartheta_2^i + (W_3^i)')\mathbf{e}_3^i)(N)& \\ = \mathbf{F}_N + m_N\ddot{\mathbf{W}}^{\bar{i}}(N),&\end{aligned} \tag{8.12}$$

while (8.11) remains the same. These are the dynamic node conditions for the Bresse–Timoshenko system.

EXAMPLE 8.3. Let us again consider the case of a carpenter's square as above. The condition (8.10) on the sum of forces at N now reads

$$\begin{cases} E_1h_1((W_1^1)' - \kappa_3^1 W_3^1) - \alpha^1\theta_3^1 = G_2h_2(\vartheta_2^2 + (W_3^2)' + \kappa_3^2 W_1^2), \\ E_2h_2((W_1^2)' - \kappa_3^2 W_3^2) - \alpha^2\theta_3^2 = -G_1h_1(\vartheta_2^1 + (W_3^1)' + \kappa_3^1 W_1^1). \end{cases} \tag{8.13}$$

The condition (8.11) is already independent of the triads $\mathbf{e}^1, \mathbf{e}^2$ and remains unchanged. From this the corresponding conditions in the isothermal situation, and also in the uncurved case, are easily obtained.

8.2.2. *Pinned Joints.* As mentioned above, in the case of pinned joints we retain (8.1), (8.2), but we do not impose (8.3) as geometrical or compatibility conditions. Also, for the Euler-Bernoulli system, the sum of shear forces at the node has to satisfy (8.4) or (8.6) in the 3–d or 2–d case, respectively. The conditions on the sum of moments, however, is replaced by the much simpler condition

$$[(E_iI_{i33}(W_3^i)'' + \alpha^i\theta_3^i)\mathbf{e}_2^i - (E_iI_{i22}(W_2^i)'' + \alpha^i\theta_2^i)\mathbf{e}_3^i](N) = \mathbf{M}_N^i, \tag{8.14}$$

in the 3–d case, or

$$[(E_iI_{i33}(W_3^i)'' + \alpha^i\theta_3^i)](N) = M_N^i \tag{8.15}$$

in the 2–d case. As for the Bresse–Timoshenko system we obtain the dynamic node conditions (8.10) together with

$$[E_iI_{i33}(\vartheta_2^i)' - \alpha^i\theta_3^i)](N) = M_N^i. \tag{8.16}$$

9. Rotating Two-link Flexible Nonlinear Shearable Beams

We consider two nonlinear shearable beams which are coupled through a revolutive joint (pinned joint) with mass m_{r2} and moment I_{r2}. This aggregate is attached to a cylinder of mass m_{r1} and moment I_{r1}. For simplicity we assume that the radius of the cylinder is zero (the more general case is also easily dealt with). Both the cylinder and the attached beams are then supposed to move rigidly in the $\mathbf{e}_1\mathbf{e}_3$ plane, i.e. there is a translation $\mathbf{y}_1$ and a rotation $\mathbf{\Pi}_1$ of the entire system. In addition we allow the second beam to rotate about the revolutive joint, the rotation being $\mathbf{\Pi}_2$. We assume, for simplicity, that the beams are uniform, and neglect warping and thermal effects as well as damping effects. All these features can be handled with some extra computational work.

In the inertial frame we have

$$\boldsymbol{\nabla}_i = \mathbf{\Pi}_i\mathbf{r}_i + \mathbf{y}_i$$

and

$$\boldsymbol{\mathcal{R}}_i = \mathbf{\Pi}_i\mathbf{R}_i + \mathbf{y}_i.$$

We write $\mathbf{R}_i$ instead of $\mathbf{R}_{0i}$. In particular $\mathbf{r}_i = (x,0,0)^T$ and $\mathbf{R}_i = \mathbf{r}_i + \mathbf{W}_i$. We require that the beams are connected at the joint, in the undeformed and the deformed configuration:

$$\mathbf{r}_1(\ell_1) = \mathbf{r}_2(0),$$
$$\boldsymbol{\mathcal{R}}_1(\ell_1) = \boldsymbol{\mathcal{R}}_2(0).$$

This amounts to

$$\mathbf{y}_2 = \mathbf{\Pi}\mathbf{R}_1(\ell_1) + \mathbf{y}_1$$

and

$$\mathbf{\Pi}\mathbf{R}_1 + \mathbf{y}_1 = \mathbf{\Pi}\mathbf{R}_2 + \mathbf{y}_2 = \mathbf{\Pi}_2\mathbf{R}_2(0) + \mathbf{\Pi}_1\mathbf{r}_1(\ell_1) + \mathbf{y}_1,$$

hence

$$\mathbf{\Pi}_1(\mathbf{R}_1 - \mathbf{r}_1)(\ell_1) = \mathbf{\Pi}_2(\mathbf{R}_2 - \mathbf{r}_2)(0) \Longleftrightarrow \mathbf{\Pi}_1\mathbf{W}_1(\ell_1) = \mathbf{\Pi}_2\mathbf{W}_2(0).$$

We proceed to write down the kinetic and strain energies. The kinetic energy is given by

$$T = \sum_{i=1}^{2} T_i = \frac{1}{2}\sum_{i=1}^{2}\{I_i\dot{\theta}^2 + m_i\|\dot{\mathbf{y}}_i\|^2 + \int_0^{\ell_i}[I(\dot{\alpha}_i^2 + \dot{\theta}_i^2) + \rho_i\|\mathbf{\Pi}_i D_i\mathbf{R}_i + \dot{\mathbf{y}}_i\|^2]dx\}.$$

The strain energy can be written as

$$U = \sum_{i=1}^{2} U_i = \frac{1}{2}\sum_{i=1}^{2}\int_0^{\ell_i} [E_iA_i(u_i' + \frac{1}{2}\alpha_i^2)^2 + E_iI_i(\alpha_i')^2 + G_iA_i(\alpha_i + w_i')^2]dx.$$

We derive the equations of motion and the dynamic node-and boundary conditions by the requirement that the total energy be conserved, i.e.

$$E := T + U, \quad \dot{E} = 0.$$

We remark, however, that using this principle one has to take special care of Coriolis forces which do not contribute to work. We first calculate $\dot{U}_i$. We have

$$\begin{aligned}
\dot{U}_i = &\int_0^{\ell_i} [E_iA_i(u_i' + \frac{1}{2}\alpha_i^2)(\dot{u}_i' + \alpha_i\dot{\alpha}_i) \\
&+ E_iI_i\alpha_i'\dot{\alpha}_i' + G_iA_i(\alpha_i + w_i')(\dot{\alpha}_i + \dot{w}_i')]dx \\
= &< (E_iA_i(u_i' + \frac{1}{2}\alpha_i^2), 0, G_iA_i(\alpha_i + w_i'))^T, \dot{\mathbf{R}}_i > |_0^{\ell_i} + E_iI_i\alpha_i'\dot{\alpha}_i|_0^{\ell_i} \\
&- \int_0^{\ell_i} < (E_iA_i(u_i' + \frac{1}{2}\alpha_i^2)', 0, G_iA_i(\alpha_i + w_i')')^T, \dot{\mathbf{R}}_i > dx \\
&+ \int_0^{\ell_i} [E_iA_i(u_i' + \frac{1}{2}\alpha_i^2)\alpha_i + G_iA_i(\alpha_i + w_i') - E_iI_i\alpha_i'']\dot{\alpha}_i dx \\
= &[< \frac{\partial U_i}{\partial \bar{\varepsilon}_i}, \dot{\mathbf{R}}_i > + \frac{\partial U_i}{\partial \bar{\kappa}_i}\dot{\alpha}_i]|_0^{\ell_i} \\
&- \int_0^{\ell_i} < \frac{\partial U_i}{\partial \mathbf{R}_i}, \dot{\mathbf{R}}_i > dx - \int_0^{\ell_i} \frac{\partial U_i}{\partial \alpha_i}\dot{\alpha}_i dx,
\end{aligned}$$

where $\frac{\partial}{\partial \mathbf{R}_i}$ etc. are Fréchet derivatives. As for the time derivative of the kinetic energy, we find

$$\begin{aligned}
\dot{T}_i = &I_{ri}\ddot{\theta}_i\dot{\theta}_i + m_{ri} < \ddot{\mathbf{y}}, \dot{\mathbf{y}} > + \int_0^{\ell_i} [I_i(\ddot{\alpha}_i + \ddot{\theta}_i)\dot{\alpha}_i + I_i(\ddot{\alpha}_i + \ddot{\theta}_i)\dot{\theta}_i]dx \\
&+ \int_0^{\ell_i} \rho_i < \dot{\theta}_i\mathbf{J}(\dot{\mathbf{R}}_i + \dot{\theta}_i\mathbf{J}\dot{\mathbf{R}}_i) + \ddot{\mathbf{R}}_i + \ddot{\theta}_i\mathbf{J}\mathbf{R}_i + \dot{\theta}_i\mathbf{J}\dot{\mathbf{R}}_i + \mathbf{\Pi}_i^T\ddot{\mathbf{y}}_i, \\
&\qquad \dot{\mathbf{R}}_i + \dot{\theta}_i\mathbf{J}\mathbf{R}_i + \mathbf{\Pi}_i^T\dot{y}_i > dx.
\end{aligned}$$

We denote the last integral, which stems from the elastic deformation, by

$\dot{T}_i^e$. By straightforward calculations we find

$$\dot{T}_i^e = \int_0^{\ell_i} \rho_i < \ddot{\mathbf{R}}_i + \ddot{\theta}_i \mathbf{J}\mathbf{R}_i + 2\dot{\theta}_i \mathbf{J}\dot{\mathbf{R}}_i - \dot{\theta}_i^2 \mathbf{R}_i + \mathbf{\Pi}^T \ddot{\mathbf{y}}_i, \dot{\mathbf{R}}_i + \dot{\theta}_i \mathbf{J}\mathbf{R}_i + \mathbf{\Pi}_i^T \dot{y}_i > dx$$

$$= \int_0^{\ell_i} \rho_i < D_i^2 \mathbf{R}_i + \mathbf{\Pi}_i^T \ddot{\mathbf{y}}_i, \dot{\mathbf{R}}_i > dx + \int_0^{\ell_i} \rho_i < \mathbf{\Pi}_i D_i^2 \mathbf{R}_i dx + \ell_i \ddot{\mathbf{y}}, \dot{\mathbf{y}} > dx + \int_0^{\ell_i} < \mathbf{R}_i, \mathbf{J}^T (D_i^2 \mathbf{R}_i + \mathbf{\Pi}_i^T \ddot{\mathbf{y}}_i) > dx\, \dot{\theta}_i.$$

We require that

$$\mathbf{R}_1(0) = 0, \quad \frac{\partial U_2}{\partial \bar{\varepsilon}_2}(\ell_2) = 0,$$
$$\alpha_1(0) = 0, \quad \alpha_2'(\ell_2) = 0,$$
$$\mathbf{\Pi}_1 \mathbf{W}_1(\ell_1) = \mathbf{\Pi}_2 \mathbf{W}_2(0).$$

Hence,

$$\dot{\mathbf{R}}_1(0) = 0, \quad \dot{\alpha}_1(0) = 0$$

and

$$\dot{\mathbf{R}}_2(0) = (\dot{\theta}_1 - \dot{\theta}_2)\mathbf{\Pi}_2^T \mathbf{\Pi}_1 \mathbf{J}\mathbf{W}_1(\ell_1) + \mathbf{\Pi}_2^T \mathbf{\Pi}_1 \dot{\mathbf{R}}_1(\ell_1).$$

Using these expressions and the boundary conditions in $\dot{E} = \sum_{i=1}^2 (\dot{T}_i + \dot{U}_i)$, we can express the nodal terms as

$$< \frac{\partial U_1}{\partial \bar{\varepsilon}_1}(\ell_1), \dot{\mathbf{R}}_1(\ell_1) > - < \frac{\partial U_2}{\partial \bar{\varepsilon}_2}(0), \dot{\mathbf{R}}_2(0) > = < \frac{\partial U_1}{\partial \bar{\varepsilon}_1}(\ell_1), \dot{\mathbf{R}}_1(\ell_1) > - < \frac{\partial U_2}{\partial \bar{\varepsilon}_2}(0), (\dot{\theta}_1 - \dot{\theta}_2)\mathbf{\Pi}_2^T \mathbf{\Pi}_1 \mathbf{J}\mathbf{W}_1(\ell_1) + \mathbf{\Pi}_2^T \mathbf{\Pi}_1 \dot{\mathbf{R}}_1(\ell_1) >$$
$$= < \frac{\partial U_1}{\partial \bar{\varepsilon}_1}(\ell_1) - \mathbf{\Pi}_1^T \mathbf{\Pi}_2 \frac{\partial U_2}{\partial \bar{\varepsilon}_2}(0), \dot{\mathbf{R}}_1(\ell_1) > - (\dot{\theta}_1 - \dot{\theta}_2) < \frac{\partial U_2}{\partial \bar{\varepsilon}_2}(0), \mathbf{\Pi}_2^T \mathbf{\Pi}_1 \mathbf{J}\mathbf{W}_1(\ell_1) > .$$

The terms involving $\dot{\theta}_1, \dot{\theta}_2$ are taken together with the corresponding terms coming from the expression for the time derivative of the kinetic energy. The equation $\dot{E} = 0$ has to hold for all test functions satisfying the boundary conditions and the continuity condition. We obtain the following set of equations governing the motion of the system. Note that $\mathbf{y}_2 = \mathbf{\Pi}_1 \mathbf{r}_1(\ell_1)$.

$$(I_{r1} + I_1\ell_1)\ddot{\theta}_1 + \int_0^{\ell_1} I_1\ddot{\alpha}_1 dx =< \mathbf{W}_1(\ell_1), \mathbf{J}^T\mathbf{\Pi}_1^T\mathbf{\Pi}_2\frac{\partial U_2}{\partial \bar{\varepsilon}_2}(0) >$$
$$- \int_0^{\ell_1} < \mathbf{R}_1, \mathbf{J}^T(D_1^2\mathbf{R}_1 + \mathbf{\Pi}_1^T\ddot{y}_1 > dx,$$

$$(I_{r2} + I_2\ell_2)\ddot{\theta}_2 + \int_0^{\ell_2} I_2\ddot{\alpha}_2 dx = - < \mathbf{W}_1(\ell_1), \mathbf{J}^T\mathbf{\Pi}_1^T\mathbf{\Pi}_2\frac{\partial U_2}{\partial \bar{\varepsilon}_2}(0) >$$
$$- \int_0^{\ell_2} < \mathbf{R}_2, \mathbf{J}^T(D_2^2\mathbf{R}_2 + \mathbf{\Pi}_2^T\ddot{\mathbf{\Pi}}_1\mathbf{R}_1(\ell_1) > dx,$$

$$(m_{r1} + \rho_0\ell_1)\ddot{\mathbf{y}}_1 = - \int_0^{\ell_1} \rho_0\mathbf{\Pi}_1 D_1^2\mathbf{R}_1 dx,$$
$$(m_{r2} + \rho_0\ell_2)\ddot{\mathbf{\Pi}}_1\mathbf{r}_1(\ell_1) = - \int_0^{\ell_1} \rho_0\mathbf{\Pi}_2 D_2^2\mathbf{R}_2 dx,$$

$$I_1(\ddot{\alpha}_1 + \ddot{\theta}_1) = \frac{\partial U_1}{\partial \alpha_1} = E_1I_1\alpha_1'' - E_1A_1(u_1' + \frac{1}{2}\alpha_1^2)\alpha_1 - G_1A_1(\alpha_1 + w_1'),$$
$$I_2(\ddot{\alpha}_2 + \ddot{\theta}_2) = \frac{\partial U_2}{\partial \alpha_2} = E_2I_2\alpha_2'' - E_2A_2(u_2' + \frac{1}{2}\alpha_2^2)\alpha_2 - G_2A_2(\alpha_2 + w_2'),$$

$$\rho_0 D_1^2\mathbf{R}_1 + \mathbf{\Pi}_1^T\ddot{y}_1 = \frac{\partial U_1}{\partial \mathbf{R}_1},$$
$$\rho_0 D_2^2\mathbf{R}_2 + \mathbf{\Pi}_2^T\ddot{\mathbf{\Pi}}_1\mathbf{r}_1(\ell_1) = \frac{\partial U_2}{\partial \mathbf{R}_2}.$$

The boundary and node conditions are given by

$$\mathbf{R}_1(0) = 0, \quad \frac{\partial U_2}{\partial \bar{\varepsilon}_2}(\ell_2) = 0,$$
$$\alpha_1(0) = 0, \quad \alpha_2'(\ell_2) = 0,$$
$$\mathbf{\Pi}_1\mathbf{W}_1(\ell_1) = \mathbf{\Pi}_2\mathbf{W}_2(0),$$
$$\mathbf{\Pi}_1\frac{\partial U_1}{\partial \bar{\varepsilon}_1}(\ell_1) = \mathbf{\Pi}_2\frac{\partial U_2}{\partial \bar{\varepsilon}_2}(0),$$
$$\alpha_1'(\ell_1) = \alpha_2'(0).$$

It is straightforward to show that the case where $\theta_1 = \theta_2$, $m_{r2} = 0$, $I_{r2} = 0$ reduces to the rotating beam model described in Section 3. One can also derive from this general set of equations an Euler-Bernoulli type rotating two-link beam model, and one with dynamic stiffening. We refer to Khorrami [42] where a rotating two-link Euler-Bernoulli beam has been developed. His equations however, do not reduce to a single rotating beam equation if the rotation angles are the same and the mass at the joint is not considered. Rather, in that situation the second beam may oscillate independent of the first in his model.

CHAPTER IV

A General Hyperbolic Model for Networks of One Dimensional Elements

In this chapter we study a general linear, hyperbolic model for vibrating networks of one-dimensional elements. This model accommodates not only the elastic string networks of Chapter II, but also applies to networks of beams governed by one or other of the linear, isothermal, hyperbolic beam models of Chapter III and to mixed networks which contain both beams and strings. The existence and regularity of solutions to the general network system is established, along lines identical to those used in Section 4 of Chapter II. Energy estimates involving characteristics are proved and then used to obtain the estimates from which exact controllability and stabilizability follow for networks containing no closed loops, again following the argument used in Chapter II.

1. The General Model

We begin by describing the network configuration and establishing some notation which generalizes or supplements that of Chapter II. We suppose as before that the network consists of n elements located in $\mathbb{R}^3$ (or $\mathbb{R}^2$ in the case of planar models) indexed by $i \in \mathcal{I}$ and parameterized by $x \in [0, l_i]$. We suppose also that these elements are joined at some of the endpoints in such a way that the resulting network is connected. The elements are associated with curves which generally describe the location of a physical object such as a string or of the centerline of a three dimensional object such as a beam. Associated with each curve we have a function $\mathbf{r}^i(x,t)$: $[0, l_i] \times [0,T] \mapsto \mathbb{R}^{p_i}$, which describes the displacements of the variables describing the configuration of the i-th element. For string networks one has $p_i = 3$ and $\mathbf{r}^i(x,t)$ describes the displacement of the i-th string at time t and at the point parameterized by the natural arc length x. More generally, some components of $\mathbf{r}^i$ may describe the spatial displacement of the centerline while others may describe changes in "internal variables" introduced into the model by a kinematic hypothesis. As in Chapter II we let $\mathbf{r}$ denote the n-tuple of functions $(\mathbf{r}^i)_{i\in\mathcal{I}}$.

In order to formulate geometric constraints on the network we associate with each node $q_j \times p_i$ matrices $\mathbf{C}_{ij}$ of rank q_j indexed by $j \in \mathcal{J}$ and $i \in \mathcal{I}_j$, where necessarily $q_j \leq p_i$. These matrices pick out that "component" of $\mathbf{r}_i$ which is to be involved in the geometric condition at the j-th node. At the multiple nodes of the network we then impose the *geometric node condition*

$$\text{for } j \in \mathcal{J}_M,\ \mathbf{C}_{ij}\mathbf{r}^i(x_{ij},t) \text{ is the same for all } i \in \mathcal{I}_j. \tag{1.1}$$

This is a continuity condition on the configuration functions at the j-th node which reflects the way in which the corresponding elements are joined together. For a given type of element there are many possible joint conditions. Sometimes, as will be illustrated in the next section, it may happen that the node conditions (1.1) indeed involve only some components of the configuration displacements (see the reference to pinned joints in Section 2.2 below). We remark also that the multiple node condition as we have formulated it is convenient in the most important applications but a more general condition is sometimes more appropriate (see Remark 1.2 and Subsection 2.5). In this section we want to walk a delicate line between generality and intelligibility; there are many physically reasonable combinations of geometric and dynamic conditions which can be accommodated within the general model at the cost of a proliferation of notation.

For each $j \in \mathcal{J}$ we define $\mathcal{D}^j\mathbf{r}$ to be the value (common value in the case of multiple nodes) of $\mathbf{C}_{ij}\mathbf{r}^i(x_{ij})$. We now suppose that certain nodes are clamped:

$$\mathcal{D}^j\mathbf{r} = \mathbf{0} \quad \text{for } j \in \mathcal{J}^C. \tag{1.2}$$

In case $\mathbf{C}_{ij}$ is injective for each $i \in \mathcal{I}_j$ this latter condition is equivalent to $\mathbf{r}^i(x_{ij}) = \mathbf{0}$ and we say that the node is *fully clamped*; otherwise it is only *partially clamped.* Other nodes may be subject to Dirichlet controls

$$\mathcal{D}^j\mathbf{r} = \mathbf{u}^j \quad \text{for } j \in \mathcal{J}^D, \tag{1.3}$$

where $\mathbf{u}^j(t)$ is a preassigned control function. We shall later also allow conditions of Neumann type. The form that these should take will emerge from the variational argument which follows.

Let the system have associated with it kinetic and potential energy functions of the form

$$\mathcal{K}(t) = \sum_{i=1}^{n} \frac{1}{2}\int_0^{l_i} |\mathbf{P}_i\dot{\mathbf{r}}^i(x,t)|^2\,dx, \tag{1.4}$$

and

$$\mathcal{U}(t) = \sum_{i=1}^{n} \frac{1}{2}\int_0^{l_i} \left[|\mathbf{Q}_i[\mathbf{r}^i_{,x} + \mathbf{R}_i\mathbf{r}^i](x,t)|^2 + [\mathbf{S}_i\mathbf{r}\cdot\mathbf{r}]\,(x,t)\right]dx, \tag{1.5}$$

where

- $\mathbf{P}_i$, $\mathbf{Q}_i$, $\mathbf{S}_i$ and $\mathbf{R}_i$ are continuous $p_i \times p_i$ matrix valued functions of x;
- $\mathbf{P}_i$ and $\mathbf{Q}_i$ are symmetric and positive definite;
- $\mathbf{S}_i$ is either positive definite symmetric or the zero matrix.

One can also introduce the work functional

$$(1.6)\qquad \mathcal{W}(t) = \sum_{i\in\mathcal{I}} \int_0^{l_i} \mathbf{f}^i(x,t)\cdot\mathbf{r}^i(x,t)\,dx + \sum_{j\in\mathcal{J}^N} \mathbf{u}^j(t)\cdot\mathcal{D}^j\mathbf{r}(t),$$

where $\mathcal{J}^N \subset \mathcal{J}\setminus[\mathcal{J}^C\cup\mathcal{J}^D]$ is the set of nodal indices at which we wish to impose a Neumann control $\mathbf{u}^j(t)$. We apply Hamilton's principle to the Lagrangian functional

$$\mathcal{L}(\mathbf{r}) = \int_0^T [\mathcal{K}(t) + \mathcal{W}(t) - \mathcal{U}(t)]\,dt,$$

allowing only variations in $\mathbf{r}$ which respect the geometric conditions (4.1), (4.2) and (4.3). First one uses, for each i, variations in $\mathbf{r}^i$ having compact support in $(0, l_i)$. In a standard way one obtains the system of equations

$$(1.7)\qquad \mathbf{P}_i^2\ddot{\mathbf{r}}^i = \left[\mathbf{Q}_i^2(\mathbf{r}^i_{,x} + \mathbf{R}_i\mathbf{r}^i)\right]_{,x} - \mathbf{R}_i^t\mathbf{Q}_i^2(\mathbf{r}^i_{,x} + \mathbf{R}_i\mathbf{r}^i) + \mathbf{S}_i\mathbf{r}_i + \mathbf{f}^i$$

(where "t" denotes the transpose of a matrix). One then considers arbitrary variations satisfying the geometric conditions and uses the equations (4.7) to get the variational identity

$$(1.8)\qquad \sum_{j\in\mathcal{J}} \int_0^T \sum_{i\in\mathcal{I}_j} \epsilon_{ij}\mathbf{Q}_i^2\left(\mathbf{r}^i_{,x} + \mathbf{R}_i\mathbf{r}^i\right)(x_{ij},t)\cdot\delta\mathbf{r}^i(x_{ij},t)\,dt$$
$$+ \sum_{j\in\mathcal{J}^N} \int_0^T \mathbf{u}^j(t)\cdot\mathcal{D}^j\delta\mathbf{r}(t)\,dt = 0.$$

In order to write down the dynamic conditions resulting from (4.8) we let $\mathbf{\Pi}_{ij}$ denote orthogonal projection onto the the kernel of the transformation defined by $\mathbf{C}_{ij}$ and $\mathbf{\Pi}_{ij}^{\perp}$ be orthogonal projection onto the orthogonal complement of that kernel. Denote by $\mathbf{C}_{ij}^{+}$ the generalized inverse of $\mathbf{C}_{ij}$. This is the $p_i \times q_j$ matrix which satisfies

$$\mathbf{C}_{ij}\mathbf{C}_{ij}^{+} = \mathbf{I}_j \quad\text{and}\quad \mathbf{C}_{ij}^{+}\mathbf{C}_{ij} = \mathbf{\Pi}_{ij}^{\perp},$$

where $\mathbf{I}_j$ denotes the $q_j \times q_j$ identity matrix. The variations $\delta\mathbf{r}^i$ can now be chosen so that

$$\delta\mathbf{r}^i(x_{ij},t) = \begin{cases} \mathbf{C}_{ij}^{+}\boldsymbol{\psi}(t) + \boldsymbol{\psi}_{ij}(t), & \text{for } j\in\mathcal{J}^N,\\ \boldsymbol{\psi}_{ij}(t), & \text{for } j\in\mathcal{J}^F,\end{cases}$$

where $\psi(t)$ and $\psi^{ij}(t)$ are arbitrary smooth functions taking their respective values in $\mathbb{R}^{q_j}$ and in $\ker(\mathbf{C}_{ij})$. Evidently $\mathcal{D}^j\delta\mathbf{r} = \psi$ or $\mathbf{0}$ corresponding to the two cases. The last variational identity then yields the following dynamic conditions:

$$\mathcal{N}^j\mathbf{r} = \begin{cases} \mathbf{u}^j & \text{for } j \in \mathcal{J}^N, \\ \mathbf{0} & \text{for } j \in \mathcal{J}^F, \end{cases} \tag{1.9}$$

with $\mathcal{J}^F = \mathcal{J} \setminus [\mathcal{J}^C \cup \mathcal{J}^D \cup \mathcal{J}^N]$, as well as

$$\mathcal{N}_i^j\mathbf{r} = \mathbf{0}, \quad \text{for all } j \in \mathcal{J} \text{ and } i \in \mathcal{I}_j, \tag{1.10}$$

where

$$\mathcal{N}^j\mathbf{r} = \sum_{i\in\mathcal{I}_j} [\mathbf{C}_{ij}^+]^t \mathbf{Q}_i^2 \left(\mathbf{r}^i_{,x} + \mathbf{R}_i\mathbf{r}^i\right)(x_{ij}, t) \;\; \text{for } j \in \mathcal{J}_M, \tag{1.11}$$

and

$$\mathcal{N}_i^j\mathbf{r} = \epsilon_{ij}\mathbf{\Pi}_{ij}\mathbf{Q}_i^2 \left(\mathbf{r}^i_{,x} + \mathbf{R}_i\mathbf{r}^i\right)(x_{ij}, t) \;\; \text{for } j \in \mathcal{J}_M \text{ and } i \in \mathcal{I}_j. \tag{1.12}$$

The condition (1.10) simply expresses the fact that the component of each $\mathbf{r}^i(x_{ij}, t)$ belonging to the kernel of $\mathbf{C}_{ij}$ is free.

REMARK 1.1. One could of course place further geometric (Dirichlet) or dynamic (Neumann) conditions on the components which belong to the kernel of $\mathbf{C}_{ij}$. For example, if we add the following expression to the work functional (1.6)

$$\int_0^T \sum_{j\in\mathcal{J}^{N_1}} \sum_{i\in\mathcal{I}_j} \mathbf{v}^{ij} \cdot \mathbf{\Pi}_{ij}\mathbf{r}^i(x_{ij}, t)\, dt$$

where $\mathcal{J}^{N_1}$ is any subset of $\mathcal{J}$, the condition (1.10) becomes

$$\mathcal{N}_i^j\mathbf{r} = \begin{cases} \mathbf{v}^{ij} & \text{for } j \in \mathcal{J}^{N_1} \text{ and } i \in \mathcal{I}_j, \\ \mathbf{0} & \text{for } j \in \mathcal{J} \setminus \mathcal{J}^{N_1}. \end{cases} \tag{1.13}$$

In fact there are many ways in which Dirichlet and Neumann conditions could be combined, also at multiple nodes. To insist on full generality would further complicate our notation and we choose not to pursue this theme further.

REMARK 1.2. A reasonable generalization of the multiple node condition (4.1) takes the form

$$\mathcal{D}^j\mathbf{r} = \sum_{i\in\mathcal{I}_j} \mathbf{D}_{ji}\mathbf{r}^i(x_{ij}, t) = \mathbf{0}, \tag{1.14}$$

where the $\mathbf{D}_{ji}$ are $r_j \times p_i$ matrices. One can then get dynamic node conditions from the variational argument if one makes use of a basis $\mathbf{v}_{jk} = (\mathbf{v}^i_{jk})$ of solutions to the homogeneous linear system

$$\sum_{i \in \mathcal{I}_j} \mathbf{D}_{ji}\mathbf{v}^i = \mathbf{0}.$$

We do not work out the details here but refer to the third variant on the example of Subsection 2.5 as well as Remark 5.3, which illustrate this generalization of the multiple node conditions.

2. Some Special Cases

After briefly describing how the string networks of Chapter II fit the general model we draw on the some of the linear, hyperbolic and isothermal beam models of Chapter III, Section 7 to specify a variety of beam networks as well as some simple networks involving both strings and beams.

2.1. String Networks. Here $p_i = 3$ for each $i \in \mathcal{I}$ and each $\mathbf{C}_{ij}$ is set equal to the 3×3 identity matrix $\mathbf{I}$. Furthermore $\mathbf{Q}_i$ is the positive definite square root of the matrix called $\mathbf{Q}_i$ in Chapter II, $\mathbf{P}_i = \sqrt{\rho_i}\mathbf{I}$ and $\mathbf{R}_i = \mathbf{S}_i = \mathbf{0}$. Everything reduces exactly to the model considered in Chapter II.

REMARK 2.1. In the above model it is natural for the matrices to have no spatial dependence. When the network is subject to gravity the equilibrium configuration will "sag" and then the matrices entering the linearized equations do depend on x (see Section 7, Chapter II and [95]).

2.2. Networks of Planar Timoshenko Beams. Such networks were considered in [49] and the treatment closely parallels that of string networks. The equations, given in Section 7, Chapter III are obtained from (1.7) if we make the following assignment of matrices to each element, for the moment suppressing the index i:

$$\text{(2.1)} \qquad \begin{cases} \mathbf{r} = (W_1, W_3, \vartheta_2), \\ \mathbf{P}^2 = \operatorname{diag}(m_0, m_0, \rho_0 I_{33}), \quad \mathbf{Q}^2 = \operatorname{diag}(Eh, Gh, EI_{33}), \\ \mathbf{R} = \begin{pmatrix} 0 & 0 & 0 \\ 0 & 0 & 1 \\ 0 & 0 & 0 \end{pmatrix}, \qquad \mathbf{S} = \mathbf{0}. \end{cases}$$

The network is planar if $\mathbf{e}_{i1}$ and $\mathbf{e}_{i3}$ span the same planar subspace of $\mathbb{R}^3$ for all $i \in \mathcal{I}$, so that all motions occur in that plane and $\mathbf{e}_{i2} = \mathbf{n}$ can be chosen independently of i as the normal to the plane of motion. The geometric constraints require continuity of the displacements $\mathbf{W}_i = W_{i1}\mathbf{e}_{i1} + W_{i3}\mathbf{e}_{i3}$ of the center lines at multiple nodes, and for "rigid joints" one also demands continuity of the rotation angles ϑ_{i2}. These conditions can be expressed in the form (1.1) if we set $\mathbf{C}_{ij}$ equal to the orthogonal matrix having as successive columns $\mathbf{e}_{i1}$, $\mathbf{e}_{i3}$ and $\mathbf{n}$. One then has $\Pi_{ij} = \mathbf{0}$ and $\mathbf{C}_{ij}^+ = \mathbf{C}_{ij}^t$. The condition (1.10) becomes vacuous and (1.9) gives

$$(2.2)\quad \mathcal{N}^j\mathbf{r} = \sum_{i\in\mathcal{I}_j} \epsilon_{ij}[E_i h_i W_{i1,x}\mathbf{e}_{i1} + G_i h_i (W_{i3,x} + \vartheta_{i2})\mathbf{e}_{i3} + E_i I_{i33}\vartheta_{i2,x}\mathbf{n}](x_{ij},t) = \begin{cases} \mathbf{u}^j & \text{for } j \in \mathcal{J}^N, \\ \mathbf{0} & \text{for } j \in \mathcal{J}^F. \end{cases}$$

These can be decoupled into planar and normal components with the aid of any pair of orthonormal vectors $\mathbf{e}_1$ and $\mathbf{e}_2$ which are chosen independently of i and span the network plane. One gets

$$(2.3)\quad \sum_{i\in\mathcal{I}_j} \epsilon_{ij}[E_i h_i W_{i1,x}\mathbf{e}_{i1} + G_i h_i (W_{i3,x} + \vartheta_{i2})\mathbf{e}_{i3}](x_{ij},t) = \begin{cases} (\mathbf{u}^j\cdot\mathbf{e}_1)\mathbf{e}_1 + (\mathbf{u}^j\cdot\mathbf{e}_2)\mathbf{e}_2 & \text{for } j \in \mathcal{J}^N, \\ \mathbf{0} & \text{for } j \in \mathcal{J}^F, \end{cases}$$

and

$$(2.4)\quad \sum_{i\in\mathcal{I}_j} \epsilon_{ij}[E_i I_{i33}\vartheta_{i2,x}](x_{ij},t) = \begin{cases} \mathbf{u}^j\cdot\mathbf{n} & \text{for } j \in \mathcal{J}^N, \\ \mathbf{0} & \text{for } j \in \mathcal{J}^F. \end{cases}$$

Physically, at multiple nodes, these two conditions assert that the resultants of bending moments and forces both are zero or balanced by those of the imposed forces.

If a multiple joint, indexed by j, is pinned rather than rigid there is no constraint on the angles $\vartheta_{i2}(x_{ij},t)$ and the appropriate form of the continuity condition is given by (1.1) with $\mathbf{C}_{ij} = \mathbf{E}^t\mathbf{E}_i$ where $\mathbf{E}_i$ is again the 3×3 matrix having $\mathbf{e}_{i1}$, $\mathbf{e}_{i3}$ and $\mathbf{n}$ as successive columns while $\mathbf{E}$ is a 2×3 matrix having as its rows the vectors $\mathbf{e}_1$ and $\mathbf{e}_2$ which were introduced above. In this case the dynamic conditions (1.9) reduce to (1.3) alone, while (1.10) is no longer vacuous and yields

$$(2.5)\quad [E_i I_{i33}\vartheta_{i2,x}](x_{ij},t) = 0 \text{ for all } i \in \mathcal{I}_j.$$

In this case at the pinned joint all the $\vartheta_{i2,x}$ are free.

It is of course possible that at some multiple nodes the joint is rigid while at others it is pinned.

Here we must set

$$
(2.6)\quad \begin{cases}
\mathbf{r} = (W_1, W_2, W_3, \vartheta_1, \vartheta_2, \vartheta_2), \\
\mathbf{P}^2 = \operatorname{diag}(m_0, m_0, m_0, \rho I, \rho I_{33}, \rho I_{22}), \\
\mathbf{Q}^2 = \operatorname{diag}(EA, GA, GA, GI, EI_{33}, EI_{22}), \\
\mathbf{R} = \begin{pmatrix} 0 & 0 & 0 & 0 & 0 & 0 \\ 0 & 0 & 0 & 0 & 0 & -1 \\ 0 & 0 & 0 & 0 & 1 & 0 \\ 0 & 0 & 0 & 0 & 0 & 0 \\ 0 & 0 & 0 & 0 & 0 & 0 \\ 0 & 0 & 0 & 0 & 0 & 0 \end{pmatrix}, \quad S = \mathbf{0}.
\end{cases}
$$

At rigid joints the geometric multiple node condition (8.1) and (8.3) of Chapter III can be written as

$$
\begin{cases}
\sum_{k=1}^{3} W_{ik}\mathbf{e}_k^i \text{ coincide for } i \in \mathcal{I}_j, \\
\sum_{k=1}^{3} \vartheta_{ik}\mathbf{e}_k^i \text{ coincide for } i \in \mathcal{I}_j.
\end{cases}
$$

These can be expressed in the form (1.1) with $\mathbf{C}_{ij} = \operatorname{diag}(\mathbf{E}^t\mathbf{E}_i, \mathbf{E}^t\mathbf{E}_i)$, where $\mathbf{E}_i$ and $\mathbf{E}$ respectively denote the 3×3 matrices having as columns the triad $\mathbf{e}_1^i, \mathbf{e}_2^i, \mathbf{e}_2^i$ associated with the i-th beam or orthonormal basis vectors $\mathbf{e}_1, \mathbf{e}_2, \mathbf{e}_2$ chosen independently of i. In this case each $\mathbf{C}_{ij}$ is invertible so condition (1.10) is degenerate. One is left with the dynamic conditions (1.9) which can be written out explicitly. We note that $[\mathbf{C}_{ij}^{+}]^t = \mathbf{C}_{ij}$. One can decouple the first three components of $\mathcal{N}^j\mathbf{r}$ from the second three. Denote the 6 components of the control vectors $\mathbf{u}^j$ by u_k^j with $k = 1, \dots, 6$. If one carefully rewrites the expression (1.11) for $\mathcal{N}^j$, the conditions (1.9) become

$$
(2.7)\quad \sum_{i\in\mathcal{I}_j} \epsilon_{ij}\Big[E_iA_iW_{i1,x}\mathbf{e}_1^i + G_iA_i(W_{i2,x} - \vartheta_{i3})\mathbf{e}_2^i + G_iA_i(W_{i3,x} + \vartheta_{i2})\mathbf{e}_3^i\Big] = \begin{cases} \sum_{k=1}^{3} u_k^j\mathbf{e}_k & \text{for } j \in \mathcal{J}^N \\ \mathbf{0} & \text{for } j \in \mathcal{J}^N \end{cases},
$$

and

$$
(2.8)\quad \sum_{i\in\mathcal{I}_j} \epsilon_{ij}[G_iI_i\vartheta_{i1,x}\mathbf{e}_1^i + E_iI_{i33}\vartheta_{i2,x}\mathbf{e}_2^i + E_iI_{i22}\vartheta_{i3,x})\mathbf{e}_3^i] = \begin{cases} \sum_{k=4}^{6} u_k^j\mathbf{e}_{k-3} & \text{for } j \in \mathcal{J}^N, \\ \mathbf{0} & \text{for } j \in \mathcal{J}^N. \end{cases}
$$

For a pin-joint (or perhaps a ball-joint is a better term for a three dimensional joint allowing all proper rotations) there is no geometric condition on the (infinitesimal) rotations and it is an easy exercise to work out what the dynamic conditions are in that case.

2.4. Networks of Initially Curved Bresse Beams. . These planar beam models are treated much like the Timoshenko beams. We have

$$(2.9)\qquad \begin{cases} \mathbf{r} = (W_1, W_3, \vartheta_2), \\ \mathbf{P}^2 = \operatorname{diag}(m_0, m_0, \rho_0 I_{33}), \quad \mathbf{Q}^2 = \operatorname{diag}(Eh, Gh, EI_{33}), \\ \mathbf{R} = \begin{pmatrix} 0 & -\kappa_3 & 0 \\ \kappa_3 & 0 & 1 \\ 0 & 0 & 0 \end{pmatrix}, \quad \mathbf{S} = \mathbf{0}. \end{cases}$$

One can use the same matrices $\mathbf{C}_{ij}$ as before and obtain the following minor variation of the conditions (2.3):

$$(2.10)\qquad \sum_{i\in\mathcal{I}_j} \epsilon_{ij} \Big[E_i h_i (W_{i1,x} - \kappa_{i3} W_{i3}) \mathbf{e}_{i1} + G_i h_i (W_{i3,x} + \kappa_{i3} W_{i1} + \vartheta_{i2}) \mathbf{e}_{i3} \Big] (x_{ij}, t) = \begin{cases} (\mathbf{u}^j \cdot \mathbf{e}_1)\mathbf{e}_1 + (\mathbf{u}^j \cdot \mathbf{e}_2)\mathbf{e}_2 & \text{for } j \in \mathcal{J}^N \\ \mathbf{0} & \text{for } j \in \mathcal{J}^F \end{cases}.$$

The condition (2.4) remains unchanged.

2.5. Beams and Strings. As an example of a structure involving elements of different natures we consider some ways in which strings and beams can be joined together. This will also illustrate the need for the generalization of the multiple node condition (1.1), which was indicated in Remark 1.2.

It is useful to recall from Chapter III that in the *planar* Timoshenko model the beam is thought of as occupying a narrow rectangle $\{x_1\mathbf{e}_1 + x_3\mathbf{e}_3 : x_1 \in (0, l), x_3 \in (-h, h)\}$. According to the kinematic hypotheses the displacement at the point corresponding to (x_1, x_3) is then given by

$$(2.11)\qquad [W_1(x_1, t) + x_2\vartheta_2(x_1, t)]\mathbf{e}_1 + W_3(x_1, t)\mathbf{e}_3.$$

In setting up the expressions for potential and kinetic energies the variable x_3 is integrated away and enters the constants occurring in the Timoshenko equations, and if we set $x = x_1$ we end up with the apparently one dimensional model of this section. However, in modeling the joints between Timoshenko beams and other objects it is helpful to think of the displacements in terms of both variables $x := x_1$ and x_3.

We shall now consider a planar network consisting of two strings indexed by 1 and 2 attached in various ways to one beam indexed by 3: since there is only one beam we do not bother to use the index for the quantities associated with it. For $i = 1, 2$ the spaces $\mathbf{V}_i$ and $\mathbf{H}_i$ as well as their inner products are chosen as in Chapter II and Subsection 2.1 above, while for the beam the choice of spaces is made in accordance with Subsection 2.2. In the absence of Neumann controls this leads to the Lagrangian function

$$
(2.12)\quad \mathcal{L}(\mathbf{r}) = \int_0^T \frac{1}{2}\Big[\sum_{i=1}^{2}\int_0^{l_i} [\rho_i|\dot{\mathbf{r}}^i|^2 - \mathbf{Q}_i\mathbf{r}^i_{,x}.\mathbf{r}^i_{,x}]\,dx
+ \int_0^{l_3} \{[m_0\dot{W}_1^2 + m_0\dot{W}_3^2 + \rho_0 I_{33}\dot{\vartheta}_2^2]
- [EhW_{1,x}{}^2 + Gh(W_{3,x}+\vartheta_2)^2 + EI_{33}\vartheta_{2,x}{}^2]\}\,dx\Big]\,dt.
$$

Now one needs to specify the nature of the joints, which will involve the endpoints of the string and endfaces of the beam. We note first that in all cases variations of $\mathbf{r}^i$ inside $(0, l_i)$ will lead to the equations (3.10) of Chapter II for the two strings and the Timoshenko system of Chapter III, Section 7 for the beam. General variations, for the moment not subject to any geometric constraints, then lead one to the variational identity

$$
(2.13)\quad 0 = \int_0^T \Big(\sum_{i=1}^{2}\Big[[\mathbf{Q}_i\mathbf{r}^i_{,x}\cdot\delta\mathbf{r}^i](x,t)\Big]_{x=0}^{x=l_i}
+ \Big[[EhW_{1,x}\delta W_1 + Gh(W_{3,x}+\vartheta_2)\delta\mathbf{W}_3 + EI_{33}\vartheta_{2,x}\delta\vartheta_2](x,t)\Big]_{x=0}^{x=l_3}\Big)\,dt.
$$

We do not write down the various controls or free conditions at the endpoints which are not joined to another element, where one has all the possibilities discussed in the previous sections.

Consider first the situation where *the two strings are attached respectively to the two endpoints of the centerline of the beam.* Here, if one takes into account the form (2.11) of the beam displacement, one has the geometric constraints

$$
(2.14)\quad \begin{cases} \mathbf{r}^1(0,t) = W_1(0,t)\mathbf{e}_1 + W_3(0,t)\mathbf{e}_3, \\ \mathbf{r}^2(0,t) = W_1(l_3,t)\mathbf{e}_1 + W_3(l_3,t)\mathbf{e}_3. \end{cases}
$$

These conditions at the two multiple nodes located at the endpoints of the beam are certainly of the form (1.1). One can use the general scheme to write down explicitly the corresponding dynamic conditions (1.9) and (1.10) but it is also instructive to derive these directly. Since the angular variable ϑ_2 is not involved in the geometric constraint, one can choose $\delta\vartheta_2$ arbitrarily while letting the other variations vanish. One obtains from (2.13) that

$$
(2.15)\quad \vartheta_{2,x}(0,t) = \vartheta_{2,x}(l_3,t) = 0.
$$

We then consider variations $\delta\mathbf{r}^1$, δW_1 and δW_3 which respect the first of the above conditions and vanish at the other endpoints and set $\delta\mathbf{r}^2 = \mathbf{0}$

and $\delta\vartheta_2 = 0$. Then, because

$$\delta\mathbf{r}^1(0,t) = \delta W_1(0,t)\mathbf{e}_1 + \delta W_3(0,t)\mathbf{e}_3,$$

(2.13) reduces to

$$[-\mathbf{Q}_1\mathbf{r}^1_{,x}.\mathbf{e}_1 - EhW_{1,x}](0,t)\delta W_1(0,t) \\ + [-\mathbf{Q}_1\mathbf{r}^1_{,x}.\mathbf{e}_3 - Gh(W_{3,x} + \vartheta_2)](0,t)\delta W_3(0,t) = 0.$$

One obtains the following two dynamic node conditions associated with the junction between the first string and the beam:

$$\begin{cases} \mathbf{Q}_1\mathbf{r}^1_{,x}(0,t)\cdot\mathbf{e}_1 + EhW_{1,x}(0,t) = 0, \\ \mathbf{Q}_1\mathbf{r}^1_{,x}(0,t)\cdot\mathbf{e}_3 + Gh[W_{3,x} + \vartheta_2](0,t) = 0. \end{cases} \tag{2.16}$$

Similarly

$$\begin{cases} \mathbf{Q}_2\partial\mathbf{r}^2(0,t)\cdot\mathbf{e}_1 - EhW_{1,x}(l_3,t) = 0, \\ \mathbf{Q}_2\mathbf{r}^2_{,x}(0,t)\cdot\mathbf{e}_3 - Gh[W_{3,x} + \vartheta_2](l_3,t) = 0. \end{cases} \tag{2.17}$$

Next we consider the situation where *the two strings are both attached to the centerline of the beam at one end*, which corresponds to the geometric constraint

$$\mathbf{r}^1(0,t) = \mathbf{r}^2(0,t) = W_1(0,t)\mathbf{e}_1 + W_3(0,t)\mathbf{e}_3. \tag{2.18}$$

Again ϑ_2 is not involved and one obtains the condition $\vartheta_{2,x}(0,t) = 0$. Then we take variations with $\delta\vartheta_2 = 0$ and $\delta\mathbf{r}^1$, $\delta\mathbf{r}^2$, δW_1 and δW_3 having support near the junction, to obtain from (2.13) that

$$\begin{cases} \left[\mathbf{Q}_1\mathbf{r}^1_{,x}(0,t) + \mathbf{Q}_2\mathbf{r}^2_{,x}(0,t)\right]\cdot\mathbf{e}_1 + EhW_{1,x}(0,t) = 0, \\ \left[\mathbf{Q}_1\mathbf{r}^1_{,x}(0,t) + \mathbf{Q}_2\mathbf{r}^2_{,x}(0,t)\right]\cdot\mathbf{e}_3 + Gh[W_{3,x} + \vartheta_2](0,t) = 0. \end{cases} \tag{2.19}$$

Finally we consider a third variant which illustrates that the continuity condition (1.1) is somewhat restrictive. Suppose that *the two strings are attached to the two extreme points of one endface of the beam.* This leads to the constraints

$$\begin{cases} \mathbf{r}^1(0,t) = [W_1(0,t) + h\vartheta_2(0,t)]\mathbf{e}_1 + W_3(0,t)\mathbf{e}_3, \\ \mathbf{r}^2(0,t) = [W_1(0,t) - h\vartheta_2(0,t)]\mathbf{e}_1 + W_3(0,t)\mathbf{e}_3. \end{cases} \tag{2.20}$$

These two multiple node conditions cannot be written in the form (1.1) in a way which satisfies the conditions on $\mathbf{C}_{ij}$ but can be expressed in the form discussed in Remark 1.2. Again we can proceed in an ad hoc way. For the variations we must require

$$\begin{cases} \delta\mathbf{r}^1(0,t) = [\delta W_1(0,t) + h\delta\vartheta_2(0,t)]\mathbf{e}_1 + \delta W_3(0,t)\mathbf{e}_3, \\ \delta\mathbf{r}^2(0,t) = [\delta W_1(0,t) - h\delta\vartheta_2(0,t)]\mathbf{e}_1 + \delta W_3(0,t)\mathbf{e}_3. \end{cases}$$

This gives rise to the dynamic conditions (2.18) as before, together with the additional condition

$$h[\mathbf{Q}_1\mathbf{r}^1_{,x}(0,t) - \mathbf{Q}_2\partial\mathbf{r}^2(0,t)]\cdot\mathbf{e}_1 + EI_{33}\vartheta_{3,x}(0,t) = 0. \tag{2.21}$$

Note that in this last situation, unlike the previous one where the strings are both joined to the center line of the beam, the variables $\mathbf{r}^1(0,t)$ and $\mathbf{r}^2(0,t)$ together determine, and are determined by, the beam variables $W_1(0,t)$, $W_3(0,t)$ and $\vartheta_2(0,t)$.

REMARK 2.2. The three different arrangements of beam and strings lead to the same equations with different geometric and dynamic coupling conditions where they are joined together. Supplementing these by boundary conditions, controls and initial conditions one is led to systems for which existence of solutions will follow from Section 3. The systems behave in different ways (for example see Remark 5.3 for comments on exact controllability).

3. Existence and Regularity of Solutions

In this section we shall sketch how the standard variational and semigroup methods used in Section 4 of Chapter II to obtain existence and regularity results for solutions of the string network system apply with only very minor changes to the general system we have derived in Section 1 of this Chapter. This general system is

$$\begin{cases} \mathbf{P}_i^2\ddot{\mathbf{r}}^i = \left[\mathbf{Q}_i^2(\mathbf{r}^i_{,x} + \mathbf{R}_i\mathbf{r}^i)\right]_{,x} \\ \qquad\qquad - \mathbf{R}_i^t\mathbf{Q}_i^2(\mathbf{r}^i_{,x} + \mathbf{R}_i\mathbf{r}^i) + \mathbf{S}_i\mathbf{r}^i + \mathbf{f}^i \ \text{ for } i \in \mathcal{I}, \\ \mathbf{r}^i(x_{ij},t) \ \text{coincide for } j \in \mathcal{J}_M,\ i \in \mathcal{I}_j, \\ \mathcal{D}^j\mathbf{r}(t) = \mathbf{u}^j(t) \text{ for } j \in \mathcal{J}^D,\ \mathcal{D}^j\mathbf{r}(t) = \mathbf{0} \text{ for } j \in \mathcal{J}^C, \\ \mathcal{N}^j\mathbf{r}(t) = \mathbf{u}^j(t) \text{ for } j \in \mathcal{J}^N,\ \mathcal{N}^j\mathbf{r}(t) = \mathbf{0} \text{ for } j \in \mathcal{J}^F, \\ \mathcal{N}_i^j\mathbf{r}(t) = \mathbf{0} \text{ for } j \in \mathcal{J} \text{ and } i \in \mathcal{I}_j, \\ \mathbf{r}(0) = \mathbf{r}_0,\quad \dot{\mathbf{r}}(0) = \dot{\mathbf{r}}_0. \end{cases} \tag{3.1}$$

The governing equations can again be written in the form

$$\frac{\partial}{\partial t}\begin{pmatrix}\mathbf{r}\\ \dot{\mathbf{r}}\end{pmatrix} = \begin{pmatrix}\mathbf{0} & \mathbf{I}\\ \mathbf{A} & \mathbf{0}\end{pmatrix}\begin{pmatrix}\mathbf{r}\\ \dot{\mathbf{r}}\end{pmatrix} + \begin{pmatrix}\mathbf{0}\\ \mathbf{f}/\rho\end{pmatrix} \tag{3.2}$$

with $\mathbf{f}/\rho = (\mathbf{f}^i/\rho_i)_{i\in\mathcal{I}}$, and

$$\mathbf{A}\mathbf{r} = \left(\mathbf{P}_i^{-2}\left[\left[\mathbf{Q}_i^2(\mathbf{r}^i_{,x} + \mathbf{R}_i\mathbf{r}^i)\right]_{,x} - \mathbf{R}_i^t\mathbf{Q}_i^2(\mathbf{r}^i_{,x} + \mathbf{R}_i\mathbf{r}^i) + \mathbf{S}_i\mathbf{r}_i\right]\right)_{i\in\mathcal{I}}. \tag{3.3}$$

As *state space* for our system we take the Hilbert space $\mathcal{H} = \mathbf{V} \times \mathbf{H}$ where

$$\mathbf{H} = \prod_{i=1}^{n} L_2(0, l_i; \mathbb{R}^{p_i})$$

and

$$\mathbf{V} = \left\{ \mathbf{r} \in \prod_{i\in\mathcal{I}} H^1(0, l_i; \mathbb{R}^{p_i}) \middle| \begin{cases} \mathbf{C}_{ij}\mathbf{r}^i(x_{ij}) \text{ coincide for } j \in \mathcal{J}_M, \, i \in \mathcal{I}_j, \\ \mathcal{D}^j\mathbf{r} = 0 \text{ for } j \in \mathcal{J}^C \end{cases} \right\}.$$

On $\mathbf{H}$ we define the inner product

$$(3.4) \qquad (\mathbf{r}, \mathbf{s})_{\mathbf{H}} = \sum_{i\in\mathcal{I}} \frac{1}{2} \int_0^{l_i} \mathbf{P}_i^2 \mathbf{r}^i \cdot \mathbf{s}^i \, dx,$$

corresponding to kinetic energy. On $\mathbf{V}$ we define

(3.5)

$$(\mathbf{r}, \mathbf{s})_{\mathbf{V}} = \sum_{i\in\mathcal{I}} \frac{1}{2} \int_0^{l_i} \left[\mathbf{Q}_i^2 [\mathbf{r}^i_{,x} + \mathbf{R}_i\mathbf{r}^i] \cdot [\mathbf{s}^i_{,x} + \mathbf{R}_i\mathbf{s}^i] + [\mathbf{S}_i\mathbf{r}^i \cdot \mathbf{s}^i] \, (x, t) \right] dx,$$

which corresponds to the potential energy, as well as $(\mathbf{r}, \mathbf{s})_{\mathbf{V},\lambda} = (\mathbf{r}, \mathbf{s})_{\mathbf{V}} + \lambda(\mathbf{r}, \mathbf{s})_{\mathbf{H}}$. It is not difficult to see that for $\lambda > 0$ the norm associated with $< \mathbf{r}, \tilde{\mathbf{r}} >_{\mathbf{V},\lambda}$ is equivalent to the standard norm. In some important cases this is also true for $\lambda = 0$. The following lemma gives some sufficient (but not necessary) conditions for this to happen.

LEMMA 3.1. *Suppose that either*

(i) *every $\mathbf{S}_i$ is positive definite; or*

(ii) *there is at least one fully clamped node, and for each element labelled i_1, say, there exists a path denoted by $[i_1, i_2, \ldots, i_r]$ leading along the network elements labelled successively by $i_1, i_2, \ldots, i_r$ and concluding at a fully clamped node with the property that if j_k is the index of the node between the i_{k-1}-th and the i_k-th element the matrix $\mathbf{C}_{i_k j_k}$ is invertible.*

Then $\|\mathbf{r}\|_{\mathbf{V}}$ is a norm on $\mathbf{V}$ and is equivalent to the product norm

$$\|\mathbf{r}\| = \left[\sum_{i=1}^{n} \int_0^{l_i} [|\mathbf{r}^i|^2 + |\mathbf{r}^i_{,x}|^2] dx \right]^{\frac{1}{2}}.$$

PROOF. The triangle property and the homogeneity property of $\|\mathbf{r}\|_{\mathbf{V}}$ are evident, so to verify that this is a norm one need only show that $\|\mathbf{r}\|_{\mathbf{V}} = 0$ implies $\mathbf{r} = \mathbf{0}$. This is obvious if each $\mathbf{S}_i$ is positive definite. Otherwise one concludes that $\mathbf{r}^i_{,x} + \mathbf{R}_i\mathbf{r}^i = \mathbf{0}$ for each i. If $\mathbf{r}^i$ vanishes at one of the endpoints of $[0, l_i]$ it follows from a uniqueness theorem for solutions of differential equations that $\mathbf{r}^i$ vanishes identically. If we now start at a fully clamped node and work our way from element to element along paths such as those described in the alternative hypothesis, using the node condition

(1.1) and the invertibility of each $\mathbf{C}_{i_k j_k}$, we can conclude that indeed each $\mathbf{r}^i$ vanishes.

From the continuity of $\mathbf{Q}_i(x)$, $\mathbf{R}_i(x)$ and $\mathbf{S}_i(x)$ it is obvious that $\|\mathbf{r}\|_{\mathbf{V}} \leq C \|\mathbf{r}\|$, for some suitable constant C. The reverse inequality is obvious in the case of the first hypothesis but more delicate in the alternative case. We again use an idea from differential equations. Let $\mathbf{T}_i(x)$ be a (nonsingular) $p_i \times p_i$ matrix-valued solution of

$$\mathbf{T}_{i,x}(x) = \mathbf{T}_i(x)\mathbf{R}_i(x).$$

Then

$$[\mathbf{T}_i\mathbf{r}^i]_{,x} = \mathbf{T}_i[\mathbf{r}^i_{,x} + \mathbf{R}_i\mathbf{r}^i].$$

Hence

$$\mathbf{T}_i(x)\mathbf{r}^i(x) = \mathbf{T}_i(0)\mathbf{r}^i(0) + \int_0^x [\mathbf{T}_i(y)\mathbf{r}^i(y)]_{,x}dy,$$

and so one gets the estimate

$$|\mathbf{T}_i(l_i)\mathbf{r}^i(l_i)|^2 + \int_0^{l_i} |\mathbf{T}_i\mathbf{r}^i|^2dx \leq C\Big[|\mathbf{T}_i(0)\mathbf{r}^i(0)|^2 + \int_0^{l_i} |\mathbf{r}^i_{,x} + \mathbf{R}_i\mathbf{r}^i|^2dx\Big].$$

Since $\mathbf{T}_i(x)$ is continuous and continuously invertible one also has, after redefinition of C,

$$|\mathbf{r}^i(l_i)|^2 + \int_0^{l_i} |\mathbf{r}^i|^2dx \leq C\Big[|\mathbf{r}^i(0)|^2 + \int_0^{l_i} |\mathbf{r}^i_{,x} + \mathbf{R}_i\mathbf{r}^i|^2dx\Big].$$

The latter holds also with l_i and 0 interchanged. One again proceeds from element to element along paths starting at a fully clamped node, getting alternately estimates on $|\mathbf{r}^i(0)|^2$ or $|\mathbf{r}^i(l_i)|^2$ and on $\int_0^{l_i} |\mathbf{r}^i|^2dx$ in terms of $\|\mathbf{r}\|_{\mathbf{V}}$. This gives the desired estimate $\|\mathbf{r}\| \leq C \|\mathbf{r}\|_{\mathbf{V}}$. □

REMARK 3.1. In most applications the elements will all be of the same nature and the conclusion of the lemma will be valid whenever all the $\mathbf{C}_{ij}$ are invertible and there is at least one fully clamped node, since then the second of the alternative hypotheses is trivially valid. For an exception see Remark 5.3 concerning the mixed string/beam network.

The argument leading to well-posedness of our model now proceeds word for word as in Section 3 of Chapter II, with only minor notational adjustments:

- The continuity condition becomes: the vectors $\mathbf{C}_{ij}\mathbf{r}^i(x_{ij}, t)$ coincide for $j \in \mathcal{J}_M$ and $i \in \mathcal{I}_j$;
- the condition $\mathcal{N}^j_i\mathbf{r} = \mathbf{0}$ for $j \in \mathcal{J}$ and $i \in \mathcal{I}_j$ needs to be stated along with the other conditions;
- $\mathbb{R}^3$ needs to be replaced by $\mathbb{R}^{p_i}$ or $\mathbb{R}^{q_j}$ as appropriate.

Exactly as before one constructs a semigroup $\mathcal{U}(t)$ with a generator $\mathcal{A}$ and uses this semigroup to solve (3.1) for initial data in dom($\mathcal{A}^\infty$) and control functions which satisfy

$$\text{(3.6)}\qquad \begin{cases} \mathbf{u}_j(t) - \mathcal{D}^j \mathbf{r}_0 \in \mathbf{C}_0^\infty((0,T]; \mathbb{R}^{q_j}) & \text{for } j \in \mathcal{J}^D, \\ \mathbf{u}_j(t) \in \mathbf{C}_0^\infty((0,T]; \mathbb{R}^{q_j}) & \text{for } j \in \mathcal{J}^N. \end{cases}$$

One then has to make a priori estimates which allow the passage to less regular data. This involves the same steps as before, but the explicit computations need to be redone. For one thing the energy density is now given by

$$e_i(x,t) = \frac{1}{2}\Big[|\mathbf{P}_i \dot{\mathbf{r}}^i|^2 + |\mathbf{Q}_i(\mathbf{r}^i_{,x} + \mathbf{R}_i \mathbf{r}^i)|^2 + \mathbf{S}_i \mathbf{r}^i \cdot \mathbf{r}^i\Big].$$

It is in fact also useful to introduce

$$\tilde{e}_i(x,t) = \frac{1}{2}\Big[|\mathbf{P}_i \dot{\mathbf{r}}^i|^2 + |\mathbf{Q}_i(\mathbf{r}^i_{,x} + \mathbf{R}_i \mathbf{r}^i)|^2\Big].$$

and

$$\tilde{\mathcal{E}}^\dagger(x; t_1, t_2) = \int_{t_1}^{t_2} \tilde{e}(x,t)\,dt.$$

We have to prove the analog of Lemma 4.2 of Chapter II which, in the prosent context, takes the following form.

LEMMA 3.2. *Let* $\mathbf{r}$ *be a smooth solution of*

$$\text{(3.7)}\qquad \left[\mathbf{P}^2 \mathbf{r}_{,t}\right]_{,t} = \left[\mathbf{Q}^2(\mathbf{r}_{,x} + \mathbf{R}\mathbf{r})\right]_{,x} - \mathbf{R}^t\mathbf{Q}^2(\mathbf{r}_{,x} + \mathbf{R}\mathbf{r}) - \mathbf{S}\mathbf{r}.,$$

where $m(x)$ *is a given smooth* multiplier function. *Then*

$$\text{(3.8)}\qquad \int_0^T \Big[\tilde{e}(x,t)m(x)\Big]_{x=0}^{x=l} dt = \int_0^T \int_0^l \tilde{e}(x,t) m_{,x}\,dx dt$$
$$+ \frac{1}{2}\int_0^T \int_0^l \Big[\mathbf{M}_1 \dot{\mathbf{r}} \cdot \dot{\mathbf{r}} + \mathbf{M}_2(\mathbf{r}_{,x} + \mathbf{R}\mathbf{r}) \cdot (\mathbf{r}_{,x} + \mathbf{R}\mathbf{r})$$
$$- m\mathbf{S}\mathbf{r} \cdot (\mathbf{r}_{,x} + \mathbf{R}\mathbf{r})\Big]\,dx dt + \int_0^l \Big[m\mathbf{P}^2 \dot{\mathbf{r}} \cdot (\mathbf{r}_{,x} + \mathbf{R}\mathbf{r})\Big]_{t=0}^{t=T} dx,$$

where

$$\mathbf{M}_1 = m(\mathbf{P}^2)_{,x} - m\mathbf{R}^t\mathbf{P}^2, \quad \mathbf{M}_2 = m\mathbf{R}^t\mathbf{Q}^2 - m(\mathbf{Q}^2)_{,x}.$$

As a consequence

$$\text{(3.9)}\qquad \tilde{\mathcal{E}}^\dagger(0;0,T) + \tilde{\mathcal{E}}^\dagger(l;0,T) \le C\left[\int_0^T \mathcal{E}(t)\,dt + \mathcal{E}(0) + \mathcal{E}(T)\right].$$

PROOF. Here we multiply the equation by $m(\mathbf{r}_{,x} + \mathbf{R}\mathbf{r})$ and integrate by parts. First we get

$$
\begin{aligned}
(3.10) \quad & \int_0^T \int_0^l \mathbf{P}^2 \ddot{\mathbf{r}} \cdot m(\mathbf{r}_{,x} + \mathbf{R}\mathbf{r})\, dx dt \\
&= - \int_0^T \int_0^l m\mathbf{P}^2 \dot{\mathbf{r}} \cdot (\dot{\mathbf{r}}_{,x} + \mathbf{R}\dot{\mathbf{r}})\, dx dt + \int_0^l \Big[m\mathbf{P}^2\dot{\mathbf{r}} \cdot (\mathbf{r}_{,x} + \mathbf{R}\mathbf{r}) \Big]_{t=0}^{t=T} dx \\
&= \frac{1}{2} \int_0^T \int_0^l [m_{,x}\mathbf{P}^2 + m(\mathbf{P}^2)_{,x} - m\mathbf{R}^t\mathbf{P}^2]\dot{\mathbf{r}} \cdot \dot{\mathbf{r}}\, dx dt \\
&\quad - \int_0^T \Big[\frac{m}{2} |\mathbf{P}\dot{\mathbf{r}}|^2 \Big]_{x=0}^{x=l} dt + \int_0^l \Big[m\mathbf{P}^2\dot{\mathbf{r}} \cdot (\mathbf{r}_{,x} + \mathbf{R}\mathbf{r}) \Big]_{t=0}^{t=T} dx,
\end{aligned}
$$

where we have used the identity

$$
m\mathbf{P}^2\dot{\mathbf{r}} \cdot \dot{\mathbf{r}}_{x} = \frac{1}{2}[m\mathbf{P}^2\dot{\mathbf{r}} \cdot \dot{\mathbf{r}}]_{,x} - \frac{1}{2}[m_{,x}\mathbf{P}^2 + m(\mathbf{P}^2)_{,x}]\dot{\mathbf{r}} \cdot \dot{\mathbf{r}}.
$$

To obtain the next identity we first note that

$$
\begin{aligned}
m\big[\mathbf{Q}^2(\mathbf{r}_{,x} + \mathbf{R}\mathbf{r})\big]_{,x} \cdot (\mathbf{r}_{,x} + \mathbf{R}\mathbf{r}) &= \frac{1}{2}\big[m\mathbf{Q}^2(\mathbf{r}_{,x} + \mathbf{R}\mathbf{r}) \cdot (\mathbf{r}_{,x} + \mathbf{R}\mathbf{r})\big]_{,x} \\
&\quad - \frac{1}{2}[m_{,x}\mathbf{Q}^2 - m(\mathbf{Q}^2)_{,x}](\mathbf{r}_{,x} + \mathbf{R}\mathbf{r}) \cdot (\mathbf{r}_{,x} + \mathbf{R}\mathbf{r}).
\end{aligned}
$$

Then we integrate by parts with respect to x to get

$$
\begin{aligned}
(3.11) \quad & - \int_0^T \int_0^l \Big[\big[\mathbf{Q}^2(\mathbf{r}_{,x} + \mathbf{R}\mathbf{r})\big]_{,x} - \mathbf{R}^t\mathbf{Q}^2(\mathbf{r}_{,x} + \mathbf{R}\mathbf{r}) \\
&\quad - \mathbf{S}\mathbf{r} \Big] \cdot m(\mathbf{r}_{,x} + \mathbf{R}\mathbf{r})\, dx dt = - \int_0^T \Big[\frac{m}{2} |\mathbf{Q}(\mathbf{r}_{,x} + \mathbf{R}\mathbf{r})|^2 \Big]_{x=0}^{x=l} dt \\
&\quad + \frac{1}{2} \int_0^T \int_0^l \Big[[m_{,x}\mathbf{Q}^2 - m(\mathbf{Q}^2)_{,x} + m\mathbf{R}^t\mathbf{Q}^2](\mathbf{r}_{,x} + \mathbf{R}\mathbf{r}) \cdot (\mathbf{r}_{,x} + \mathbf{R}\mathbf{r}) \\
&\quad - m\mathbf{S}\mathbf{r} \cdot (\mathbf{r}_{,x} + \mathbf{R}\mathbf{r}) \Big]\, dx dt.
\end{aligned}
$$

By adding identities (3.10) and (3.11) and then rearranging we get (3.8). The estimate (3.9) then follows by setting $m(x) = -1 + 2x/l$ as before. □

The next step is to prove the a priori estimate analogous to that of Lemma 4.3 of Chapter II. We restate the result.

LEMMA 3.3. $\mathbf{r}$ *be a smooth solution of* (3.1). *Then*

$$
(3.12) \quad \mathcal{E}(t) \le C \left[\mathcal{E}(0) + \int_0^t \Big[\sum_{j \in \mathcal{J}^D} |\dot{\mathbf{u}}^j(\tau)|^2 + \sum_{j \in \mathcal{J}^N} |\mathbf{u}^j(\tau)|^2 \Big] d\tau \right].
$$

PROOF. The proof follows exactly the same lines as before. The following calculation has to be made:

$$
\begin{aligned}
\frac{d\mathcal{E}}{dt}(\tau) &= \sum_{i\in\mathcal{I}} \int_0^{l_i} \left[\mathbf{P}_i^2 \dot{\mathbf{r}}^i \cdot \ddot{\mathbf{r}}^i + \mathbf{Q}_i^2[\mathbf{r}^i_{,x} + \mathbf{R}_i\mathbf{r}^i] \cdot [\dot{\mathbf{r}}^i_{,x} + \mathbf{R}_i\dot{\mathbf{r}}^i]\right](x,\tau)\,dx \\
&= \sum_{i\in\mathcal{I}} \left[\mathbf{Q}_i^2[\mathbf{r}^i_{,x} + \mathbf{R}_i\mathbf{r}^i] \cdot \dot{\mathbf{r}}^i\right]_{x=0}^{x=l_i} \\
&= \sum_{j\in\mathcal{J}} \sum_{i\in\mathcal{I}_j} \epsilon_{ij} [\mathbf{Q}_i^2[\mathbf{r}^i_{,x} + \mathbf{R}_i\mathbf{r}^i] \cdot \dot{\mathbf{r}}^i](x_{ij},\tau).
\end{aligned}
$$

Now

$$
\begin{aligned}
\dot{\mathbf{r}}^i(x_{ij},\tau) &= \mathbf{\Pi}_{ij}^{\perp}\dot{\mathbf{r}}^i(x_{ij},\tau) + \mathbf{\Pi}_{ij}\dot{\mathbf{r}}^i(x_{ij},\tau) = \mathbf{C}_{ij}^{+}\mathbf{C}_{ij}\dot{\mathbf{r}}^i(x_{ij},\tau) + \mathbf{\Pi}_{ij}\dot{\mathbf{r}}^i(x_{ij},\tau) \\
&= \mathbf{C}_{ij}^{+}\boldsymbol{\mathcal{D}}^j\dot{\mathbf{r}}(\tau) + \mathbf{\Pi}_{ij}\dot{\mathbf{r}}^i(x_{ij},\tau)
\end{aligned}
$$

Substituting this back into the previous identity and taking into account the form of the various operators involved we obtain

$$
\begin{aligned}
\frac{d\mathcal{E}}{dt}(\tau) &= \sum_{j\in\mathcal{J}} \boldsymbol{\mathcal{N}}^j\mathbf{r} \cdot \boldsymbol{\mathcal{D}}^j\dot{\mathbf{r}} \\
&= \sum_{j\in\mathcal{J}^N} \boldsymbol{\mathcal{N}}^j\mathbf{r} \cdot \dot{\mathbf{u}}^j(\tau) + \sum_{j\in\mathcal{J}^N} \boldsymbol{\mathcal{D}}^j\dot{\mathbf{r}} \cdot \mathbf{u}^j(\tau).
\end{aligned}
\tag{3.13}
$$

Since

$$
\begin{aligned}
|\boldsymbol{\mathcal{D}}^j\dot{\mathbf{r}}(\tau)|^2 &\le C\tilde{e}_i(x_{ij},t) \quad \text{for any } i \in \mathcal{I}_j, \\
|\boldsymbol{\mathcal{N}}^j\mathbf{r}(\tau)|^2 &\le C\sum_{i\in\mathcal{I}_j} \tilde{e}_i(x_{ij},t),
\end{aligned}
$$

the remainder of the argument, using the estimate (3.9), goes through as before. □

One is led to the following general existence theorem:

THEOREM 3.1. *Suppose that*
(i) $\mathbf{r}_0 \in \mathbf{V}$, $\dot{\mathbf{r}}_0 \in \mathbf{H}$, $\mathbf{f} \in L_2(0,T;\mathbf{H})$;
(ii) $\mathbf{u}^j \in H^1(0,T : \mathbb{R}^{q_j})$ *and satisfies* $\mathbf{u}^j(0) = \boldsymbol{\mathcal{D}}^j\mathbf{r}_0$ *for each* $j \in \mathcal{J}^D$;
(iii) $\mathbf{u}^j(t) \in L_2(0,T : \mathbb{R}^{q_j})$ *for each* $j \in \mathcal{J}^N$.
Then there exists a unique $\mathbf{r} \in C([0,T];\mathbf{V}) \cap C^1([0,T];\mathbf{H})$ *such that*

$$\boldsymbol{\mathcal{D}}^j\mathbf{r}(t) = \mathbf{u}^j(t) \quad \text{for } j \in \mathcal{J}^D$$

and which satisfies the remaining requirements of (4.1) *in the following weak sense: for any* $\phi \in \mathbf{V}_0$ *the function* $(\mathbf{r}(t),\phi)_{\mathbf{H}}$ *has an absolutely continuous first derivative in* t *and almost everywhere satisfies*

$$
\frac{d^2}{dt^2}(\mathbf{r}(t),\phi)_{\mathbf{H}} + (\mathbf{r}(t),\phi)_{\mathbf{V}} = (\mathbf{f}(t)/\rho,\phi)_{\mathbf{H}} + \sum_{j\in\mathcal{J}^N} \mathbf{u}^j(t) \cdot \boldsymbol{\mathcal{D}}^j\phi
\tag{3.14}
$$

as well as the initial conditions

$$(3.15)\qquad (\mathbf{r}(0),\phi)_{\mathbf{H}} = (\mathbf{r}_0,\phi)_{\mathbf{H}},\ \frac{d}{dt}(\mathbf{r}(t),\phi)_{\mathbf{H}}\Big|_{t=0} = (\dot{\mathbf{r}}_0,\phi)_{\mathbf{H}}.$$

The solution satisfies the estimate

$$(3.16)\quad \mathcal{E}(t) \le C\left[\mathcal{E}(0) + \int_0^t \Big[\sum_{j\in\mathcal{J}^D} |\dot{\mathbf{u}}^j(\tau)|^2 + \sum_{j\in\mathcal{J}^N} |\mathbf{u}^j(\tau)|^2\Big]\, d\tau\right].$$

4. Energy Estimates for Hyperbolic Systems

In this section we derive the energy estimates which are needed to prove results on controllability and stabilizability.

We consider a single system

$$(4.1)\qquad \left[\mathbf{P}^2\mathbf{r}_{,t}\right]_{,t} = \left[\mathbf{Q}^2(\mathbf{r}_{,x}+\mathbf{R}\mathbf{r})\right]_{,x} - \mathbf{R}^t\mathbf{Q}^2(\mathbf{r}_{,x}+\mathbf{R}\mathbf{r}) - \mathbf{S}\mathbf{r},$$

which is associated with the energy density

$$(4.2)\qquad e(x,t) = \frac{1}{2}\left[|\mathbf{P}\mathbf{r}_{,t}|^2 + |\mathbf{Q}(\mathbf{r}_{,x}+\mathbf{R}\mathbf{r})|^2 + \mathbf{S}\mathbf{r}\cdot\mathbf{r}\right].$$

Here we find it convenient to use "$_{,t}$" as an alternative to the "$\dot{}$" notation for partial derivatives with respect to t. For the purposes of this section we shall suppose

- $\mathbf{P}$ and $\mathbf{Q}$ are smooth, positive definite and symmetric $p\times p$ matrix valued functions of x and t;
- $\mathbf{S}$ is either a positive definite and symmetric $p\times p$ matrix valued function of x and t or, alternatively, it is the zero matrix;
- $\mathbf{R}$ is a $p\times p$ matrix valued functions of x and possibly, in case $\mathbf{S}$ is positive definite, of t as well.

The dependence on t is not needed for the applications we have in mind here but may be useful in other contexts and does not introduce additional difficulties.

We shall need the following two identities

$$(4.3)\qquad e_{,t} - [\mathbf{Q}^2(\mathbf{r}_{,x}+\mathbf{R}\mathbf{r})\cdot\mathbf{r}_{,t}]_{,x} = q_1(\mathbf{r},\mathbf{r}_{,t},\mathbf{r}_{,x})$$

and

$$(4.4)\qquad e_{,x} - [\mathbf{P}^2\mathbf{r}_{,t}\cdot(\mathbf{r}_{,x}+\mathbf{R}\mathbf{r})]_{,t} = q_2(\mathbf{r},\mathbf{r}_{,t},\mathbf{r}_{,x}),$$

where

$$(4.5)\quad q_1(\mathbf{r},\mathbf{r}_{,t},\mathbf{r}_{,x}) = \frac{1}{2}[\mathbf{P}^2]_{,t}\mathbf{r}_{,t}\cdot\mathbf{r}_{,t} + \frac{1}{2}[\mathbf{Q}^2]_{,t}(\mathbf{r}_{,x}+\mathbf{R}\mathbf{r})\cdot(\mathbf{r}_{,x}+\mathbf{R}\mathbf{r}) + \frac{1}{2}\mathbf{S}_{,t}\mathbf{r}\cdot\mathbf{r} + \mathbf{Q}^2\mathbf{R}_{,t}\mathbf{r}\cdot(\mathbf{r}_{,x}+\mathbf{R}\mathbf{r}).$$

and

$$
\begin{aligned}
(4.6)\quad q_2(\mathbf{r},\mathbf{r}_{,t},\mathbf{r}_{,x}) &= \frac{1}{2}[\mathbf{P}^2]_{,x}\mathbf{r}_{,t}\cdot\mathbf{r}_{,t} - \frac{1}{2}[\mathbf{Q}^2]_{,x}(\mathbf{r}_{,x}+\mathbf{R}\mathbf{r})\cdot(\mathbf{r}_{,x}+\mathbf{R}\mathbf{r}) \\
&+ \frac{1}{2}\mathbf{S}_{,x}\mathbf{r}\cdot\mathbf{r} - \mathbf{P}^2\mathbf{r}_{,t}\cdot\mathbf{R}_{,t}\mathbf{r} + (\mathbf{r}_{,x}+\mathbf{R}\mathbf{r})\cdot\mathbf{R}^t\mathbf{Q}^2(\mathbf{r}_{,x}+\mathbf{R}\mathbf{r}) \\
&\qquad + (\mathbf{r}_{,x}+\mathbf{R}\mathbf{r})\cdot\mathbf{S}\mathbf{r} + \mathbf{S}\mathbf{r}\cdot\mathbf{r}_{,x} - \mathbf{P}^2\mathbf{r}_{,t}\cdot\mathbf{R}\mathbf{r}_{,t}.
\end{aligned}
$$

Under the given hypotheses on the matrices it is easily verified that

$$(4.7)\qquad |q_i| \le Ce \quad \text{for } i = 1, 2.$$

We note that

$$
(4.8)\quad \begin{cases}
\mathbf{Q}^2(\mathbf{r}_{,x}+\mathbf{R}\mathbf{r})\cdot\mathbf{r}_{,t} = \mathbf{Q}(\mathbf{r}_{,x}+\mathbf{R}\mathbf{r})\cdot\mathbf{Q}\mathbf{P}^{-1}[\mathbf{P}\mathbf{r}_{,t}] = \frac{1}{2}\mathbf{M}\mathbf{w}\cdot\mathbf{w}, \\
\mathbf{P}^2\mathbf{r}_{,t}\cdot(\mathbf{r}_{,x}+\mathbf{R}\mathbf{r}) = \mathbf{P}\mathbf{r}_{,t}\cdot\mathbf{P}\mathbf{Q}^{-1}[\mathbf{Q}(\mathbf{r}_{,x}+\mathbf{R}\mathbf{r})] = \frac{1}{2}\mathbf{N}\mathbf{w}\cdot\mathbf{w}, \\
e = \frac{1}{2}\left[|\mathbf{w}|^2 + \mathbf{S}\mathbf{r}\cdot\mathbf{r}\right],
\end{cases}
$$

where

$$\mathbf{w} = \begin{pmatrix} \mathbf{P}\mathbf{r}_{,t} \\ \mathbf{Q}(\mathbf{r}_{,x}+\mathbf{R}\mathbf{r}) \end{pmatrix}$$

and

$$\mathbf{M} = \begin{pmatrix} 0 & \mathbf{P}^{-1}\mathbf{Q} \\ \mathbf{Q}\mathbf{P}^{-1} & 0 \end{pmatrix}, \quad \mathbf{N} = \begin{pmatrix} 0 & \mathbf{P}\mathbf{Q}^{-1} \\ \mathbf{Q}^{-1}\mathbf{P} & 0 \end{pmatrix}.$$

One easily verifies that

$$(4.9)\qquad \mathbf{w}_{,t} = \mathbf{M}\mathbf{w}_{,x} + \mathbf{l}(\mathbf{r},\mathbf{r}_{,t},\mathbf{r}_{,x}),$$

where $\mathbf{l}(\mathbf{r},\mathbf{r}_{,t},\mathbf{r}_{,x})$ is equal to

$$\begin{pmatrix} -\mathbf{P}^{-1}[\mathbf{R}^t\mathbf{Q}^2(\mathbf{r}_{,x}+\mathbf{R}\mathbf{r}) + \mathbf{S}\mathbf{r} + \mathbf{P}_{,t}\mathbf{P}\mathbf{r}_{,t} - \mathbf{Q}_{,x}\mathbf{Q}(\mathbf{r}_{,x}+\mathbf{R}\mathbf{r})] + \mathbf{P}_{,t}\mathbf{r}_{,t} \\ \mathbf{Q}\mathbf{R}_{,t}\mathbf{r} + \mathbf{Q}\mathbf{R}\mathbf{r}_{,t} + \mathbf{Q}_{,t}(\mathbf{r}_{,x}+\mathbf{R}\mathbf{r}) - \mathbf{Q}\mathbf{P}^{-1}\mathbf{P}_{,x}\mathbf{r}_{,t} \end{pmatrix}.$$

Then $|\mathbf{l}(\mathbf{r},\mathbf{r}_{,t},\mathbf{r}_{,x})|^2 \le Ce$.

The estimates we wish to obtain involve the characteristics, which are defined in terms of the eigenvalues of $\mathbf{M}$. It is an easy exercise to show that these eigenvalues have the form $\pm\omega_k$ where $\{\omega_k^2\}_{k=1}^p$ are the eigenvalues of the positive definite matrix $\mathbf{P}^{-1}\mathbf{Q}^2\mathbf{P}^{-1}$ and the ω_k are positive. The eigenvalues will in general be functions of x and t. We denote the largest and the smallest of the ω_k by $\overline{\omega}$ and $\underline{\omega}$ respectively. We let $C_\pm^k(x_0, t_0)$ denote the characteristic curve $x = x_\pm^k(t)$ obtained by solving the system

$$(4.10)\qquad \frac{dx}{dt} = \pm\omega_k(x,t), \quad x(t_0) = x_0.$$

We let $\overline{C}_\pm(x_0,t_0)$ and $\underline{C}_\pm(x_0,t_0)$ denote the characteristics corresponding to $\pm\overline{\omega}$ and $\pm\underline{\omega}$ respectively; these are described by functions $\overline{x}_\pm(t)$ and

$\underline{x}_{\pm}(t)$. Note that the characteristic curves can also be described in terms of graphs $t = t_{\pm}^{k}(x)$ obtained by solving

$$\frac{dt}{dx} = \pm\frac{1}{\omega_k(x,t)}, \quad t(x_0) = t_0. \tag{4.11}$$

We shall obtain estimates by using the divergence theorem over certain regions in the x, t–plane bounded by characteristics and time-like or space-like intervals. Let us recall the notation

$$\mathcal{E}(t; x_1, x_2) = \int_{x_1}^{x_2} e(x,t)\,dx \qquad \mathcal{E}^{\dagger}(x; t_1, t_2) = \int_{t_1}^{t_2} e(x,t)\,dt.$$

We can formulate and prove various estimates in terms of these energy integrals. The first is a very familiar hyperbolic energy estimate.

LEMMA 4.1. *Let* $\mathbf{r}(x,t) \in W^2((0,l)\times(0,T); I\!R^p)$ *be a solution of* (4.1). *Suppose that* $0 \le x_1 \le x_2 \le l$. *Then if* $\overline{x}_+(t)$ *and* $\overline{x}_-(t)$ *correspond to the characteristics* $\overline{C}_+(x_1, 0)$ *and* $\overline{C}_-(x_2, 0)$, *respectively, one has, as long as* $\overline{x}_+(t) < \overline{x}_-(t)$,

$$\mathcal{E}(t; \overline{x}_+(t), \overline{x}_-(t)) \le C\mathcal{E}(0; x_1, x_2). \tag{4.12}$$

PROOF. We apply the divergence theorem to the equation (4.3) over the region $\Delta(t)$ bounded above and below by the intervals $\{t\}\times[\overline{x}_+(t), \overline{x}_-(t)]$ and $\{0\}\times[x_1, x_2]$ and laterally by the characteristics $\overline{C}_+(x_1, 0)$ and $\overline{C}_-(x_2, 0)$. One has

$$\iint\limits_{\Delta(t)} q_1(\mathbf{r}, \mathbf{r}_{,t}, \mathbf{r}_{,x})\,dx\,d\tau = \oint\limits_{\partial\Delta(t)} \boldsymbol{\nu}\cdot(-\frac{1}{2}\mathbf{M}\mathbf{w}\cdot\mathbf{w}, e)ds, \tag{4.13}$$

where $\boldsymbol{\nu}$ denotes the outward pointing unit normal to the boundary curve $\partial\Delta(t)$ and ds denotes the element of (positive) arc length along the curve. Along the top bounding interval one has $\boldsymbol{\nu} = (0,1)$, while along the base interval $\boldsymbol{\nu} = (0,-1)$. Along the lateral characteristics one has, respectively, $\boldsymbol{\nu} = [1+\overline{\omega}^2]^{-1/2}(\mp 1, \overline{\omega})$ and therefore, noting the third identity of (4.8), $\boldsymbol{\nu}\cdot(-\frac{1}{2}\mathbf{M}\mathbf{w}\cdot\mathbf{w}, e)$ is a positive multiple of

$$[\pm\frac{1}{2}\mathbf{M}\mathbf{w}\cdot\mathbf{w} + \overline{\omega}e] = \frac{1}{2}[\pm\mathbf{M}\mathbf{w}\cdot\mathbf{w} + \overline{\omega}|\mathbf{w}|^2 + \overline{\omega}\mathbf{S}\mathbf{w}\cdot\mathbf{w}].$$

By use of Schwarz's inequality, the fact that $|\mathbf{M}\mathbf{w}\cdot\mathbf{w}| \ge \overline{\omega}|\mathbf{w}|^2$ and the hypothesis that $\mathbf{S}$ vanishes or is positive definite, one concludes that along the lateral characteristic curve segments $\boldsymbol{\nu}\cdot(-\frac{1}{2}\mathbf{M}\mathbf{w}\cdot\mathbf{w}, e)$ is positive. If we write out the line integral in (4.15) as the sum of four integrals corresponding to the four different curve segments we can discard the positive contributions along the characteristic curves and obtain

$$\mathcal{E}(t;\overline{x}_+(t),\overline{x}_-(t)) \le \mathcal{E}(0;x_1,x_2) + \iint\limits_{\Delta(t)} q_1(\mathbf{r},\mathbf{r}_{,t},\mathbf{r}_{,x})\,dx\,d\tau.$$

In view of (4.7) it follows that

$$\mathcal{E}(t;\overline{x}_+(t),\overline{x}_-(t)) \le \mathcal{E}(0;x_1,x_2) + C\int_0^t \mathcal{E}(\tau;\overline{x}_+(\tau),\overline{x}_-(\tau))\,d\tau$$

for some suitable constant C. The estimate (4.14) then follows from Gronwall's lemma. □

To formulate the next technical lemma we need to introduce the orthogonal projections Π_+ and Π_- onto the direct sums of the eigenspaces corresponding, respectively, to positive and negative eigenvalues of the symmetric matrix $\mathbf{M}$. These are smooth functions of x and t. To see this one notes that the eigenvalues of $\mathbf{M}$ never vanish so that one can, for given x and t, use the representation

$$\Pi_\pm(x,t) = \frac{1}{2\pi i}\int_{\gamma_\pm} [z\mathbf{I} - \mathbf{M}(x,t)]^{-1}\,dz,$$

where $\gamma_\pm$ are contours surrounding the positive and negative eigenvalues of $\mathbf{M}$ respectively. The smoothness of the eigenvalues follows easily from this. We also introduce $\mathbf{w}_\pm = \Pi_\pm \mathbf{w}$, which have the same regularity as $\mathbf{w}$.

LEMMA 4.2. *Let $\mathbf{r}(x,t) \in W^2((0,l)\times(0,T);\mathbb{R}^p)$ be a solution of (4.1). Suppose that if $\mathbf{S}=0$, $\mathbf{P}$ is independent of t. Let $a \in [0,l]$ and $\overline{C}_\pm(a,T)$ be associated with the functions $\overline{x}_\pm(t)$ respectively. If a is equal to 0 or l only one of these characteristics is defined. If $\overline{x}_+(t)$ meets the initial interval $\{0\}\times[0,l]$ one has the estimates*

$$\text{(4.14)}\quad \int_0^T |\mathbf{w}_+(a,t)|^2\,dt \;\le\; C\left[\mathcal{E}(0;\overline{x}_+(0),a) + \int_0^T \mathcal{E}(\tau;\overline{x}_+(\tau),a)\,d\tau\right]$$

and

$$\text{(4.15)}\quad \int_0^T [\mathbf{S}\mathbf{r}\cdot\mathbf{r}](a,t)\,dt \le C\left[\mathcal{E}(0;\overline{x}_+(0),a) + \int_0^T \mathcal{E}(\tau;\overline{x}_+(\tau),a)\,d\tau\right].$$

Similarly, if $\overline{x}_-(t)$ meets the initial interval $\{0\}\times[0,l]$ one has

$$\text{(4.16)}\quad \int_0^T |\mathbf{w}_-(a,t)|^2\,dt \;\le\; C\left[\mathcal{E}(0;a,\overline{x}_-(0)) + \int_0^T \mathcal{E}(\tau;a,\overline{x}_-(\tau))\,d\tau\right]$$

and

$$\int_0^T [\mathbf{Sr}\cdot\mathbf{r}](a,t)\,dt \le C\left[\mathcal{E}(0;a,\overline{x}_-(0)) + \int_0^T \mathcal{E}(\tau;a,\overline{x}_-(\tau))\,d\tau\right]. \tag{4.17}$$

PROOF. It follows from (4.9) that

$$\frac{\partial \mathbf{w}_\pm}{\partial t} = \mathbf{M}\frac{\partial \mathbf{w}_\pm}{\partial x} + \mathbf{l}_\pm(\mathbf{r},\mathbf{r}_{,t},\mathbf{r}_{,x}), \tag{4.18}$$

with

$$\mathbf{l}_\pm = \Pi_\pm \mathbf{l} + \frac{\partial}{\partial t}\,\Pi_\pm \mathbf{w} - \frac{\partial}{\partial x}\,\Pi_\pm \mathbf{w}.$$

Now

$$\frac{\partial}{\partial t}|\mathbf{w}_\pm|^2 = \frac{\partial}{\partial x}[\mathbf{M}\mathbf{w}_\pm\cdot\mathbf{w}_\pm] + q_\pm(\mathbf{r},\mathbf{r}_{,t},\mathbf{r}_{,x}), \tag{4.19}$$

where $q_\pm = 2\mathbf{w}_\pm\cdot\mathbf{l}_\pm - \mathbf{M}_{,x}\mathbf{w}_\pm\cdot\mathbf{w}_\pm$ are quadratic functions of the arguments $\mathbf{r},\mathbf{r}_{,t}$ and $\mathbf{r}_{,x}$ satisfying $|q_\pm| \le Ce$.

To obtain the estimate (4.14) we integrate $q_+(\mathbf{r},\mathbf{r}_{,t},\mathbf{r}_{,x})$ over the region Δ_+ bounded by the intervals $[\overline{x}_+(0),a]\times 0$ and $a\times[0,T]$ along with the characteristic curve $\mathcal{C}_+(a,T)$. By using (4.19) and the divergence theorem we obtain

$$\iint_{\Delta_+} q_+(\mathbf{r},\mathbf{r}_{,t},\mathbf{r}_{,x})\,dx\,d\tau = \oint_{\partial\Delta_+} \boldsymbol{\nu}\cdot(-\mathbf{M}\mathbf{w}_+\cdot\mathbf{w}_+, |\mathbf{w}_+|^2)ds. \tag{4.20}$$

There are three contributions to the line integral. The contribution corresponding to the characteristic curve $\mathcal{C}_+(a,T)$ is positive since along that curve $\boldsymbol{\nu} = [1+\overline{\omega}^2]^{-1/2}(-1,\overline{\omega})$, so that $\boldsymbol{\nu}\cdot(-\mathbf{M}\mathbf{w}_+\cdot\mathbf{w}_+, |\mathbf{w}_+|^2)$ is a positive multiple of the positive function

$$(-1,\overline{\omega})\cdot(-\mathbf{M}\mathbf{w}_+\cdot\mathbf{w}_+, |\mathbf{w}_+|^2) = \mathbf{M}\mathbf{w}_+\cdot\mathbf{w}_+ + \overline{\omega}|\mathbf{w}_+|^2.$$

Discarding this term one is left with

$$\int_0^T [\mathbf{M}\mathbf{w}_+\cdot\mathbf{w}_+](a,t)\,dt - \int_{\overline{x}_+(0)}^a |\mathbf{w}_+(x,0)|^2\,dx \le \iint_{\Delta_+} q_+(\mathbf{r},\mathbf{r}_{,t},\mathbf{r}_{,x})\,dx\,d\tau.$$

Then (4.14) follows from $|\mathbf{w}_+|^2 \le e$, $|q_\pm| \le Ce$ and $\underline{\omega}|\mathbf{w}_+|^2 \le \mathbf{M}\mathbf{w}_+\cdot\mathbf{w}_+$.

The estimate (4.15) is vacuous if $\mathbf{S} = 0$; otherwise it is obtained by applying the divergence theorem over Δ_+ to

$$\frac{\partial}{\partial x}[\epsilon\mathbf{Sr}\cdot\mathbf{r}] + \frac{\partial}{\partial t}[\mathbf{Sr}\cdot\mathbf{r}] = q(\mathbf{r},\mathbf{r}_{,t},\mathbf{r}_{,x}),$$

where ϵ is chosen so that $\epsilon \le \overline{\omega}$. One has $q \le Ce$. Along the characteristic, $\boldsymbol{\nu}\cdot(\epsilon\mathbf{Sr}\cdot\mathbf{r}, \mathbf{Sr}\cdot\mathbf{r})$ is a positive multiple of the positive function $(-\epsilon+\overline{\omega})\mathbf{Sr}\cdot\mathbf{r}$.

Discarding the line integral over the characteristic curve segment one is left with

$$\epsilon \int_0^T [\mathbf{Sr}\cdot\mathbf{r}](a,t)\,dt - \int_{\overline{x}_+(0)}^{a} [\mathbf{Sr}\cdot\mathbf{r}](x,0)\,dx \le \iint_{\Delta_+} q(\mathbf{r},\mathbf{r}_{,t},\mathbf{r}_{,x})\,dx\,d\tau.$$

The estimate (4.15) is an immediate consequence.

The estimates (4.16) and (4.17) are proved in an analogous way. □

We can combine our previous results to obtain the following theorem.

THEOREM 4.1. *Let* $\mathbf{r}(x,t) \in W^2((0,l)\times(0,T);\mathbb{R}^p)$ *be a solution of* (4.1). *Let* $0 \le x_1 \le x_2 \le l$. *Let* $\overline{x}_+(t)$ *and* $\overline{x}_-(t)$ *correspond to the characteristics* $\overline{C}_+(x_1,0)$ *and* $\overline{C}_-(x_2,0)$, *respectively, and suppose that these characteristics do not cross for* $0 \le t < T$. *Let* $\overline{x}_+(T) \le a \le \overline{x}_+(T)$. *Then*

$$\mathcal{E}(T;\overline{x}_+(T),\overline{x}_-(T)) + \mathcal{E}^{\uparrow}(a;0,T) \le C\mathcal{E}(0;x_1,x_2). \tag{4.21}$$

PROOF. In view of lemma 4.1 it remains only to prove that

$$\mathcal{E}^{\uparrow}(a;0,T) \le C\mathcal{E}(0;x_1,x_2). \tag{4.22}$$

We note that $e = |\mathbf{w}_+|^2 + |\mathbf{w}_-|^2 + \mathbf{Sr}\cdot\mathbf{r}$. Note also that the characteristics $\overline{C}_\pm(a,T)$ lie between $\overline{C}_+(x_1,0)$ and $\overline{C}_-(x_2,0)$ so that it follows from the estimates (4.14), (4.15) and (4.16) that

$$\mathcal{E}^{\uparrow}(a;0,T) \le C\left[\mathcal{E}(0;x_1,x_2) + \int_0^T \mathcal{E}(\tau;\overline{x}_+(\tau),\overline{x}_-(\tau))\,d\tau\right].$$

Now (4.21) follows from (4.12). □

Note that the roles of x and t can be reversed by rewriting (4.9) as

$$\frac{\partial \mathbf{w}}{\partial x} = \mathbf{N}\frac{\partial \mathbf{w}}{\partial x} - \mathbf{Nl}(\mathbf{r},\mathbf{r}_{,t},\mathbf{r}_{,x}), \tag{4.23}$$

where we have used the easily verified fact that $\mathbf{N}$ is the inverse of $\mathbf{M}$. The latter fact also implies that the eigenvalue of $\mathbf{N}$ are the reciprocals of those of $\mathbf{M}$, so that, in particular, the largest eigenvalue of $\mathbf{N}$ is $1/\underline{\omega}$. The characteristics are also the same. We rewrite Theorem 4.1 in the form that is needed for proofs of exact controllability.

THEOREM 4.2. *Let* $\mathbf{r}(x,t) \in W^2((0,l)\times(0,T);\mathbb{R}^p)$ *be a solution of* (4.1). *Let* $0 \le t_1 \le t_2 \le T$. *Let* $\underline{x}_+(t)$ *and* $\underline{x}_-(t)$ *correspond to the characteristics* $\underline{C}_+(l,t_2)$ *and* $\underline{C}_-(l,t_1)$ *respectively and suppose that these characteristics do not cross for* $0 \le x < l$. *Let* $\underline{x}_-(0) \le a \le \overline{x}_+(0)$. *Then*

$$\mathcal{E}^{\uparrow}(0;\overline{x}_+(T),\overline{x}_-(T)) + \mathcal{E}(a;0,l) \le C\mathcal{E}^{\uparrow}(l;t_1,t_2). \tag{4.24}$$

A similar estimate holds if we interchange the roles of 0 *and* l *throughout and use the characteristics* $\underline{C}_+(0,t_1)$ *and* $\underline{C}_-(0,t_2)$.

In case $\mathbf{P}$ and $\mathbf{Q}$ are independent of t, both $\mathbf{N}$ and its eigenvalues $1/\omega_k$ are functions of x alone. The characteristics $\underline{\mathcal{C}}_+(l, t_2)$ and $\underline{\mathcal{C}}_-(l, t_1)$ are then given by

$$t = t_2 - \int_x^l \frac{1}{\underline{\omega}(y)}\,dy \quad \text{and} \quad t = t_1 + \int_x^l \frac{1}{\underline{\omega}(y)}\,dy.$$

Taking this into account we obtain immediately a variant on the previous theorem.

THEOREM 4.3. *Let* $\mathbf{r}(x,t) \in W^2((0,l)\times(0,T); I\!R^p)$ *be a solution of* (4.1), *where* $\mathbf{P}$ *and* $\mathbf{Q}$ *are assumed to be independent of* t. *Suppose that* $t_2 - t_1 \geq 2S_\star$, *where*

$$S_\star = \int_0^l \frac{1}{\underline{\omega}(x)}\,dx. \tag{4.25}$$

Then for any $t \in [t_1 + S_\star, t_2 - S_\star]$,

$$\mathcal{E}^\dagger(0; t_1 + S_\star, t_2 - S\star) + \mathcal{E}(t; 0, l) \leq C\mathcal{E}^\dagger(l; t_1, t_2). \tag{4.26}$$

The same estimate holds with the roles of 0 *and* l *reversed.*

We can use our methods to derive another estimate which plays a complementary role to the estimates of the previous two theorems.

THEOREM 4.4. *Let* $\mathbf{r}(x,t) \in W^2((0,l)\times(0,T); I\!R^p)$ *be a solution of the equation* (4.1) *which satisfies the boundary conditions*

$$\begin{cases} \dot{\mathbf{r}}(0,t) = \mathbf{0} & \text{or} \quad [\mathbf{r}_{,x} + \mathbf{R}\mathbf{r}](0,t) = \mathbf{0}, \\ \dot{\mathbf{r}}(l,t) = \mathbf{0} & \text{or} \quad [\mathbf{r}_{,x} + \mathbf{R}\mathbf{r}](l,t) = \mathbf{0} \end{cases}$$

for $t \in (0,T)$. *Then*

$$\mathcal{E}^\dagger(0; 0, T) + \mathcal{E}^\dagger(l; 0, T) \leq C\mathcal{E}(0). \tag{4.27}$$

PROOF. This theorem depends on Lemma 4.2 and is somewhat subtle.

First we note that an easy computation, taking into account the boundary conditions, gives

$$\frac{d}{dt}\mathcal{E}(t) \leq C\mathcal{E}(t),$$

from which it follows that

$$\mathcal{E}(t) \leq C\mathcal{E}(0) \quad \text{for } t \in [0,T]. \tag{4.28}$$

It is therefore enough to show that

$$\mathcal{E}^\dagger(0; t, t+\delta) + \mathcal{E}^\dagger(l; t, t+\delta) \leq C\mathcal{E}(t) \tag{4.29}$$

for each t and some corresponding positive δ. We do this for $t = 0$. In fact we shall estimate the term $\mathcal{E}^\dagger(l; 0, \delta)$; the estimate of the term at the endpoint 0 is similar. Now in Lemma 4.2 we set $a = l$ and $T = \delta$ chosen sufficiently small to ensure that $\overline{C}_+(l,\delta)$ meets the initial interval. Then it follows from the inequalities (4.14), (4.15) and (4.28) that

$$\int_0^\delta \left[|\mathbf{w}_+(l,t)|^2 + \mathbf{Sr}\cdot\mathbf{r}\right] dt \le C\mathcal{E}(0). \tag{4.30}$$

The final step is to see how $|\mathbf{w}_+(l,t)|^2$ is related to $|\mathbf{w}(l,t)|^2 = |\mathbf{Pr}_{,t}|^2 + |\mathbf{Q}(\mathbf{r}_{,x}+\mathbf{Rr})|^2$. Note first that, because of the boundary conditions, one or other of the terms in the last expression vanishes. We have to look more closely at the projection defining $\mathbf{w}_+ = \Pi_+\mathbf{w}$. In order to do this pick an orthonormal basis $\{\mathbf{f}^k\}_{k=1}^p$ of $I\!R^p$ consisting of eigenvectors of $\mathbf{P}^{-1}\mathbf{Q}^2\mathbf{P}^{-1}$ corresponding to the eigenvalues $\{\omega_k^2\}_{k=1}^p$. Then one easily verifies that one has an orthonormal basis of eigenvectors $\{\mathbf{F}_+^k\}_{k=1}^p \cup \{\mathbf{F}_-^k\}_{k=1}^p$, where $F_\pm^k = \alpha_k(\mathbf{f}_k, \pm\omega_k\mathbf{Q}^{-1}\mathbf{P}\mathbf{f}_k)\}_{k=1}^p$ with suitable normalizing constants α_k. The first set of eigenfunctions corresponds to the positive eigenvalues ω_k while the second set corresponds to negative eigenvalues $-\omega_k$. Now, if we assume the boundary condition $\mathbf{w}(l,t) = \mathbf{0}$ and remember that $\mathbf{w} = (\mathbf{Pr}_{,t}, \mathbf{Q}(\mathbf{r}_{,x}+\mathbf{Rr}))$ we find

$$\begin{aligned}
\mathbf{w}_+(l,t) = \Pi_+\mathbf{w}(l,t) &= \sum_{k=1}^p \left[\mathbf{w}(l,t)\cdot\mathbf{F}^k\right]\mathbf{F}^k \\
&= \sum_{k=1}^p \left[\alpha_k\omega_k[\mathbf{Q}(\mathbf{r}_{,x}+\mathbf{Rr})](l,t)\cdot\mathbf{Q}^{-1}\mathbf{P}\mathbf{f}_k\right]\mathbf{F}^k \\
&= \sum_{k=1}^p \left[\alpha_k\omega_k[\mathbf{P}(\mathbf{r}_{,x}+\mathbf{Rr})](l,t)\cdot\mathbf{f}_k\right]\mathbf{F}^k.
\end{aligned}$$

It follows that

$$\begin{aligned}
|\mathbf{w}_+(l,t)|^2 &= \sum_{k=1}^p |\alpha_k\omega_k[\mathbf{P}(\mathbf{r}_{,x}+\mathbf{Rr})](l,t)\cdot\mathbf{f}_k|^2 \\
&\ge c|[\mathbf{Q}(\mathbf{r}_{,x}+\mathbf{Rr})](l,t)|^2 = c|\mathbf{w}(l,t)|^2
\end{aligned} \tag{4.31}$$

where

$$c = \min_{1\le k\le p} |\alpha_k\omega_k|^2 \;\; \|\mathbf{QP}^{-1}\|^2,$$

with the norm being the operator norm on matrices. From (4.30) and (4.31) we get the desired estimate on $\mathcal{E}^\dagger(l:0,\delta)$, which completes the proof. □

REMARK 4.1. When the coefficients are not time dependent this result, as well as a corresponding result for networks, follows directly from Lemma 3.3. When needed, the above result is also readily generalized to networks and allows for time dependence of the coefficients. Theorems 4.2 and 4.4 together provide the estimates which should allow the application of the Hilbert Uniqueness method (as described in Lions [60]) to hyperbolic systems with time dependent coefficients.

5. Exact Controllability of the Network Model.

The primal problem we wish to consider is the following:

$$
(5.1)\quad \begin{cases}
\mathbf{P}_i^2\ddot{\mathbf{r}}^i = \left[\mathbf{Q}_i^2(\mathbf{r}^i_{,x} + \mathbf{R}_i\mathbf{r}^i)\right]_{,x} \\
\qquad\qquad - \mathbf{R}_i^t\mathbf{Q}_i^2(\mathbf{r}^i_{,x} + \mathbf{R}_i\mathbf{r}^i) + \mathbf{S}_i\mathbf{r}^i \;\text{ for } i \in \mathcal{I}, \\
\mathbf{r}^i(x_{ij}, t) \text{ coincide for } j \in \mathcal{J}_M,\ i \in \mathcal{I}_j, \\
\mathcal{D}^j\mathbf{r}(t) = \mathbf{u}^j(t) \text{ for } j \in \mathcal{J}^D, \quad \mathcal{D}^j\mathbf{r}(t) = \mathbf{0} \text{ for } j \in \mathcal{J}^C, \\
\mathcal{N}^j\mathbf{r}(t) = \mathbf{u}^j(t) \text{ for } j \in \mathcal{J}^N, \quad \mathcal{N}^j\mathbf{r}(t) = \mathbf{0} \text{ for } j \in \mathcal{J}^F, \\
\mathcal{N}_i^j\mathbf{r}(t) = \mathbf{0} \text{ for } j \in \mathcal{J} \text{ and } i \in \mathcal{I}_j, \\
\mathbf{r}(0) = \mathbf{0}, \ \dot{\mathbf{r}}(0) = \mathbf{0}.
\end{cases}
$$

Here we make the same assumptions on the matrix coefficients as in Section 2.3; in particular they do not depend on t. We use the same notation for the spaces of controls as in Section 5, Chapter II, namely,

$$\mathbf{U}^D = \prod_{j\in\mathcal{J}^D} H^1((0,T];\mathbb{R}^{q_j}), \ \text{ and } \ \mathbf{U}^N = \prod_{j\in\mathcal{J}^N} H^1((0,T];\mathbb{R}^{q_j}).$$

We denote elements of these spaces by $\mathbf{u}^D$ and $\mathbf{u}^N$ respectively, and use the inner products

$$(\mathbf{u}^D, \mathbf{v}^D)_{\mathbf{U}^D} = \sum_{j\in\mathcal{J}^D} \int_0^T \dot{\mathbf{u}}^j(t)\cdot\dot{\mathbf{v}}^j(t)\,dt$$

and

$$(\mathbf{u}^N, \mathbf{v}^N)_{\mathbf{U}^N} = \sum_{j\in\mathcal{J}^N} \int_0^T \mathbf{u}^j(t)\cdot\mathbf{v}^j(t)\,dt.$$

As before we let $\mathbf{U} = \mathbf{U}^D \times \mathbf{U}^N$ with elements $\mathbf{u} = \mathbf{u}^D \times \mathbf{u}^N$.

Because of Theorem 3.1 we can then define the control to state map $\mathcal{C}_T : \mathbf{U} \mapsto \mathcal{H}$ by $\mathcal{C}_T\mathbf{u} = \mathbf{r}(T) \times \dot{\mathbf{r}}(T)$ where $\mathbf{r}$ is the solution of (5.1). The surjectivity of the bounded linear operator $\mathcal{C}_T$ (i.e. the exact controllability of the networks) is then equivalent to the following estimate on the adjoint operator $\mathcal{C}_T^\star : \mathcal{H} \mapsto \mathbf{U}$:

$$(5.2)\quad \|\boldsymbol{\lambda}_0 \times \dot{\boldsymbol{\lambda}}_0\|_{\mathcal{H}} \le C\,\|\mathcal{C}_T^\star[\boldsymbol{\lambda}_0 \times \dot{\boldsymbol{\lambda}}_0]\|_{\mathbf{U}}, \quad \text{for all } \boldsymbol{\lambda}_0 \times \dot{\boldsymbol{\lambda}}_0 \in \mathcal{H}.$$

As in Section 3, Chapter II, $\mathcal{H}$ and $\mathbf{U}$ have been identified with their dual spaces. We need to identify $\mathcal{C}_T^\star[\boldsymbol{\lambda}_0 \times \dot{\boldsymbol{\lambda}}_0]$, which depends on the of inner products for $\mathbf{U}$ and $\mathcal{H}$.

The adjoint system will be

$$
\text{(5.3)}\qquad
\begin{cases}
\mathbf{P}_i^2\ddot{\boldsymbol{\lambda}}^i = \left[\mathbf{Q}_i^2(\boldsymbol{\lambda}^i_{,x} + \mathbf{R}_i\boldsymbol{\lambda}^i)\right]_{,x} \\
\qquad\qquad - \mathbf{R}_i^t\mathbf{Q}_i^2(\boldsymbol{\lambda}^i_{,x} + \mathbf{R}_i\boldsymbol{\lambda}^i) + \mathbf{S}_i\boldsymbol{\lambda}_i \ \text{ for } i \in \mathcal{I}, \\
\boldsymbol{\lambda}^i(x_{ij}, t) \ \text{ coincide for } j \in \mathcal{J}_M,\ i \in \mathcal{I}_j, \\
\mathcal{D}^j\boldsymbol{\lambda}(t) = \mathbf{0} \ \text{ for } j \in \mathcal{J}^D \cup \mathcal{J}^C, \\
\mathcal{N}^j\boldsymbol{\lambda}(t) = \mathbf{0}(t) \ \text{ for } j \in \mathcal{J}^N \cup \mathcal{J}^F, \\
\mathcal{N}_i^j\boldsymbol{\lambda}(t) = \mathbf{0} \ \text{ for } j \in \mathcal{J} \text{ and } i \in \mathcal{I}_j, \\
\boldsymbol{\lambda}(T) = \boldsymbol{\lambda}_0, \ \ \dot{\boldsymbol{\lambda}}(T) = \dot{\boldsymbol{\lambda}}_0.
\end{cases}
$$

We then have the following characterization of $\mathcal{C}_T^\star$.

PROPOSITION 5.1. *Suppose that $\|\cdot\|_{\mathbf{V}}$ is a norm on $\mathbf{V}$ (which is guaranteed, for example, by the hypotheses of Lemma 3.1). Let $\boldsymbol{\lambda}(t)$ be the solution of* (4.3) *corresponding to terminal data $\boldsymbol{\lambda}_0 \times \dot{\boldsymbol{\lambda}}_0 \in \mathcal{H}$. Then*

$$
\text{(5.4)}\qquad \mathcal{C}_T^\star[\boldsymbol{\lambda}_0 \times \dot{\boldsymbol{\lambda}}_0] = \left\{\int_0^t \mathcal{N}^j\boldsymbol{\lambda}(\tau)\,d\tau\right\}_{j\in\mathcal{J}^D} \times \left\{\mathcal{D}^j\dot{\boldsymbol{\lambda}}(t)\right\}_{j\in\mathcal{J}^N},
$$

where $\boldsymbol{\lambda}$ is the solution of the adjoint system (5.3).

PROOF. The hypothesis allows us to use $(\cdot,\cdot)_{\mathbf{V}}$ as inner product on $\mathbf{V}$-components of elements of $\mathcal{H}$. As in the proof of Lemma 3.3, we evaluate

$$
\begin{aligned}
&\frac{d}{dt}\left[(\mathbf{r},\boldsymbol{\lambda})_{\mathbf{V}} + (\dot{\mathbf{r}},\dot{\boldsymbol{\lambda}})_{\mathbf{H}}\right] \\
&\quad = \sum_{i\in\mathcal{I}} \int_0^{l_i} \frac{1}{2}\Big[\mathbf{P}_i^2\ddot{\mathbf{r}}^i \cdot \dot{\boldsymbol{\lambda}}^i + \mathbf{Q}_i^2(\mathbf{r}^i_{,x} + \mathbf{R}_i\mathbf{r}^i)\cdot(\dot{\boldsymbol{\lambda}}^i_{,x} + \mathbf{R}_i\dot{\boldsymbol{\lambda}}^i) + \mathbf{S}_i\mathbf{r}^i\cdot\dot{\boldsymbol{\lambda}}^i \\
&\qquad\qquad + \mathbf{P}_i^2\dot{\mathbf{r}}^i\cdot\ddot{\boldsymbol{\lambda}}_i + \mathbf{Q}_i^2(\dot{\mathbf{r}}^i_{,x} + \mathbf{R}_i\dot{\mathbf{r}}^i)\cdot(\boldsymbol{\lambda}^i_{,x} + \mathbf{R}_i\boldsymbol{\lambda}_i) + \mathbf{S}_i\dot{\mathbf{r}}^i\cdot\boldsymbol{\lambda}_i\Big]\,dx \\
&\quad = \sum_{i\in\mathcal{I}} \left[\mathbf{Q}_i^2(\mathbf{r}^i_{,x} + \mathbf{R}_i\mathbf{r}^i)\cdot\dot{\boldsymbol{\lambda}}^i + \dot{\mathbf{r}}^i\cdot\mathbf{Q}_i^2(\boldsymbol{\lambda}^i_{,x} + \mathbf{R}_i\boldsymbol{\lambda}^i)\right]_{x=0}^{x=l_i}
\end{aligned}
$$

where we have integrated by parts with respect to x. Thus, taking into account

$$
\begin{aligned}
\dot{\mathbf{r}}^i(x_{ij},\tau) &= \mathbf{C}_{ij}^+\mathcal{D}^j\dot{\mathbf{r}}(\tau) + \boldsymbol{\Pi}_{ij}\dot{\mathbf{r}}^i(x_{ij},\tau), \\
\dot{\boldsymbol{\lambda}}^i(x_{ij},\tau) &= \mathbf{C}_{ij}^+\mathcal{D}^j\dot{\boldsymbol{\lambda}}(\tau) + \boldsymbol{\Pi}_{ij}\dot{\boldsymbol{\lambda}}^i(x_{ij},\tau),
\end{aligned}
$$

along with the various nodal conditions in (5.1) and (5.3), we obtain

$$
\begin{aligned}
\frac{d}{dt}(\mathbf{r} \times \dot{\mathbf{r}}, \boldsymbol{\lambda} \times \dot{\boldsymbol{\lambda}})_{\mathcal{H}} &= \sum_{j \in \mathcal{J}} \left[\mathcal{D}^j \dot{\boldsymbol{\lambda}} \cdot \mathcal{N}^j \mathbf{r} + \mathcal{N}^j \boldsymbol{\lambda} \cdot \mathcal{D}^j \dot{\mathbf{r}} \right] \\
&= \sum_{j \in \mathcal{J}^D} \mathcal{N}^j \boldsymbol{\lambda} \cdot \dot{\mathbf{u}}^j + \sum_{j \in \mathcal{J}^N} \mathcal{D}^j \dot{\boldsymbol{\lambda}} \cdot \mathbf{u}^j .
\end{aligned} \tag{5.5}
$$

Upon integrating with respect to t from 0 to T one finds

$$
(\mathcal{C}_T \mathbf{u}, \boldsymbol{\lambda}_0 \times \dot{\boldsymbol{\lambda}}_0)_{\mathcal{H}} = \int_0^T \Big[\sum_{j \in \mathcal{J}^D} \mathcal{N}^j \boldsymbol{\lambda} \cdot \dot{\mathbf{u}}^j + \sum_{j \in \mathcal{J}^N} D^j \dot{\boldsymbol{\lambda}} \cdot \mathbf{u}^j \Big] \, dt.
$$

Hence (5.4) follows by Riesz identification. □

Because of this lemma the estimate (5.2), characterizing exact controllability in time T, becomes

$$
\| \boldsymbol{\lambda}_0 \times \dot{\boldsymbol{\lambda}}_0 \|_{\mathcal{H}}^2 \leq C \int_0^T \Big[\sum_{j \in \mathcal{J}^D} |\mathcal{N}^j \boldsymbol{\lambda}(t)|^2 + \sum_{j \in \mathcal{J}^N} |\mathcal{D}^j \dot{\boldsymbol{\lambda}}(t)|^2 \Big] \, dt. \tag{5.6}
$$

This nontrivial estimate will again be established, as in Section 5, Chapter II, for tree networks with one fully clamped node and for T larger than a certain critical time $T_\star$, for which we will give an explicit expression. The counterexamples of that section of course tell us that we cannot expect a better result.

We shall need two lemmas. The first one is a reformulation of Theorem 4.3 with $S_\star$ replaced by $S_\star^i$, $t = T/2$, $t_1 = S$ and $t_2 = T - S$.

LEMMA 5.1. *Let $\mathbf{r}^i \in W^2((0, l_i) \times (0, T); \mathbb{R}^{p_i})$ be a solution of the equation of the form (4.1) for the i-th string and suppose that $T \geq T_\star^i = 2S_\star^i$, where*

$$
S_\star^i = \int_0^{l_i} \frac{1}{\underline{\omega}_i(x)} dx. \tag{5.7}
$$

Then there exists a constant C_i such that

$$
E_i(\frac{T}{2}) + E_i^\uparrow(0; S + S_\star^i, T - S - S_\star^i) \leq C_i E_i^\uparrow(l_i; S, T - S) \tag{5.8}
$$

for all nonnegative S satisfying $2S + T_\star^i \leq T$. The same inequality holds with 0 and l_i interchanged.

A second lemma is a consequence of the geometric and dynamic conditions which hold at multiple nodes.

LEMMA 5.2. *Suppose that $q_j = p_i = p$ for all $i = 1, \cdots, n$ and $j \in \mathcal{J}_M$, in which case the matrices $\mathbf{C}_{ij}$ are invertible. Let $\mathbf{r}$ be a solution of the system (5.1) with the initial condition replaced by the requirement $\mathbf{r}(0) \times \dot{\mathbf{r}}(0) \in \mathcal{H}$. Then for $j \in \mathcal{J}_M$ and $i_0 \in \mathcal{I}_j$ one has the estimate*

$$E_{i_0}^{\dagger}(x_{i_0 j}; t_1, t_2) \le C \sum_{i \in \mathcal{I}_j - \{i_0\}} E_i^{\dagger}(x_{ij}; t_1, t_2), \tag{5.9}$$

where C is a suitable constant.

PROOF. Under the hypothesis of the lemma the node condition (8) is redundant and, in view of (7) and the positive definiteness of each $\mathbf{Q}_i$, it follows that

$$|\mathbf{Q}_{i_0}[\mathbf{r}_{i_0}' + \mathbf{R}_{i_0}\mathbf{r}_{i_0}](x_{i_0 k}, t)|^2 \le C \sum_{i \in \mathcal{I}_j - \{i_0\}} |\mathbf{Q}_i^2[\mathbf{r}_i' + \mathbf{R}_i\mathbf{r}_i](x_{ij}, t)|^2.$$

From (1) it follows that

$$|\mathbf{P}_{i_0}\dot{\mathbf{r}}^{i_0}(x_{i_0 j}, t)|^2 \le C|\mathbf{P}_i\dot{\mathbf{r}}^i(x_{ij}, t)|^2 \text{ for any } i \in \mathcal{I}_j - \{i_0\}.$$

From these estimates (49) is immediate. □

REMARK 5.1. The estimate (5.9) is valid under the assumption that $q_j = p_{i_0} \le p_i$ for $i \in \mathcal{I}_j$.

The statement and proof of the next main theorem is very close to that of Theorem 5.2, Chapter II and depends on the previous two lemmas.

THEOREM 5.1. *Suppose that the network contains no closed circuit, that all multiple nodes and no simple nodes are free, that exactly one simple node (indexed by $j = m$) is fully clamped and that the remaining simple nodes are controlled either geometrically or dynamically. In the interests of simplicity suppose also that $p_i = q_j$ for all i and j, in which case the matrices $\mathbf{C}_{ij}$ are all invertible. Let $\mathbf{r}$ be a solution of the system (5.1) with the homogeneous initial conditions replaced by the requirement $\mathbf{r}_0 \times \dot{\mathbf{r}}_0 \in \mathcal{H}$. Then*

$$\mathcal{E}(\frac{T_\star}{2}) = \sum_{i=1}^{n} \mathcal{E}_i(\frac{T_\star}{2}) \le C \sum_{j \in \mathcal{J}_S \setminus \{m\}} \mathcal{E}_{i_j}^{\dagger}(x_j; 0, T_\star), \tag{5.10}$$

where

$$T_\star = \max\{T_{i_1} + T_{i_1} + \cdots + T_{i_r} : [i_1, i_2, \ldots, i_r] \textit{ leads from the clamped node to a controlled node}\}. \tag{5.11}$$

REMARK 5.2. The result also holds for more general networks. For example it also holds if for each path $[i_1, i_2, \ldots, i_r]$ which leads from the clamped node to a controlled node the following condition is satisfied: letting j be the index of the multiple node between the i_k-th and the i_{k+1}-th elements, one has $q_j = p_{i_j} \le p_{i_{j+1}}$.

Finally we state the result on exact controllability, which, as before, follows immediately from the previous theorem.

THEOREM 5.2. *Under the hypotheses of the previous theorem, the network is exactly controllable for each time $T \geq T_\star$.*

REMARK 5.3. The general model allows for elements of different nature such as the networks of one beam and two strings introduced in Subsection 2.5. We can adapt the reasoning of this section to consider the controllability of those networks. If one end of the beam is clamped and the strings are joined to the extreme points of the other endface, the motion of the whole system can indeed be exactly controlled from the remaining endpoints of the strings. In the other cases the beam cannot be controlled by control of the strings. The key element is the way in which energy is transmitted through the multiple nodes; only in the one indicated case is the conclusion of Lemma 5.2 valid; the main argument in deriving the energy estimate goes through easily once that lemma is established.

6. Stabilizability of the Network Model.

We return to the question of nonlinear feedbacks with a view to stabilizing the vibrations of a general network as modeled in this chapter. We make some simplifying, but unessential assumptions. In particular we assume that $q_j = p_i = p$ for all i and j and that the m-th node is clamped. We then consider the following *nonlinear velocity feedback* system:

$$(6.1)\qquad \begin{cases} \mathbf{P}_i^2\ddot{\mathbf{r}}^i = \left[\mathbf{Q}_i^2(\mathbf{r}^i_{,x} + \mathbf{R}_i\mathbf{r}^i)\right]_{,x} \\ \qquad\qquad - \mathbf{R}_i^t\mathbf{Q}_i^2(\mathbf{r}^i_{,x} + \mathbf{R}_i\mathbf{r}^i) + \mathbf{S}_i\mathbf{r}_i \;\text{ for } i \in \mathcal{I}, \\ \mathbf{r}^i(x_{ij}, t) \;\text{ coincide for } j \in \mathcal{J}_M,\ i \in \mathcal{I}_j, \\ \boldsymbol{\mathcal{D}}^m\mathbf{r}(t) = \mathbf{0}, \\ -\boldsymbol{\mathcal{N}}^j\mathbf{r}(t) \in \mathbf{g}^j(\boldsymbol{\mathcal{D}}^j\dot{\mathbf{r}}(t)) \text{ for } j \in \mathcal{J}_S \setminus \{m\}, \\ \boldsymbol{\mathcal{N}}^j\mathbf{r}(t) = \mathbf{0} \text{ for } j \in \mathcal{J}_M, \\ \mathbf{r}(0) = \mathbf{r}_0,\ \ \dot{\mathbf{r}}(0) = \dot{\mathbf{r}}_0. \end{cases}$$

Here each $\mathbf{g}^j(\mathbf{u})$ will be a maximal, monotone function obtained as the subdifferential $\partial\mathbf{G}^j(\mathbf{u})$ of a convex function $\mathbf{G}^j(\mathbf{u}) : I\!R^p \mapsto I\!R$ taking its minimum value at $\mathbf{0}$. From (3.12) it then follows that for any sufficiently regular solution of (6.1)

$$(6.2)\qquad \begin{aligned} \frac{d\mathcal{E}}{dt}(t) &= \sum_{j\in\mathcal{J}_S\setminus\{m\}} \boldsymbol{\mathcal{N}}^j\mathbf{r}(t)\cdot\boldsymbol{\mathcal{D}}^j\dot{\mathbf{r}}(t) \\ &\quad \text{with } -\boldsymbol{\mathcal{N}}^j\mathbf{r}(t) \in \mathbf{g}^j(\boldsymbol{\mathcal{D}}^j\dot{\mathbf{r}}(t)). \end{aligned}$$

Since necessarily $\mathbf{0} \in \mathbf{g}^j(\mathbf{0})$, monotonicity implies

$$[-\mathcal{N}^j \mathbf{r}(t) - \mathbf{0}] \cdot [\mathcal{D}^j \dot{\mathbf{r}}(t) - \mathbf{0}] \geq 0,$$

so that it follows from (6.2) that the energy is decreasing monotonely.

Our two main tasks are to establish the existence of solutions to the feedback system (6.) and to find conditions on the functions $\mathbf{g}^j$ which imply that the energy actually decays to zero, preferably with a specified decay rate. We follow closely Conrad and Pierre [20] who provide an abstract setting for such problems (see also Lasiecka [53]). For general facts concerning convex functions we refer to Ekeland and Temam [28], while extensive information on maximal monotone operators and contraction semigroups may be found in Brezis [11] (noting in particular [11, Theorem 3.1]).

THEOREM 6.1. *Let the maximal monotone functions* $\mathbf{g}^j$ *be as described above. Let* $\operatorname{dom}(\mathcal{A}_s) \subset \mathbf{V} \times \mathbf{H}$ *consist of all* $\mathbf{r} \times \dot{\mathbf{r}}$ *with* $\dot{\mathbf{r}} \in \mathbf{V}$ *and* $\mathbf{r}$ *satisfying*

$$(6.3) \qquad \begin{cases} \textit{there exists } \mathbf{f} \in \mathbf{H} \textit{ such that} \\ (\mathbf{r}, \phi)_{\mathbf{V}} = (\mathbf{f}, \phi)_{\mathbf{H}} + \sum_{j \in \mathcal{J}^S \setminus \{m\}} \mathbf{u}^j \cdot \mathcal{D}^j \phi \\ \textit{for all } \phi \in \mathbf{V}, \textit{ with } -\mathbf{u}^j \in \mathbf{g}^j(\mathcal{D}^j \dot{\mathbf{r}}). \end{cases}$$

For $\mathbf{r} \times \dot{\mathbf{r}} \in \operatorname{dom}(\mathcal{A}_s)$ *define* $\mathcal{A}_s$ *by setting* $\mathcal{A}_s[\mathbf{r} \times \dot{\mathbf{r}}] = -\dot{\mathbf{r}} \times \mathbf{f}$. *Then* $\mathcal{A}_s$ *is a single valued, maximal monotone operator on* $\mathcal{H}$ *which has dense domain. Thus* $-\mathcal{A}_s$ *generates a contraction semigroup* $\{\mathcal{U}_s(t)\}_{t \geq 0}$ *on* $\mathcal{H}$ *so that* $\mathcal{U}_s(t)[\mathbf{r}_0 \times \dot{\mathbf{r}}_0]$ *provides a solution to the system* (6.1). *Moreover,* $\mathcal{A}_s$ *has compact resolvent.*

PROOF. It helps ones intuition to note that (6.3) is just the weak form of

$$-\mathbf{A}\mathbf{r} = \mathbf{f}, \quad \mathcal{D}^m \mathbf{r} = \mathbf{0}, \quad -\mathcal{N}^j \mathbf{r} \ (= -\mathbf{u}^j) \ \in \mathbf{g}^j(\mathcal{D}^j \dot{\mathbf{r}}) \text{ for } j \in \mathcal{J}^S \setminus \{m\}.$$

For a given $\mathbf{r} \times \dot{\mathbf{r}}$ there can exist at most one $\mathbf{f}$ in condition (6.3), so that $\mathcal{A}_s$ is well defined as a single valued operator. To see this, let $\mathbf{V}_0$ denote the subspace of $\mathbf{V}$ whose elements satisfy $\mathcal{D}^j \mathbf{r} = \mathbf{0}$ for $j \in \mathcal{J}^S \setminus \{m\}$. Then $\mathbf{V}_0$ is dense in $\mathbf{H}$. If $\mathbf{f}$ and $\tilde{\mathbf{f}}$ both satisfy the conditions of (6.3), the difference of the two weak identities with $\phi \in \mathbf{V}_0$ gives $(\mathbf{f} - \tilde{\mathbf{f}}, \phi)_{\mathbf{H}} = 0$, from which it follows that $\mathbf{f} = \tilde{\mathbf{f}}$. We usually write $-\mathbf{A}\mathbf{r}$ in place of $\mathbf{f}$ so that, in particular, $\mathcal{A}_s[\mathbf{r} \times \dot{\mathbf{r}}] = -\dot{\mathbf{r}} \times (-\mathbf{A}\mathbf{r})$.

The monotonicity of $\mathcal{A}_s$ is immediate: for $\mathbf{r} \times \dot{\mathbf{r}}$ and $\mathbf{s} \times \dot{\mathbf{s}}$ in $\operatorname{dom}(\mathcal{A}_s)$

one has

$$\begin{aligned}
(\mathcal{A}_s[\mathbf{r}\times\dot{\mathbf{r}}]-&\mathcal{A}_s[\mathbf{s}\times\dot{\mathbf{s}}],\mathbf{r}\times\dot{\mathbf{r}}-\mathbf{s}\times\dot{\mathbf{s}})_{\mathcal{H}}\\
&=(-\dot{\mathbf{r}}+\dot{\mathbf{s}},\mathbf{r}-\mathbf{s})_{\mathbf{V}}+(-\mathbf{A}\mathbf{r}+\mathbf{A}\mathbf{s},\dot{\mathbf{r}}-\dot{\mathbf{s}})_{\mathbf{H}}\\
&=-\sum_{j\in\mathcal{J}^S\setminus\{m\}}[\mathbf{u}^j-\mathbf{v}^j]\cdot\mathcal{D}^j[\dot{\mathbf{r}}-\dot{\mathbf{s}}]\\
&=\sum_{j\in\mathcal{J}^S\setminus\{m\}}[(-\mathbf{u}^j)-(-\mathbf{v}^j)]\cdot[\mathcal{D}^j\dot{\mathbf{r}}-\mathcal{D}^j\dot{\mathbf{s}}]\\
&\geq 0,
\end{aligned}$$

where we have used the weak identity in (6.3) for $\mathbf{r}$ and $\mathbf{s}$ with $\phi=\dot{\mathbf{r}}-\dot{\mathbf{s}}$, the fact that $-\mathbf{u}^j\in\mathbf{g}^j(\mathcal{D}^j\dot{\mathbf{r}})$ and $-\mathbf{v}^j\in\mathbf{g}^j(\mathcal{D}^j\dot{\mathbf{s}})$ as well as the monotonicity of each $\mathbf{g}^j$.

To prove the maximal monotonicity it is enough to prove that the operator $\mathbf{I}+\mathcal{A}_s$ is surjective (see Proposition 2.2 of [11]). The equation $[\mathbf{I}+\mathcal{A}_s](\mathbf{r}\times\dot{\mathbf{r}})=\mathbf{f}\times\dot{\mathbf{f}}$ needs to be solved for each $\mathbf{f}\times\dot{\mathbf{f}}\in\mathbf{V}\times\mathbf{H}$. The equation translates into

$$\mathbf{r}-\dot{\mathbf{r}}=\mathbf{f},\quad \dot{\mathbf{r}}-\mathbf{A}\mathbf{r}=\dot{\mathbf{f}},\quad -\mathcal{N}^j\mathbf{r}\in\mathbf{g}^j(\mathcal{D}^j\dot{\mathbf{r}})\text{ for } j\in\mathcal{J}^S\setminus\{m\},$$

which can be rewritten in the form

$$\dot{\mathbf{r}}=\mathbf{r}-\mathbf{f},\quad \begin{cases}\mathbf{r}-\mathbf{A}\mathbf{r}=\mathbf{f}+\dot{\mathbf{f}},\\ -\mathcal{N}^j\mathbf{r}\in\mathbf{g}^j(\mathcal{D}^j[\mathbf{r}-\mathbf{f}])\text{ for } j\in\mathcal{J}^S\setminus\{m\}\end{cases}. \tag{6.4}$$

It is enough to solve the latter equations (which no longer involve $\dot{\mathbf{r}}$); the first equation then determines $\dot{\mathbf{r}}$. The weak form of the equations for $\mathbf{r}$ is

$$\begin{cases}(\mathbf{r},\phi)_{\mathbf{H}}+(\mathbf{r},\phi)_{\mathbf{V}}=(\mathbf{f}+\dot{\mathbf{f}},\phi)_{\mathbf{H}}+\sum_{j\in\mathcal{J}^S\setminus\{m\}}\mathbf{u}^j\cdot\mathcal{D}^j\phi\\ \text{for all }\phi\in\mathbf{V},\text{ with }-\mathbf{u}^j\in\mathbf{g}^j(\mathcal{D}^j[\mathbf{r}-\mathbf{f}]),\end{cases} \tag{6.5}$$

The solution of this equation is obtained by showing that the following convex functional on $\mathbf{V}$ achieves its minimum at an element $\mathbf{r}$ which necessarily satisfies (6.5):

$$\begin{aligned}\mathcal{G}(\mathbf{r})=\frac{1}{2}(\mathbf{r},\mathbf{r})_{\mathbf{H}}+\frac{1}{2}(\mathbf{r},\mathbf{r})_{\mathbf{V}}-(\mathbf{f}+\dot{\mathbf{f}},\mathbf{r})_{\mathbf{H}}&\\ +\sum_{j\in\mathcal{J}^S\setminus\{m\}}\mathbf{G}^j(\mathcal{D}^j[\mathbf{r}-\mathbf{f}]).&\end{aligned} \tag{6.6}$$

This functional does achieve a minimum. To see this note first that

$$-(\mathbf{r},\mathbf{f})_{\mathbf{H}}+\sum_{j\in\mathcal{J}^S\setminus\{m\}}\mathbf{G}^j(\mathcal{D}^j[\mathbf{r}-\mathbf{f}])\geq a+b\,\|\mathbf{r}\|_{\mathbf{H}}+c\,\|\mathbf{r}\|_{\mathbf{V}},$$

where to estimate the convex terms we use the fact that each $\mathbf{G}^j$ is bounded below by an affine function, as well as the continuity of $\mathcal{D}^j:\mathbf{V}\mapsto\mathbb{R}^p$. It is then evident that $\mathcal{G}(\mathbf{r}) \to \infty$ as $\|\mathbf{r}\|_{\mathbf{V}} \to \infty$. The existence of a minimizing element, which for the moment we denote by $\tilde{\mathbf{r}}$, is then a consequence of a standard result on convex functions (see [28, Proposition 1.2, Chapter II]). It follows easily from

$$\mathcal{G}(\tilde{\mathbf{r}}) \leq \mathcal{G}(\tilde{\mathbf{r}} + \epsilon[\mathbf{r} - \tilde{\mathbf{r}}]) \text{ for all } \mathbf{r} \in \mathbf{V}$$

that

$$\begin{aligned} -\epsilon[(\tilde{\mathbf{r}}, \mathbf{r} - \tilde{\mathbf{r}})_{\mathbf{V}} + (\tilde{\mathbf{r}}, \mathbf{r} - \tilde{\mathbf{r}})_{\mathbf{H}} - (\mathbf{f} + \dot{\mathbf{f}}, \mathbf{r} - \tilde{\mathbf{r}})_{\mathbf{H}}] - \epsilon^2[\|\mathbf{r} - \tilde{\mathbf{r}}\|_{\mathbf{H}}^2 + \|\mathbf{r} - \tilde{\mathbf{r}}\|_{\mathbf{V}}^2] \\ \leq \sum_{j \in \mathcal{J}^S \setminus \{m\}} \Big[\mathbf{G}^j(\mathcal{D}^j[\tilde{\mathbf{r}} + \epsilon(\mathbf{r} - \tilde{\mathbf{r}}) - \mathbf{f}]) - \mathbf{G}^j(\mathcal{D}^j[\tilde{\mathbf{r}} - \mathbf{f}])\Big] \\ \leq \sum_{j \in \mathcal{J}^S \setminus \{m\}} \epsilon\Big[\mathbf{G}^j(\mathcal{D}^j[\mathbf{r} - \mathbf{f}]) - \mathbf{G}^j(\mathcal{D}^j[\tilde{\mathbf{r}} - \mathbf{f}])\Big]. \end{aligned}$$

In the last step we used the following consequence of the convexity of $\mathbf{G}^j$:

$$\mathbf{G}^j(\mathcal{D}^j[\tilde{\mathbf{r}} + \epsilon(\mathbf{r} - \tilde{\mathbf{r}}) - \mathbf{f}]) \leq (1-\epsilon)\mathbf{G}^j(\mathcal{D}^j[\tilde{\mathbf{r}} - \mathbf{f}]) + \epsilon\mathbf{G}^j(\mathcal{D}^j[\tilde{\mathbf{r}} + \epsilon(\mathbf{r} - \mathbf{f}]).$$

Dividing by ϵ and letting ϵ go to zero we get

$$\begin{aligned} -[(\tilde{\mathbf{r}}, \mathbf{r} - \tilde{\mathbf{r}})_{\mathbf{V}} + (\tilde{\mathbf{r}}, \mathbf{r} - \tilde{\mathbf{r}})_{\mathbf{H}} - (\mathbf{f} + \dot{\mathbf{f}}, \mathbf{r} - \tilde{\mathbf{r}})_{\mathbf{H}}] \\ \leq \sum_{j \in \mathcal{J}^S \setminus \{m\}} \Big[\mathbf{G}^j(\mathcal{D}^j[\mathbf{r} - \mathbf{f}]) - \mathbf{G}^j(\mathcal{D}^j[\tilde{\mathbf{r}} - \mathbf{f}])\Big]. \end{aligned} \tag{6.7}$$

Thus the linear functional $\mathbf{l}(\boldsymbol{\phi}) = (\tilde{\mathbf{r}}, \boldsymbol{\phi})_{\mathbf{V}} + (\tilde{\mathbf{r}}, \boldsymbol{\phi})_{\mathbf{H}} - (\mathbf{f} + \dot{\mathbf{f}}, \boldsymbol{\phi})_{\mathbf{H}}$ satisfies

$$-\mathbf{l}(\boldsymbol{\phi}) \leq \sum_{j \in \mathcal{J}^S \setminus \{m\}} \Big[\mathbf{G}^j(\mathcal{D}^j[\tilde{\mathbf{r}} - \mathbf{f} + \boldsymbol{\phi}]) - \mathbf{G}^j(\mathcal{D}^j[\tilde{\mathbf{r}} - \mathbf{f}])\Big]. \tag{6.8}$$

so that $-\mathbf{l}$ belongs to the subdifferential at $\tilde{\mathbf{r}}$ of the convex function

$$\mathcal{G}(\mathbf{r}) = \sum_{j \in \mathcal{J}^S \setminus \{m\}} \mathbf{G}^j(\mathcal{D}^j[\mathbf{r} - \mathbf{f}]).$$

To complete the proof of existence one must show that

$$l(\boldsymbol{\phi}) = \sum_{j \in \mathcal{J}^S \setminus \{m\}} \mathbf{u}^j \cdot \mathcal{D}^j \boldsymbol{\phi},$$

with $-\mathbf{u}^j \in \mathbf{g}^j(\mathcal{D}^j[\mathbf{r} - \mathbf{f}])$. Since the righthand side of (6.8) vanishes for $\boldsymbol{\phi} \in \mathbf{V}_0$ (the subspace introduced above), it follows (by replacing $\boldsymbol{\phi}$ with $-\boldsymbol{\phi}$ in the inequality) that $l(\boldsymbol{\phi}) \equiv 0$ on $\mathbf{V}_0$. Now let $\mathbf{V}_1$ be the orthogonal complement of $\mathbf{V}_0$ with respect to the inner product $(\cdot,\cdot)_{\mathbf{V}}$, and let $\boldsymbol{\phi} = \boldsymbol{\phi}_0 + \boldsymbol{\phi}_1$ be the corresponding decomposition of elements. It is easy to see that the map $\phi_1 \mapsto \left(\mathcal{D}^j\phi_1\right)_{j\in\mathcal{J}^S\setminus\{m\}}$ is a bijective map of $\mathbf{V}_1$ onto $[\mathbb{R}^p]^{|\mathcal{J}^S|-1}$ and thus one can conclude that

$$\mathbf{l}(\phi) = \mathbf{l}(\phi_1) = \sum_{j\in\mathcal{J}^S\setminus\{m\}} \mathbf{u}^j \cdot \mathcal{D}^j\phi.$$

That $-\mathbf{u}^j$ belongs to $\mathbf{g}^j(\mathcal{D}^j[\tilde{\mathbf{r}} - \mathbf{f}])$ for each j also follows easily from (6.8).

This completes the proof of existence. We now turn to the proof of the compactness property. It is enough to show that $(\mathbf{I} + \mathcal{A}_s)^{-1} : \mathbf{V}\times\mathbf{H} \mapsto \mathbf{V}\times\mathbf{H}$ is compact. Since $[\mathbf{I}+\mathcal{A}_s](\mathbf{r}\times\dot{\mathbf{r}}) = \mathbf{f}\times\dot{\mathbf{f}}$ is equivalent to (6.4) (with the weak formulation (6.5)), it is enough by Rellich's compactness theorem to show that solutions of the latter system satisfy the a priori estimate

$$\sum_{i\in\mathcal{I}}\left[\|\mathbf{r}^i\|_{W^2(0,l_i;\mathbb{R}^p)} + \|\dot{\mathbf{r}}\|_{W^1(0,l_i;\mathbb{R}^p)}\right] \le C\left[\|\mathbf{f}\|_{\mathbf{V}} + \|\dot{\mathbf{f}}\|_{\mathbf{H}}\right]. \tag{6.9}$$

The first step is to set $\phi = \mathbf{r} - \mathbf{f}$ in (6.5). After simple rearrangement of terms one gets

$$\|\mathbf{r}-\mathbf{f}\|_{\mathbf{H}}^2 + \|\mathbf{r}\|_{\mathbf{V}}^2 = (\mathbf{r},\mathbf{f})_{\mathbf{V}} + (\dot{\mathbf{f}},\mathbf{r}-\mathbf{f})_{\mathbf{H}} + \sum_{j\in\mathcal{J}^S\setminus\{m\}} \mathbf{u}^j\cdot\mathcal{D}^j[\mathbf{r}-\mathbf{f}].$$

Since $-\mathbf{u}^j \in \mathbf{g}^j(\mathcal{D}^j[\mathbf{r}-\mathbf{f}])$ and $\mathbf{0}\in\mathbf{g}^j(\mathbf{0})$ it follows from the monotonicity of $\mathbf{g}^j$ that

$$\sum_{j\in\mathcal{J}^S\setminus\{m\}} \mathbf{u}^j\cdot\mathcal{D}^j[\mathbf{r}-\mathbf{f}] = -\sum_{j\in\mathcal{J}^S\setminus\{m\}} (-\mathbf{u}^j)\cdot\mathcal{D}^j[\mathbf{r}-\mathbf{f}] \le 0.$$

From the previous identity we get the estimate

$$\|\mathbf{r}-\mathbf{f}\|_{\mathbf{H}}^2 + \|\mathbf{r}\|_{\mathbf{V}}^2 \le \epsilon\,\|\mathbf{r}-\mathbf{f}\|_{\mathbf{H}}^2 + \|\mathbf{r}\|_{\mathbf{V}}^2 + \frac{1}{4\epsilon}\left[\|\dot{\mathbf{f}}\|_{\mathbf{H}}^2 + \|\mathbf{f}\|_{\mathbf{V}}^2\right],$$

so that

$$\|\mathbf{r}-\mathbf{f}\|_{\mathbf{H}}^2 + \|\mathbf{r}\|_{\mathbf{V}}^2 \le C\left[\|\dot{\mathbf{f}}\|_{\mathbf{H}}^2 + \|\mathbf{f}\|_{\mathbf{V}}^2\right]. \tag{6.10}$$

Now we are essentially finished. Since

$$\|\dot{\mathbf{r}}\|_{\mathbf{V}} = \|\mathbf{r}-\mathbf{f}\|_{\mathbf{V}} \le \|\mathbf{r}\|_{\mathbf{V}} + \|\mathbf{f}\|_{\mathbf{V}},$$

if we take into account Lemma 3.1 the estimate (6.10) implies that part of (6.9) which pertains to $\dot{\mathbf{r}}$. To get the compactness estimate on $\mathbf{r}$ we note that the norm defined by

$$\|\mathbf{r}\|^2 = \|\mathbf{r}\|_V^2 + \sum_{i\in\mathcal{I}}\int_0^{l_i} |\mathbf{r}^i_{,xx}|^2\,dx$$

is equivalent to the norm

$$\sum_{i\in\mathcal{I}} \| \mathbf{r}^i \|_{W^2(0,l_i;\mathbb{R}^p)} \cdot$$

Therefore (6.10) implies estimate (6.9) for $\mathbf{r}$ since

$$\| \mathbf{Ar} \|_H = \| \mathbf{r} - \mathbf{f} - \dot{\mathbf{f}} \|_H \leq \| \mathbf{r} - \mathbf{f} \|_{\mathbf{H}} + \| \dot{\mathbf{f}} \|_{\mathbf{H}}$$

and

$$\sum_{i\in\mathcal{I}} \int_0^{l_i} |\mathbf{r}^i_{,xx}|^2 \, dx \leq C\Big[\| \mathbf{Ar} \|_H^2 + \| \mathbf{r} \|_V^2\Big].$$

□

We now prove a theorem which guarantees that for any initial state the solution to the feedback system decays asymptotically to the zero energy state.

THEOREM 6.2. *Suppose that the network has no closed circuits and that the hypotheses of the previous existence theorem are satisfied. Suppose also that the minimum of each* $\mathbf{G}^j(\mathbf{u})$ *is achieved only at* $\mathbf{0}$ *and that*

$$\mathbf{g}^j(\mathbf{0}) = \partial \mathbf{G}^j(\mathbf{0}) = \{\mathbf{0}\} \text{ for all } j \in \mathcal{J} \setminus \{m\}. \tag{6.11}$$

Then

$$\lim_{t\to\infty} \|\boldsymbol{\mathcal{U}}_s(t)[\mathbf{r}_0 \times \dot{\mathbf{r}}_0]\|_{\boldsymbol{\mathcal{H}}} = 0 \text{ for all } \mathbf{r}_0 \times \dot{\mathbf{r}}_0 \in \boldsymbol{\mathcal{H}}. \tag{6.12}$$

PROOF. The proof follows standard lines. From the compactness of the resolvent of $\boldsymbol{\mathcal{A}}_s$ it follows that the trajectories $\{\boldsymbol{\mathcal{U}}_s(t)[\mathbf{r}_0 \times \dot{\mathbf{r}}_0]\}_{t\geq 0}$ are relatively compact in $\boldsymbol{\mathcal{H}}$ and hence that the ω-limit set $\omega(\mathbf{r}_0 \times \dot{\mathbf{r}}_0)$ is non-empty. We shall see that, under the stated hypotheses, it consists of the zero state alone. Since the energy norm is nonincreasing along each trajectory a standard proof shows that the energy is constant on $\omega(\mathbf{r}_0 \times \dot{\mathbf{r}}_0)$. Moreover it is easily seen that if $\mathbf{s}_0 \times \dot{\mathbf{s}}_0 \in \omega(\mathbf{r}_0 \times \dot{\mathbf{r}}_0)$ so does the whole trajectory $[\mathbf{s} \times \dot{\mathbf{s}}](t) = \boldsymbol{\mathcal{U}}_s(t)[\mathbf{s}_0 \times \dot{\mathbf{s}}_0]$. Consequently, energy is conserved along the trajectory. It follows from (6.2) that for all positive t,

$$\sum_{j\in\mathcal{J}_S\setminus\{m\}} \boldsymbol{\mathcal{N}}^j \mathbf{s}(t) \cdot \boldsymbol{\mathcal{D}}^j \dot{\mathbf{s}}(t) = 0, \text{ with } -\boldsymbol{\mathcal{N}}^j \mathbf{s}(t) \in \mathbf{g}^j(\boldsymbol{\mathcal{D}}^j \dot{\mathbf{s}}(t)).$$

It follows from the monotonicity of each $\mathbf{g}^j$ that

$$\boldsymbol{\mathcal{N}}^j \mathbf{s}(t) \cdot \boldsymbol{\mathcal{D}}^j \dot{\mathbf{s}}(t) = 0.$$

Since

$$\mathbf{G}^j(\mathbf{0}) \geq \mathbf{G}^j(\boldsymbol{\mathcal{D}}^j \dot{\mathbf{s}}(t)) - \boldsymbol{\mathcal{N}}^j \mathbf{s}(t) \cdot [\mathbf{0} - \boldsymbol{\mathcal{D}}^j \dot{\mathbf{s}}(t)] = \mathbf{G}^j(\boldsymbol{\mathcal{D}}^j \dot{\mathbf{s}}(t)),$$

we conclude that $\mathbf{G}^j(\mathbf{0}) = \mathbf{G}^j(\boldsymbol{\mathcal{D}}^j \dot{\mathbf{s}}(t))$. From the hypothesis it follows that $\boldsymbol{\mathcal{D}}^j \dot{\mathbf{s}}(t) = \mathbf{0}$ and $-\boldsymbol{\mathcal{N}}^j \mathbf{s}(t) \in \mathbf{g}^j(\mathbf{0})$; consequently we also have $\boldsymbol{\mathcal{N}}^j \mathbf{s}(t) = \mathbf{0}$.

Therefore for each $j \in \mathcal{J} \setminus \{m\}$ one has $e_j(x_j, t) = 0$ and it follows from the estimate of Theorem 5.1 that the solution $\mathbf{s}(t) \times \dot{\mathbf{s}}(t)$ has zero energy for all t so that, in particular, $\mathbf{s}_0 \times \dot{\mathbf{s}}_0$ is the zero state. □

REMARK 6.1. The assumption (1.11) is in fact a necessary condition for the conclusion of the theorem to hold. In order to see this suppose that, in contradiction to (6.11), some $\mathbf{g}^k(\mathbf{0})$ contains a non-zero vector $\mathbf{u}^k$. Then there exists a non-zero solution of

$$\text{(6.13)} \quad \begin{cases} \left[\mathbf{Q}_i^2(\mathbf{r}^i_{,x} + \mathbf{R}_i\mathbf{r}^i)\right]_{,x} - \mathbf{R}_i^t\mathbf{Q}_i^2(\mathbf{r}^i_{,x} + \mathbf{R}_i\mathbf{r}^i) + \mathbf{S}_i\mathbf{r}_i = \mathbf{0} \text{ for } i \in \mathcal{I}, \\ \mathbf{r}^i(x_{ij}) \text{ coincide for } j \in \mathcal{J}_M,\ i \in \mathcal{I}_j, \\ \mathcal{N}^j\mathbf{r} = \mathbf{0} \text{ for } j \in \mathcal{J}_M \\ \mathcal{D}^m\mathbf{r} = \mathbf{0}, \\ \mathcal{N}^k\mathbf{r} = \mathbf{u}^k, \quad \mathcal{N}^j\mathbf{r} = \mathbf{0} \text{ for } j \in \mathcal{J}_S \setminus \{m, k\}, \end{cases}$$

Consequently, since $\mathbf{0} \in \mathbf{g}^j(\mathbf{0})$ and $\mathbf{u}^k \in \mathbf{g}^k(\mathbf{0})$, $\mathbf{r} \times \mathbf{0}$ is a non-trivial stationary solution of (6.1); its energy does not decay to zero.

On the other hand the condition that the minimizing set $\min(\mathbf{G}^j)$ of $\mathbf{G}^j$ contain only $\mathbf{0}$ is unnecessarily stringent. Decay is, for example, guaranteed if $\min(\mathbf{G}^j) \cap \min(\partial\mathbf{G}^j) = \{\mathbf{0}\}$. For details see [20], where one can also find a condition which together with (1.11) will be necessary and sufficient for the conclusion (1.12) to hold. That condition again involves the non-existence of an overdetermined eigenvector.

We now use Theorem 5.1 to prove a result on the exponential decay of energy for the feedback system (6.1) in the case of a tree network.

THEOREM 6.3. *Suppose that the network contains no closed circuits and that the functions $\mathbf{G}^j$ are continuously differentiable, so that the $\mathbf{g}^j$ are continuous, single valued and maximal monotone. Suppose also that*

$$\text{(6.14)} \qquad \mathbf{g}_j(\mathbf{u}) \cdot \mathbf{u} \geq \delta[|\mathbf{u}|^2 + |\mathbf{g}(\mathbf{u})|^2].$$

Then the feedback semi-group decays exponentially: there exist constants β and C such that

$$\|\mathcal{U}_s(t)[\mathbf{r}_0 \times \dot{\mathbf{r}}_0]\|_{\mathcal{H}} \leq C\exp(-\beta t)\ \|\mathbf{r}_0 \times \dot{\mathbf{r}}_0\|_{\mathcal{H}}\ .$$

PROOF. From (6.2) it follows that

$$
\begin{aligned}
&\|\mathcal{U}_s(T_\star)[\mathbf{r}_0\times\dot{\mathbf{r}}_0]\|^2_{\mathcal{H}}\\
&=\|\mathbf{r}_0\times\dot{\mathbf{r}}_0\|^2_{\mathcal{H}}-\sum_{j\in\mathcal{J}_S\setminus\{m\}}\int_0^{T_\star}\mathbf{g}^j(\mathcal{D}^j\dot{\mathbf{r}}(t))\cdot\mathcal{D}^j\dot{\mathbf{r}}(t)\,dt\\
&\le\|\mathbf{r}_0\times\dot{\mathbf{r}}_0\|^2_{\mathcal{H}}-\delta\sum_{j\in\mathcal{J}_S\setminus\{m\}}\int_0^{T_\star}\left[|\mathcal{D}^j\dot{\mathbf{r}}(t)|^2+|\mathbf{g}^j(\mathcal{D}^j\dot{\mathbf{r}}(t)|^2\right]dt \qquad (6.15)\\
&=\|\mathbf{r}_0\times\dot{\mathbf{r}}_0\|^2_{\mathcal{H}}-\delta\sum_{j\in\mathcal{J}_S\setminus\{m\}}\int_0^{T_\star}\left[|\mathcal{D}^j\dot{\mathbf{r}}(t)|^2+|\mathcal{N}^j\mathbf{r}(t)|^2\right]dt.
\end{aligned}
$$

Since

$$e_{i_j}(x_j,t)\le C\left[|\mathcal{D}^j\dot{\mathbf{r}}(t)|^2+|\mathcal{N}^j\mathbf{r}(t)|^2\right],$$

it follows from (6.15) that

$$(6.16)\qquad \|\mathcal{U}_s(T_\star)[\mathbf{r}_0\times\dot{\mathbf{r}}_0]\|^2_{\mathcal{H}}\le\|\mathbf{r}_0\times\dot{\mathbf{r}}_0\|^2_{\mathcal{H}}-\delta C^{-1}\sum_{j\in\mathcal{J}_S\setminus\{m\}}\mathcal{E}^{\dagger}_{i_j}(x_j,0,T_\star).$$

If we note that $\|\mathcal{U}_s(T_\star)[\mathbf{r}_0\times\dot{\mathbf{r}}_0]\|^2_{\mathcal{H}}\le\|\mathcal{U}_s(T_\star/2)[\mathbf{r}_0\times\dot{\mathbf{r}}_0]\|^2_{\mathcal{H}}$ and then use the estimate of Theorem 5.1, we get from (6.17)

$$(6.17)\quad \|\mathcal{U}_s(T_\star)[\mathbf{r}_0\times\dot{\mathbf{r}}_0]\|^2_{\mathcal{H}}\le\|\mathbf{r}_0\times\dot{\mathbf{r}}_0\|^2_{\mathcal{H}}-C\,\|\mathcal{U}_s(T_\star)[\mathbf{r}_0\times\dot{\mathbf{r}}_0]\|^2_{\mathcal{H}}\,.$$

where the constant C has been redefined. It follows that

$$\|\mathcal{U}_s(T_\star)[\mathbf{r}_0\times\dot{\mathbf{r}}_0]\|^2_{\mathcal{H}}\le(1+C)^{-1}\,\|\mathbf{r}_0\times\dot{\mathbf{r}}_0\|^2_{\mathcal{H}};$$

the exponential decay follows from this. □

Finally we discuss feedbacks yielding only polynomial decay rates. This requires a much more subtle argument and depends on the following two lemmas.

The first lemma is a consequence of Theorem 5.2.

LEMMA 6.1. *Under the hypotheses of Theorem 5.1 we have for all* $T>0$

$$(6.18)\qquad \int_0^T\mathcal{E}(t)\,dt\le C\left[\sup_{0\le t\le T}\mathcal{E}(t)+\sum_{j\in\mathcal{J}_S\setminus\{m\}}\int_0^T e_{i_j}(x_j,t)\,dt\right].$$

PROOF. It is obviously enough to prove this when $T>T_\star=2S_\star$. From Theorem 5.2 we can conclude, after a time shift, that for any t

$$(6.19)\qquad \mathcal{E}(t)\le C\sum_{j\in\mathcal{J}_S\setminus\{m\}}\int_{t-S_\star}^{t+S_\star}e_{i_j}(x_j,\tau)\,d\tau.$$

Now

$$\int_0^T \mathcal{E}(t)\,dt = \int_0^{S_\star} \mathcal{E}(t)\,dt + \int_{T-S_\star}^{T} \mathcal{E}(t)\,dt + \int_{S_\star}^{T-S_\star} \mathcal{E}(t)\,dt.$$

The sum of the first two integrals on the right is bounded by

$$T_\star \sup_{0\le t\le T} \mathcal{E}(t),$$

while, by (6.5),

$$\begin{aligned}\int_{S_\star}^{T-S_\star} \mathcal{E}(t)\,dt &\le \sum_{j\in\mathcal{J}_S\setminus\{m\}} \int_{S_\star}^{T-S_\star} \left[\int_{t-S_\star}^{t+S_\star} e_{i_j}(x_j,\tau)\,d\tau\right] dt \\ &\le T_\star \sum_{j\in\mathcal{J}_S\setminus\{m\}} \int_0^T e_{i_j}(x_j,\tau)\,d\tau,\end{aligned}$$

where we have changed the order of integration. The result follows from the latter estimates. □

The next lemma, due to Lasiecka and Tataru [54], will be stated without proof.

LEMMA 6.2. *Let h_1 be a positive, increasing function with $h_1(0) = 0$ Suppose that a sequence of positive numbers s_k satisfies the estimate*

$$s_{k+1} + h_1(s_{k+1}) \le s_k. \tag{6.20}$$

Then $s_k \le S(k)$ where $S(t)$ is the solution of

$$\frac{d}{dt}S(t) + h(S(t)) = 0, \quad S(0) = s_0, \tag{6.21}$$

with h a positive, increasing function satisfying $h(0) = 0$ defined by $h = id - [id + h_1]^{-1}$, "id" denoting the identity map.

We can now prove the following theorem.

THEOREM 6.4. *Suppose that the network contains no closed circuits and that the functions $\mathbf{G}^j$ are continuously differentiable, so that the $\mathbf{g}^j$ are continuous, single valued and maximal monotone. Suppose also that*

$$\mathbf{g}_j(\mathbf{u})\cdot\mathbf{u} \ge \begin{cases} \delta[|\mathbf{u}|^2 + |\mathbf{g}(\mathbf{u})|^2] & \text{for } |\mathbf{u}| > 1, \\ \delta[|\mathbf{u}|^2 + |\mathbf{g}(\mathbf{u})|^2]^p & \text{for } |\mathbf{u}| \le 1, \end{cases} \tag{6.22}$$

where $p > 1$ and $\delta > 0$. Then the feedback semi-group has decay-rate

$$\|\boldsymbol{\mathcal{U}}_s(t)[\mathbf{r}_0 \times \dot{\mathbf{r}}_0]\|_{\mathcal{H}} = O\left(t^{1/(1-p)}\right). \tag{6.23}$$

PROOF. Since the energy is decreasing we get from (6.18) that

$$
\begin{aligned}
T\mathcal{E}(T) &\le C\left[\mathcal{E}(0) + \sum_{j\in\mathcal{J}_S\setminus\{m\}} \int_0^T e_{i_j}(x_j,t)\,dt\right] \\
&\le C\left[\mathcal{E}(0) + \sum_{j\in\mathcal{J}_S\setminus\{m\}} \int_0^T \left[|\mathcal{D}^j\dot{\mathbf{r}}|^2 + |\mathbf{g}^j(\mathcal{D}^j\dot{\mathbf{r}})|^2\right] dt\right],
\end{aligned}
\tag{6.24}
$$

where here and below we repeatedly adjust the value of C. Let

$$
I_j = \{t > 0 \,|\, |\mathcal{D}^j\dot{\mathbf{r}}(t)| \le 1\}, \quad J_j = \{t > 0 \,|\, |\mathcal{D}^j\dot{\mathbf{r}}(t)| > 1\}.
$$

Then, taking into account the hypothesis (6.22),

$$
\begin{aligned}
&\sum_{j\in\mathcal{J}_S\setminus\{m\}} \int_0^T \left[|\mathcal{D}^j\dot{\mathbf{r}}|^2 + |\mathbf{g}^j(\mathcal{D}^j\dot{\mathbf{r}})|^2\right] dt \\
&\quad \le C \sum_{j\in\mathcal{J}_S\setminus\{m\}} \left[\int_{I_j} \mathbf{g}^j(\mathcal{D}^j\dot{\mathbf{r}})\cdot\mathcal{D}^j\dot{\mathbf{r}}\,dt + \int_{J_j} \left[\mathbf{g}^j(\mathcal{D}^j\dot{\mathbf{r}})\cdot\mathcal{D}^j\dot{\mathbf{r}}\right]^{1/p} dt\right] \\
&\quad \le C\Big[\sum_{j\in\mathcal{J}_S\setminus\{m\}} \int_0^T \mathbf{g}^j(\mathcal{D}^j\dot{\mathbf{r}})\cdot\mathcal{D}^j\dot{\mathbf{r}}\,dt \\
&\qquad + T^{1/q}\Big[\sum_{j\in\mathcal{J}_S\setminus\{m\}} \int_0^T \mathbf{g}^j(\mathcal{D}^j\dot{\mathbf{r}})\cdot\mathcal{D}^j\dot{\mathbf{r}}\,dt\Big]^{1/p}\Big],
\end{aligned}
$$

where we have used Hölder's inequality. We substitute this, along with the identity

$$
\mathcal{E}(T) = \mathcal{E}(0) - \sum_{j\in\mathcal{J}_S\setminus\{m\}} \int_0^T \mathbf{g}^j(\mathcal{D}^j\dot{\mathbf{r}})\cdot\mathcal{D}^j\dot{\mathbf{r}}\,dt \tag{6.25}
$$

(which follows from (6.2)) into (6.24) to get

$$
\begin{aligned}
T\mathcal{E}(T) \le C\Big[\mathcal{E}(T) + 2 &\sum_{j\in\mathcal{J}_S\setminus\{m\}} \int_0^T \mathbf{g}^j(\mathcal{D}^j\dot{\mathbf{r}})\cdot\mathcal{D}^j\dot{\mathbf{r}}\,dt \\
&+ T^{1/q}\Big[\sum_{j\in\mathcal{J}_S\setminus\{m\}} \int_0^T \left[\mathbf{g}^j(\mathcal{D}^j\dot{\mathbf{r}})\cdot\mathcal{D}^j\dot{\mathbf{r}}\right] dt\Big]^{1/p}\Big].
\end{aligned}
\tag{6.26}
$$

We can rearrange this, again using (6.25), to obtain

$$
\mathcal{E}(T) \le \frac{C}{T-C}\left[2\big(\mathcal{E}(0) - \mathcal{E}(T)\big) + T^{1/q}\big(\mathcal{E}(0) - \mathcal{E}(T)\big)^{1/p}\right]
$$

which for T sufficiently large gives

$$\begin{aligned}\mathcal{E}(T) &\leq \left[(\mathcal{E}(0)-\mathcal{E}(T)) + (\mathcal{E}(0)-\mathcal{E}(T))^{1/p}\right] \\ &= [\mathrm{id}+h_0](\mathcal{E}(0)-\mathcal{E}(T)),\end{aligned} \tag{6.27}$$

where $h_0(s) = s^{1/p}$. If we then set $h_1 = [\mathrm{id}+h_0]^{-1}$ it follows that

$$\mathcal{E}(T) + h_1(\mathcal{E}(T)) \leq \mathcal{E}(0). \tag{6.28}$$

Let $s_k = \mathcal{E}(kT)$. Then it follows from the previous estimate, which is valid over the interval $[kT, kT+T]$, that the hypothesis (6.20) of Lemma 6.2 holds. This allows us to prove the decay rate. We define $h(s)$ as in the lemma. It is then an easy exercise to show that the two identities

$$h_1(s+s^{1/p}) = s \quad \text{and} \quad h(s+h_1(s)) = h_1(s)$$

together with the hypothesis $p > 1$ imply

$$\lim_{s\to 0}\frac{h_1(s)}{s^p} = 1 \quad \text{and} \quad \lim_{s\to 0}\frac{h(s)}{s^p} = 1. \tag{6.29}$$

Since Theorem 6.2 implies that $\lim_{k\to\infty} s_k = 0$ we can assume s_0 to be as small as we please. The solution $S(t)$ of (6.21) satisfies $S(t) \leq s_0$ for $t > 0$. Given $\epsilon \in (0,1)$ s_0 can be assumed so small that $|h(s) - s^p| < \epsilon s^p$ for $s \leq s_0$. Then, in particular it follows from the equation (6.21) that

$$\frac{d}{dt}S(t) + (1-\epsilon)S(t)^p \leq 0.$$

It follows that $S(t) = O(t^{1/(1-p)}$ so that $\mathcal{E}(kT) = O([kT]^{-1/(p-1)})$. Since $\mathcal{E}(t)$ is decreasing this implies the decay estimate (6.23). □

CHAPTER V

Spectral Analysis and Numerical Simulation of 1-d Networks

1. Preliminaries

In this and in the following section we shall consider the problem of determining the frequency spectrum and the corresponding modes of a given network of elastic strings or elastic beams. This problem is of great importance for purposes of obtaining Fourier series expansions (Galerkin approximations) of solutions to the dynamic equations of motion and also for certain approaches to the theories of control and optimal design. There is a huge literature devoted to this problem, within which we can identify two groups of papers. The first group, comprising the majority of these papers, is concerned with a finite element approximation of the underlying system and with the generalized eigenvalue problem associated with that approximation. The latter problem is then solved using a suitable numerical procedure, such as the shifted Q-R algorithm or Jacobi's method, if the dimension of the approximating space is comparatively small, or by other methods such as the Lancos algorithm, the Raleigh-Ritz method, etc., if one is interested in obtaining a particular subset of eigenvalues. For repetitive structures, which are typical in network problems, yet another method has been employed with great success, namely, the method of *slicing the spectrum*, which is based on Sylvester's theorem of invariance of the number of positive, vanishing and negative eigenvalues under similarity transformations. This method has been developed for repetitive structures by Wittrick and Williams [110], [111] and others [103]. The underlying idea is that of substructuring, that is, of dividing the structure into smaller identical pieces. The analysis is then performed on the level of the substructure and the results are *read into* the master structure in an appropriate way.

That this strategy is not only admissible for finite dimensional approximations but also for the infinite dimensional full problem has been observed by Destyunder [27] and, very recently, by Balakrishnan [7], who exploited the ideas of [110], [111]. These papers are among of the second group attacking the full infinite dimensional problem. A conceptually quite different subset of this second group contains the works of Nicaise, von Below, Ali Mehmeti and others ([75], [77], [76], [106], [105], [67], [69], [71], [68]). The latter papers are essentially devoted to the eigenvalue problem for the Laplace- Beltrami operator on smooth, connected graphs . In particular, for one-dimensional elastic links the method of von Below [106] is closest to the ideas which we will pursue here. The basis of this method is the possibility of rescaling the spatial variables associated with each member of the structure to the uniform interval $[0,1]$. Once such a rescaled formulation and a transformation to a first order system is obtained, some classical results on the spectrum of general first order systems can be used. See the work of Schäfke and Schneider and references in the survey article by Schneider [98]. See also von Below's Habilitation thesis [107]. We will follow [106], where a slightly more direct and constructive approach is taken: the rescaled system is kept as a second order system and then explicitly solved using Hadamard's operational calculus and an appropriate version of the variation of constants formula. The eigenvalues can then, in principle, be computed from this representation. We will also provide some numerical evidence based on a slicing technique similar to the one by Wittrick and Williams [110].

We remark that the following sections are very much in the spirit of von Below's works, even though the problems encountered here are more difficult in that the models we study deal with in-plane or even fully 3-d displacements rather than scalar functions which are exclusively considered in von Below's work and, incidentally, also in the work of Nicaise [75], [77], etc. However, for the sake of completeness, the arguments will be provided in some detail.

1.1. Notation. We consider a nonempty, finite, simple and connected graph $\mathbf{G}$ in $I\!R^2$ with n vertices $V(\mathbf{G}) := \{v_i,\ i = 1,...,n\}$, and N edges $E(\mathbf{G}) := \{\mathbf{k}_j,\ j = 1,...,N\}$. The edges $\mathbf{k}_j$ are parameterized by $\pi_j : [0,\ell_j] \mapsto I\!R^2$, where the running variable $x_j \in [0,\ell_j]$ represents the arc length. The maps π_j are assumed to be C^2-smooth. For $i,j \in V(\mathbf{G})$ we set $\mathbf{k}_{s(i,h)} := v_iv_j = v_jv_i$, where v_iv_j signifies the edge joining the vertices v_i, v_j, if $v_iv_j \in E(\mathbf{G})$. In fact, as we will be concerned here only with networks consisting of straight edges, we assign to each edge $\mathbf{k}_{s(i,h)}$ the unit edge vector $\mathbf{e}_{s(i,h)}$, directed along the edge, and its orthogonal complement $\mathbf{e}^{\perp}_{s(i,h)}$. We also introduce the unit vector $\mathbf{n}$ normal to the plane of the graph. To each triad $\mathbf{e}_{s(i,h)}$, $\mathbf{e}^{\perp}_{s(i,h)}$, $\mathbf{n}$ we associate the corresponding set of orthogonal projections $\mathbf{P}_{s(i,h)}$, $\mathbf{P}^{\perp}_{s(i,h)}$, $\mathbf{P}_{\mathbf{n}}$. Further, we introduce the incidence matrix

$$d_{ij} = \begin{cases} 1 & \text{if } \pi_j(\ell_j) = v_i, \\ -1 & \text{if } \pi_j(0) = v_i, \\ 0 & \text{else,} \end{cases} \tag{1.1}$$

and the adjacency matrix

$$e_{ih} = \begin{cases} 1 & \text{if } v_i v_h \in E(\mathbf{G}), \\ 0 & \text{else.} \end{cases} \tag{1.2}$$

The individual length $\ell_{s(i,h)}$ of the edge joining v_i, v_h can then be expressed by $\ell_{ih} := \ell_{s(i,h)} e_{ih}$, where may set $s(i,h) = 1$ if $e_{ih} = 0$. This notation makes it possible to normalize the description of the running variable in each individual edge as follows: for $x \in [0,1]$,

$$\begin{aligned} \gamma_{ih}(x) :=& \ell_{ih} \left(\frac{1 + d_{is(i,h)}}{2} - x d_{is(i,h)} \right) \\ =& \begin{cases} \ell_{s(i,h)}(1-x) & \text{if } \pi_{s(i,h)}(\ell_{s(i,h)}) = v_i,\ e_{ih} = 1, \\ \ell_{s(i,h)} x & \text{if } \pi_{s(i,h)}(0) = v_i,\ e_{ih} = 1, \\ 0 & \text{if } e_{ih} = 0. \end{cases} \end{aligned}$$

For each vertex we define the set $\Gamma(v_i) := \{v_h \in V(\mathbf{G}) | e_{ih} = 1\}$ of all vertices adjacent to the vertex v_i, and $d(v_i) := |\Gamma(v_i)|$ the edge degree of the vertex v_i. We can now characterize the network as follows:

$$\begin{aligned} \mathbf{G} =& \cup_{j=1}^{N} \mathbf{k}_j, \\ \partial V(\mathbf{G}) =& \{v_i \in V(\mathbf{G}) |\, d(v_i) = 1\}, \\ \overset{\circ}{V}(\mathbf{G}) =& \{v_i \in V(\mathbf{G}) |\, d(v_i) > 1\}. \end{aligned}$$

Obviously, $\partial V(\mathbf{G})$ signifies the set of simple nodes, where only one edge starts or ends, while $\overset{\circ}{V}(\mathbf{G})$ signifies the set of multiple nodes, where two or more edges meet. A function $\mathbf{r} : \mathbf{G} \mapsto \mathbb{R}^r$ can then be viewed as a collection of functions with values along the individual edges given by

$$\mathbf{r}_j := \mathbf{r}(\pi_j(x)), \quad x \in [0, \ell_j].$$

It is, in some circumstances, more convenient to express those functions $\mathbf{r}_j$ in the normalized form introduced above, i.e.,

$$\mathbf{r}_{ih}(x) := e_{ih} \mathbf{r}_{s(i,h)}[\gamma_{ih}(x)], \quad x \in [0,1]. \tag{1.3}$$

With this definition we find

$$\begin{aligned}\mathbf{r}_{ih}(0) =& e_{ih}\begin{cases}\mathbf{r}_{s(i,h)}(\ell_{s(i,h)}) & \text{if } \pi_{s(i,h)}(\ell_{s(i,h)}) = v_i,\\ \mathbf{r}_{s(i,h)}(0) & \text{if } \pi_{s(i,h)}(0) = v_i,\end{cases}\\ =& e_{ih}\mathbf{r}_{s(i,h)}(\pi^{-1}_{s(i,h)}(v_i)) = e_{ih}\mathbf{r}(v_i),\end{aligned}$$

and also

$$\mathbf{r}_{hi}(x) = \mathbf{r}_{ih}(1-x).$$

This notation is adopted from [106].

1.2. Networks of Strings. In this section we consider a network of strings which, in its reference configuration, coincides with a graph $\mathbf{G}$ as described in the last section. The equations governing the motion are formally derived in Chapter II. The model we consider here, however, is also easily developed along the lines of Chapter III on the modeling of beam networks. Let $\mathbf{r}$ denote the deviation from the reference configuration. By the definition above, $\mathbf{r}_j$ denotes the displacement of the string (edge) with label j. We define

$$\begin{aligned}u_j\mathbf{e}_j &:= <\mathbf{r}_j, \mathbf{e}_j> \mathbf{e}_j = P_j\mathbf{r}_j,\\ w_j\mathbf{e}_j^\perp &:= <\mathbf{r}_j, \mathbf{e}_j^\perp> \mathbf{e}_j^\perp = P_j^\perp\mathbf{r}_j,\end{aligned}$$

so that u_j, w_j represent the longitudinal and vertical displacements of the jth string, respectively. Let p_j^2, q_j^2 denote the longitudinal and flexural rigidities. Furthermore, let $\mathbf{K}_j := p_j^2P_j + q_j^2P_j^\perp$ be the stiffness operator associated with jth string. Then the equations governing the motion of the network of strings can be written as follows:

(1.4)

$$\begin{cases}\ddot{\mathbf{r}}_j - \mathbf{K}_j\mathbf{r}_j'' = 0, \quad j = 1, ..., N,\\ \mathbf{r}_j(v) = 0, \quad \forall v \in \partial_0 V(\mathbf{G}),\\ \mathbf{K}_j\mathbf{r}_j'(v) = f_v, \quad \forall v \in \partial_1 V(\mathbf{G}),\\ \mathbf{r}_i(v_h) = \mathbf{r}_j(v_h), \quad \forall i,j \in \operatorname{ind}\Gamma(v_h),\ v_h \in \overset{\circ}{V}(\mathbf{G}),\\ \sum_{h:v_h\in\Gamma(v_i)} d_{is(i,h)}\mathbf{K}_{s(i,h)}\mathbf{r}'_{s(i,h)}(\pi^{-1}_{s(i,h)}(v_i)) = \mathbf{g}_v, \quad \forall i : v_i \in \overset{\circ}{V}(\mathbf{G}),\\ \mathbf{r}_j(\cdot,0) = \mathbf{r}_{j0}, \quad \dot{\mathbf{r}}_j(\cdot,0) = \mathbf{r}_{j1}.\end{cases}$$

Here $\partial_0 V(\mathbf{G}) \subset \partial V(\mathbf{G})$ denotes the set of simple nodes with clamped boundary conditions, $\partial_1 V(\mathbf{G})$ is the set of stress free (or externally loaded as is the case when $\mathbf{f}_v \neq 0$) simple nodes and $\operatorname{ind}\Gamma(V_h)$ is the index set of $\Gamma(v_h)$. The third equation expresses the continuity of the network, or its connectedness also in the deformed configuration, while the fourth equation represents the balance of forces (with a possible external loading modeled through the functions $\mathbf{g}_v$). These equations may be expressed in terms of the functions $\mathbf{r}_{ih}$ defined above as follows:

$$\begin{cases} \ddot{\mathbf{r}}_{ih} - l_{ih}^{(-2)}\mathbf{K}_{s(i,h)}\mathbf{r}''_{ih} = 0, \\ \mathbf{r}_{hi}(x) = \mathbf{r}_{ih}(1-x),\ x \in [0,1], \end{cases} \tag{1.5}$$

$$\exists \phi \in \mathbb{R}^{2N} \,|\, \mathbf{r}_{ih}(0) = e_{ih}\phi(i), \tag{1.6}$$

$$\sum_h \ell_{ih}^{(-1)}\mathbf{K}_{s(i,h)}\mathbf{r}'_{s(i,h)}(0) = \mathbf{g}_i, \quad \forall i : v_i \in V(\mathbf{G}), \tag{1.7}$$

where $\phi(i) \in \mathbb{R}^2$ in (1.6) represents $\mathbf{r}(v_i)$. If $v_i \in \partial_0 V(\mathbf{G})$, then $\phi_i = 0$. If v_i in (1.7) is in $\partial_1 V(\mathbf{G})$, then there is exactly one vertex in $\Gamma(v_i)$, and (1.7), for that i, expresses the Neumann boundary condition for that node. The balance condition (1.7) is the analog of the Kirchhoff law in the context of electrical networks. Of course, $\ell_{ih}^{(-1)}$ means $\ell_{s(i,h)}^{-1}e_{ih}$.

The above setup is in complete analogy with the one to be found in [106], where a static $\mathbb{R}$-valued Neumann eigenvalue problem is considered in the context of diffusion equations on a network.

1.3. Networks of Timoshenko Beams. The situation is notationally very similar to the last section but mathematically more involved in the case where each edge is considered as a Timoshenko beam in its reference configuration. In addition to the longitudinal and vertical displacements (u_j, w_j) of the centerline of the beam we also have a rotation $\psi_j := \langle \mathbf{P}_\mathbf{n}\mathbf{r}_j, \mathbf{n} \rangle$ of the cross section out of its position perpendicular to the centerline, i.e., we have to account for shearing. We denote the shear modulus of the j th beam by q_j^2, its flexural rigidity by m_j^2 and, as before, p_j^2 denotes its longitudinal stiffness. We only consider here an initially straight and untwisted linear Timoshenko-Bresse system as described above:

$$\begin{cases} \ddot{u}_j - p_j^2 u''_j = 0, \\ \ddot{w}_j - q_j^2(w'_j + \psi_j)' = 0, \\ \ddot{\psi}_j - m_j^2\psi''_j + q_j^2(w'_j + \psi_j) = 0, \end{cases} \tag{1.8}$$

$$\mathbf{r}_j(v_i) = \mathbf{r}_h(v_i), \quad \forall h, j \in \operatorname{ind}\Gamma(v_i), \tag{1.9}$$

$$\begin{aligned} \sum_h d_{is(i,h)}\{p_{s(i,h)}^2 u'_{s(i,h)}\mathbf{e}_{s(i,h)} + q_{s(i,h)}^2(w'_{s(i,h)} + \psi_{s(i,h)})\mathbf{e}_{s(i,h)}^{\perp} \\ + m_{s(i,h)}^2\psi'_{s(i,h)}\mathbf{n}\}(v_i) = \mathbf{g}_i,\ \forall v_i \in \overset{\circ}{V}(\mathbf{G}), \end{aligned} \tag{1.10}$$

$$
(1.11) \qquad \begin{cases} \mathbf{r}_i(v) = 0, \quad \forall v \in \partial_0 V(\mathbf{G}), \\ \{p_j^2 u_j' \mathbf{e}_j + q_j^2(w_j' + \psi_j)\mathbf{e}_j^\perp\}(v_i) = \mathbf{f}_i, \\ m_j^2 \psi_j'(v_i) = h_i, \quad \forall v_i \in \partial_1 V(\mathbf{G}). \end{cases}
$$

In analogy to the previous section we define the operators

$$
\begin{aligned}
\mathbf{K}_j &:= p_j^2 P_j + q_j^2 P_j^\perp + m_j^2 \mathbf{P_n}, \\
\mathbf{M}_j &:= q_i^2 < \mathbf{P_n}(\cdot), \mathbf{n} > \mathbf{e}_j^\perp, \\
\mathbf{N}_j &:= q_j^2 < P_j^\perp(\cdot), \mathbf{e}_j^\perp > \mathbf{n}, \\
\mathbf{S}_j &:= \mathbf{N}_j - \mathbf{M}_j \Rightarrow \mathbf{S}_j^* = -\mathbf{S}_j.
\end{aligned}
$$

Then the state equations (1.8) can be written as

$$
(1.12) \qquad \ddot{\mathbf{r}}_j - \mathbf{K}_j \mathbf{r}_j'' + \mathbf{S}_j \mathbf{r}_j' + q_j^2 \mathbf{P_n} \mathbf{r}_j = 0.
$$

The continuity condition is exactly of the form as in (1.4), with the only difference that the vectors have now three components. Let us consider the homogeneous situation. The balance of forces requirement can then be expressed as

$$
(1.13) \qquad \sum_h d_{is(i,h)}(\mathbf{K}_{s(i,h)} \mathbf{r}_{s(i,h)}' + \mathbf{M}_{s(i,h)} \mathbf{r}_{s(i,h)})(\pi_{s(i,h)}^{-1})(v_i) = 0,
$$
$$
\forall v_i \in V(\mathbf{G}) \setminus \partial_0 V(\mathbf{G}).
$$

We can as well express these equations in terms of matrix entries as follows:

$$
(1.14) \qquad \begin{cases} \ddot{\mathbf{r}}_{ih} - \ell_{ih}^{(-2)} \mathbf{K}_{s(i,h)} \mathbf{r}_{ih}'' + \ell_{ih}^{(-1)} \mathbf{S}_{s(i,h)} \mathbf{r}_{ih}' + q_{s(i,h)}^2 \mathbf{P_n} \mathbf{r}_{ih} = 0, \\ \exists \phi : \mathbf{r}_{ih}(0) = e_{ih}\phi(i); \;\; \phi(i) = 0, \quad \forall v_i \in \partial_0 V(\mathbf{G}), \\ \sum_h [\ell_{ih}^{(-1)} \mathbf{K}_{s(i,h)} \mathbf{r}_{ih}'(0) + d_{is(i,h)} \mathbf{M}_{s(i,h)} \mathbf{r}_{ih}(0)] = 0, \\ \qquad\qquad \forall v_i \in V(\mathbf{G}) \setminus \partial_0 V(\mathbf{G}), \\ \mathbf{r}_{hi}(x) = \mathbf{r}_{ih}(1-x), \;\; \mathbf{r}_{ih}(\cdot, 0) = \mathbf{r}_{ih0}, \;\; \dot{\mathbf{r}}_{ih}(\cdot, 0) = \mathbf{r}_{ih1}. \end{cases}
$$

1.4. Networks of Euler-Bernoulli Beams. Here we consider the case where all edges are taken to represent Euler-Bernoulli beams. By this we mean that shear strains are not taken into account or, what amounts to the same thing, that the cross sections of the deformed beam stay perpendicular to the deformed centerline. As we are going to discuss planar networks only, we have again two displacements, the longitudinal, u_j, and the vertical, w_j. We denote the longitudinal and flexural stiffnesses by p_j^2, q_j^2, respectively. We will consider only homogeneous boundary and nodal conditions. We use the notation developed in section 1.1. By setting

$$
\mathbf{r}_j = u_j \mathbf{e}_j + w_j \mathbf{e}_j^\perp, \quad K_j = q_j^2 P_j, \quad H_j = p_j^2 P_j,
$$

we can rewrite the state equations

$$\ddot{u}_j - p_j^2 u_j'' = 0, \quad \ddot{w}_j + q_j^2 w_j'''' = 0, \tag{1.15}$$

as

$$\ddot{\mathbf{r}}_j - \mathbf{H}_j \mathbf{r}_j'' + \mathbf{K}_j \mathbf{r}_j'''' = 0. \tag{1.16}$$

The continuity (connectedness) condition again reads

$$\mathbf{r}_j(v_h) = \mathbf{r}_i(v_h), \quad \forall v_h \in \overset{\circ}{V}(\mathbf{G}). \tag{1.17}$$

This time, however, depending on the particular nature of the joint, we have other geometric or compatibility conditions. In the case of rigid joints, the case of most interest, we have

$$\mathbf{P}_j^{\perp} \mathbf{r}_j'(v_h) = \mathbf{P}_i^{\perp} \mathbf{r}_i'(v_h), \quad \forall v_h \in \overset{\circ}{V}(\mathbf{G}), \tag{1.18}$$

and in the case of pinned joints we do not have any additional condition of the type (1.18).

Let us now look at the dynamical node conditions or, in the terminology of Lumer, at the connection/interaction operators. These are

$$\sum_h d_{is(i,h)}(\mathbf{K}_{s(i,h)} \mathbf{r}_{s(i,h)}''' - \mathbf{H}_{s(i,h)} \mathbf{r}_{s(i,h)}')(\pi_{s(i,h)}^{-1}(v_i)) = 0, \tag{1.19}$$

$$\sum_h d_{is(i,h)} \mathbf{K}_{s(i,h)} \mathbf{r}_{s(i,h)}''(\pi_{s(i,h)}^{-1}(v_i)) = 0, \quad \forall v_i \in \overset{\circ}{V}(\mathbf{G}), \tag{1.20}$$

where the summation is taken over all $h \in \operatorname{ind}\Gamma(v_i)$. In the case of a pinned joint (1.20) is replaced by the stronger condition

$$\mathbf{K}_{s(i,h)} \mathbf{r}_{s(i,h)}''(\pi_{s(i,h)}^{-1}(v_i)) = 0, \quad \forall h \in \operatorname{ind}\Gamma(v_i). \tag{1.21}$$

Of course we also have initial conditions

$$\mathbf{r}_j(\cdot, 0) = \mathbf{r}_{0j}, \quad \dot{\mathbf{r}}_j(\cdot, 0) = \mathbf{r}_{1j}. \tag{1.22}$$

We are now going to express this system in terms of the matrix notation developed earlier. We have

$$\ddot{\mathbf{r}}_{ih} + \ell_{ih}^{(-4)} \mathbf{K}_{s(i,h)} \mathbf{r}_{ih}'''' - \ell_{ih}^{(-2)} \mathbf{H}_{s(i,h)} \mathbf{r}_{ih}'' = 0; \tag{1.23}$$

$$\begin{cases} \exists \phi \in R^{2N} : \ \mathbf{r}_{ih}(0) = e_{ih}\phi(i), \\ \phi(i) = 0, \ \forall v_i \in \partial_0 V(\mathbf{G}); \end{cases} \tag{1.24}$$

$$\begin{cases} \exists \xi \in R^N : \ \ell_{ih}^{(-1)} d_{is(i,h)} \mathbf{P}_{s(i,h)}^{\perp} \mathbf{r}_{ih}'(0) = e_{ih}\xi_i, \\ \xi_i = 0, \ \forall v_i \in \partial_0 V(\mathbf{G}); \end{cases} \tag{1.25}$$

$$\sum_h \{\ell_{ih}^{(-3)} \mathbf{K}_{s(i,h)} \mathbf{r}_{ih}'''(0) - \ell_{ih}^{(-1)} \mathbf{H}_{s(i,h)} \mathbf{r}_{ih}'(0)\} = 0; \tag{1.26}$$

$$\sum_h d_{is(i,h)} \ell_{ih}^{(-2)} \mathbf{K}_{s(i,h)} \mathbf{r}_{ih}''(0) = 0, \quad v_h \in V(\mathbf{G}) \setminus \partial_0 V(\mathbf{G}); \tag{1.27}$$

$$\mathbf{r}_{hi}(x) = \mathbf{r}_{ih}(1-x). \tag{1.28}$$

As remarked earlier, in the case of pinned joints we have to neglect (1.25) and instead have to strengthen (1.27) to

$$\mathbf{K}_{s(i,h)} \mathbf{r}_{ih}''(0) = 0, \quad \forall v_h \in V(\mathbf{G}) \setminus \partial_0 V(\mathbf{G}). \tag{1.29}$$

2. Eigenvalue Problems for Networks of 1–d Elements

2.1. Introduction. As in [106] we introduce a matrix notation which provides a general framework for eigenvalue problems on multiconnected domains. The notation is slightly different from the one introduced by Nicaise [75] who, independently, discussed the eigenvalue problem for the second derivative operator on homogeneous networks. Given two matrices A, B we denote by $A \cdot B$ the matrix with entries $(a_{ih} b_{ih})$ (coordinatewise multiplication). It is understood that the same notation is used also in the case where the entries of B are vectors (in which case we write $\mathbf{B}$ in place of B) while those of A are scalars : i.e. $A \cdot \mathbf{B} = (a_{ih} \mathbf{b}_{ih})$. One can also introduce the following functional calculus with matrices: given a function $f(\cdot)$ and a matrix A we define $f(A)$ to be the matrix $(f(a_{ih}))$. This notation can be extended to the case of vector-valued matrix entries in the following way. If $\mathbf{B} = (\mathbf{b})_{ih} = (b_{ih} \mathbf{e}_{s(i,h)} + b_{ih}^{\perp} \mathbf{e}_{s(i,h)}^{\perp})$ then $f(\mathbf{B}) = (f(b_{ih}) \mathbf{e}_{s(i,h)} + f(b_{ih}^{\perp}) \mathbf{e}_{s(i,h)}^{\perp})$. Let us also introduce the vector $e := (1, \dots, 1)^*$ such that $Ae = \sum_h a_{ih}$; an analogous notation is used for vector-valued matrices. Without loss of generality we may choose the indices of the set $\partial_1 V(\mathbf{G})$ to be $i = k+1, \dots, n$. We introduce the diagonal matrix $\mathcal{Z}$ with zeros in the first k entries and ones in the others. We will in general have to consider a function $\mathbf{r} : \mathbf{G} \mapsto \mathbb{R}^r$ describing the deformation of the mechanical system which, in its rest configuration, covers the graph $\mathbf{G}$. The matrix $R := (\mathbf{r}_{ih})$ is then defined following the procedure outlined in Section 1.1. The general structure of an eigenvalue problem associated with a network of the kind described above will then be of the following type:

$$\begin{cases} \mathcal{L}^{(-2)} \cdot \mathcal{K} \cdot R'' + \mathcal{L}^{(-1)} \cdot \mathcal{S} \cdot R' + \mathcal{Q} \cdot R = -\lambda^2 R, \\ \exists \phi : R'(0) = \phi e^* \mathcal{E} =: \Phi; \quad \phi_i = 0, \quad \forall i : v_i \in \partial_0 V(\mathbf{G}), \\ \mathcal{Z}(\mathcal{L}^{(-1)} \cdot \mathcal{K} \cdot R'(0) + \mathcal{D} \cdot \mathcal{M} R(0)) e = 0, \\ R^*(x) = R(1-x), \quad \forall x \in [0,1]. \end{cases} \tag{2.1}$$

The matrices $\mathcal{E}, \mathcal{D}$ are the adjacency and incidence matrix of the underlying graph as defined by (1.2) and (1.1), respectively. For the case of scalar valued functions r and vanishing matrices $\mathcal{S}, \mathcal{M}$ this system has been studied by von Below [106] and, based on a slightly different matrix setup, by Nicaise [75]. It should be noted that the restriction to the second order case is just a matter of convenience and that higher order problems can be cast into a similar form. This is particularly important for networks composed of Euler-Bernoulli beams.

2.2. General String Networks. We use the notation above with $\mathcal{K}$ given as in Section 2. We obtain the following problem

$$(2.2) \qquad \begin{cases} \mathcal{L}^{(-2)} \cdot \mathcal{K} \cdot R''(x) = -\lambda^2 R(x), \\ \exists \phi : R(0) = \phi e^* \cdot \mathcal{E}; \quad \phi_i = 0, \quad \forall i : v_i \in \partial_0 V(\mathbf{G}), \\ \mathcal{Z}(\mathcal{L}^{(-1)} \cdot \mathcal{K} \cdot R'(0))e = 0, \\ R^*(x) = R(1-x), \quad \forall x \in [0,1]. \end{cases}$$

We define $\mathcal{B} := \mathcal{L} \cdot \mathcal{K}^{(-1/2)}$. Then using the functional calculus (which has been attributed to Hadamard) introduced above, we may define

$$(2.3) \qquad R(x) = \cos(x\lambda\mathcal{B}) \cdot \Phi + \frac{1}{\lambda}\mathcal{B}^{(-1)} \cdot \sin(x\lambda\mathcal{B}) \cdot \Psi,$$

with the initial matrices $\Phi = \phi e^* \cdot \mathcal{E}$, $\Psi = R'(0)$. It is easy to see that $R(x)$ given by (2.3) is the general solution to the system of equations (1.5) in which the inertia term is replaced by $-\lambda^2 \mathbf{r}_{ih}$. The problem then consists of finding matrices Φ, Ψ such that the conditions (1.6) and (1.7), transformed into matrix notation as in the second and third equation of (2.2), are satisfied. The possibility of representing the general solution to the matrix-differential equation in a comprehensible form is a crucial step in this particular approach to the analysis of the eigenvalues. The analogous representation in the case of a network of Timoshenko beams is more involved and far less obvious. However, the representation (2.3) can be utilized in essentially the same way it was done in [106] (see also [75] for the case of homogeneous networks). Here we give the necessary conditions for λ to be an eigenvalue.

THEOREM 2.1. *If λ is an eigenvalue of the system (2.2), then λ satisfies one of the following conditions:*

(i) $\lambda = 0$ *(iff $\partial_0 V(\mathbf{G}) = \emptyset$);*

(ii) $\lambda = \ell^{-1}_{s(i,h)} p_{s(i,h)} \pi k$ *for some $i, h = 1, \ldots, n$; $k \in \mathbf{Z}$;*

(iii) $\lambda = \ell^{-1}_{s(i,h)} q_{s(i,h)} \pi k$ *for some $i, h = 1, \ldots, n$; $k \in \mathbf{Z}$;*

(iv) λ *is such that* $\ker \mathcal{Z}(\mathcal{C} - \operatorname{diag} H_i) \neq 0$, *where*

$$C_{ih} := \ell_{s(i,h)} p_{s(i,h)} (\sin(\lambda \ell_{s(i,h)} p_{s(i,h)}^{-1}))^{-1} e_{ih} \mathbf{P}_{s(i,h)} + \ell_{s(i,h)} q_{s(i,h)} (\sin(\lambda q_{s(i,h)}^{-1}))^{-1} e_{ih} \mathbf{P}_{s(i,h)}^{\perp},$$

$$H_i := \sum_h e_{ih} \ell_{s(i,h)} (p_{s(i,h)} (\sin(\lambda p_{s(i,h)}^{-1}))^{-1} \cos(\lambda \ell_{s(i,h)} p_{s(i,h)}^{-1}) \mathbf{P}_{s(i,h)} + q_{s(i,h)} (\sin(\lambda q_{s(i,h)}^{-1}))^{-1} \cos(\lambda \ell_{s(i,h)} q_{s(i,h)}^{-1}) \mathbf{P}_{s(i,h)}^{\perp}).$$

The eigenelements, which are characterized through the nodal vectors ϕ, Ψ, *then satisfy*

(i) $\phi_i = \phi_1$, $\forall i = 2, ..., n$; $\Psi_{ih} = 0$, $\forall i, h = 1, \dots, n$;

(ii) $$\begin{cases} \mathbf{P}_{s(i,h)} \phi_h = (-1)^k \mathbf{P}_{s(i,h)} \phi_i, & \mathbf{P}_{s(i,h)} \Psi_{ih} = -(-1)^k \mathbf{P}_{s(i,h)} \Psi_{hi}, \\ \mathbf{P}_{s(i,h)}^{\perp} \phi_j = \mathbf{P}_{s(i,h)}^{\perp} \Psi_{jk} = 0, & \forall j, k = 1, \dots, n; \end{cases}$$

(iii) $$\begin{cases} \mathbf{P}_{s(i,h)}^{\perp} \phi_h = (-1)^k \mathbf{P}_{s(i,h)}^{\perp} \phi_i, & \mathbf{P}_{s(i,h)}^{\perp} \Psi_{ih} = -(-1)^{-1} \mathbf{P}_{s(i,h)}^{\perp} \Psi_{hi}, \\ \mathbf{P}_{s(i,h)} \phi_j = \mathbf{P}_{s(i,h)}^{\perp} \Psi_{jk} = 0, & \forall j, k = 1, \dots, n; \end{cases}$$

(iv) $\phi \in \ker \mathcal{Z}(\mathcal{C} - \operatorname{diag} H_i)$, *and*

$$\Psi_{ih} = \lambda e_{ih} \ell_{s(i,h)} \{ p_{s(i,h)}^{-1} (\sin(\lambda \ell_{s(i,h)} p_{s(i,h)}^{-1}))^{-1} (\mathbf{P}_{s(i,h)} (\Phi_{ih}^* - \cos(\lambda \ell_{s(i,h)} p_{s(i,h)}^{-1}) \Phi_{ih}) + q_{s(i,h)}^{-1} (\sin(\lambda q_{s(i,h)}^{-1}))^{-1} (\mathbf{P}_{s(i,h)}^{\perp} (\Phi_{ih}^* - \cos(\lambda \ell_{s(i,h)} q_{s(i,h)}^{-1}) \Phi_{ih} \},$$

respectively.

PROOF. The proof is given in the spirit of [106]. However, the situation here is slightly more complicated in that we have vector-valued functions and also allow for Dirichlet conditions in the simple nodes. Therefore, we provide the arguments.

(i) We first deal with the case where λ is zero. Suppose that $\partial_0 V(\mathbf{G}) = \emptyset$. Then (2.2) reduces to

$$\begin{gathered} \mathcal{L}^{(-2)} \cdot \mathcal{K} \cdot R''(x) = 0, \\ \exists \phi : R(0) = \phi e^* \cdot \mathcal{E}; \ \phi_i = 0, \ i : v_i \in \partial_0 V(\mathbf{G}), \\ (\mathcal{L}^{(-1)} \cdot \mathcal{K} \cdot R'(0)) z = 0, \\ R^*(x) = R(1-x), \quad \forall x \in [0,1]. \end{gathered}$$

The general solution to the first equation, under the requirement given by the last equation, yields

$$R(x) = \Phi + x(\Phi^* - \Phi),$$

whereas the third equation leads to

$$\mathcal{L}^{(-1)} \cdot \mathcal{K} \cdot (\Phi^* - \Phi)e = 0.$$

We recall the meaning of $\Phi^* = \mathcal{E} \cdot e\phi^*$ and obtain

$$(\mathcal{L}^{(-1)} \cdot \mathcal{K})\phi = \operatorname{diag}((\mathcal{L}^{(-1)} \cdot \mathcal{K})e)\phi.$$

In contrast to [106], we cannot invert the operator on the right side of this equation. Nevertheless,

$$\operatorname{diag}((\mathcal{L}^{(-1)} \cdot \mathcal{K})e) = \operatorname{diag}\big(\sum_h \ell^{-1}_{s(i,h)} p^2_{s(i,h)} e_{ih} \mathbf{P}_{s(i,h)} + \sum_h \ell^{-1}_{s(i,h)} q^2_{s(i,h)} e_{ih} \mathbf{P}^{\perp}_{s(i,h)}\big),$$

hence the balance condition reduces to

$$\sum_h \ell^{-1}_{s(i,h)} p^2_{s(i,h)} e_{ih} \mathbf{P}_{s(i,h)}(\phi_h) + \sum_h \ell^{-1}_{s(i,h)} q^2_{s(i,h)} e_{ih} \mathbf{P}^{\perp}_{s(i,h)}(\phi_h)$$
$$= \big(\sum_h \ell^{-1}_{s(i,h)} p^2_{s(i,h)} e_{ih} \mathbf{P}_{s(i,h)} + \sum_h \ell^{-1}_{s(i,h)} q^2_{s(i,h)} e_{ih} \mathbf{P}^{\perp}_{s(i,h)}\big)\phi_i.$$

This equation is obviously satisfied if all nodal deformations ϕ_i are the same, which means that the graph as a whole is shifted by a vector $\phi = \phi_i$, $\forall i$.

Let now $\partial_0 V(\mathbf{G}) \neq \emptyset$. It is sufficient to consider the case $\partial_0 V(\mathbf{G}) = \{\phi_1\}$. Then, however, the representation of the solution above gives $\mathbf{r}_{1,2}(x) \equiv 0$ which then implies $\phi_2 = 0$ and so on. Hence, zero cannot be an eigenvalue if one exterior node is clamped. (Note that we do not consider a possibly rigid rotation around node 1 of the whole graph in this model.)

(ii) We are now going to demonstrate the occurrence of the eigenvalues of each individual string under some periodic boundary conditions, dictated by the "connectivity" conditions.

Let us consider the case where $\lambda_{j,k} = \lambda_{s(i,h),k} = \ell^{-1}_{s(i,h)} p_{s(i,h)} \pi k$, with $j = s(i,h)$ for some pair (i,h); the case corresponding to (iii) in the statement of the theorem is done in the same way. We set to zero all initial data $\Phi_{m,n}, \Psi_{m,n}$ such that $\lambda_{s(i,h),k} \neq \ell^{-1}_{s(m,n)} p_{s(m,n)} \pi k$. Then we are left with

$$\mathbf{r}_{ih}(x) = e_{ih}[\cos(xk\pi)\Phi_{ih} + \frac{1}{k\pi}\sin(xk\pi)\Psi_{ih}]$$

and the reflection condition

$$\mathbf{r}_{ih}(1-x) = e_{ih}((-1)^k \cos(k\pi x)\Phi_{ih} - (-1)^k \frac{1}{k\pi}\sin(xk\pi)\Psi_{ih}) \stackrel{!}{=} \mathbf{r}_{hi}(x).$$

This implies (ii) as stated in the theorem. In addition, if v_i is in $V(\mathbf{G}) \setminus \partial_0 V(\mathbf{G})$ we have to satisfy the balance condition, the second equation in (2.2), which now reads

$$\sum_m \ell_{s(i,m)}^{-1} e_{im} p_{s(i,m)}^2 \Psi_{im} = 0.$$

Note that if the coefficients of the $s(i,m)$-th string joining the i-th and m-th vertices are different from the ones of the $s(i,h)$-th string, then by assumption the corresponding $\Psi_{im} = 0$. For the others the condition above has to be taken into account. (We do not claim that the conditions (ii) in the theorem are sufficient.) Note also that if one end of the j-th string ($j = s(i,h)$) is clamped, the balance condition above is not effective for the corresponding vertex, so that the vertex adjacent to the clamped end stays fixed and the derivatives have to satisfy (ii).

(iii) is handled as (ii).

(iv) We now turn to the case where λ is neither zero nor of the form discussed in (ii) or (iii):

$$\lambda \neq \ell_{s(i,h)}^{-1} p_{s(i,h)} \pi k, \quad \lambda \neq \ell_{s(i,h)}^{-1} q_{s(i,h)} \pi k, \quad \forall i, h, k.$$

We will extensively use the reflection formula:

$$\begin{aligned}
\mathbf{r}_{ih}(1) = \mathbf{P}_{s(i,h)} e_{ih} [&\cos(\lambda \ell_{s(i,h)} p_{s(i,h)}^{-1}) \mathbf{\Phi}_{ih} \\
&+ \frac{1}{\lambda} \ell_{s(i,h)}^{-1} p_{s(i,h)} \sin(\lambda \ell_{s(i,h)} p_{s(i,h)}^{-1}) \Psi_{ih}] \\
&+ \mathbf{P}_{s(i,h)}^{\perp} e_{ih} [\cos(\lambda \ell_{s(i,h)} q_{s(i,h)}^{-1}) \mathbf{\Phi}_{ih} \\
&+ \frac{1}{\lambda} \ell_{s(i,h)}^{-1} q_{s(i,h)} \sin(\lambda \ell_{s(i,h)} q_{s(i,h)}^{-1}) \Psi_{ih}] = \mathbf{r}_{hi}(0) = \mathbf{\Phi}_{hi}.
\end{aligned}$$

The strategy is in principal taken from [106], but again we cannot invert operators here. The idea is to eliminate Ψ_{ih} from the balance condition and thereby obtain a system of linear equations in the unknowns ϕ_i. We recall that

$$\mathbf{\Phi}_{ih} = u_{ih0} \mathbf{e}_{s(i,h)} + w_{ih0} \mathbf{e}_{s(i,h)}^{\perp}, \quad \Psi_{ih} = u'_{ih0} e_{s(i,h)} + w'_{ih0} \mathbf{e}_{s(i,h)}^{\perp},$$

and rewrite the reflection condition obtained above as

$$\begin{aligned}
u'_{ih0} &= e_{ih} \lambda p_{s(i,h)}^{-1} \sin(\lambda \ell_{s(i,h)} p_{s(i,h)}^{-1})^{-1} (u^*_{ih0} - \cos(\lambda \ell_{s(i,h)} p_{s(i,h)}^{-1}) u_{ih0}), \\
w'_{ih0} &= e_{ih} \lambda q_{s(i,h)}^{-1} \sin(\lambda \ell_{s(i,h)} q_{s(i,h)}^{-1})^{-1} (w^*_{ih0} - \cos(\lambda \ell_{s(i,h)} q_{s(i,h)}^{-1}) w_{ih0}).
\end{aligned}$$

This result is transferred back to the old notation

$$\begin{aligned}\Psi_{ih} = e_{ih}\lambda\ell_{s(i,h)}p_{s(i,h)}^{-1}\sin(\lambda\ell_{s(i,h)}p_{s(i,h)}^{-1})^{-1}\mathbf{P}_{s(i,h)}\\ (\Phi_{ih}^* - \cos(\lambda\ell_{s(i,h)}p_{s(i,h)}^{-1})\Phi_{ih})\\ + e_{ih}\lambda\ell_{s(i,h)}q_{s(i,h)}^{-1}\sin(\lambda\ell_{s(i,h)}q_{s(i,h)}^{-1})^{-1}\\ \mathbf{P}_{s(i,h)}^{\perp}(\Phi_{ih}^* - \cos(\lambda\ell_{s(i,h)}q_{s(i,h)}^{-1})\Phi_{ih}).\end{aligned}$$

We are now in the position to eliminate Ψ_{ih} from the balance condition $(2.2)_2$. The result is

$$\begin{aligned}\sum_h\{e_{ih}\ell_{s(i,h)}p_{s(i,h)}\sin(\lambda\ell_{s(i,h)}p_{s(i,h)}^{-1})^{-1}\\ \mathbf{P}_{s(i,h)}(\Phi_{ih}^* - \cos(\lambda\ell_{s(i,h)}p_{s(i,h)}^{-1})\Phi_{ih})\\ + e_{ih}\ell_{s(i,h)}q_{s(i,h)}\sin(\lambda\ell_{s(i,h)}q_{s(i,h)}^{-1})^{-1}\\ \mathbf{P}_{s(i,h)}^{\perp}(\Phi_{ih}^* - \cos(\lambda\ell_{s(i,h)}q_{s(i,h)}^{-1})\Phi_{ih}) = 0\end{aligned}$$

for all i such that $v_i \in \overset{\circ}{V}(\mathbf{G}) \cup \partial_1 V(\mathbf{G})$. In order to make this equation a bit more transparent we introduce the operators $\mathcal{C}, H_i$ as in statement (iv) of the theorem and verify that, in fact, the balance equation reduces to the equation given in (iv).

This concludes the proof of the theorem. □

We are now going to give sufficient conditions in the special case where ratios of the physical and geometrical quantities are uniform. In particular, we will be looking at completely uniform networks.

2.3. Homogeneous String Networks. We consider the special case where the quantities $\ell_{s(i,h)}p_{s(i,h)}^{-1}$, $\ell_{s(i,h)}q_{s(i,h)}^{-1}$ are equal to a single number c, implying that $p_{s(i,h)} = q_{s(i,h)}$ and hence $\ell_{s(i,h)}p_{s(i,h)} = \ell_{s(i,h)}q_{s(i,h)}$ is a songle number b. Hence, the operator $\mathcal{L}^{(-2)} \cdot \mathbf{K}$ reduces to cI and (2.2) reduces to

$$\begin{aligned}R''(x) = -(c\lambda)^2 R(x),\\ \exists\phi:\ R(0) = \phi e^* \cdot \mathcal{E};\ \ \phi_i = 0,\ \ \forall i : v_i \in \partial_0 V(\mathbf{G}),\\ \mathcal{Z}(\mathcal{L} \cdot R'(0)e) = 0,\\ R^*(x) = R(1-x).\end{aligned}$$

(I) We shall first discuss the Neumann case, $\partial_0 V(\mathbf{G}) = \emptyset$. Here we can fully adopt the results of von Below [106] and Nicaise [75]. In the case where $\sin(c\lambda) \neq 0$ we find

$$\mathbf{r}'_{ih}(0) = \Psi_{ih} = \frac{c\lambda}{\sin(c\lambda)}(e\phi^* - \cos(c\lambda)\phi e^*) \cdot \mathcal{E}$$

and

$$(2.4)\qquad (\mathbf{r}_{ih})_{ih} = \cos(c\lambda x)\mathcal{E}\cdot\phi e^* + \frac{\sin(c\lambda x)}{\sin(c\lambda)}(e\phi^* - \cos(c\lambda)\phi e^*)\cdot\mathcal{E},$$

which are used in the balance conditions to obtain

$$(\mathcal{L}\cdot(e\phi^* - \cos(c\lambda)\phi e^*)\cdot\mathcal{E})e = 0.$$

This, in turn, can be expressed as

$$\mathcal{L}\phi = \cos(c\lambda)\mathrm{diag}\,(\mathcal{L}e)\phi.$$

Upon introducing the matrix

$$(2.5)\qquad Z := (\mathrm{diag}\,(\mathcal{L}e))^{-1}\mathcal{L}$$

we see that the equation to be satisfied by ϕ is in fact an eigenvalue problem for Z:

$$Z\phi = \mu\phi$$

if we put $\lambda = \arccos(\mu)/c$. It turns out that the case where $\sin(c\lambda) = 0$, i.e., where $\lambda = c^{-1}k\pi$, corresponds to the eigenvalues $1, -1$ of Z. Therefore, all possible eigenvalues of (2.2) can be discussed entirely in terms of the matrix Z. For easier reference, we display the results obtained in [106] or in equivalent form in [75]. First it is shown in [106] that Z has eigenvalues $\mu_i : i = 1, \dots, n$ satisfying $1 = \mu_1 > \mu_2 \geq \cdots \geq \mu_n \geq -1$, with corresponding eigenvectors $\xi_1 = e, \xi_2, \dots, \xi_n$, respectively. In particular, -1 is an eigenvalue if and only if the graph $\mathbf{G}$ contains no odd cycle, which is the case if and only if it is bipartite (i.e., the vertex set $V(\mathbf{G})$ can be separated into two disjoint sets such that each edge in $E(\mathbf{G})$ has one end point in either set). Using these eigenvalues and eigenelements, von Below derives the following characterization of the eigenvalues and eigenelements of (2.2). Label the strings of eigenvalues by λ_{sk}, where s denotes the number of a string of eigenvalues while $k \in N$ is the running index in that string. The corresponding eigenelement is denoted by ϕ_{sk}. For simplicity, we change $c\lambda$ to λ. Let α be a generic real number.

(1) $\lambda_{10} = 0, \quad \phi_{10} = \alpha e$;
(2) $\lambda_{sk} = -\arccos(\mu_s) + (k+1)\pi$ if $k \equiv 1 \bmod 2$,
$2 \leq s \leq n$ and $-1 < \mu_s$, $k \in N$, $\phi_{sk} = \alpha\xi_s$;
(3) $\lambda_{sk} = \arccos(\mu_s) + k\pi$ if $k \equiv 0 \bmod 2$,
$2 \leq s \leq n$ and $-1 < \mu_s$, $k \in N$, $\phi_{sk} = \alpha\xi_s$;
(4) $\lambda_{1k} = 2\pi k,\ k \neq 0,\ \phi_{1k} = \alpha e,\ \Psi^* = -\Psi,\ (\mathcal{L}\cdot\Psi)e = 0$;
(5) $\lambda_{n+1,k} = (2k+1)\pi,\ k \in N,\ \phi_{n+1,k} = 0$,
$\Psi^* = \Psi$, $(\mathcal{L}\cdot\Psi)e = 0$ if $\mathbf{G}$ is *not* bipartite;
(6) $\lambda_{nk} = (2k+1)\pi,\ k \in N,\ \phi_{nk} = \alpha\xi_n$,
$\Psi^* = \Psi$, $(\mathcal{L}\cdot\Psi)e = 0$ if $\mathbf{G}$ is bipartite.

(II) Let us now discuss the case where $\partial_0 V(\mathbf{G}) \neq \emptyset$. The only difference from the previous case is that the balance condition now is

$$\mathcal{Z}(\mathcal{L} \cdot (e\phi^* - \cos(c\lambda)\phi e^*) \cdot \mathcal{E})e = 0.$$

By a remark in [106], it is sufficient to consider the case $\mathcal{L} = \mathcal{E}$. The system then reduces to

$$\mathcal{Z}\mathcal{E}\phi = \cos(\lambda)\text{diag}\,(\mathcal{E}e)\phi.$$

Now the application of $\mathcal{Z}$ just deletes the first k rows (which are to be viewed as rows with 2-d columns) of $\mathcal{E}$. But also the first k entries (which in fact are (2,1)-vectors themselves) in ϕ are zero. Therefore, the system is n-k square. However, a complete characterisation as above does not seem to be known even in the scalar case.

In the purely serial case (collinear or not) this means that the multiplicities of the former eigenvalues are always equal to one.

2.3.1. *Examples.* We discuss the simplest situation in which all constants are taken to be equal to one. Then the spectrum of (2.2) can be determined as above from the eigenvalue problem

$$\mathcal{E}\phi = \text{diag}(d(v_i))\cos(\lambda)\phi. \tag{2.6}$$

Recall that $d(v_i)$ is the edge degree of the vertex v_i. The simplest case, in which the eigenvalues of type (iv) can be determined by the eigenvalues of the adjacency matrix alone, is the case where $d(v_i) \equiv d$, $\forall i = 1, ..., n$. The latter is true for a simple closed loop, where the edge degree is equal to 2 for all vertices. We have

$$\mathcal{E}\phi = d\cos(\lambda)\phi. \tag{2.7}$$

EXAMPLE 2.1. *The unit square.*

In this case $\mathbf{G}$ is bipartite and 2-regular in the sense that the edge degree is 2 for all vertices, i.e., $d(v_i) = 2$. Here $\mathcal{E}$ is given by

$$\mathcal{E} = \begin{pmatrix} 0 & I & 0 & I \\ I & 0 & I & 0 \\ 0 & I & 0 & I \\ I & 0 & I & 0 \end{pmatrix}.$$

The eigenvalues of $\mathcal{E}$ are $0, 2, -2$ and those of Z are $0, 1, -1$ modulo multiplicities, hence $\lambda = k\pi$ including $k = 0$, and $\lambda = (k + \frac{1}{2})\pi$. This shows that only the value $\pi/2$ gives rise to a *new* string of eigenvalues, i.e., eigenvalues which are not among those of an individual string of length 1, with multiplicity 4. We have chosen this example because the eigenvalues can also be obtained by elementary means from classical results on the spectrum of the second order derivative operator with constant coefficients and with periodic boundary conditions. Indeed, if all the material properties are one and the same constant, so that the governing operator is the same on all links, then the continuity and equilibrium conditions at the nodes are equivalent to periodic boundary conditions. The spectrum of this operator is exactly as above. This analogy is not, however, apparent when the material constants differ.

There are four different configurations of nodal values according to the four eigenvalues $\pi/2$. These four nodal configurations are precisely the four corresponding eigenvectors ϕ^i of the adjacency matrix:

$$\phi^1 = \begin{pmatrix} -0.0940 \\ -0.6897 \\ 0.0157 \\ 0.1233 \\ 0.0940 \\ 0.6897 \\ -0.0157 \\ -0.1233 \end{pmatrix}, \quad \phi^2 = \begin{pmatrix} -0.5659 \\ 0.1421 \\ -0.1256 \\ 0.3792 \\ 0.5659 \\ -0.1421 \\ 0.1256 \\ -0.3792 \end{pmatrix},$$

$$\phi^3 = \begin{pmatrix} -0.3713 \\ -0.0110 \\ 0.4481 \\ -0.4015 \\ 0.3713 \\ 0.0110 \\ -0.4481 \\ 0.4015 \end{pmatrix}, \quad \phi^4 = \begin{pmatrix} -0.1818 \\ -0.0631 \\ -0.5321 \\ -0.4240 \\ 0.1818 \\ 0.0631 \\ 0.5321 \\ 0.4240 \end{pmatrix}.$$

It is easy to interpret these node configurations geometrically; we omit the details. Note that in the notation developed here the component ϕ^1_1 of the first eigenvector consists of the first two entries in ϕ^1 above and signifies the deformed position of the first node, and so on. To be very explicit, let us identify the deformation matrix $R = \mathbf{r}$ and then let us go back to the original setup in which each individual string is labeled counterclockwise along the square. We display only nonzero entries.

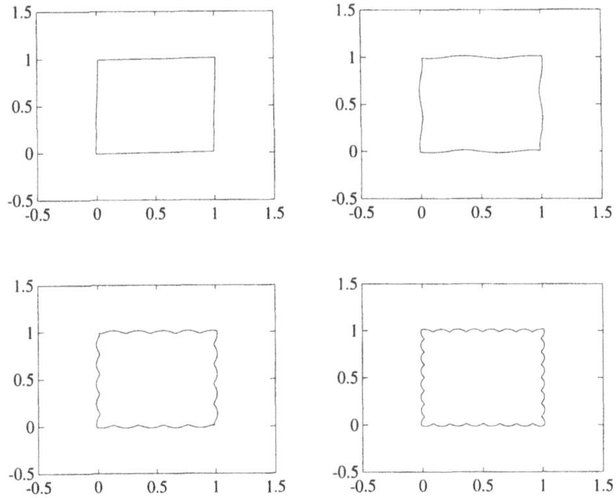

FIGURE V.1. Some eigenmodes of the unit square

$$
\begin{aligned}
\mathbf{r}_{12}(x) &= \cos(x\frac{\pi}{2})\phi^1 + \sin(x\frac{\pi}{2})\phi^2 = \mathbf{r}_2(x),\\
\mathbf{r}_{14}(x) &= \cos(x\frac{\pi}{2})\phi^1 + \sin(x\frac{\pi}{2})\phi^4 = \mathbf{r}_1(x),\\
\mathbf{r}_{21}(x) &= \cos(x\frac{\pi}{2})\phi^2 + \sin(x\frac{\pi}{2})\phi^1 = \mathbf{r}_2(1-x),\\
\mathbf{r}_{23}(x) &= \cos(x\frac{\pi}{2})\phi^2 + \sin(x\frac{\pi}{2})\phi^3 = \mathbf{r}_3(1-x),\\
\mathbf{r}_{32}(x) &= \cos(x\frac{\pi}{2})\phi^3 + \sin(x\frac{\pi}{2})\phi^2 = \mathbf{r}_3(x),\\
\mathbf{r}_{34}(x) &= \cos(x\frac{\pi}{2})\phi^3 + \sin(x\frac{\pi}{2})\phi^4 = \mathbf{r}_4(x),\\
\mathbf{r}_{41}(x) &= \cos(x\frac{\pi}{2})\phi^4 + \sin(x\frac{\pi}{2})\phi^1 = \mathbf{r}_1(1-x),\\
\mathbf{r}_{43}(x) &= \cos(x\frac{\pi}{2})\phi^4 + \sin(x\frac{\pi}{2})\phi^3 = \mathbf{r}_4(1-x).
\end{aligned}
$$

It is easy to check the continuity and balance conditions in either setting directly using this table. We display some typical configurations in Figure V.1.

We now want to use the matrix in (iv) of Theorem 2.1 directly in order to compute eigenvalues and corresponding multiplicities even in the case where the material properties are not assumed be equal to a single constant. In order to obtain numerical evidence, we plot the determinant of the matrix $C - \text{diag}(H_i)$, chopping off all values beyond 10. See the first graph of Figure V.2. The zeros indicate eigenvalues of the entire graph. In the second graph of Figure V.2 we show the multiplicities. This is done in the following way. We perform a Doolittle L-R decomposition of the

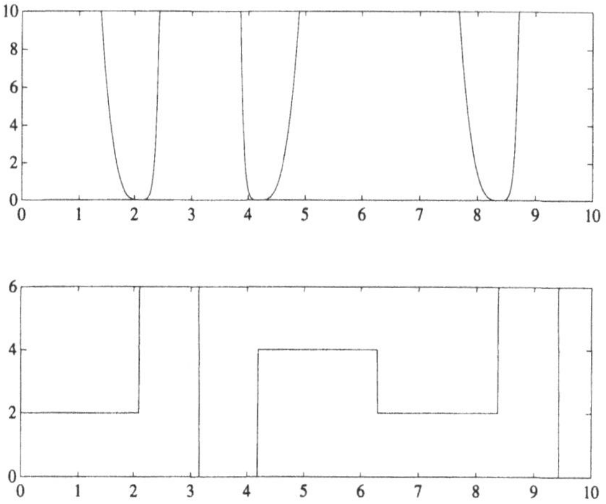

FIGURE V.2. Eigenvalues of unit square

matrix above and then invoke Sylvester's theorem on the invariance of the number of positive, vanishing and negative eigenvalues. These numbers can be recovered from the diagonal of the upper triangular matrix R. This is essentially the argument of Wittrick and Williams [110]. However, in their work it is applied to the finite element approximation, whereas here we apply it to the fully infinite dimensional system by using the exact solution. Note that zero is a two-fold eigenvalue and that at individual eigenvalues there is pole of the determinant of $C - \operatorname{diag}(H_i)$. At odd multiples of π we recognize a jump by four indicating, as was proved above, the multiplicity four. In passing the poles the multiplicity reduces by four, indicating that also the individual eigenvalues have multiplicity four.

This analysis becomes particularly interesting when applied to the case where the material properties are not equal to the same constant. This is the case where we do not have explicit knowledge of the eigenvalue distribution. We concentrate on the first eigenvalue coming from the graph (that is, the first one related to the vanishing of the determinant of $C - \operatorname{diag}(H_i)$). In order to be able to compare to the previous case, we take as a base the same constant 1 for all the material constants. Then, however, we perturb these by $0.1\times$random numbers. The result is displayed in Figure V.3. We clearly observe the splitting of the spectrum into simple eigenvalues. We remark that, at this time, we do not have a proof of this natural result in a general context.

EXAMPLE 2.2. *The carpenter's square.*

We have two links connected together in a right angle. We label the only interior node with #1, and the two simple nodes by #2 and #3. Again

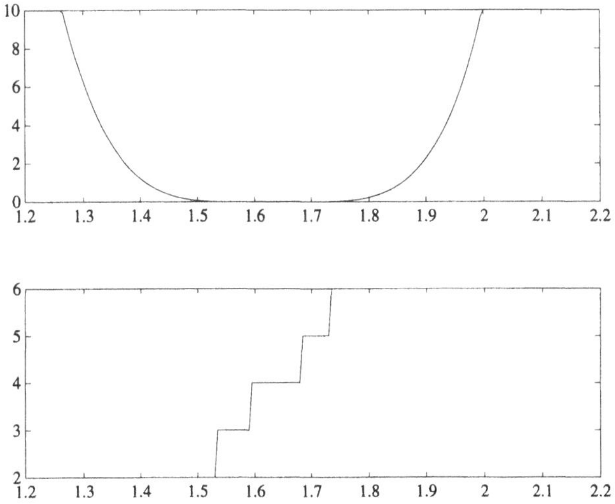

FIGURE V.3. Splitting of eigenvalues

the graph is bipartite, but not regular, and does not contain a circuit. The adjacency matrix is given by

$$\mathcal{E} = \begin{pmatrix} 0 & I & I \\ I & 0 & 0 \\ I & 0 & 0 \end{pmatrix}.$$

We first consider the case where both exterior nodes satisfy the Neumann condition. Since this time the edge degree is not uniform, we have to consider the generalized eigenvalue problem

$$\mathcal{E}\phi = \mu \mathrm{diag}\,(2I, I, I)\phi$$

and then take $\lambda = \arccos(\mu)$. We obtain the eigenvalues $\lambda = \pi/2$ plus integer multiples of π. We also have all integer multiples of π including zero. The same is true for the serial beam, where the angle between the strings is 180 degrees rather than 90 degrees. In fact, our analysis clearly shows that the eigenvalues (eigenfrequencies) are independent of the angle at the joint. In general, we can show that the serial string (collinear or non-collinear) having m elements possesses all multiples of π/m plus multiples of π as eigenvalues (if, as assumed, the exterior nodes satisfy Neumann conditions).

In the case of the carpenter's square as above but with both exterior nodes satisfying Dirichlet conditions we find

$$\sum_h e_{1h}\phi_h = \phi_2 + \phi_3 = 2\mu\phi_1$$

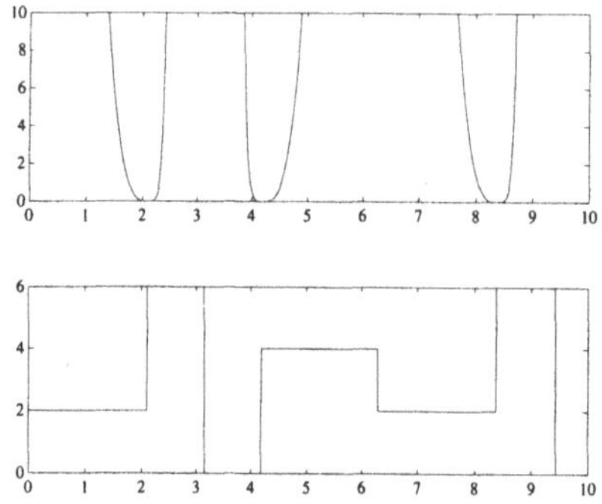

FIGURE V.4. Eigenvalues of a triangle

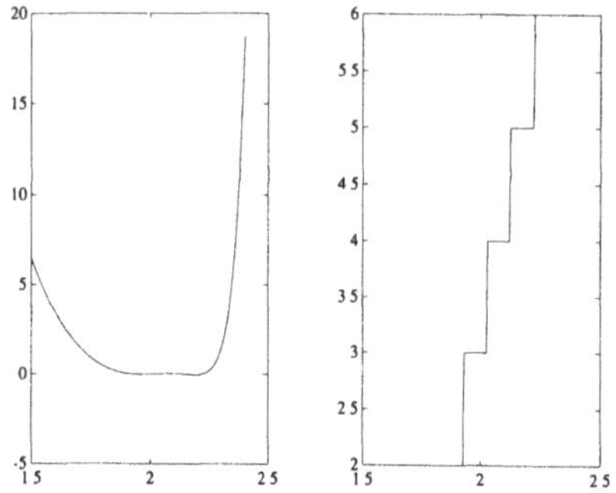

FIGURE V.5. Splitting of the spectrum

and $\phi_2 = \phi_3 = 0$, so that $\mu = 0$ and hence $\cos(\lambda) = 0$ and, therefore, $\pi/2$ plus all integer multiples of π are eigenvalues. In addition, we have all integer multiples of π with the exception of 0.

If we take the same model, but instead of two Dirichlet nodes we take one node satisfying the Dirichlet and the other the Neumann condition we obtain the equations

$$\sum_h e_{1h}\phi_h = \phi_2 + \phi_3 = 2\mu\phi_1, \quad \sum_h e_{3,h}\phi_h = \phi_1 = \mu\phi_3$$

which, because of $\phi_2 = 0$, reduce to $\phi_1 = 2\mu^2\phi_1$ or to $\mu = \pm 1/\sqrt{2}$ and imply that $\lambda = \pi/4, 3\pi/4$ plus all integer multiples of π.

This example shows how the "softening" of boundary conditions (i.e., going from Dirichlet to Neumann conditions) leads to a decrease of the lowest eigenvalue, as it should be.

EXAMPLE 2.3. *The equal sided triangle.*

This case is also easily treated. This is the simplest case of a non-bipartite graph. Also $N = n$, and $\mathbf{G}$ is 2-regular. In fact this graph is also complete in the sense that $\mathcal{E} = ee^* - I$:

$$\mathcal{E} = \begin{pmatrix} 0 & I & I \\ I & 0 & I \\ I & I & 0 \end{pmatrix}.$$

Here we obtain $\lambda = (k + 1/3)\pi$ for odd k and $(k + 2/3)\pi$ for even k. In addition, we obtain 0 and all *even* multiples of π. We display the same numerical evidence as in the first example: in the first graph of Figure V.4 we see the plot of the determinant of $\mathcal{C} - \text{diag}(H_i)$ and in the second we show the corresponding multiplicities, these being in agreement with the theory.

We also perform the perturbation of constants analogous to the one in Example 2.1. The result is displayed in Figure V.5. Again we observe the splitting of multiple eigenvalues.

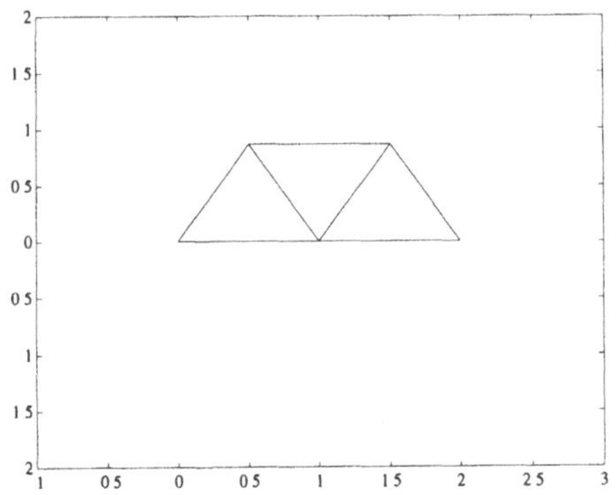

FIGURE V.6. A typical frame

EXAMPLE 2.4. *The triangle with appendices.*

We now take the triangle of the last example and attach a string at each node. The exterior nodes are first assumed to satisfy Neumann conditions. We label the multiple nodes as 1, 2, 3 and the simple nodes of the attachments as 4, 5, 6, respectively. We still have $N = n$, but now the graph is not regular (and, of course, not complete). Here the adjacency matrix takes on the form

$$\mathcal{E} = \begin{pmatrix} 0 & I & I & I & 0 & 0 \\ I & 0 & I & 0 & I & 0 \\ I & I & 0 & 0 & 0 & I \\ I & 0 & 0 & 0 & 0 & 0 \\ 0 & I & 0 & 0 & 0 & 0 \\ 0 & 0 & I & 0 & 0 & 0 \end{pmatrix}.$$

As a result we obtain from Z three different strings of eigenvalues with

$$\arccos(\mu_s)/3 = 0.3570\pi, \quad 0.6082\pi, \quad 0.7785\pi.$$

The other eigenvalues are 0 and again the *even* multiples of π.

If we now consider the three attached strings as being clamped, then the adjacency matrix of the sub-graph problem, obtained by deleting the rows associated with clamped vertices, is the same as in the previous example. However, the edge degree still is 3 rather than 2 as in the previous example. In this case we do not have 0 but rather the multiples of π and the eigenvalues associated with $\arccos(\mu)/4 = .6082\pi,\ .2677\pi$.

EXAMPLE 2.5. *Outlook for general frames.*

In the general case we have to deal with repetitive structures as in Wittrick and Williams [110], [111], Simpson and Tabarrok [103]. For homogeneous networks the eigenvalues can be given explicitly according to the theory. This can also be done in the 3-dimensional case. We close the collection of examples with a suggestive picture which appears to be typical in this context; see Figure V.6. In this situation we obtain evidence on the distribution of eigenvalues and their multiplicities; see Figure V.7.

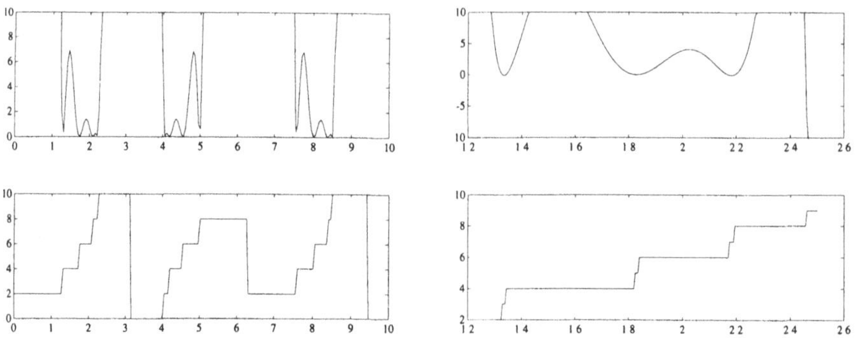

FIGURE V.7. Distribution of eigenvalues

2.4. Networks of Timoshenko Beams. We are now going to discuss the eigenvalue problem for the Timoshenko system (1.8), (1.9), (1.10) and (1.11), i.e., the system

$$\text{(2.8)}\quad \begin{cases} \ell_{ih}^{(-2)}\mathbf{K}_{s(i,h)}\mathbf{r}_{ih}'' + \ell_{ih}^{(-1)}\mathbf{S}_{s(i,h)}\mathbf{r}_{ih}' + q_{s(i,h)}^2\mathbf{P}_{\mathbf{n}}\mathbf{r}_{ih} = -\lambda^2\mathbf{r}_{ih},\\ \sum_h[\ell_{ih}^{(-1)}\mathbf{K}_{s(i,h)}\mathbf{r}_{ih}'(0) + d_{is(i,h)}\mathbf{M}_{s(i,h)}\mathbf{r}_{ih}(0)] = 0,\\ \exists\phi : \mathbf{r}_{ih}(0) = e_{ih}\phi(i),\\ \mathbf{r}_{hi}(x) = \mathbf{r}_{ih}(1-x). \end{cases}$$

As in the preceding case, we want to derive a general solution formula for the Cauchy problem. In order to do this it is convenient to eliminate the shear angle as usual. To this end we have to write down the equations of motion in terms of $u_{ih}, w_{ih}, \psi_{ih}$ as follows:

$$\begin{aligned} \ell_{s(i,h)}^{-2}p_{s(i,h)}^2u_{ih}'' &= -\lambda^2u_{ih},\\ \ell_{s(i,h)}^{-2}q_{s(i,h)}^2w_{ih}'' + \ell_{s(i,h)}^{-1}q_{s(i,h)}^2\psi_{ih}' &= -\lambda^2w_{ih},\\ \ell_{s(i,h)}^{-2}\gamma_{s(i,h)}^2\psi_{ih}'' - \ell_{s(i,h)}^{-1}q_{s(i,h)}^2w_{ih}' - q_{s(i,h)}^2\psi_{ih} &= -\lambda^2\psi_{ih}. \end{aligned}$$

We differentiate the second equation twice and the third equation once. Also, we express ψ_{ih}' in terms of the vertical displacement and its second derivative, using the second equation. The result is

$$\begin{aligned} \ell_{s(i,h)}^{-2}q_{s(i,h)}^2w_{ih}'''' + \ell_{s(i,h)}^{-1}q_{s(i,h)}^2\psi_{ih}''' &= -\lambda^2w_{ih}'',\\ \ell_{s(i,h)}^{-2}\gamma_{s(i,h)}^2\psi_{ih}''' - \ell_{s(i,h)}^{-1}q_{s(i,h)}^2w_{ih}'' - q_{s(i,h)}^2\psi_{ih}' &= -\lambda^2\psi_{ih}',\\ \psi_{ih}' = -\ell_{s(i,h)}^{-1}w_{ih}'' - \lambda^2\ell_{s(i,h)}q_{s(i,h)}^{-2}w_{ih}&. \end{aligned}$$

We can now use the second and the third equation to eliminate ψ_{ih}. We obtain

$$\text{(2.9)}\quad w_{ih}'''' + \left(1 + \frac{\ell_{s(i,h)}^2}{\gamma_{s(i,h)}^2}\right)\lambda^2w_{ih}'' - \lambda^2\ell_{s(i,h)}^4\frac{q_{s(i,h)}^2}{\gamma_{s(i,h)}^2}(q_{s(i,h)}^2 - \lambda^2)w_{ih} = 0.$$

Since we want to solve the initial value problem, we have to use the compatibility conditions

$$
(2.10)\quad \begin{cases}
w_{ih}(0) = w_{ih0}, \quad w'_{ih}(0) = w'_{ih0}, \\
w''_{ih}(0) = -\lambda^2 \ell^2_{s(i,h)} q^{-2}_{s(i,h)} w_{ih0} - \ell_{s(i,h)} \psi'_{ih0}, \\
w'''_{ih}(0) = -\ell^2_{s(i,h)} \left(\lambda^2 q^{-2}_{s(i,h)} + \dfrac{q^2_{s(i,h)}}{\gamma^2_{s(i,h)}} \right) w'_{ih0} \\
\qquad\qquad + \ell^3_{s(i,h)} \gamma^{-2}_{s(i,h)} (\lambda^2 - q^2_{s(i,h)}) \psi_{ih0}).
\end{cases}
$$

The characteristic equation for the fourth order equation (2.9) is

$$
r^4 - \left(1 + \frac{\ell^2_{s(i,h)}}{\gamma^2_{s(i,h)}}\right) \lambda^2 r^2 - \ell^4_{s(i,h)} \frac{q^2_{s(i,h)}}{\gamma^2_{s(i,h)}} \lambda^2 (q^2_{s(i,h)} - \lambda^2) = 0
$$

and it has the roots

$$
r_{\pm} = \frac{1}{2} \left(1 + \left(\frac{\ell_{s(i,h)}}{\gamma_{s(i,h)}}\right)^2\right) \lambda^2
$$
$$
\pm \left[\frac{1}{4} \left(1 + \left(\frac{q_{s(i,h)}}{\gamma_{s(i,h)}}\right)^2\right)^2 \lambda^4 + \ell^4_{s(i,h)} \left(\frac{q_{s(i,h)}}{\gamma_{s(i,h)}}\right)^2 \lambda^2 (q^2_{s(i,h)} - \lambda^2) \right]^{1/2}.
$$

Note that, in order to obtain oscillatory solutions, λ has to be larger than $q_{s(i,h)}$ if it is not equal to zero. We define $\mu_{ih1} := \sqrt{r_+}$, $\mu_{ih2} := \sqrt{r_-}$, $\mu_{ih3} := -\mu_{ih1}$, $\mu_{ih4} := -\mu_{ih2}$. We can now write down the general solution to the initial value problem (2.9), (2.10). To this end we define

$$\begin{aligned}
\rho_{ih}^{00} &:= \frac{1}{\mu_{ih1}^2 - \mu_{ih2}^2}(\lambda^2 \ell_{s(i,h)}^2 q_{s(i,h)}^{-2} - \mu_{ih2}^2),\\
\rho_{ih}^{10} &:= \frac{1}{\mu_{ih1}} \frac{1}{\mu_{ih1}^2 - \mu_{ih2}^2} \left(\lambda^2 \ell_{s(i,h)}^2 q_{s(i,h)}^{-2} - \mu_{ih2}^2 + \ell_{s(i,h)}^2 \frac{q_{s(i,h)}^2}{\gamma_{s(i,h)}^2}\right),\\
\rho_{ih}^{01} &:= -\frac{1}{\mu_{ih1}^2 - \mu_{ih2}^2}(\lambda^2 \ell_{s(i,h)}^2 q_{s(i,h)}^{-2} - \mu_{ih1}^2),\\
\rho_{ih}^{11} &:= -\frac{1}{\mu_{ih2}} \frac{1}{\mu_{ih1}^2 - \mu_{ih2}^2} \left(\lambda^2 q_{s(i,h)}^{-2} - \mu_{ih1}^2 + \ell_{s(i,h)}^2 \left(\frac{q_{s(i,h)}}{\gamma_{s(i,h)}}\right)^2\right),\\
\theta_{ih}^{00} &:= \frac{1}{\mu_{ih1}} \frac{1}{\mu_{ih1}^2 - \mu_{ih2}^2} \ell_{s(i,h)}^3 \gamma_{s(i,h)}^{-2} (q_{s(i,h)}^2 - \lambda^2),\\
\theta_{ih}^{10} &:= \frac{1}{\mu_{ih1}^2 - \mu_{ih2}^2} \ell_{s(i,h)}\\
\theta_{ih}^{01} &:= -\frac{1}{\mu_{ih2}} \frac{1}{\mu_{ih1}^2 - \mu_{ih2}^2} \ell_{s(i,h)}^3 \gamma_{s(i,h)}^{-2} (q_{s(i,h)}^2 - \lambda^2),\\
\theta_{ih}^{11} &:= -\frac{1}{\mu_{ih1}^2 - \mu_{ih2}^2} \ell_{s(i,h)}.
\end{aligned}$$

Then w_{ih} is given by

$$\begin{aligned}
w_{ih}(x) = (\rho_{ih}^{00} \cos(\mu_{ih1} x) &+ \rho_{ih}^{01} \cos(\mu_{ih2} x)) w_{ih0}\\
&+ (\rho_{ih}^{10} \sin(\mu_{ih1} x) + \rho_{ih}^{11} \sin(\mu_{ih2} x)) w_{ih0}'\\
&+ (\theta_{ih}^{10} \cos(\mu_{ih1} x) + \theta_{ih}^{11} \cos(\mu_{ih2} x)) \psi_{ih0}'\\
&\qquad + (\theta_{ih}^{00} \sin(\mu_{ih1} x) + \theta_{ih}^{01} \sin(\mu_{ih2} x)) \psi_{ih0}.
\end{aligned}$$

We use this representation to solve the equation

$$\psi_{ih}'' + \left(\frac{\ell_{s(i,h)}}{\gamma_{s(i,h)}}\right)^2 (\lambda^2 - q_{s(i,h)}^2) \psi_{ih} = \ell_{s(i,h)} \left(\frac{q_{s(i,h)}}{\gamma_{s(i,h)}}\right)^2 w_{ih}'$$

governing ψ_{ih} by the method of variation of parameters. Let us introduce

$$\xi_{ih} := \frac{\ell_{s(i,h)}}{\gamma_{s(i,h)}} \sqrt{\lambda^2 - q_{s(i,h)}^2}.$$

Then ψ_{ih} is given by

(2.12)

$$
\begin{aligned}
\psi_{ih} = \psi_{ih0} & \left[\cos(\xi_{ih} x) \left(1 - \ell_{s(i,h)} \left(\frac{q_{s(i,h)}}{\gamma_{s(i,h)}} \right)^2 \left(\frac{\theta^{00}_{ih}\mu_{ih1}}{\xi^2_{ih} - \mu^2_{ih1}} + \frac{\theta^{01}_{ih}\mu_{ih2}}{\xi^2_{ih} - \mu^2_{ih2}} \right) \right) \right. \\
& + \cos(\mu_{ih1} x) \left(\ell_{s(i,h)} \frac{q_{s(i,h)}}{\gamma_{s(i,h)}} \theta^{00}_{ih} \mu_{ih1} \frac{1}{\xi^2_{ih} - \mu^2_{ih1}} \right) \\
& \left. + \cos(\mu_{ih2} x) \left(\ell_{s(i,h)} \frac{q_{s(i,h)}}{\gamma_{s(i,h)}} \theta^{01}_{ih} \mu_{ih2} \frac{1}{\xi^2_{ih} - \mu^2_{ih2}} \right) \right] \\
+ \psi'_{ih0} & \left[\frac{1}{\xi_{ih}} \sin(\xi_{ih} x) \left(1 + \ell_{s(i,h)} \left(\frac{q_{s(i,h)}}{\gamma_{s(i,h)}} \right)^2 \left(\frac{\theta^{10}_{ih}\mu^2_{ih1}}{\xi^2_{ih} - \mu^2_{ih1}} + \frac{\theta^{11}_{ih}\mu^2_{ih2}}{\xi^2_{ih} - \mu^2_{ih2}} \right) \right) \right. \\
& - \frac{1}{\mu_{ih1}} \sin(\mu_{ih1} x) \left(\ell_{s(i,h)} \left(\frac{q_{s(i,h)}}{\gamma_{s(i,h)}} \right)^2 \frac{\theta^{10}_{ih}\mu^2_{ih1}}{\xi^2_{ih} - \mu^2_{ih1}} \right) \\
& \left. - \frac{1}{\mu_{ih2}} \sin(\mu_{ih2} x) \left(\ell_{s(i,h)} \left(\frac{q_{s(i,h)}}{\gamma_{s(i,h)}} \right)^2 \frac{\theta^{11}_{ih}\mu^2_{ih2}}{\xi^2_{ih} - \mu^2_{ih2}} \right) \right] \\
+ w_{ih0} & \left[\frac{1}{\xi_{ih}} \sin(\xi_{ih} x) \left(1 - \ell_{s(i,h)} \left(\frac{q_{s(i,h)}}{\gamma_{s(i,h)}} \right)^2 \left(\frac{\rho^{00}_{ih}\mu^2_{ih1}}{\xi^2_{ih} - \mu^2_{ih1}} + \frac{\rho^{01}_{ih}\mu^2_{ih2}}{\xi^2_{ih} - \mu^2_{ih2}} \right) \right) \right. \\
& - \frac{1}{\mu_{ih1}} \sin(\mu_{ih1} x) \left(\ell_{s(i,h)} \left(\frac{q_{s(i,h)}}{\gamma_{s(i,h)}} \right)^2 \frac{\rho^{00}_{ih}\mu^2_{ih1}}{\xi^2_{ih} - \mu^2_{ih1}} \right) \\
& \left. - \frac{1}{\mu_{ih2}} \sin(\mu_{ih2} x) \left(\ell_{s(i,h)} \left(\frac{q_{s(i,h)}}{\gamma_{s(i,h)}} \right)^2 \frac{\rho^{01}_{ih}\mu^2_{ih2}}{\xi^2_{ih} - \mu^2_{ih2}} \right) \right] \\
+ w'_{ih0} & \left[\cos(\xi_{ih} x) \left(1 - \ell_{s(i,h)} \left(\frac{q_{s(i,h)}}{\gamma_{s(i,h)}} \right)^2 \left(\frac{\rho^{10}_{ih}\mu_{ih1}}{\xi^2_{ih} - \mu^2_{ih1}} + \frac{\rho^{11}_{ih}\mu_{ih2}}{\xi^2_{ih} - \mu^2_{ih2}} \right) \right) \right. \\
& + \cos(\mu_{ih1} x) \left(\ell_{s(i,h)} \left(\frac{q_{s(i,h)}}{\gamma_{s(i,h)}} \right)^2 \frac{\rho^{10}_{ih}\mu_{ih1}}{\xi^2_{ih} - \mu^2_{ih1}} \right) \\
& \left. + \cos(\mu_{ih2} x) \left(\ell_{s(i,h)} \left(\frac{q_{s(i,h)}}{\gamma_{s(i,h)}} \right)^2 \frac{\rho^{11}_{ih}\mu_{ih2}}{\xi^2_{ih} - \mu^2_{ih2}} \right) \right].
\end{aligned}
$$

We also have the longitudinal deformation

$$
(2.13) \qquad u_{ih} = \cos\left(\lambda \frac{\ell_{s(i,h)}}{p_{s(i,h)}} \right) u_{ih0} + \frac{p_{s(i,h)}}{\ell_{s(i,h)}\lambda} \sin\left(\lambda \frac{\ell_{s(i,h)}}{p_{s(i,h)}} \right) u'_{ih0}.
$$

Given the deformations (2.11), (2.12) and (2.13) we can go back to the matrix formulation and write

$$
\mathbf{r}_{ih} = u_{ih}\mathbf{P}_{s(i,h)} + w_{ih}\mathbf{P}^{\perp}_{s(i,h)} + \psi_{ih}\mathbf{P}_{\mathbf{n}}.
$$

We want to express this in terms of the initial values. To this end we define

$$\begin{aligned} u_{ih}(x) &=: U^0_{ih}(\lambda,x)u_{ih0} + U^1_{ih}(\lambda,x)u'_{ih0},\\ w_{ih}(x) &=: W^{00}_{ih}(\lambda,x)w_{ih0} + W^{01}_{ih}(\lambda,x)w'_{ih0}\\ &\qquad\qquad + W^{10}_{ih}(\lambda,x)\psi_{ih0} + W^{11}_{ih}(\lambda,x)\psi'_{ih0},\\ \psi_{ih}(x) &=: Z^{00}_{ih}(\lambda,x)w_{ih0} + Z^{01}_{ih}(\lambda,x)w'_{ih0}\\ &\qquad\qquad + Z^{10}_{ih}(\lambda,x)\psi_{ih0} + Z^{11}_{ih}(\lambda,x)\psi'_{ih0}, \end{aligned}$$

where the functions on the right hand side are defined using the formulas (2.11), (2.12) and (2.13). Thus

$$\begin{aligned} r(x) = &\begin{pmatrix} U^0(\lambda,x)\mathbf{P}_{s(i,h)} & O & O\\ O & W^{00}(\lambda,x)\mathbf{P}^{\perp}_{s(i,h)} & W^{10}(\lambda,x)\mathbf{P}^{\perp}_{s(i,h)}\\ O & Z^{00}(\lambda,x)\mathbf{P}_{\mathbf{n}} & Z^{10}(\lambda,x)\mathbf{P}_{\mathbf{n}} \end{pmatrix}\Phi\\ &+ \begin{pmatrix} U^1 & O & O\\ O & W^{01}(\lambda,x)\mathbf{P}^{\perp}_{s(i,h)} & W^{11}(\lambda,x)\mathbf{P}^{\perp}_{s(i,h)}\\ O & Z^{01}(\lambda,x)\mathbf{P}_{\mathbf{n}} & Z^{11}(\lambda,x)\mathbf{P}_{\mathbf{n}} \end{pmatrix}\Psi. \end{aligned}$$

Another way of writing this is

$$\begin{aligned} \mathbf{r}_{ih}(x) = &\begin{pmatrix} U^0_{ih}(\lambda,x) & O & O\\ O & W^{00}_{ih}(\lambda,x) & W^{10}_{ih}(\lambda,x)\\ O & Z^{00}_{ih}(\lambda,x) & Z^{10}_{ih}(\lambda,x) \end{pmatrix}_{s(i,h)} \begin{pmatrix} u_{ih0}\\ w_{ih0}\\ \psi_{ih0} \end{pmatrix}_{s(i,h)}\\ &+ \begin{pmatrix} U^1_{ih} & O & O\\ O & W^{01}_{ih}(\lambda,x) & W^{11}_{ih}(\lambda,x)\\ O & Z^{01}_{ih}(\lambda,x) & Z^{11}_{ih}(\lambda,x) \end{pmatrix}_{s(i,h)} \begin{pmatrix} u'_{ih0}\\ w'_{ih0}\\ \psi'_{ih0} \end{pmatrix}_{s(i,h)}, \end{aligned}$$

where the matrices and the vectors are given in the base $\mathbf{e}_{s(i,h)}$, $\mathbf{e}^{\perp}_{s(i,h)}$, $\mathbf{n}$, as indicated by the suffix $s(i,h)$. This representation is particularly useful if one wants to have a compact notation. However, one has to keep in mind that for each different pair of indices i,h one is then talking about a different meaning of the component form. We recall

$$\begin{aligned} \mathbf{r}_{ih}(0) &= \Phi_{ih} = u_{ih0}\mathbf{e}_{s(i,h)} + w_{ih0}\mathbf{e}^{\perp}_{s(i,h)} + \psi_{ih0}\mathbf{n},\\ \mathbf{r}'_{ih}(0) &= \Psi_{ih} = u'_{ih0}\mathbf{e}_{s(i,h)} + w'_{ih0}\mathbf{e}^{\perp}_{s(i,h)} + \psi'_{ih0}\mathbf{n}. \end{aligned}$$

We may now define

$$\begin{aligned} \Theta^0_{ih}(\lambda,x) = U^0_{ih}(\lambda,x)\mathbf{P}_{s(i,h)} &+ W^{00}_{ih}(\lambda,x)\mathbf{P}_{s(i,h)}\\ &+ W^{10}_{ih}(\lambda,x)\mathbf{M}_{s(i,h)} + Z^{00}_{ih}\mathbf{N}_{s(i,h)} + Z^{10}_{ih}\mathbf{P}_{\mathbf{n}}, \end{aligned}$$

$$\Theta^1_{ih}(\lambda,x) = U^1_{ih}(\lambda,x)\mathbf{P}_{s(i,h)} + W^{01}_{ih}(\lambda,x)\mathbf{P}_{s(i,h)} + W^{11}_{ih}(\lambda,x)\mathbf{M}_{s(i,h)} + Z^{01}_{ih}(\lambda,x)\mathbf{N}_{s(i,h)} + Z^{11}_{ih}\mathbf{P}_{\mathbf{n}}.$$

Then the solution to the Cauchy problem can be written as

$$\mathbf{r}_{ih}(x) = \Theta^0_{ih}(\lambda,x)\cdot\Phi_{ih} + \Theta^1_{ih}(\lambda,x)\cdot\Psi_{ih}. \tag{2.14}$$

2.4.1. *The Case Where* $\lambda = 0$. In the case where $\lambda = 0$ we have to solve

$$\ell^{(-2)}_{ih}\mathbf{K}_{s(i,h)}\mathbf{r}''_{ih} + \ell^{(-1)}_{ih}\mathbf{S}_{s(i,h)}\mathbf{r}'_{ih} + q^2_{s(i,h)}\mathbf{P}_{\mathbf{n}}\mathbf{r}_{ih} = 0,$$

or, what amounts to be same thing,

$$\ell^{-2}_{s(i,h)}p^2_{s(i,h)}u''_{ih} = 0,$$
$$\ell^{-2}_{s(i,h)}q^2_{s(i,h)}w''_{ih} + \ell^{-1}_{s(i,h)}q^2_{s(i,h)}\psi'_{ih} = 0,$$
$$\ell^{-2}_{s(i,h)}\gamma^2_{s(i,h)}\psi''_{ih} - \ell^{-1}_{s(i,h)}q^2_{s(i,h)}w'_{ih} - q^2_{s(i,h)}\psi_{ih} = 0.$$

This implies

$$u_{ih}(x) = u_{ih0} + x(u^*_{ih0} - u_{ih0}).$$

Furthermore, $w''_{ih}(x) = -\ell_{s(i,h)}\psi'_{ih}(x)$ implies $\psi'''_{ih}(x) \equiv 0$. Hence

$$\psi_{ih}(x) = \psi_{ih0} + x\psi_{ih1} + x^2\psi_{ih2}$$

for some ψ_{ih2}. This implies

$$w''_{ih}(x) = -\psi_{ih1} - 2x\psi_{ih2}.$$

The identity $(w^*_{ih})''(x) = w''_{ih}(1-x)$ implies

$$\psi_{ih2} = \frac{1}{2}(\psi^*_{ih1} - \psi_{ih1}).$$

However, we also have $\psi^*_{ih}(x) = \psi_{ih}(1-x)$ and $(\psi'_{ih})^*(x) = -\psi'_{ih}(1-x)$, which implies that $\psi^*_{ih1} = \psi_{ih1}$. Thus, $\psi_{ih}(x)$ is reduced to

$$\psi_{ih}(x) = \psi_{ih0} + x\psi_{ih1},$$

which implies

$$\psi_{ih}(x) = \psi_{ih0} + x(\psi^*_{ih0} - \psi_{ih0}).$$

But then

$$w''_{ih}(x) = -(\psi^*_{ih0} - \psi_{ih0})\ell_{s(i,h)}$$

and hence

$$w_{ih}(x) = -\frac{1}{2}x^2(\psi^*_{ih0} - \psi_{ih0})\ell_{s(i,h)} + w_{ih1}x + w_{ih0}.$$

Again, by reflection at 1 we deduce $\psi^*_{ih0} = \psi_{ih0}$ and $w_{ih1} = w^*_{ih0} - w_{ih0}$. This reduces $\psi_{ih}(x)$ to a constant

$$\psi_{ih}(x) = \psi_{ih0} = \psi^*_{ih0}$$

and w_{ih} to

$$w_{ih}(x) = w_{ih0} + x(w^*_{ih0} - w_{ih0}).$$

With this information we go back to the differential equation for ψ_{ih}, which now reads

$$w'_{ih}(x) = \ell_{s(i,h)}\psi_{ih0} = w^*_{ih0} - w_{ih0}.$$

But since $\psi^*_{ih0} = \psi_{ih0}$, we conclude that $w^*_{ih0} = w_{ih0}$ which, in turn, reduces $w_{ih}(x)$ to a constant

$$w_{ih}(x) = w_{ih0} = w^*_{ih0}$$

and, in addition, $\psi_{ih}(x)$ to zero. Therefore, the necessary conditions for zero to be an eigenvalue are

$$\begin{aligned} u_{ih}(x) =& u_{ih0} + x(u^*_{ih0} - u_{ih0}), \\ w_{ih}(x) =& w_{ih0} = w^*_{ih0}, \\ \psi_{ih}(x) =& 0. \end{aligned}$$

As $u_{ih}(x)$ is the only variable with a non-vanishing derivative, the balance equation reduces to

$$\sum_h d_{is(i,h)}(p^2_{s(i,h)}u'_{ih}(0)\mathbf{e}_{s(i,h)}) = 0,$$

which turns out to be equivalent to

$$\sum_h d_{is(i,h)}(p^2_{s(i,h)}u^*_{ih0}\mathbf{e}_{s(i,h)}) = \sum_h d_{is(i,h)}(p^2_{s(i,h)}u_{ih0}\mathbf{e}_{s(i,h)}).$$

By the same argument as in [106] we deduce $u^*_{ih0} = u_{ih0}$, which then reduces $u_{ih}(x)$ to a constant. *This shows that zero is an eigenvalue if and only if no vertex in $\partial V(\mathbf{G})$ is clamped.*

2.4.2. *The Case Where λ Belongs to an Individual Beam.* Suppose we keep all the nodes fixed, i.e., $\Phi_{ih} = 0$. Then we can use the reflection formula $\mathbf{r}^*_{ih}(0) = \mathbf{r}_{ih}(1)$ in order to obtain

$$0 = \Theta^1_{ih}\Psi$$

or, in matrix form,

$$0 = \begin{pmatrix} U^1_{ih} & O & O \\ O & W^{01}_{ih}(\lambda, x) & W^{11}_{ih}(\lambda, x) \\ O & Z^{01}_{ih}(\lambda, x) & Z^{11}_{ih}(\lambda, x) \end{pmatrix}_{s(i,h)} \begin{pmatrix} u'_{ih0} \\ w'_{ih0} \\ \psi'_{ih0} \end{pmatrix}_{s(i,h)}.$$

Since in this equation u'_{ih0} is uncoupled from w'_{ih0}, ψ'_{ih0}, we can set the latter equal to zero and choose the eigenvalue λ to be a multiple of π.

Likewise, we can take u'_{ih0} to be zero and look for values λ such that the sub-matrix

$$V_{ih}^1 := \begin{pmatrix} W_{ih}^{01}(\lambda,1) & W_{ih}^{11}(\lambda,1) \\ Z_{ih}^{01}(\lambda,1) & Z_{ih}^{11}(\lambda,1) \end{pmatrix}$$

has determinant equal to zero. Going back to the formula for the vertical displacement, we see that this determinant condition describes precisely the eigenvalues of a single Timoshenko beam. Unfortunately, there is no simple explicit representation of those values. Note that we have dispersion in the Timoshenko model. The determinant of the matrix V_{ih}^1 can be computed more easily in the case of homogeneous networks. That case will be studied later. Let us display the formula for the determinant in the case where all coefficients are equal to one:

$$\det(V_{ih}^1) = \det(V^1) = \frac{1}{2\lambda^2}(1 - \cos(\mu_1)\cos(\mu_2)) + \frac{1}{2\lambda\sqrt{\lambda^2-1}}\sin(\mu_1)\sin(\mu_2).$$

The roots of the equation $\det(V^1) = 0$ can be calculated to be asymptotically equal to the integer multiples of π. Again, the fact that they are not exactly (possibly non-integer) multiples of π reflects the phenomenon of dispersion of bending and shear waves in this model.

2.4.3. *Eigenvalues for the Entire Graph.* In this section we assume that the eigenvalue λ is neither zero nor of the form just discussed. We will use the general solution formula (2.14), as before. Recall

$$\Theta_{ih}^1(\lambda,x)\cdot\Psi_{ih} = \begin{pmatrix} U_{ih}^1 & O & O \\ O & W_{ih}^{01}(\lambda,x) & W_{ih}^{11}(\lambda,x) \\ O & Z_{ih}^{01}(\lambda,x) & Z_{ih}^{11}(\lambda,x) \end{pmatrix}_{s(i,h)} \begin{pmatrix} u'_{ih0} \\ w'_{ih0} \\ \psi'_{ih0} \end{pmatrix}_{s(i,h)}$$

and

$$\Theta_{ih}^0(\lambda,x)\cdot\Psi_{ih} = \begin{pmatrix} U_{ih}^0 & O & O \\ O & W_{ih}^{00}(\lambda,x) & W_{ih}^{10}(\lambda,x) \\ O & Z_{ih}^{00}(\lambda,x) & Z_{ih}^{10}(\lambda,x) \end{pmatrix}_{s(i,h)} \begin{pmatrix} u_{ih0} \\ w_{ih0} \\ \psi_{ih0} \end{pmatrix}_{s(i,h)}$$

Then, by the reflection property, we have the identity

$$\Theta_{ih}^1(\lambda,1)\cdot\Psi_{ih} = \Phi_{ih}^* - \Theta_{ih}^0(\lambda,1)\cdot\Phi_{ih} \tag{2.15}$$

which, in matrix form, is

$$\begin{pmatrix} U^1_{ih} & O & O \\ O & W^{01}_{ih}(\lambda,x) & W^{11}_{ih}(\lambda,x) \\ O & Z^{01}_{ih}(\lambda,x) & Z^{11}_{ih}(\lambda,x) \end{pmatrix}_{s(i,h)} \begin{pmatrix} u'_{ih0} \\ w'_{ih0} \\ \psi'_{ih0} \end{pmatrix}_{s(i,h)} =$$

$$\begin{pmatrix} u^*_{ih0} \\ w^*_{ih0} \\ \psi^*_{ih0} \end{pmatrix}_{s(i,h)} - \begin{pmatrix} U^0_{ih} & O & O \\ O & W^{00}_{ih}(\lambda,x) & W^{10}_{ih}(\lambda,x) \\ O & Z^{00}_{ih}(\lambda,x) & Z^{10}_{ih}(\lambda,x) \end{pmatrix}_{s(i,h)} \begin{pmatrix} u_{ih0} \\ w_{ih0} \\ \psi_{ih0} \end{pmatrix}_{s(i,h)} .$$

By assumption, the matrices Θ^1_{ih} are invertible. Therefore, we obtain for Ψ_{ih} the equation

$$\begin{pmatrix} u'_{ih0} \\ w'_{ih0} \\ \psi'_{ih0} \end{pmatrix}_{s(i,h)} = (\Theta^1_{ih})^{-1} \begin{pmatrix} u^*_{ih0} \\ w^*_{ih0} \\ \psi^*_{ih0} \end{pmatrix}_{s(i,h)} - A \begin{pmatrix} u_{ih0} \\ w_{ih0} \\ \psi_{ih0} \end{pmatrix}_{s(i,h)} ,$$

where A is given by

$$A = \begin{pmatrix} \dfrac{U^0_{ih}}{U^1_{ih}} & O & O \\ O & \dfrac{1}{\det}(W^{00}_{ih} Z^{11}_{ih} - W^{11}_{ih} Z^{00}_{ih}) & \dfrac{1}{\det}(W^{10}_{ih} Z^{11}_{ih} - W^{11}_{ih} Z^{10}_{ih}) \\ O & \dfrac{1}{\det}(Z^{00}_{ih} W^{01}_{ih} - Z^{01}_{ih} W^{00}_{ih}) & \dfrac{1}{\det}(Z^{10}_{ih} W^{01}_{ih} - Z^{01}_{ih} W^{10}_{ih}) \end{pmatrix}_{s(i,h)}$$

We have suppressed the argument $(\lambda, 1)$ and introduced the notation det $=$ $\det(\Theta^1_{ih}(\lambda,1))$. We are going to use this expression for Ψ in terms of Φ in the balance equations in order to obtain a system of equations involving just Φ. The idea is the same as in the string case; however, the system is much more complicated. We have

$$\text{(2.16)} \quad \sum_h (\ell^{-1}_{s(i,h)} e_{ih} \operatorname{diag}(p^2_{s(i,h)}, q^2_{s(i,h)}, \gamma^2_{s(i,h)})) \left[(\Theta^1_{ih})^{-1} \begin{pmatrix} u_{hi0} \\ w_{hi0} \\ \psi_{hi0} \end{pmatrix}_{s(i,h)} \right.$$

$$\left. - (\Theta^1_{ih})^{-1} \Theta^0_{ih} \begin{pmatrix} u_{ih0} \\ w_{ih0} \\ \psi_{ih0} \end{pmatrix}_{s(i,h)} \right] = \sum_k d_{is(i,h)} M_{s(i,h)} \begin{pmatrix} u_{ih0} \\ w_{ih0} \\ \psi_{ih0} \end{pmatrix}_{s(i,h)} ,$$

for all i such that $v_i \in \overset{\circ}{V}(\mathbf{G}) \cup \partial_1 V(\mathbf{G})$. Even though the equation (2.16) has a very compact form, one has to keep in mind that the components are taken with respect to the individual bases associated with each edge. Therefore (2.16) is not suitable for computations.

We collect the results in a theorem. In contrast to the first theorem, we dispense with the characterization of the corresponding eigenstates. Those can be seen from the preceding discussion.

THEOREM 2.2. *If λ^2 is an eigenvalue of* (2.8), *then λ satisfies one of the following conditions:*
(i) $\lambda = 0$ *(iff $\partial_0 V(\mathbf{G}) = \emptyset$).*
(ii) *λ is a solution of $U^1_{ih} = 0$ or of $\det(V^1_{ih}) = 0$ for some pair i, h.*
(iii) *λ is the solution of* (2.16).

2.5. Homogeneous Timoshenko Networks. In the case of the completely homogeneous network all quantities $\det(V^1_{ih})$, Z^{kl}_{ih}, W^{kl}_{ih} are independent of i, h. We have

$$\det(V^1) = \frac{1}{2\lambda^2}(1 - \cos(\mu_1)\cos(\mu_2)) + \frac{1}{2\lambda\sqrt{\lambda^2-1}}\sin(\mu_1)\sin(\mu_2),$$

$$\begin{aligned}
U^0(\lambda, x) &= \cos(\lambda x),\\
U^1(\lambda, x) &= \frac{1}{\lambda}\sin(\lambda x),\\
Z^{00}(\lambda, x) &= \frac{1}{2\mu_1}\lambda\sin(\mu_1 x) - \frac{1}{2\mu_2}\lambda\sin(\mu_2 x),\\
Z^{10}(\lambda, x) &= \frac{\lambda-1}{2\lambda}\cos(\mu_1 x) + \frac{\lambda+1}{2\lambda}\cos(\mu_2 x),\\
Z^{01}(\lambda, x) &= -\frac{1}{2\lambda}\cos(\mu_1 x) + \frac{1}{2\lambda}\cos(\mu_2 x),\\
Z^{11}(\lambda, x) &= \frac{1}{2\mu_1}\sin(\mu_1) + \frac{1}{2\mu_2}\sin(\mu_2 x),
\end{aligned}$$

$$\begin{aligned}
W^{00}(\lambda, x) &= \frac{1}{2}\cos(\mu_1 x) + \frac{1}{2}\cos(\mu_2 x),\\
W^{01}(\lambda, x) &= \frac{\lambda+1}{\sqrt{\lambda^2+\lambda}}\frac{1}{2\lambda}\sin(\mu_1 x) + \frac{\lambda-1}{\sqrt{\lambda^2-1}}\frac{1}{2\lambda}\sin(\mu_2 x),\\
W^{10}(\lambda, x) &= -\frac{\lambda^2-1}{2\lambda\sqrt{\lambda^2+\lambda}}\sin(\mu_1 x) + \frac{\lambda^2-1}{2\lambda\sqrt{\lambda^2-\lambda}}\sin(\mu_2 x),\\
W^{11}(\lambda, x) &= \frac{1}{2\lambda}\cos(\mu_1 x) - \frac{1}{2\lambda}\cos(\mu_2 x).
\end{aligned}$$

In the situation under consideration equation (2.16) can be rewritten in terms of the projections $\mathbf{P}_{s(i,h)}$, $\mathbf{P}_{s(i,h)}^{\perp}$, $\mathbf{P}_{\mathbf{n}}$ as

$$
\begin{aligned}
(2.17)\quad & \sum_h \Big\{ \frac{1}{U^1}\mathbf{P}_{s(i,h)}\phi_h + \frac{1}{\det}(Z^{11}\mathbf{P}_{s(i,h)}^{\perp}\phi_h - W^{11}\mathbf{M}_{s(i,h)}\phi_h) \\
& \qquad + \frac{1}{\det}(-Z^{01}\mathbf{N}_{s(i,h)}\phi_h + W^{01}\mathbf{P}_{\mathbf{n}}\phi_h)\Big\} \\
& = \sum_h \Big\{ \frac{U^0}{U^1}\mathbf{P}_{s(i,h)} + \frac{1}{\det}(W^{00}Z^{11} - W^{11}Z^{00})\mathbf{P}_{s(i,h)}^{\perp} \\
& + \frac{1}{\det}(W^{10}Z^{11} - W^{11}Z^{10})\mathbf{M}_{s(i,h)} + \frac{1}{\det}(Z^{00}W^{01} - Z^{01}W^{00})\mathbf{N}_{s(i,h)} \\
& \qquad + \frac{1}{\det}(Z^{01}W^{10} - Z^{10}W^{01})\mathbf{P}_{\mathbf{n}}\Big\}\phi_i + \Big(\sum_h d_{is(i,h)}\mathbf{M}_{s(i,h)}\Big)\phi_i.
\end{aligned}
$$

The matrix elements W^{kl}, Z^{kl} are evaluated at $(\lambda, 1)$. It is clear that even in this simple situation equation (2.17) is too complicated to analyze in general. It is possible, however, to compute the eigenvalues from (2.17) for very simple configurations such as serial collinear beams or certain non-colinear networks. But the computations are enormous, even if only few edges are involved. Things turn out to be much simpler if we concentrate on the asymptotic behavior of the eigenvalues, that is, the behavior of large λ. If all constants are equal to one, we have

$$\mu_1 = \sqrt{\lambda^2 + \lambda}, \quad \mu_2 = \sqrt{\lambda^2 - \lambda}.$$

Consequently $\mu_k \approx \lambda$ for large λ. We find, therefore, the asymptotic formulas

$$
\begin{gathered}
\det \approx \frac{1}{\lambda^2}\sin^2\lambda, \\
Z^{00}(\lambda,1) \approx 0, \quad Z^{10}(\lambda,1) \approx \cos(\lambda), \\
Z^{01}(\lambda,1) = o(\frac{1}{\lambda}), \quad Z^{11}(\lambda,1) \approx \frac{1}{\lambda}\sin(\lambda), \\
W^{00}(\lambda,1) \approx \cos(\lambda), \quad W^{10}(\lambda,1) \approx 0, \\
W^{01}(\lambda,1) \approx \frac{1}{\lambda}\sin(\lambda), \quad W^{11}(\lambda,1) = o(\frac{1}{\lambda}).
\end{gathered}
$$

We multiply equation (2.17) by $\det^{1/2} = (\sin\lambda)/\lambda$ to obtain the asymptotic equation

$$
\begin{aligned}
(2.18)\quad & \sum_h e_{ih}\{\mathbf{P}_{s(i,h)}\phi_h + \mathbf{P}_{s(i,h)}^{\perp}\phi_h + \mathbf{P}_{\mathbf{n}}\phi_h\} \\
& \qquad = \sum_h e_{ih}\{U^0(\lambda,1)\mathbf{P}_{s(i,h)} + \cos(\lambda)\mathbf{P}_{s(i,h)}^{\perp} + \cos(\lambda)\mathbf{P}_{\mathbf{n}}\}\phi_i.
\end{aligned}
$$

Equation (2.18) obviously reduces to

$$\mathcal{E}\phi = \cos(\lambda)\text{diag}\,(d_i)\phi, \tag{2.19}$$

which we recognize as the same equation as in the string case. This, however, comes as no surprise if one takes into account that, by assumption, all wave speeds with respect to the longitudinal, vertical and shearing motions, respectively, are the same and that the effect of dispersion is only present in the lower modes.

3. Numerical Simulations of Controlled 1–d Networks

3.1. Introductory Remarks. We do not attempt to give an *overview* of numerical methods in the general context of multilink flexible structures. There is a vast engineering literature on this topic. Instead, we want to concentrate on the very specific problems treated in this monograph. Indeed, we shall only discuss one dimensional links and leave the higher dimensional case for future study. Within the context of one dimensional links such as strings and beams, we focus on the simulation of the transient behavior of the solutions. From the point of view of open loop control concepts such as approximate, spectral or exact controllability, it is exactly the transient behavior which is subjected to controls. This is also true in the context of stabilizability by means of feedback controls. In particular, the total energy as a function of time is the object of interest. We therefore need numerical schemes which *conserve the total energy* of the approximated system as long as no controls are applied to the system. That is to say, we have to rule out methods which exhibit *algorithmic damping.* In addition, the frequencies of the approximating system should be close to those of the original problem or, in other words, there should only be a *small amount of numerical dispersion.* These requirements, taken together, are very restrictive if large structures are involved. In this respect, numerical simulations of the kind considered here are quite different from those pursued in the engineering sciences as far as large flexible structures are concerned. In the engineering context one is not interested in energy conserving methods but rather in schemes *guaranteeing* a certain degree of algorithmic damping. This is due to the fact that the generalized eigenvalues corresponding to the finite element approximation (which is used almost exclusively) usually differ quite substantially from the corresponding eigenvalues of the original system if the number of modes considered is large. In particular, the largest eigenvalues of the finite element approximation sometimes have little to do with the spectrum of the original problem. Therefore, damping of high modes is a sought after property of a numerical scheme. Also, because the condition numbers of the *condensed* stiffness and mass matrices literally explode with the number of modes, iterative schemes are mostly used to solve the corresponding algebraic problem after complete discretization.

As we are mainly interested in giving numerical support to theoretical results which have been obtained in the preceding sections and because of the above requirements and their natural consequences, we will discuss examples of low dimension, that is networks with only a small number of flexible elements (say ≤ 200 DOF). As a result, we do not have to worry too much about CPU time and we can resort to classical techniques which have been described in the literature. The schemes which we will consider are:

- The $(\frac{1}{2}, \frac{1}{4})$- Newmark scheme applied to the semi-discrete equations resulting from the finite element procedure.
- The Lax-Friedrichs scheme for the fully discretized finite difference approach.
- High order implicit Runge-Kutta schemes for semi-discretized systems and differential-algebraic systems.

The calculations have been carried out on a 486MS-DOS PC and on a HP 9000 Apollo work station using MATLAB and also FORTRAN codes. We are not going to be any more specific about source codes or CPU times.

3.2. Networks of Strings.

3.2.1. *Absorbing Controls.* In order to give an exemplaric introduction, we will consider the numerical realization of stabilizing feedback controls. These controls are of more practical interest than open loop controls. In particular, the feedback controls we discuss are of the absorbing type. They can be viewed as being at the interface between controllability concepts and stabilizability concepts. Loosely speaking, an absorbing control is one which passes the entire energy flux in one direction. The history of these controls goes back to the beginning of control theory of distributed parameter systems and is originally connected with D.L. Russell; see [92] for an excellent review. From the mathematical viewpoint it is most easily explained for the 1-d wave equation in its characteristic description as in [92].

EXAMPLE 3.1. *The 1-d wave equation.*
Let us write down the scaled classical 1-d-wave equation as follows:

$$\begin{cases} \ddot{w}(x,t) = w''(x,t), & (x,t) \in (0,1) \times (0,T), \\ w(0,t) = 0, \;\; w'(1,t) = \phi, & t \in (0,T), \\ w(x,0) = w_0(x), \;\; \dot{w}(x,0) = w_1(x), & x \in [0,1]. \end{cases}$$

Under the transformations

$$u_1 := \dot{w}, \; u_2 := w',$$

$$v_1 := \frac{1}{\sqrt{2}}(-u_1 + u_2), \; v_2 = \frac{1}{\sqrt{2}}(u_1 + u_2),$$

we have the first order system

$$\dot{\mathbf{v}} = \Gamma \mathbf{v}', \quad \Gamma = \begin{pmatrix} -1 & 0 \\ 0 & 1 \end{pmatrix},$$
$$v_1(0) = v_2(0), \quad v_1(1) + v_2(1) = \sqrt{2}\phi.$$

In this simple case the total energy and its time derivative are given by

$$E = \int_0^1 \|\mathbf{v}\|^2 dx, \quad \dot{E} = \frac{1}{2}\{-v_1(1)^2 + v_2(1)^2\}.$$

It is obvious that the optimal choice of a control is the one leading to $v_2(1) = 0$ for all $t \geq 0$. In fact, it is readily seen that this choice amounts to canceling the outgoing signals at the boundary point $x = 1$. This means that the control mimics the response of the half-infinite string. This is Russell's philosophy in [92]: "let nature take its course". In terms of feedback controls the goal of canceling outgoing signals can be achieved by letting $\phi(t) = v_1(1,t)/\sqrt{2}$, which is equivalent to the closed loop boundary condition $w'(1,t) = -\dot{w}(1,t)$ which, in turn, is a continous system analogue to the classical *dead beat control.* It is also possible to interpret this in terms of an open loop control as follows:

$$\phi(t) = \frac{1}{2}\begin{cases} w_1(t) - w_0'(t) & t \in (0,1), \\ -(w_0'(1-t) + w_1(t)), & t \in (1,2). \end{cases} \tag{3.1}$$

The controls just derived steer the initial conditions to rest in time $T = 2$. It is a classical result that these controls are also time optimal and, on the optimal time interval, norm-minimal. Similar arguments can be applied to the Timoshenko system. We note, however, that because of the dispersion of waves in that case, we do not have exact cancellation of outgoing signals at the boundary with the consequence that exact controllability by velocity feedback controls cannot be achieved. Nevertheless, exact controllability can clearly be proved for open-loop controls. It is apparent that besides the necessity of discretizing the state equation and the boundary conditions in an appropriate way, the total energy as a function of time plays a key role. In particular, it appears to be necessary to use a numerical scheme which conserves the total energy of the discretized system as long as no controls are applied. This can, for instance, be achieved by the classical *Lax-Friedrichs scheme* discussed by Le Roux [91]. See also Halpern [59] for various schemes simulating absorbing boundaries. This scheme is applied to the system above in terms of the variables u_i, and it is also appropriate to quasi-linear models. To avoid unnecessary indexing, we write $u = w'$, $v = \dot{w}$, where w solves

$$\ddot{w} = (\sigma(w'))'$$

for a suitable function σ such that

$$\dot{u} = v', \quad \dot{v} = u'.$$

Upon introducing the grid $x_i = i \cdot \Delta x$, $i = 0, ..., n$, $t_j = j \cdot \Delta t$, $j = 0, ...m$, where $\Delta x = 1/n$, $\Delta t = q\Delta x$, q being the appropriate mesh ratio, we have (see [91])

$$(3.2) \qquad \begin{cases} u_{i+1}^{j+1} = \frac{1}{2}\{(u_{i+1}^j + u_{i-1}^j) + \frac{q}{2}(v_{i+1}^j - v_{i-1}^j)\}, \\ v_{i+1}^{j+1} = \frac{1}{2}\{(v_{i+1}^j + v_{i-1}^j) + \frac{q}{2}(\sigma(u_{i+1}^j) - \sigma(u_{i-1}^j))\}. \end{cases}$$

The boundary conditions are represented by

$$(3.3) \qquad \begin{cases} v_0^{j+1} = 0, \quad u_n^{j+1} = 0, \\ u_0^{j+1} = u_1^j + q \cdot v_1^j, \quad v_n^{j+1} = v_{n-1}^{j+1} - q \cdot u_{n-1}^j. \end{cases}$$

The Courant-Friedrichs-Lewy condition requires that $q \leq 1$; we always use $q = 1$. This setup is particularly useful if the boundary feedback above is to be implemented, because it prescribes the strain at the boundary $x = 1$. We simply have to set $u_n^{j+1} = -v_n^{j+1}$. For further information, in particular on other realizations of absorbing boundary conditions , see Halpern [59].

We first look at the uncontrolled string. We take 45 spatial mesh points and 225 time steps, which amounts to time span of $[0, 5]$. As initial conditions we choose $w(x, 0) = \sin(3\pi x^2)$, $\dot{w}(x, 0) = 0$. The reason for this choice is that we do not want to look at eigenmodes but rather at the response to sufficiently rich data; see Figure V.8. *Note that we display the evolution of strain*, first because of the choice of variables (indeed we could have integrated the strain in order to obtain the displacement), but secondly because the strain is subjected to controls.

We also look at the contour plot for the uncontrolled motion; see figure V.9. This plot clearly shows the propagation along the characteristics.

The total energy, which is also easily computed numerically using the trapezoidal rule, is displayed in Figure V.10.

We are now going to apply the absorbing boundary control at the boundary $x = 1$. The result is shown in Figure V.11

The total energy is displayed in Figure V.12.

These plots serve as evidence for the remarks above. The numerical procedure can be (and has been) applied also to planar strings and planar Timoshenko beams. For similar simulations using, however, the classical direct discretization of the wave equation, see J. Schmidt [97]. □

Now that the control strategy is essentially settled, we are going to attack some exemplaric multi-string models. We will, however, only consider out-of-plane displacement models in this section and leave the planar models to

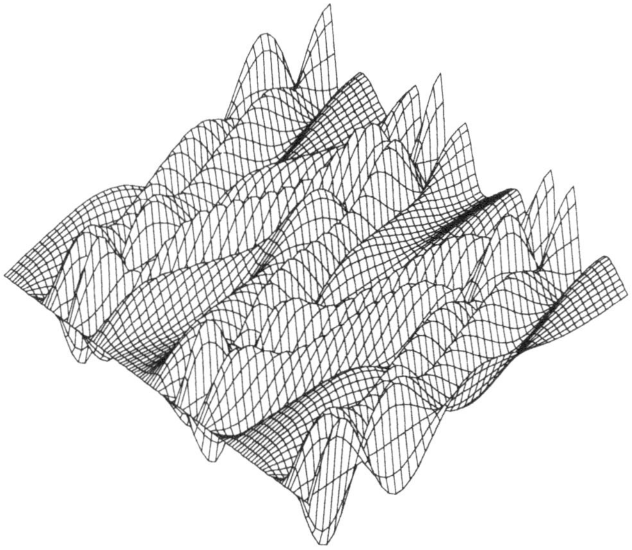

FIGURE V.8. Uncontrolled motion of a string

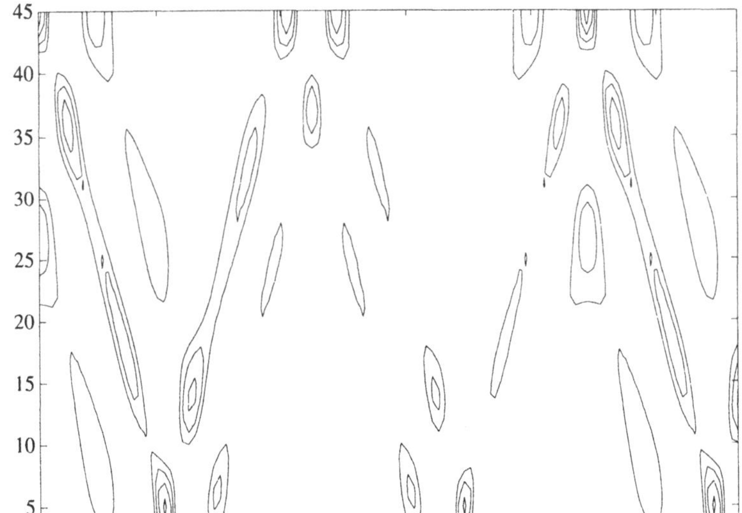

FIGURE V.9. Contour plot

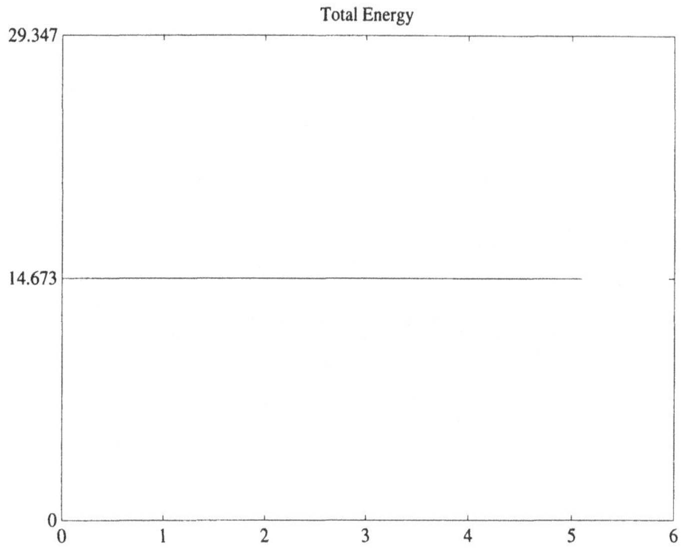

FIGURE V.10. Conservation of energy

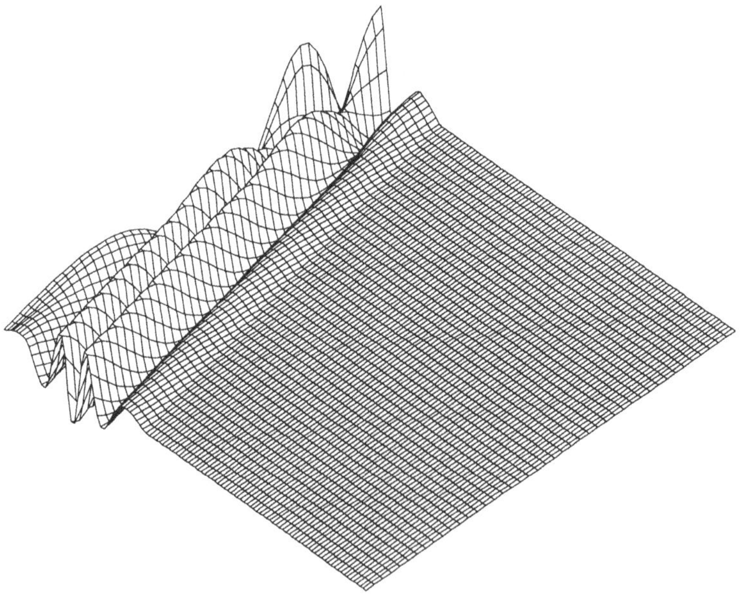

FIGURE V.11. Absorbing boundary control

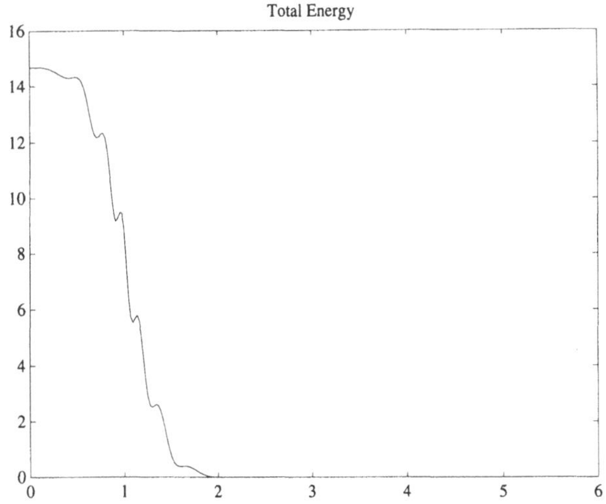

FIGURE V.12. Energy plot of disappearing solution

the finite element section. The first multilink system we look at is a system with three strings coupled together at a multiple node of vertex degree 3.

EXAMPLE 3.2. *Three strings.*

We do not write down the equations of motion, the boundary and node conditions for this model again. Rather, we just want to show the numerical simulation which is performed as in Example 3.1. Two situations are considered:

(a) two exterior nodes are clamped, the remaining one is controlled;
(b) one exterior node is clamped, one is free and the remaining one is controlled.

We label the strings such that the first and the second string are always uncontrolled, while string #3 is subjected to an absorbing control at its end as described above. We have put the initial displacement into the first string. See Figure V.13 for the first case and Figure V.14 for the second case.

Figure V.13 clearly reveals the fact that the system is not even approximately controllable: energy which initially is located in the first string enters the second and the third string through the multiple node. Then, energy is absorbed at the boundary of the third string. Indeed, energy is absorbed there until the motion in strings #1 and #2 settles to a residual motion which vanishes at the multiple node.

Figure V.14 shows to what extent the previous situation is sensitive to changes in the boundary conditions. This case parallels the two-string model with in-span control discussed by Chen et al [15]. There it was shown that for two Dirichlet (Neumann) boundary conditions and control at the mid-span, one does not have approximate controllability, while if

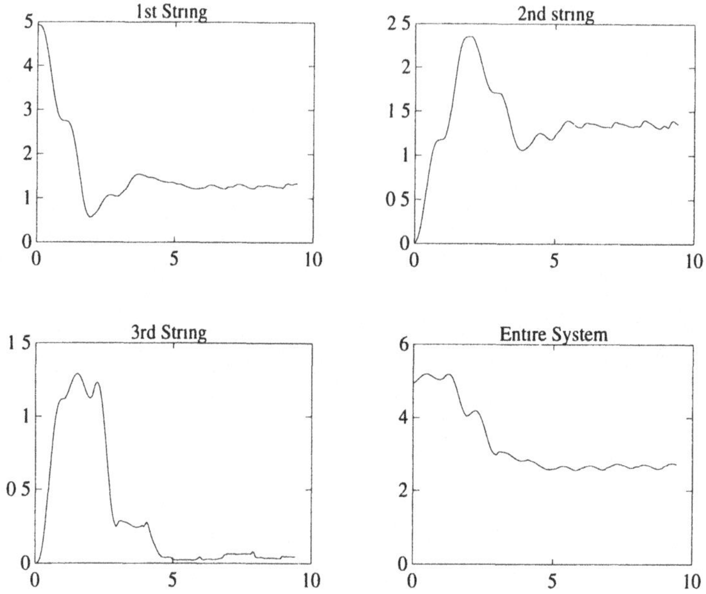

FIGURE V.13. Case (a)

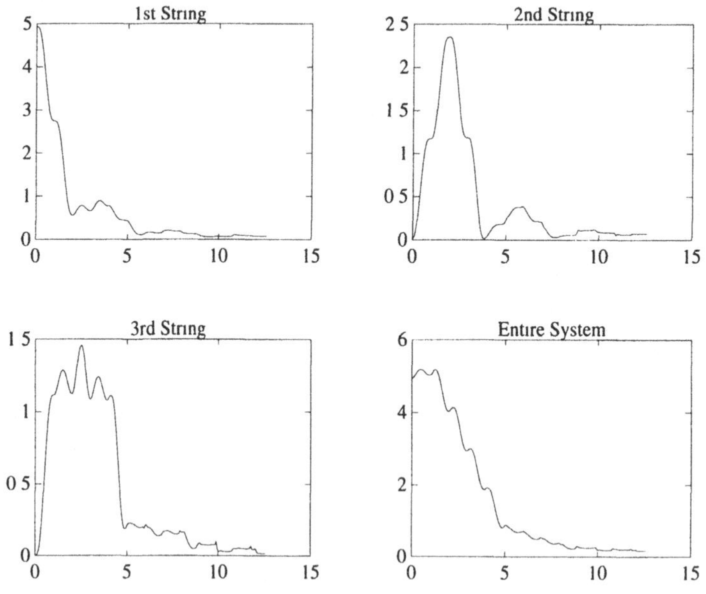

FIGURE V.14. Case (b)

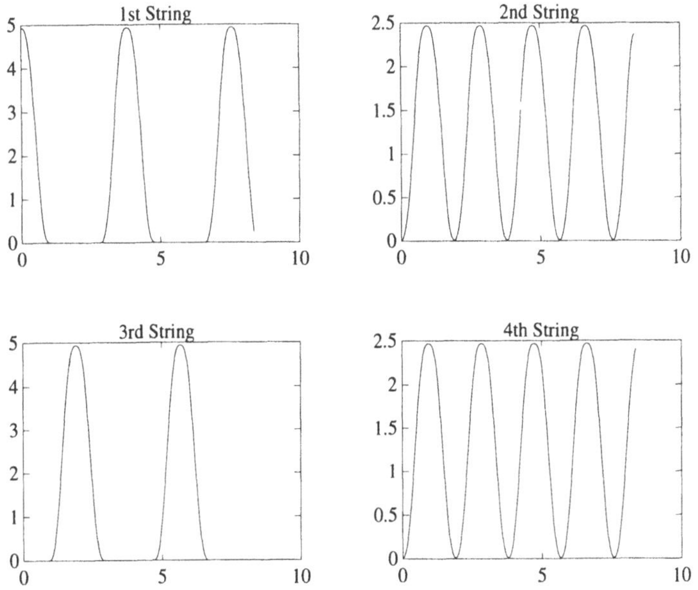

FIGURE V.15. Transfer of energy through multiple joints

one boundary condition changes from Dirichlet to Neumann (Neumann to Dirichlet) exact controllability results. In our three string model such a mathematical analysis does not seem to be known. The numerical simulations, however, strongly support an affirmative conjecture. □

EXAMPLE 3.3. *The unit square.*

We now turn to yet another example, which we have seen before: the unit square. We know that this system is not approximately controllable with controls at the joints. But in this example we do not dwell on numerical evidence concerning controllability, which we defer until the finite element section. Instead, we want to use this model to demonstrate the correctness of our modeling. We consider the unit square with initial displacement in the first (bottom) string. As in the previous cases, we take 20 spatial mesh points for each string. Figure V.15 nicely represents the physically evident flow pattern: motion penetrates from the bottom string to the two parallel upright strings #2, #4 (with half of the energy in each) and crosses into the top string #3 (which had been at rest for one time unit). At time 2 the energy is concentrated entirely in the top string, and the reverse pattern takes place. Hage See Figure V.15 □

3.2.2. *Directing Controls.* We pursue the concept of absorbing controls a bit further and consider now three identical strings and out-of-plane displacement. The fact that we concentrate on this model, rather than on the planar system, is only for the sake of a convenient display of the plots. We will consider planar systems later.

EXAMPLE 3.4. *Three serial strings.*
We have the following system:

$$\begin{cases} \ddot{w}^i = (w^i)'' \quad \text{on} \quad (0,1)\times(0,T), \\ w^1(0)=0, \;\; w^1(1)=w^2(1), \;\; (w^1)'(1)+(w^2)'(1)=f^1, \\ w^3(1)=0, \;\; w^2(0)=w^3(0), \;\; (w^2)'(0)+(w^3)'(0)=f^2, \end{cases} \tag{3.4}$$

together with initial conditions. This system describes the motion of three strings with obvious orientation of the running variable x. The orientation is chosen in order to avoid mixed boundary conditions. We perform a succession of transformations as above, namely,

$$\begin{gathered} u_1^i := \dot{w}^i, \;\; u_2^i := (w^i)', \\ v_1^i := \frac{1}{\sqrt{2}}(-u_1^i + u_2^i), \;\; v_2^i := \frac{1}{\sqrt{2}}(u_1^i + u_2^i), \\ z_i := v_1^i, \; i = 1,2,3, \;\; z_i := (v_2^{i-3})', \; i = 4,5,6. \end{gathered}$$

With these transformations we are able to write the system as

$$\begin{gathered} \dot{z}^i = -(z_i)', \; i = 1,2,3, \\ \dot{z}^i = (z_i)', \; i = 4,5,6, \\ z_1(0) = z_4(0), \;\; z_1(1) - z_4(1) = z_2(1) - z_5(1), \\ z_3(1) = z_6(1), \;\; z_2(0) - z_5(0) = z_3(0) - z_6(0), \\ z_4(1) + z_2(1) = \frac{1}{\sqrt{2}} f^1, \;\; z_5(0) + z_3(0) = \frac{1}{\sqrt{2}} f^2, \end{gathered}$$

together with initial conditions. Our goal now is to steer the energy of the entire system to the second string by applying absorbing controls at the two multiple joints connecting the second string with the first and the third strings. We adopt the strategy of P. Hagedorn and J. Schmidt, see [97]. To this end, we define

$$f^1 := \sqrt{2} z_2(1), \;\; f^2 := \sqrt{2} z_5(0). \tag{3.5}$$

By this choice the outgoing signal $z_3(0)$ at $x = 0$ and $z_4(1)$ at the boundary $x = 1$ are canceled. For the sake of convenience, we write down the boundary conditions in a matrix form reflecting incoming and outgoing signals as in Russell [92]:

$$
\begin{aligned}
\begin{pmatrix} z_1 \\ z_2 \\ z_3 \end{pmatrix}(0) &= \begin{pmatrix} 1 & 0 & 0 \\ 0 & 1 & 1 \\ 0 & 0 & 0 \end{pmatrix} \begin{pmatrix} z_4 \\ z_5 \\ z_6 \end{pmatrix}(0), \\
\begin{pmatrix} z_4 \\ z_5 \\ z_6 \end{pmatrix}(1) &= \begin{pmatrix} 0 & 0 & 0 \\ -1 & 1 & 0 \\ 0 & 0 & 1 \end{pmatrix} \begin{pmatrix} z_1 \\ z_2 \\ z_3 \end{pmatrix}(1).
\end{aligned} \tag{3.6}
$$

We consider a particular path of characteristics. For $t = 0$ we start with the three outgoing signals z_1, z_2, z_3 moving to the boundary $x = 1$ with characteristic speed equal to 1. These characteristics are exactly the incoming signals at that boundary. According to the reflection condition (3.6) at $x = 1$, z_4 is canceled at $x = 1$. Now, the triple z_4, z_5, z_6 constitutes the set of outgoing signals at $x = 1$ and, accordingly, the set of incoming signals at $x = 0$. As z_4 is already equal to zero at $x = 0$ (because of properties of characteristics), the set of nonvanishing outgoing signals at $x = 0$ reduces to the variable z_2, according to the reflection condition (3.6). Therefore, after reflection at $x = 0$ the only nonvanishing incoming signal at $x = 1$ is z_2 which, in turn, is reflected to z_5 as the only nonvanishing outgoing signal at $x = 1$. This, however, closes the cycle, because z_5 is converted to z_2 upon reflection at $x = 0$, and so on. We depict this situation in the following diagram:

$$
\begin{pmatrix} z_1 \\ z_2 \\ z_3 \end{pmatrix} \xrightarrow{\text{at } x=1} \begin{pmatrix} 0 \\ z_5 \\ z_6 \end{pmatrix} \xrightarrow{\text{at } x=0} \begin{pmatrix} 0 \\ z_2 \\ 0 \end{pmatrix} \xrightarrow{\text{at } x=1} \begin{pmatrix} 0 \\ z_5 \\ 0 \end{pmatrix} \xrightarrow{\text{at } x=0} \begin{pmatrix} 0 \\ z_2 \\ 0 \end{pmatrix}.
$$

This shows that after two reflections at the boundaries the only residual motion left is in the second string. This is what is meant by directing or diode-type controls. This concept is due to P. Hagedorn and J. Schmidt [97], who suggested the use of this kind of control in flexible multi-structures. The paper [97], however, considers only the one-link case for out-of-plane displacements (which for two such controls can also be viewed as a three-link situation). We will demonstrate that concept for *planar* frames shortly. Let us first, however, provide some numerical evidence for the situation described above. In Figure V.16 we take an initial condition distributed along the three strings which are chosen to have individual lengths $\frac{1}{3}$, so that the length of the entire system is one. That the energy remains constant after two reflections (that is at $t = \frac{2}{3}$) is seen in Figure V.17.

In the context of the original system (3.4) the controls are chosen to be as follows:

$$
f^1 = -\dot{w}^2(1) + (w^2)'(1), \quad f^2 = \dot{w}^2(0) + (w^2)'(0),
$$

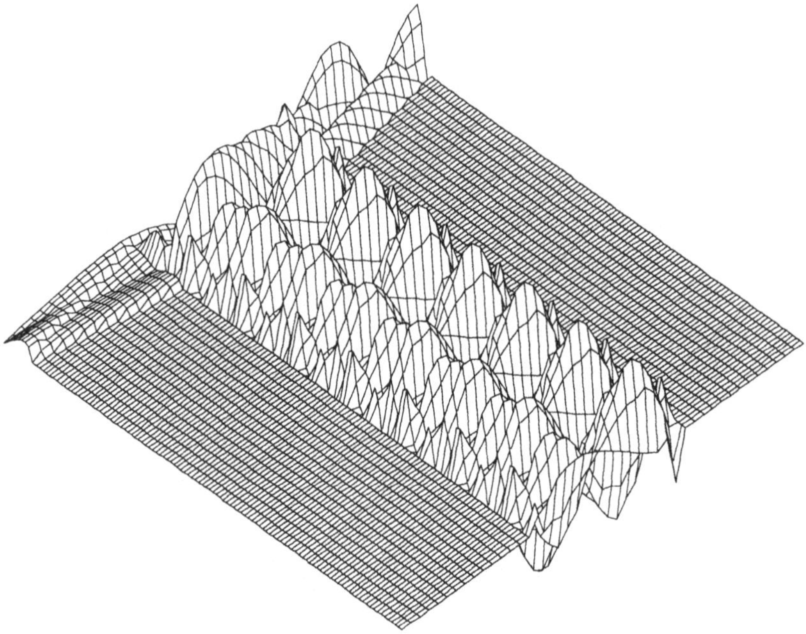

FIGURE V.16. Directing controls

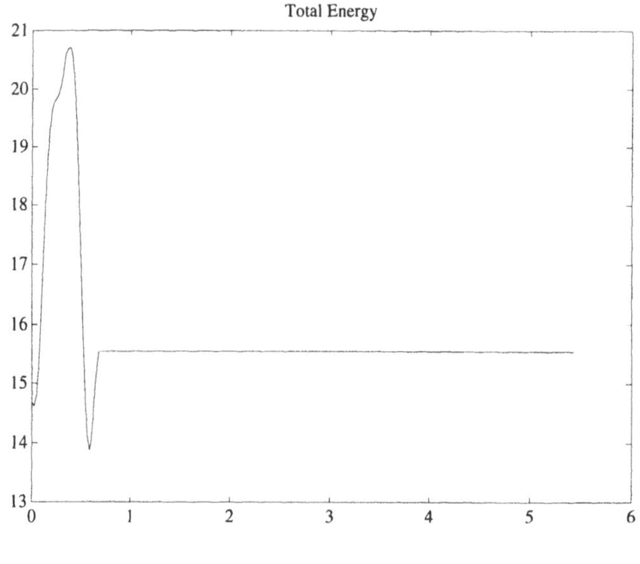

FIGURE V.17. Energy of entire system

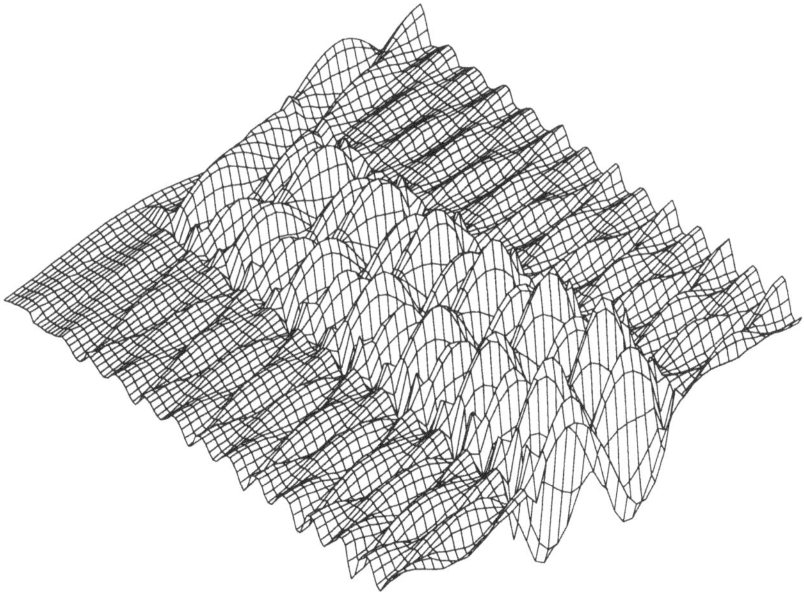

FIGURE V.18. Effect of time delay

which results in the feedback controls

$$(w^1)'(1) = -\dot{w}^1(1), \quad (w^3)'(0) = \dot{w}^3(0).$$

It has been shown in [25] that absorbing boundary controls for the 1-dimensional wave equation are sensitive to time delays in the velocity feedback. In fact, it has subsequently been shown by R. Datko [23], [24] that this phenomenon is not peculiar to strings and that it is not even circumvented by the introduction of severe damping. This instability for our three string model (3.4) is seen in Figures V.18 and V.19.

Even though this instability is rather discouraging, the situation in a real structure is not as bad as this seems to suggest. First of all, some of the energy absorbing feedback controls can be implemented as passive devices, for which a time lag will not be present. But even if the feedback law is to be realized by numerical computation (in real time), the time-lag due to the action of the microprocessor will not be a constant, but, rather, distributed by a certain stochastic process. There is, however, another remedy to this problem: the growing instability of the solutions under a feedback with time-lag is caused by feeding back an increasing amount of control energy to the system, once it is out-of-phase. In a real structure, however, a controller will have to respect certain a priori given bounds. There is an extensive literature devoted to the problem of controllability/stabilizability by bounded controls. We do not wish to dwell on this question in this monograph. We refer the reader to a recent paper by Rao [86] and to the

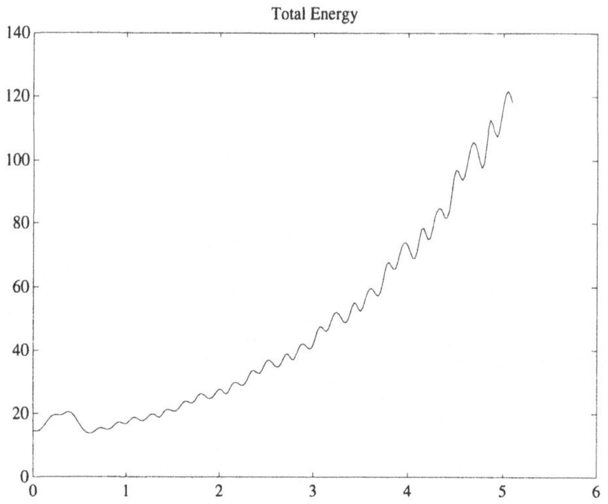

FIGURE V.19. Instability due to time delay

classical work of Haraux [35], [36] and also to Cabannes [12], [13]. Without going into the theoretical detail, we present the response of the system (3.4) without time delay but with respect to the nonlinear boundary controls:

$$f^1 = -\mathrm{sgn}(\dot{w}^2(1)) + (w^2)'(1), \quad f^2 = \mathrm{sgn}(\dot{w}^2(0)) + (w^2)'(0)$$

See Figures V.20 and V.21. It is apparent that these controls lead to considerably improved decay to a residual motion in the second string. It is interesting to observe the pattern of the strain in that string. If we now take these nonlinear boundary controls and if we account for a small time delay in the control (say one time step), then no instability is observed. See Figure V.22 and Figure V.23. Whether this observed robustness is actually true for the infinite dimensional system appears to be an open problem. The kind of nonlinearity of boundary controls just considered is in fact derived from a physical principle, namely *dry friction*. We will come back to this point later when we give an example of implicit Runge-Kutta methods applied to differential algebraic equations. □

In the next example we want to demonstrate how directing controls can be used to improve the system response by "directing" the energy to a "safe" part of the structure. This is of great practical importance because large flexible structures usually contain many uncontrollable circuits. Even if one puts the classical dissipative controls at each node, there will be a residual vibration in the system which will not decay to zero. Use of "directing" controls, however, can serve to "channel" the energy through the network to a particular set of nodes where either the energy influx is not important or where incoming energy is more easily absorbed.

EXAMPLE 3.5. *Five strings containing a circuit.*

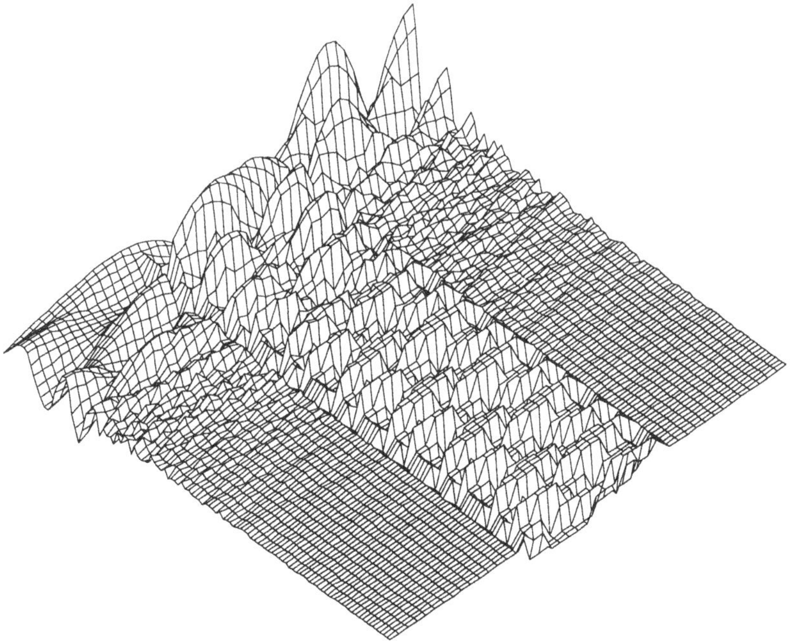

FIGURE V.20. Bang-bang control

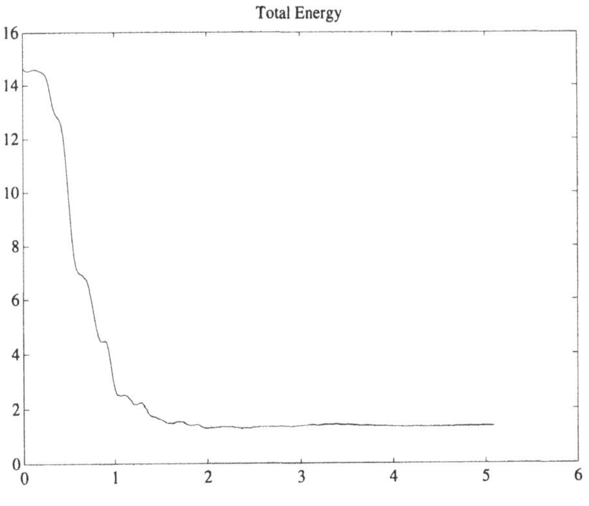

FIGURE V.21. Improved dissipation

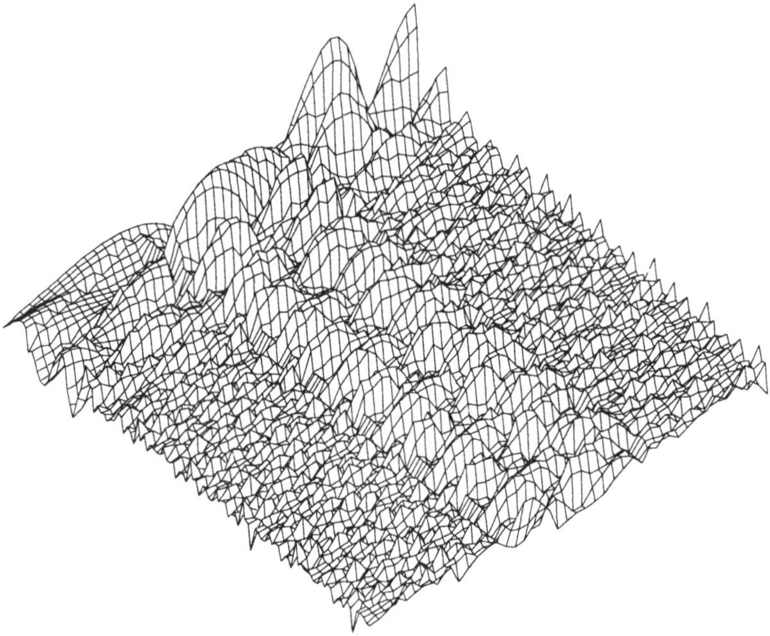

FIGURE V.22. Robustness of bang-bang control

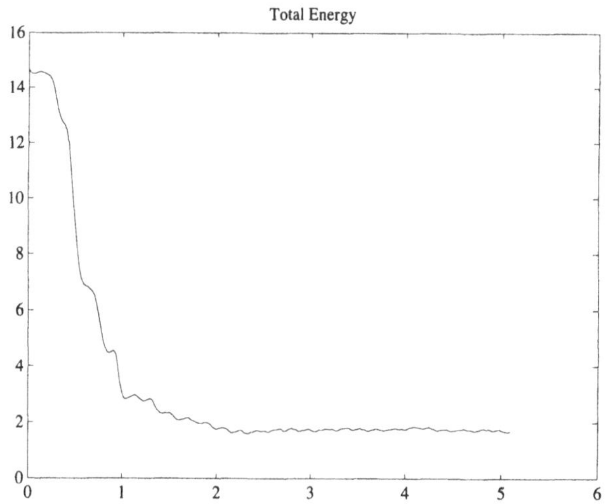

FIGURE V.23. Nonlinear dissipation with time delay

This model consists of a unit square (labels 1,2,3,4) with another string (#5) attached to the node connecting string #2 and string #3. We have an initial displacement in the first string only. In the first experiment we apply an absorbing control only at the free end of the attached string. Figures V.24 and V.25 reveal what one expects. Energy is taken out of the circuit until only a residual motion remains. In particular, the plot of the total energy of the 5th string shows that with each cycle less energy can be absorbed from the circuit because of the gradual diminishing of the nodal excitation. The situation completely changes once "directing" controls are introduced around the vertex connecting strings #2, #3 and #5. See Figures V.26 and V.27. □

4. Finite Element Approximations of Timoshenko Networks

We consider a 3-d network of isothermal Timoshenko beams without warping. We recall that the total strain energy (2.3) reflecting the additional restrictions is given by

$$W_I = \int_0^\ell U dx_1 = \int_0^\ell \{\frac{EA}{2}\bar{\varepsilon}_{11}^2 + 2GA(\bar{\varepsilon}_{12}^2 + \bar{\varepsilon}_{13}^2) + \frac{EI_{22}}{2}\tilde{\kappa}_2^2 + \frac{EI_{33}}{2}\tilde{\kappa}_3^2 + \frac{GI}{2}\tilde{\tau}^2\}dx.$$

The equations of motion are given by (2.10) which, in the present context, reduce to

$$\begin{aligned} F_{1,1} + P_1 &= 0,\\ F_{2,1} + P_2 &= 0,\\ F_{3,1} + P_3 &= 0,\\ M_{1,1} + C_1 &= 0,\\ M_{2,1} - F_3 + C_2 &= 0,\\ M_{3,1} + F_2 + C_3 &= 0. \end{aligned}$$

Under our assumptions, (2.15) now reads

$$\begin{aligned} F_1 &= EA\bar{\varepsilon}_{11} = EAW_{1,1},\\ F_2 &= 2GA\bar{\varepsilon}_{12} = GA(W_{2,1} - \vartheta_3),\\ F_3 &= 2GA\bar{\varepsilon}_{13} = GA(W_{3,1} + \vartheta_2),\\ M_1 &= GI\tilde{\tau} = GI\vartheta_1,\\ M_2 &= -EI_{33}\tilde{\kappa}_3 = EI_{33}\vartheta_{2,1},\\ M_3 &= EI_{22}\tilde{\kappa}_2 = EI_{22}\vartheta_{3,1}. \end{aligned}$$

Sometimes the area A appearing in the shear strains is replaced by "effective shear areas A_i^s". The continuity condition and the balance laws

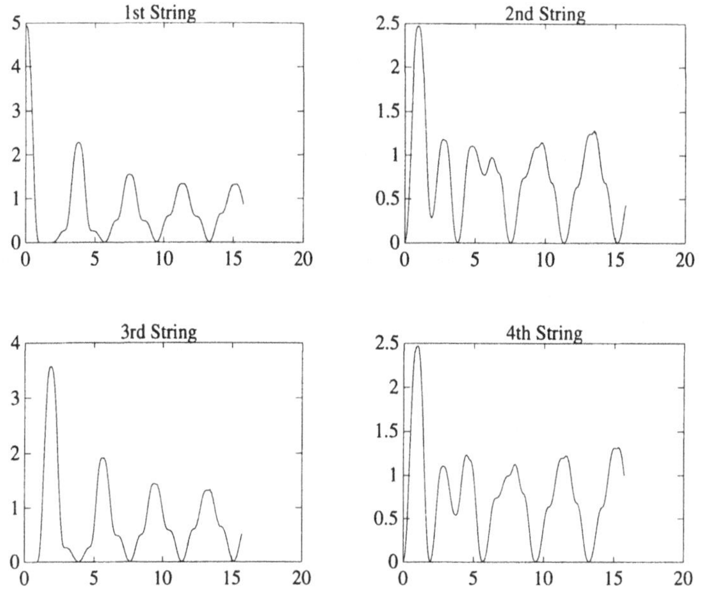

FIGURE V.24. Energy in individual strings

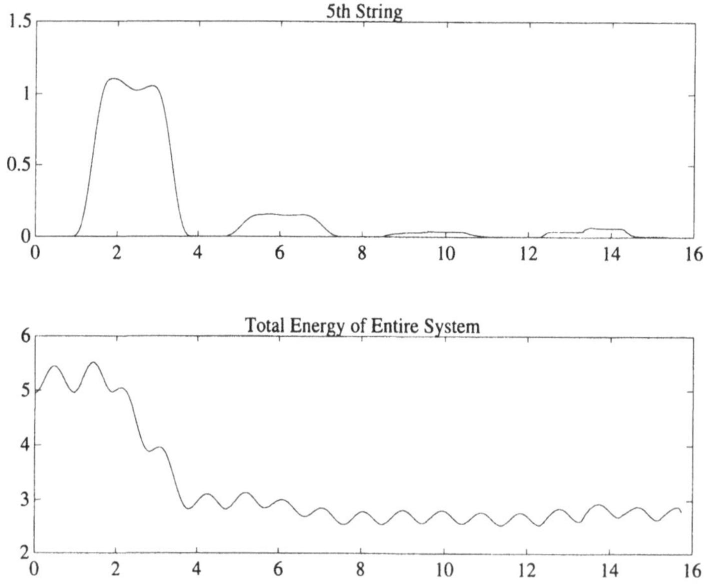

FIGURE V.25. Energy in 5th string and entire system

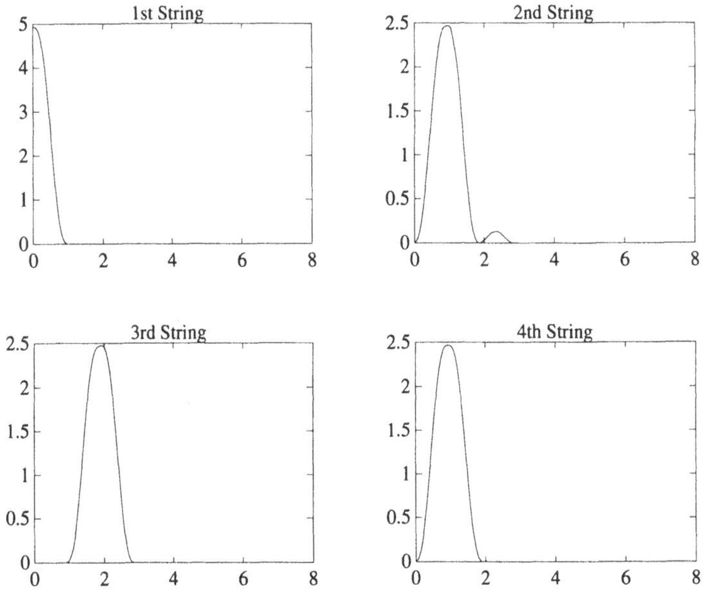

FIGURE V.26. Directing controls

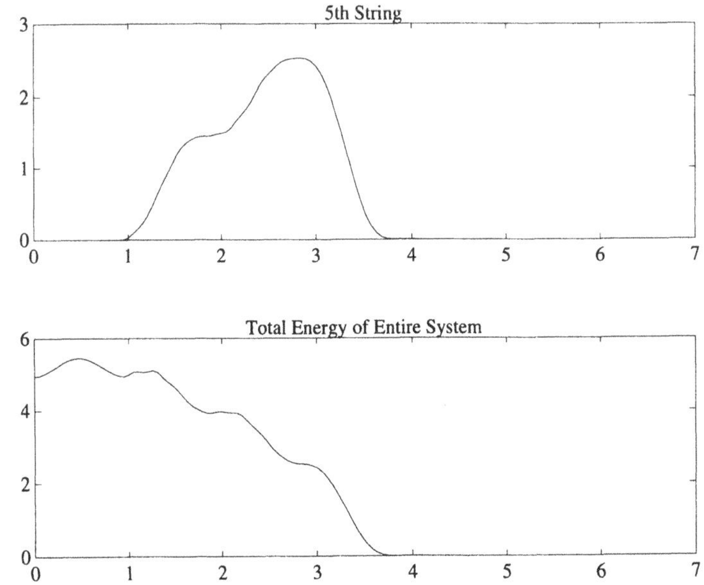

FIGURE V.27. Directing controls

of force and momentum at the nodes can be read into the global reference frame by way of the rotations Γ_i as follows: in the global frame $\mathbf{r}_i := \Gamma_i \mathbf{W}^i$ is the representation of $\mathbf{W}^i$. If there is no deflection and rotation prescribed and if there are prescribed moments $\mathbf{M}_N$ and forces $\mathbf{F}_N$ at the node N, we have

$$\Gamma_i \mathbf{W}^i(N) = \Gamma_j \mathbf{W}^j(N), \quad \forall i,j \in \mathcal{E}(N),$$
$$\sum_i \varepsilon_i(N)\Gamma_i \mathbf{F}^i + \mathbf{F}_N = 0,$$
$$\sum_i \varepsilon_i(N)\Gamma_i \mathbf{M}^i + \mathbf{M}_N = 0.$$

If we introduce now the vector quantities

$$\mathbf{u} = \begin{pmatrix} W_1 \\ W_2 \\ W_3 \\ \vartheta_1 \\ \vartheta_2 \\ \vartheta_3 \end{pmatrix}, \quad \mathbf{f} = \begin{pmatrix} P_1 \\ P_2 \\ P_3 \\ C_1 \\ C_2 \\ C_3 \end{pmatrix}, \quad \mathbf{h} = \begin{pmatrix} F_1 \\ F_2 \\ F_3 \\ M_1 \\ M_2 \\ M_3 \end{pmatrix},$$

together with their barred counterparts, we arrive at the standard problem

$$a(\bar{\mathbf{u}}, \mathbf{u}) = (\bar{\mathbf{u}}, \mathbf{f}) + (\bar{\mathbf{u}}, \mathbf{h})_\Gamma, \tag{4.1}$$

where the form $a(\cdot,\cdot)$ and the other products are given by

$$a(\bar{\mathbf{u}}, \mathbf{u}) = \sum_{i=1}^{n} \int_0^{\ell_i} [\bar{\boldsymbol{\gamma}}^T \mathbf{D}^s \boldsymbol{\gamma} + \bar{\boldsymbol{\kappa}}^T \mathbf{D}^b \boldsymbol{\kappa} + \bar{\varepsilon}_{11} EA \varepsilon_{11} + \bar{\vartheta}_1 GI \vartheta_1] dx_1,$$
$$(\bar{\mathbf{u}}, \mathbf{f}) = \int_0^{\ell_i} [\bar{\mathbf{W}}^T \mathbf{P} + \bar{\mathbf{d}}^T \mathbf{C}] dx_1,$$
$$(\bar{\mathbf{u}}, \mathbf{h})_\Gamma = \sum_i [(\bar{\mathbf{W}}^i)^T \mathbf{F}_N + (\mathbf{d}^i)^T \mathbf{M}_N].$$

Here

$$\mathbf{D}^s = \begin{pmatrix} GA_2 & 0 \\ 0 & GA_3 \end{pmatrix}, \quad \mathbf{D}^b = \begin{pmatrix} EI_{22} & 0 \\ 0 & EI_{33} \end{pmatrix}$$

are the shear and bending matrices and

$$\boldsymbol{\kappa} = \begin{pmatrix} \kappa_2 \\ \kappa_3 \end{pmatrix}, \quad \boldsymbol{\gamma} = \begin{pmatrix} \bar{\varepsilon}_{12} \\ \bar{\varepsilon}_{13} \end{pmatrix}$$

are the bending and shearing strains, respectively. The quantity $\mathbf{u}$ is sought such that (4.1) holds for all $\bar{\mathbf{u}}$ in a given space of test functions satisfying homogeneous boundary and node conditions. In order to derive the finite element approximation, we follow the presentation in Hughes [39]. We are looking for an isoparametric description, that is we assume that the displacements and the rotations are represented by nodal values via the same *shape* functions. We divide the elements of the frame in finite pieces, i.e., the finite elements, labelled by the superscript e to be distinguished from the frame element number j. For each finite element we have a number en of element nodes with displacements w^h_{ia}, and rotations ϑ^h_{ia}, $a = 1, ..., en$; the superscript h indicates the dependence on the "mesh", that is, on the number of finite elements per structural element. We interpolate between the nodal data using some shape function N_a (linear, quadratic or cubic functions are commonly used). Thus we write

$$W_i^h = \sum_{a=1}^{en} N_a w^h_{ia}, \quad \vartheta_i^h = \sum_{a=1}^{en} N_a \vartheta^h_{ia}.$$

At each node there are six degrees of freedom. We may express $\boldsymbol{\kappa}$ and $\boldsymbol{\gamma}$ by the corresponding nodal values as follows:

$$\boldsymbol{\kappa} = \begin{pmatrix} \kappa_2 \\ \kappa_3 \end{pmatrix} = \begin{pmatrix} \vartheta_{3,1} \\ -\vartheta_{2,1} \end{pmatrix} = \sum_{a=1}^{en} \begin{pmatrix} N'_a \vartheta^h_{3a} \\ -N'_a \vartheta^h_{2a} \end{pmatrix}$$

$$= \sum_{a=1}^{en} \begin{pmatrix} 0 & 0 & 0 & 0 & 0 & N'_a \\ 0 & 0 & 0 & 0 & -N'_a & 0 \end{pmatrix} \begin{pmatrix} w^h_{1a} \\ w^h_{2a} \\ w^h_{3a} \\ \vartheta^h_{1a} \\ \vartheta^h_{2a} \\ \vartheta^h_{3a} \end{pmatrix},$$

$$\boldsymbol{\gamma} = \begin{pmatrix} \gamma_2 \\ \gamma_3 \end{pmatrix} = \begin{pmatrix} W_{2,1} - \vartheta_{3,1} \\ W_{3,1} + \vartheta_{2,1} \end{pmatrix} = \sum_{a=1}^{en} \begin{pmatrix} N'_a (w^h_{2a} - \vartheta^h_{3a}) \\ -N'_a (w^h_{3a} + \vartheta^h_{2a}) \end{pmatrix}$$

$$= \sum_{a=1}^{en} \begin{pmatrix} 0 & N'_a & 0 & 0 & 0 & -N_a \\ 0 & 0 & -N'_a & 0 & N_a & 0 \end{pmatrix} \begin{pmatrix} w^h_{1a} \\ w^h_{2a} \\ w^h_{3a} \\ \vartheta^h_{1a} \\ \vartheta^h_{2a} \\ \vartheta^h_{3a} \end{pmatrix}.$$

We define

$$\mathbf{B}^b_a = \begin{pmatrix} 0 & 0 & 0 & 0 & 0 & N'_a \\ 0 & 0 & 0 & 0 & -N'_a & 0 \end{pmatrix},$$

$$\mathbf{B}^s_a = \begin{pmatrix} 0 & N'_a & 0 & 0 & 0 & -N_a \\ 0 & 0 & -N'_a & 0 & N_a & 0 \end{pmatrix},$$

$$\mathbf{B}^{ax}_a = [N'_a, 0, 0, 0, 0, 0], \quad \mathbf{B}^{to}_a = [0, 0, 0, N'_a, 0, 0],$$

$$\mathbf{B}^b = [\mathbf{B}^b_1, ..., \mathbf{B}^b_{en}], \quad \mathbf{B}^s = [\mathbf{B}^s_1, ..., \mathbf{B}^s_{en}],$$

$$\mathbf{B}^{ax} = [\mathbf{B}_1^{ax}, ..., \mathbf{B}_{en}^{ax}], \quad \mathbf{B}^{to} = [\mathbf{B}_1^{to}, ..., \mathbf{B}_{en}^{to}].$$

The superscripts *ax*, *to* stand for axial and torsional quantities, respectively. Then the element stiffness matrix $\mathbf{k}^e$ can be written as

$$\mathbf{k}^e = \mathbf{k}_b^e + \mathbf{k}_s^e + \mathbf{k}_{ax}^e + \mathbf{k}_{to}^e,$$

where

$$\mathbf{k}_b^e = \int_0^{\ell_i} \mathbf{B}^{bT}\mathbf{D}^b\mathbf{B}^b dx_1,$$

$$\mathbf{k}_s^e = \int_0^{\ell_i} \mathbf{B}^{sT}\mathbf{D}^s\mathbf{B}^s dx_1,$$

$$\mathbf{k}_{ax}^e = \int_0^{\ell_i} \mathbf{B}^{axT} EA\mathbf{B}^{ax} dx_1,$$

$$\mathbf{k}_{to}^e = \int_0^{\ell_i} \mathbf{B}^{toT} GI\mathbf{B}^{to} dx_1.$$

Similar arguments lead to the representation of the element mass matrix. In most of the calculations we used a lumped mass matrix and linear shape functions such that a one-point Gauss quadrature exactly integrates the bending stiffness and appropriately underintegrates the shear stiffness to avoid *shear locking*. Also we use a so-called residual bending flexibility to update the effective shear area according to

$$\bar{GA}_i^s = (\frac{1}{GA_i^s} + \frac{h^2}{12EI_{ii}})^{-1}.$$

Using this technique, linear functions can be used to achieve the accuracy of cubic shape functions in bending; see Hughes [39]. Once the local stiffness and mass matrices are defined, they are read into the global stiffness and global mass matrices using the local rotations Γ^i. We do not go into detail, because it is a classical procedure.

Let, finally, $\mathbf{K}$, $\mathbf{C}$, $\mathbf{M}$ denote the global stiffness, global damping and global mass matrices, respectively. Then the system of equations governing the evolution of the time dependent nodal variables can be written as

$$\mathbf{M}\ddot{\mathbf{U}} + \mathbf{C}\dot{\mathbf{U}} + \mathbf{K}\mathbf{U} = \mathbf{F}, \tag{4.2}$$

where $\mathbf{U}$ is the vector containing the nodal variables of all nodes and $\mathbf{F}$ represents all applied forces. We recall the family of Newmark schemes, which are of the predictor-corrector type. Again we follow Hughes [39]. To this end we put

$$\mathbf{a}_n = \ddot{\mathbf{U}}(t_n), \quad \mathbf{v}_n = \dot{\mathbf{U}}(t_n), \quad \mathbf{d}_n = \mathbf{U}(t_n) \tag{4.3}$$

for the acceleration, velocity and displacement, respectively. We then define the *predictors* by

$$
(4.4) \qquad \begin{cases} \tilde{\mathbf{d}}_{n+1} = \mathbf{d}_n + \Delta t \mathbf{v}_n + \Delta \frac{t^2}{2}(1 - 2\beta)\mathbf{a}_n, \\ \tilde{\mathbf{v}}_{n+1} = \mathbf{v}_n + (1 - \gamma)\Delta t \mathbf{a}_n. \end{cases}
$$

As for starting values, we have

$$
\mathbf{M}\mathbf{a}_0 = \mathbf{F} - \mathbf{C}\mathbf{v}_0 - \mathbf{K}\mathbf{d}_0.
$$

This equation can be solved for $\mathbf{a}_0$. At step $n+1$ the acceleration $\mathbf{a}_{n+1}$ is calculated from previous quantities and a priori known data according to the recursion

$$
(4.5) \qquad (\mathbf{M} + \gamma \Delta t \mathbf{C} + \beta \Delta t^2 \mathbf{K})\mathbf{a}_{n+1} = \mathbf{F}_{n+1} - \mathbf{C}\tilde{\mathbf{v}}_{n+1} - \mathbf{K}\tilde{\mathbf{d}}_{n+1}.
$$

Once $\mathbf{a}_{n+1}$ is known from (4.5), we update the predictors via the *correctors*

$$
(4.6) \qquad \begin{cases} \mathbf{d}_{n+1} = \tilde{\mathbf{d}}_{n+1} + \beta \Delta t^2 \mathbf{a}_{n+1}, \\ \mathbf{v}_{n+1} = \tilde{\mathbf{v}}_{n+1} + \gamma \Delta t \mathbf{a}_{n+1}. \end{cases}
$$

The result of (4.6) is then inserted into (4.4) which, in turn, will be used in (4.5) and so on. The two Newmark parameters can be chosen in various ways and lead to different properties of the scheme. We will always make the choice $\beta = 1/4$, $\gamma = 1/2$. This choice is known to lead to conservation of the energy, as discussed above.

In the calculations which follow we confine attention to planar frames of Timoshenko beams. In that case the local stiffness matrix $\mathbf{k}^e$ is a 6×6 matrix since there are vertical and longitudinal displacements and one rotation in the plane for each of the two nodes of a single element. For the sake of completeness we write $\mathbf{k}^e$ out explicitly:

$$
(4.7) \qquad \mathbf{k}^e = \begin{pmatrix} \frac{EA}{\ell} & 0 & 0 & -\frac{EA}{\ell} & 0 & 0 \\ 0 & \frac{A^s}{\ell} & \frac{A^s}{2} & 0 & -\frac{A^s}{\ell} & \frac{A^s}{2} \\ 0 & \frac{A^s}{\ell} & \frac{A^s\ell}{4} + \frac{EI}{2} & 0 & -\frac{A^s}{2} & \frac{A^s\ell}{4} - \frac{EI}{2} \\ -\frac{EA}{2} & 0 & 0 & \frac{EA}{\ell} & 0 & 0 \\ 0 & -\frac{A^s}{\ell} & -\frac{A^s}{2} & 0 & \frac{A^s}{\ell} & -\frac{A^s}{2} \\ 0 & \frac{A^s}{\ell} & \frac{A^s\ell}{4} - \frac{EI}{2} & 0 & -\frac{A^s}{2} & \frac{A^s\ell}{4} + \frac{EI}{2} \end{pmatrix}
$$

As for the lumped mass matrix we take

$$\text{(4.8)} \qquad \mathbf{m}^e = \frac{m_0 A\ell}{2} \begin{pmatrix} 1 & 0 & 0 & 0 & 0 & 0 \\ 0 & 1 & 0 & 0 & 0 & 0 \\ 0 & 0 & \frac{I}{A} & 0 & 0 & 0 \\ 0 & 0 & 0 & 1 & 0 & 0 \\ 0 & 0 & 0 & 0 & 1 & 0 \\ 0 & 0 & 0 & 0 & 0 & \frac{I}{A} \end{pmatrix};$$

see [39, p. 515], for the planar beam without longitudinal displacement. With the bar wave velocity $c = \sqrt{E/m_0}$ and the beam shear wave velocity $c_s = (GA^s)/(m_0 A)$, the critical time step Δt can be estimated as

$$\Delta t \le \min\left\{\frac{h}{c}, \left(\frac{h}{c_s}\right)\left(1 + \frac{A}{I}\left(\frac{h}{2}\right)^2\right)^{-\frac{1}{2}}\right\}.$$

EXAMPLE 4.1. *A single planar Timoshenko beam.*

In this example we want to consider very briefly a single planar Timoshenko beam which is clamped at $x = 0$ and subject to controls at the free end at $x = 1$. We implement a dissipative feedback with impedance 1 in the shear and axial forces as well as in the rotation. The system is known to be exponentially stabilized by usual velocity type feedback controls. Because of dispersion of waves, however, exact cancellation as in the string case does not take place. Questions which naturally arise are whether the feedback gains can be chosen in an optimal way and whether an optimal feedback may also contain the full state feedback rather than a pure velocity feedback. To give at least a little insight into these interesting issues we compare, without going into detail, the effects of typical absorbing controls versus those of an LQR-optimal control. In fact, we use as a cost functional the total energy plus the $L_2(0, \infty)$- norm of the control. We computed the LQR-feedback operator using the MATLAB control toolbox; see Figure V.28. It is interesting to note that the effects of LQR-controls are almost identical to those obtained using classical feedback controls. □

While it is fairly simple to set up finite difference schemes, such as the Lax-Friedrichs scheme, for planar systems of multi-link strings and Timoshenko beams as long as the number of structural elements is small and the graph is very simple, this method is unsuited for more complicated networks, for which the finite element method seems more appropriate. As remarked earlier, we consider our numerical simulations as tutorial and, therefore, we restrict our attention to some simple models which are nonetheless rich enough to reveal interesting properties.

EXAMPLE 4.2. *The unit square of planar Timoshenko beams*

We first consider the uncontrolled system. Again there is an initial displacement only in the bottom beam (#1). The difficulty in exhibiting

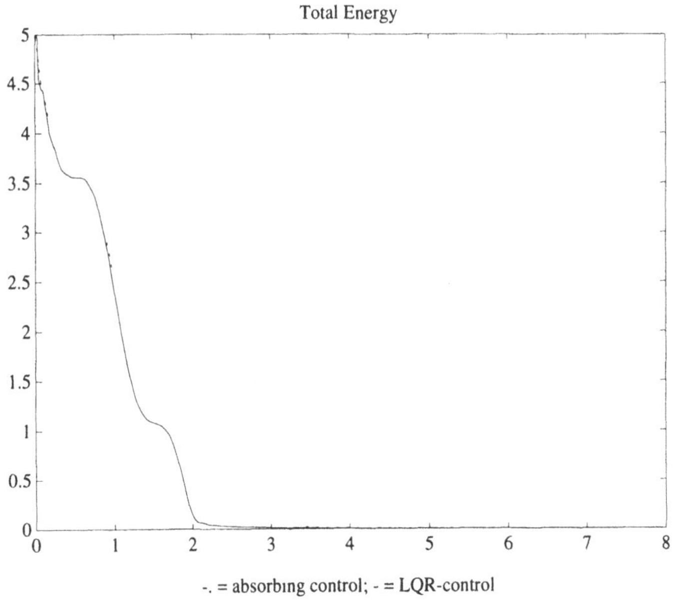

FIGURE V.28. A comparison of controls

graphically all of the displacements and rotations is obvious. At the nodes, local vertical motion of beam # 1 converts to local longitudinal motion of beams #2 and #4 which, in turn, is converted to local vertical motion in the top beam (#3). Because in the finite element approach we automatically have the local rotations Γ_i available, we can easily go back to the global reference frame. In this way the locally longitudinal motion of the 2nd and 4th beam is, indeed, globally vertical; see Figure V.29. This plot clearly shows the collision of waves in the 3rd beam after travelling safely through beams #2 and #4. The initial bump next to the corner at the bottom indicates the location of the first beam. Figure V.30 shows the conservation of energy.

In the second experiment we apply dissipative controls at all four nodes; see Figure V.31 and Figure V.32. Figure V.31 clearly shows how the motion settles to a residual oscillation which vanishes at the vertices. Accordingly, the energy settles to a constant value; see Figure V.32.

5. Implicit Runge-Kutta Method: Dry Friction at Joints

In deriving the semi-discrete approximation to the infinite dimensional distributed parameter system by the finite element approach in the last section, we have used a condensed description in the sense that only independent nodal variables have been used. This means that, for instance at a multiple node where, because of the continuity requirement, some of the displacements of one structural element can be expressed by corresponding

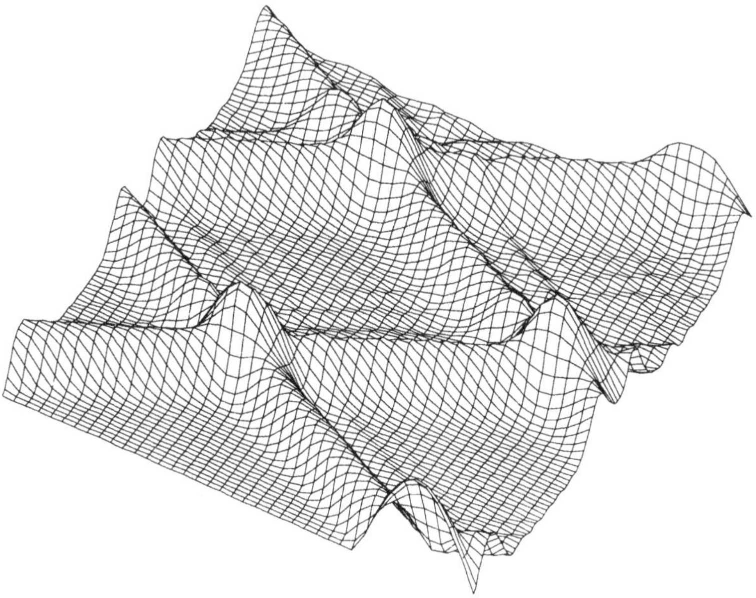

FIGURE V.29. Unit square: uncontrolled motion

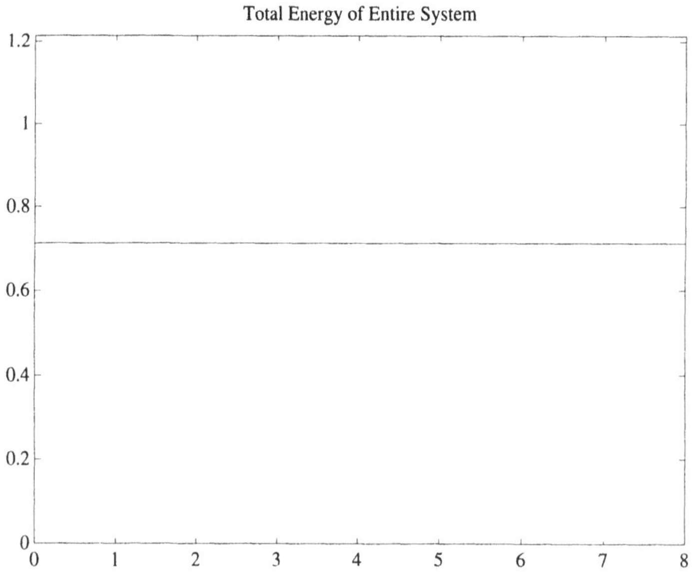

FIGURE V.30. Unit square: conservation of energy

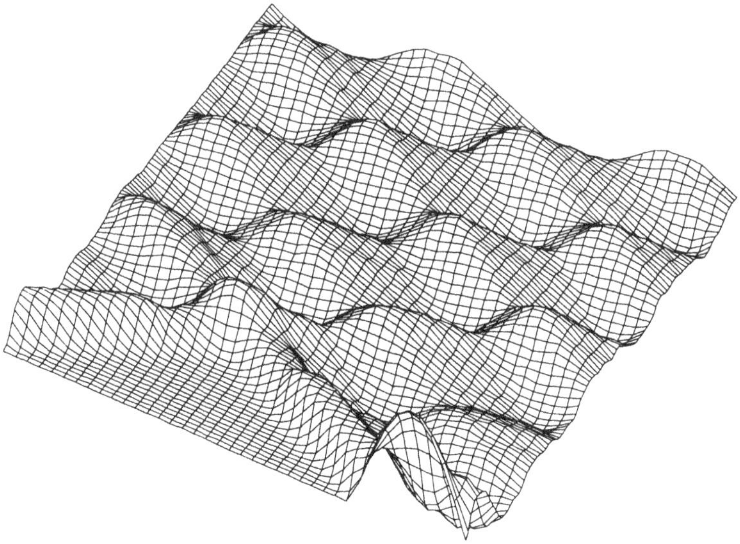

FIGURE V.31. Unit square: controlled motion

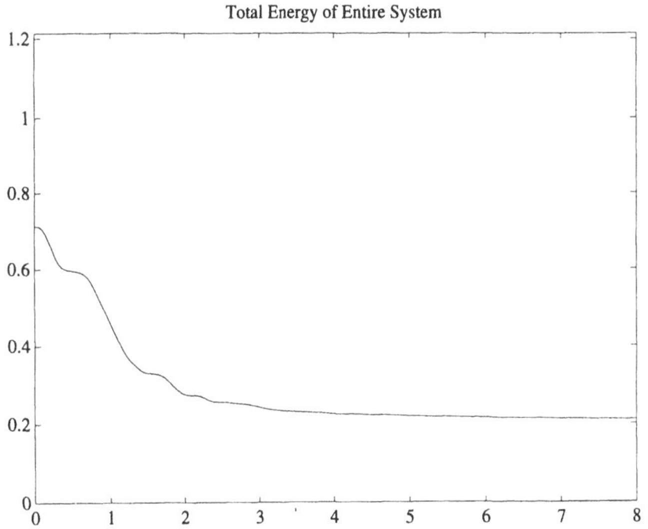

FIGURE V.32. Unit square: saturation of energy

ones of another element, only the minimal number of variables are retained. It is actually a major part of the finite element implementation to accommodate such a reduced setup. Only after this reduction is a numerical treatment as described in the last section meaningful. There is a recent tendency in the literature to treat boundary and node conditions in numerical simulations of multi-body structures as algebraic constraints. One of the reasons for this methodology is to retain the simple band structure of the stiffness and mass matrices. To make this more transparent, consider a network with a circuit. Then, because of the fact that the first and the last node of that circuit coincide, the part of the global stiffness matrix containing that circuit will have nonzero entries far away from the typical band. This makes the procedure of solving equations apparently more complicated. It appears, therefore, to be appealing to keep a simpler structure of the global mass and stiffness matrices and account for the couplings by way of algebraic constraints. This approach is taken in Shabana [99] for multilink *flexible* body dynamics, including rotating structures, and has also be used in the context of multilink *rigid* body dynamics by Hairer and Wanner [33] and Lubich [64], [65]. We will not dwell on this point further.

In this section we are mainly interested in applying the methods of differential algebraic equations to a particular problem which naturally arises in the context of damping of mechanical structures. The phenomenon we want look at is *dry friction at the joints.* It has long been known by mechanical and civil engineers that dry friction in the joints can be a particularly helpful source of damping of residual modes in structures. The idea is very simple. If one has a sufficient amount of dry (Coulomb) friction, say at a joint, then a considerable amount of energy is necessary to move that joint or to bring the element 'behind' that joint into oscillatory motion. It is reasonable to expect that under certain circumstances it will be possible to achieve "locking" of elements, that is, prevent certain elements from oscillating, by applying dry friction.

We begin by considering a simple model for this phenomenon, which we take from Cabannes [12], [13], and modify it later on. In [12], Cabannes introduced the following model of a string with dry friction. Let w denote the displacement. Again, we consider the spatial interval to be $[0,1]$. The model under consideration is

$$\begin{cases} \ddot{w} - w'' = -4\operatorname{sgn}(\dot{w}) & \text{if } \dot{w} \neq 0, \\ |\ddot{w} - w''| \leq 4 & \text{if } \dot{w} = 0. \end{cases}$$

This system is complemented by the boundary and initial conditions

$$w(0,t) = w(1,t) = 0, \;\; w(x,0) = w_0(x), \;\; \dot{w}(x,0) = w_1(x).$$

Cabannes showed that if $w_1 \equiv 0$, $w_0(0) = w_0(1) = 0$ and $w_0''(x) \leq 4$ for all $x \in [0,1]$, the string does not move. He also studied cases in which the initial curvature is beyond the value 4 and he showed that if the curvature is in the interval $[8n, 8(n+1)]$ the motion will stop after time $t = n$. He used the method of characteristics to prove his startling results. The model was later generalized to the 2-d wave equation by Haraux [35], [36], who provided a treatment of the asymptotic behavior of solutions using the method of multi-valued monotone operators. The locking phenomenon, however, was not pursued by Haraux in those papers and, in fact, *appears to be an open problem*, at least for the 2-d situation. In the 1-d case we consider the problem of dry friction at a point of the string rather than everywhere within the string. This model is also a modification of one considered in Brasey and Hairer [8] of a violin string struck by a violin bow locally within the string. The model in [8] is

$$\text{(5.1)} \qquad \begin{cases} \ddot{w} - w'' = -\rho_s \chi(x) \operatorname{sgn}(\dot{w} - v) & \text{if } \dot{w} \neq v, \\ |\ddot{w} - w''| \leq \rho_a \chi(x) & \text{if } \dot{w} = v. \end{cases}$$

Here v denotes the velocity at which the bow is moved across the string, ρ_s is proportional to the shear force, ρ_a is proportional to the adhesive forces (Coulomb frictional forces); these two constants are not necessarily the same, which gives some control on the string. The function χ is assumed to have local support. In our application it is taken to be a point mass. By this choice we intend to model dry friction at a joint. To this end we consider two identical strings of length one joined at the ends where $x = 1$ on both strings. We assume clamped boundaries at $x = 0$ for both strings and also assume dry friction at the common point $x = 1$. The resulting model is as follows:

$$\text{(5.2)} \qquad \begin{cases} \ddot{w}^i = (w^i)'', \quad i = 1, 2, \\ w^1(0) = w^2(0) = 0, \ w^1(1) = w^2(1), \\ (w^1)'(1) + (w^2)'(1) = -\rho_s z, \\ w^1(x,0) = w_0^1(x), \ \ w^2(x,0) = 0, \ \ \dot{w}^1(x,0) = \dot{w}^2(x,0) = 0, \end{cases}$$

where

$$z = \begin{cases} \operatorname{sgn}(\dot{w}^1(1)) & \text{if } \dot{w}^1(1) \neq 0, \\ \in [-\frac{\rho_a}{\rho_s}, \frac{\rho_a}{\rho_s}] & \text{else.} \end{cases}$$

We perform the familiar transformations

$$\begin{aligned} u_1 &:= \dot{w}^1, \ \ u_2 := \dot{w}_2, \ \ u_3 := (w^1)', \ \ u_4 := (w^2)', \\ v_1 &:= \frac{1}{\sqrt{2}}(-u_1 + u_3), \ \ v_3 := \frac{1}{\sqrt{2}}(u_1 + u_3), \\ v_2 &:= \frac{1}{\sqrt{2}}(-u_2 + u_4), \ \ v_4 := \frac{1}{\sqrt{2}}(u_2 + u_4). \end{aligned}$$

We then have the system

$$\dot{\mathbf{v}} = \mathbf{\Gamma}\mathbf{v}', \quad \mathbf{\Gamma} = \operatorname{diag}([-1,-1,1,1]),$$

with boundary and node conditions

$$(v_3 - v_1)(0) = 0, \quad (v_3 - v_1)(1) = (v_4 - v_2)(1),$$
$$(v_4 - v_2)(0) = 0, \quad (v_3 + v_1)(1) + (v_4 + v_2)(1) = -\rho_s\sqrt{2}z$$

with

$$z = \begin{cases} \frac{1}{\sqrt{2}}(v_3 - v_1)(1) & \text{if } (v_3 - v_1)(1) \neq 0, \\ \in [-\frac{\rho_a}{\rho_s}, \frac{\rho_a}{\rho_s}] & \text{else.} \end{cases}$$

The last set of equations can be rewritten as

$$\begin{pmatrix} v_1 \\ v_2 \end{pmatrix}(0) = \begin{pmatrix} v_3 \\ v_4 \end{pmatrix}(0), \quad \begin{pmatrix} v_3 \\ v_4 \end{pmatrix}(1) = -\begin{pmatrix} 0 & 1 \\ 1 & 0 \end{pmatrix}\begin{pmatrix} v_1 \\ v_2 \end{pmatrix}(1) - \rho_s\frac{1}{\sqrt{2}}z\begin{pmatrix} 1 \\ 1 \end{pmatrix}.$$

As a consequence of our special choice of initial conditions (we could consider more general data but wish to treat this problem only as an example), we have $v_2(\cdot, 0) = v_4(\cdot, 0) = 0$.

We will consider two cases:

(i) $|(w^1)'(\cdot, 0)| \leq \rho_a/\sqrt{2}$;
(ii) $|(w^1)'(\cdot, 0)| > \rho_a\sqrt{2}$.

We claim that in case (i), no motion takes place in the second string for $t \geq 0$.

PROOF. We have

$$(v_3 - v_1)(1) = 0 \text{ at } t = 0.$$

Take

$$z = -\frac{\sqrt{2}}{\rho_s}v_1(1,t) \in [-\frac{\rho_a}{\rho_s}, \frac{\rho_a}{\rho_s}], \quad t \in [0,1].$$

We follow the reflections at $x = 1$ and subsequently at $x = 0$, i.e.

$$\begin{aligned} v_3(1,t) &= -v_{20}(1-t) + v_{10}(1-t) = v_{10}(1-t), \\ v_4(1,t) &= 0, \quad t \in [0,1], \\ v_1(0, 1+t) &= v_{10}(1-t), \\ v_2(0, 1+t) &= 0. \end{aligned}$$

Hence, upon reflection, v_1 does not change magnitude and, with z as above, v_4 remains at zero, causing v_2 to also equal zero.

The analogous case starting with v_{30}, v_{40} and traveling first towards the boundary at $x = 0$ is settled in a similar way. The main point is that the variable z (representing the adhesive force) always compensates the shear force ('kills' the outgoing signal $v_4(1,t)$) at the boundary $x = 1$ which, in our model, represents the multiple node. □

We claim that in case (ii), the motion propagates into the second string.

PROOF. We have initially $(v_3 - v_1)(1,0) = 0$. Assume now that

$$(v_3 - v_1)(1,t) = 0,\ t \in (0,\varepsilon).$$

We have

$$z(t) \in [-\frac{\rho_a}{\rho_s}, \frac{\rho_a}{\rho_s}] \Rightarrow z(t) \neq -\frac{\sqrt{2}}{\rho_s} v_1(1,t) \text{ on } (0,\varepsilon)$$
$$\Rightarrow v_4(1,t) \neq 0 \text{ on } (0,\varepsilon) \Rightarrow v_2(0,1+t) \neq 0.$$

The latter inclusion says that there is nonzero motion in the second string. The other cases are handled in a similar way. □

We are going to give numerical support to these remarks. We consider the mesh points $x_i = i/N$, $N = 40$, and use a classical finite difference approximation to the second order operator (which, incidentally, equals the corresponding finite element stiffness matrix). Indeed, we can view this process as occuring in one string with dry friction at the midpoint. We have the following system:

$$(5.3)\quad \ddot{w}_i - N^2\{w_{i-1}(t) - 2\, w_i(t) + w_{i+1}(t)\} = \begin{cases} -\rho_s z & \text{if } i = \frac{N}{2}, \\ 0 & \text{else,} \end{cases}$$

where now z is described by

$$(5.4)\quad \begin{cases} z = \operatorname{sgn}(\dot{w}_{\frac{N}{2}}(t)) & \text{if } \dot{w}_{\frac{N}{2}}(t) \neq 0, \\ \rho_s z \in \rho_a[-1,1] & \text{if } \dot{w}_{\frac{N}{2}}(t) = 0. \end{cases}$$

We also have the boundary and initial conditions

$$(5.5)\quad w_0(t) = 0,\ \ w_N(t) = 0,\ \ w_i(0) = w_i^0,\ \ \dot{w}_i(0) = 0.$$

In the above system the variable z can be viewed as a control variable which forces the solution of the system (5.3) to satisfy $\dot{w}_{\frac{N}{2}} = 0$. In order to do that the variable z, which physically represents the adhesive force, must vary in time. In order to make the role of z more transparent, we consider a general differential algebraic equation of the following kind:

$$\begin{cases} \mathbf{y}' = \mathbf{f}(\mathbf{y},\mathbf{z}), \\ 0 = \mathbf{g}(\mathbf{y}), \end{cases}$$

where $\mathbf{f} : \mathbb{R}^n \times \mathbb{R}^m \mapsto \mathbb{R}^n$, $\mathbf{g} : \mathbb{R}^n \mapsto \mathbb{R}^n$ are C^1 and C^2 functions, respectively. Denote the Jacobian of $\mathbf{f}$ with respect to $\mathbf{y}$, $\mathbf{z}$ by $\mathbf{f_y}$, $\mathbf{f_z}$. If we differentiate $(5.6)_2$ with respect to the variable x we obtain

$$0 = \mathbf{g_y}(\mathbf{y})\mathbf{f}(\mathbf{y}, \mathbf{z}).$$

Another differentiation with respect to x yields

$$(5.6) \quad 0 = \mathbf{g_{yy}}(\mathbf{y})\mathbf{f}(\mathbf{y}, \mathbf{z}) + \mathbf{g_y}(\mathbf{y})\mathbf{f_y}(\mathbf{y}, \mathbf{z}))\mathbf{f}(\mathbf{y}, \mathbf{z}) + \mathbf{g_y}(\mathbf{y})\mathbf{f_z}(\mathbf{y}, \mathbf{z})\mathbf{z}'.$$

Obviously, if $\mathbf{g_y}(\mathbf{y})\mathbf{f_z}(\mathbf{y}, \mathbf{z})$ is invertible in a neighborhood of a solution, say $(\bar{\mathbf{y}}, \bar{\mathbf{z}})$, then (5.6) is equivalent to an ordinary differential equation. In this case for the problem

$$\mathbf{G}(\mathbf{y}, \mathbf{z}) := \mathbf{g_y}(\mathbf{y})\mathbf{f}(\mathbf{y}, \mathbf{z}) = 0, \quad \mathbf{G}(\bar{\mathbf{y}}, \bar{\mathbf{z}}) = 0,$$

there exists, by the implicit function theorem, a neighborhood $\mathcal{U}(\bar{\mathbf{y}}, \bar{\mathbf{z}})$ and a solution $\mathbf{z} = \mathbf{z}(\mathbf{y})$ of $\mathbf{G}(\mathbf{y}, \mathbf{z}) = 0$ such that (5.6) holds. See Hairer and Wanner [33] for the theory of differential algebraic equations. With this remark in mind we go back to our system, where the case just considered corresponds to (5.3), $(5.4)_2$.

If, now, z is in mode two of (5.4) and reaches the value ρ_a/ρ_s, it has to switch to the first case in (5.4) which, together with (5.3), then reduces to an ordinary differential equation. As a result, we have to switch between the differential algebraic system (5.3), $(5.4)_2$ and (5.5) and the ordinary differential equation (initial value problem) (5.3), $(5.4)_1$ and (5.5). Switching problems of this kind (in particular (5.1)) have been treated by Brasey and Hairer [8]. They used the half-explicit 5-stage Runge-Kutta method of order 4. We used the Radau5 implementation of the Radau IIa fully implicit Runge-Kutta method of order 5, also given by Hairer and Wanner in [33]. The corresponding driver was originally written by J. Verscht (Bayreuth) for the problem (5.1). We made the appropriate changes to use it for the problem under consideration and also for beam applications (which we do not reproduce here). The problem of switching between the two models is a major difficulty which is usually solved by the use of "dense output formulae" and some kind of extrapolation to find the switching manifold $\mathbf{g}(\mathbf{y}) = 0$. See Figures V.33 and V.34.

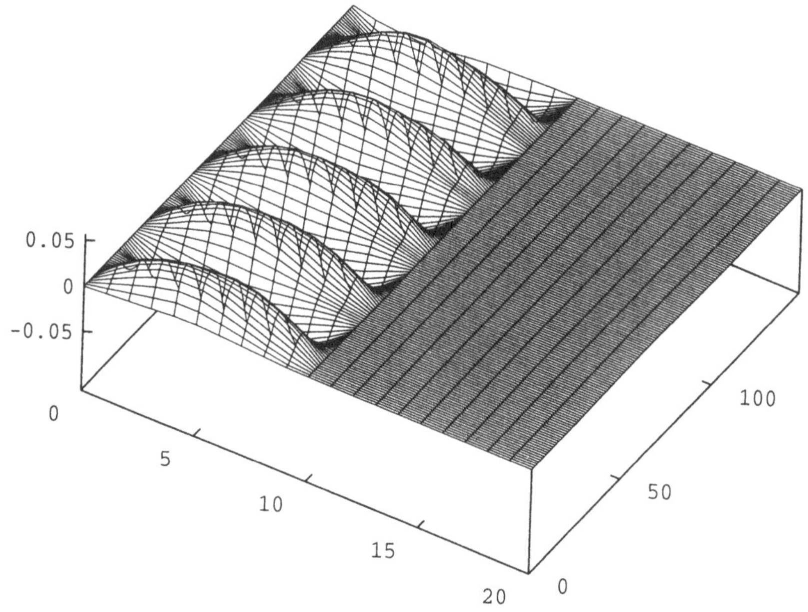

FIGURE V.33. Dry friction: locking

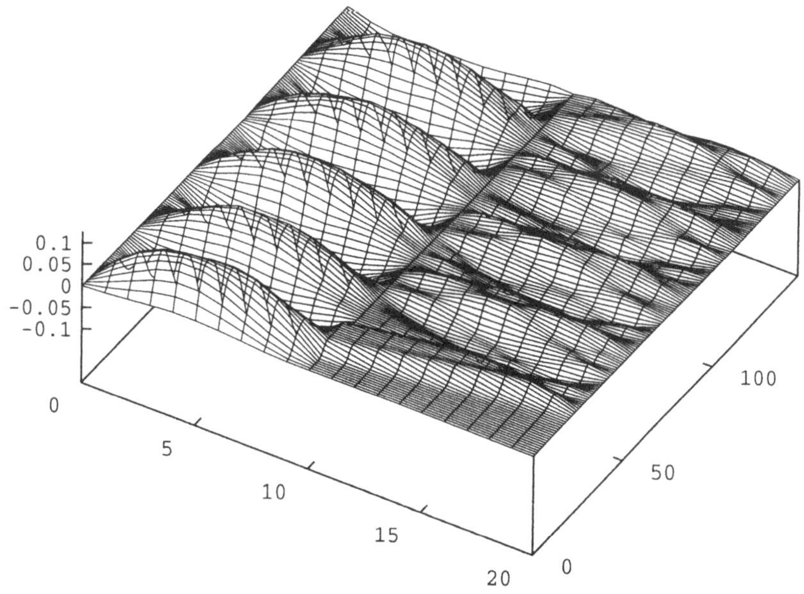

FIGURE V.34. Dry friction: spillover

CHAPTER VI

Modeling and Controllability of Interconnected Membranes

We shall consider in this Chapter the motion of an elastic body consisting of several interconnected elastic membranes. Our goals are to obtain the dynamic equations of motion of such a configuration, especially to elucidate the geometric and dynamic conditions which must hold in junction regions where two or more such elastic elements are joined together, and to study the controllability properites of the resulting (linearized) system of equations. The modeling proceeds through two stages. In Section 1 the equations of motion and boundary conditions of a general nonlinear membrane and the corresponding Hamilton's principle will be derived, assuming a linear stress-strain relation. The dynamic equations of a system of interconnected membranes are derived in Section 2 by minimizing an appropriate Lagrangian over the class of deformations which satisfies certain geometric constraints. In particular, the geometric constraint imposed in the junction regions is the very natural one that the initially connected system of membranes remains connected throughout the deformation process. When this constraint is imposed on the admissible deformations, the optimal motion is forced to satisfy a second, dynamic constaint in the junction regions that has the interpretation of a balance of forces law there. Both the equations of motion and the dynamic junction conditions involve nonlinear couplings among the components of the displacement vector. When these are linearized around the trivial equilibrium the familiar equations of motion of a single elastic membrane, in which in-plane displacements are not coupled to transverse motion, are obtained for the displacement of each individual element. However, all displacements remain coupled through the geometric and (linearized) dynamic junction conditions. The question of exact controllability of this linearized coupled system is then considered in Section 3.

Let us first fix the notation.

- Roman indices $i, j, \ldots$ take values 1,2,3.
- Greek indices $\alpha, \beta, \ldots$ take values 1,2.
- Summation convention is assumed unless explicitly noted otherwise.
- A vector $\mathbf{x} \in \mathbb{R}^3$ has coordinates (x_1, x_2, x_3) with respect to the natural basis of $\mathbb{R}^3$.

1. Modeling of Dynamic Nonlinear Elastic Membranes

Let Ω be a bounded, open, connected set in $\mathbb{R}^2$ with Lipschitz continuous boundary Γ, and let $\mathbf{a}_1, \mathbf{a}_2, \mathbf{a}_3$ be a fixed, right- handed orthonormal basis for $\mathbb{R}^3$. We consider an infinite elastic cylinder in $\mathbb{R}^3$ whose reference configuration is

$$\mathcal{B} = \{\mathbf{p}_0 + x_i \mathbf{a}_i | \, (x_1, x_2) \in \Omega, \ -\infty < x_3 < \infty\},$$

where $\mathbf{p}_0$ is a fixed vector in $\mathbb{R}^3$. We set

$$\mathbf{r}(x_1, x_2, x_3) = \mathbf{p}_0 + x_i \mathbf{a}_i.$$

It is assumed that the motion of the cylinder obeys the following hypothesis.

Basic kinematic hypothesis. The position vector at time t to the displaced location of the particle situated at $\mathbf{r}(x_1, x_2, x_3)$ in the reference configuration is

$$\mathbf{R}(x_1, x_2, x_3, t) = \mathbf{r}(x_1, x_2, x_3) + \mathbf{W}(x_1, x_2, t). \tag{1.1}$$

Hypothesis (1.1) means that the deformation of a cross-section orthogonal to the axis of the cylinder is independent of the variable x_3 along the axis of the cylinder. Thus the deformation of the cylinder is completely determined by the deformation of any of its cross-sections. The latter is specified by the vector function $\mathbf{W}$ which measures the *displacement* of a cross-section. Let $\{W_i\}$ denote the components of $\mathbf{W}$ in the $\{\mathbf{a}_i\}$ basis: $\mathbf{W} = W_i \mathbf{a}_i$. We refer to a cross-section as an *elastic membrane* and to the planar region

$$\mathcal{P} = \{\mathbf{p}_0 + x_\alpha \mathbf{a}_\alpha : \ (x_1, x_2) \in \Omega\}$$

as the *reference cross-section.*

Set $\mathbf{G}_i = \mathbf{R}_{,i}$, where a subscript after a comma means differentiation with respect to the indicated spatial variable. The components ε_{ij} of the *strain tensor* are given by

$$\varepsilon_{ij} = \frac{1}{2}(\mathbf{G}_i \cdot \mathbf{G}_j - \delta_{ij}), \tag{1.2}$$

It follows from (1.1) and (1.2) that

$$
(1.3) \qquad \begin{cases} \varepsilon_{\alpha\beta} = \dfrac{1}{2}(W_{\alpha,\beta} + W_{\beta,\alpha} + \mathbf{W}_{,\alpha} \cdot \mathbf{W}_{,\beta}), \\ \varepsilon_{\alpha 3} = \dfrac{1}{2} W_{3,\alpha}, \\ \varepsilon_{33} = 0. \end{cases}
$$

1.1. Equations of Motion. Let s^{ij} denote the components of the *second Piola-Kirchhoff stress tensor.* This symmetric tensor relates the *stress vectors* $\mathbf{t}^i$ to the quantities $\mathbf{G}_i$ through

$$
(1.4) \qquad \mathbf{t}^i = s^{ij}\mathbf{G}_j
$$

(see [108, Chapter 3]). If $\mathbf{F}$ denotes the body force per unit of undeformed volume, the three-dimensional conditions for the balance of forces is

$$
(1.5) \qquad \mathbf{t}^i_{,i} + \mathbf{F} = \rho \mathbf{R}_{,tt},
$$

where ρ is the mass of the body per unit of reference volume. Use of (1.1) and (1.4) in (1.5) allows that equation to be written

$$
(1.6) \qquad s^{ij}_{,i}\mathbf{a}_j + s^{i\beta}_{,i}\mathbf{W}_{,\beta} + s^{\alpha\beta}\mathbf{W}_{,\alpha\beta} + \mathbf{F} = \rho \mathbf{W}_{,tt}.
$$

In order to be consistent with our basic kinematic hypothesis, it is assumed that $\mathbf{F}$ and ρ are independent of x_3. Let us further assume that the material in question is *linearly elastic* (hookean), i.e.,

$$
(1.7) \qquad s^{ij} = C^{ijkl}\varepsilon_{kl},
$$

where the coefficients of elasticity C^{ijkl} *depend only on* x_1 *and* x_2. The symmetry of s^{ij} and ε_{kl} requires that

$$
(1.8) \qquad C^{ijkl} = C^{jikl} = C^{ijlk},
$$

and we further *assume* that

$$
(1.9) \qquad C^{ijkl} = C^{klij}
$$

Then, in particular, the stresses s^{ij} do not depend on x_3 and (1.6) reduces to

$$
(1.10) \qquad s^{\alpha j}_{,\alpha}\mathbf{a}_j + (s^{\alpha\beta}\mathbf{W}_{,\beta})_{,\alpha} + \mathbf{F} = \rho \mathbf{W}_{,tt}.
$$

This system comprises three scalar equations for the scalar components W_i of $\mathbf{W}$.

1.2. Edge Conditions. Let Λ denote the two-dimensional boundary of the reference cross-section $\mathcal{P}$ and assume that

$$\Lambda = \overline{\Lambda}^C \cup \overline{\Lambda}^D, \tag{1.11}$$

where Λ^C, Λ^D are mutually disjoint, relatively open subsets of Λ and $\Lambda^D \neq \emptyset$. (The superscripts C and D are used to indicate "clamped" and "dynamic," respectively.) The partition (1.11) induces a similar partition of $\Gamma := \partial\Omega$ into

$$\Gamma = \overline{\Gamma}^C \cup \overline{\Gamma}^D \tag{1.12}$$

through the mapping

$$\Gamma \mapsto \Lambda : (x_1, x_2) \mapsto \mathbf{p}_0 + x_\alpha \mathbf{a}_\alpha.$$

Let (ν_1, ν_2) denote a unit exterior normal vector and $(-\nu_2, \nu_1)$ a unit tangent vector at a point (x_1, x_2) of Γ. Since the mapping $(x_1, x_2, x_3) \mapsto x_i \mathbf{a}_i$ is orthogonal with determinant unity, $\boldsymbol{\nu} := \nu_1 \mathbf{a}_1 + \nu_2 \mathbf{a}_2$ and $\boldsymbol{\tau} := -\nu_2 \mathbf{a}_1 + \nu_1 \mathbf{a}_2$ are vectors which are normal and tangent, respectively, to Λ at the point $\mathbf{p}_0 + x_\alpha \mathbf{a}_\alpha$ with $\boldsymbol{\nu}$ pointing out of $\mathcal{P}$ and $\boldsymbol{\tau}$ positively oriented with respect to $\mathbf{a}_3$, i.e., $\boldsymbol{\tau} \times \mathbf{a}_3 = \boldsymbol{\nu}$.

The edge conditions along the part of $\partial\mathcal{B}$ corresponding to Λ^C are described by

$$\mathbf{W} = 0 \quad \text{on } \Gamma^C,\ t > 0. \tag{1.13}$$

Let $\mathbf{f}$ denote a given distribution of forces along $\Lambda^D + x_3 \mathbf{a}_3$, $-\infty < x_3 < +\infty$ which is *independent of* x_3. The condition for the balance of forces along this part of $\partial\mathcal{B}$ is

$$\nu_\alpha \mathbf{t}^\alpha = \mathbf{f}$$

which, in view of our previous hypotheses, may be written

$$\nu_\alpha s^{\alpha j} \mathbf{a}_j + \nu_\alpha s^{\alpha\beta} \mathbf{W}_{,\beta} = \mathbf{f} \quad \text{on } \Gamma^D,\ t > 0. \tag{1.14}$$

1.3. Hamilton's Principle. Let $\mathbf{W}$ satisfy (1.10) and (1.14), and let $\hat{\mathbf{W}} = \hat{W}_j \mathbf{a}_j$ be a smooth function defined on $\overline{\Omega} \times [0, T]$ such that $\hat{\mathbf{W}}|_{t=0} = \hat{\mathbf{W}}|_{t=T} = 0$, $\hat{\mathbf{W}}|_{\Gamma^C \times (0,T)} = 0$. We form the scalar produce in $\mathbb{R}^3$ of (1.10) with $\hat{\mathbf{W}}$ and integrate over $\Omega \times (0, T)$. After some integrations by parts we obtain

$$\int_0^T \int_\Omega (\rho \mathbf{W}_{,t} \cdot \hat{\mathbf{W}}_{,t} - s^{\alpha j} \hat{W}_{j,\alpha} - s^{\alpha\beta} \mathbf{W}_{,\beta} \cdot \hat{\mathbf{W}}_{,\alpha})\, d\Omega + \int_0^T \int_\Omega \mathbf{F} \cdot \hat{\mathbf{W}}\, d\Omega + \int_0^T \int_{\Gamma^D} \mathbf{f} \cdot \hat{\mathbf{W}}\, d\Gamma = 0. \tag{1.15}$$

Conversely, if $\mathbf{W}$ satisfies (1.15) for all such test functions $\hat{\mathbf{W}}$, then $\mathbf{W}$ satisfies (in an appropriate sense) (1.10) and (1.14). The variational equation

(1.15) is known as the *principle of virtual work* for the membrane problem under consideration.

We now recast (1.15) into a form known as *Hamilton's principle.* Let δ denote the first Fréchet derivative with respect to $\mathbf{W}$. We may write

$$s^{\alpha j}\hat{W}_{j,\alpha} + s^{\alpha\beta}\mathbf{W}_{,\beta}\cdot\hat{\mathbf{W}}_{,\alpha} = s^{ij}\langle\delta\varepsilon_{ij},\hat{\mathbf{W}}\rangle, \tag{1.16}$$

where $\langle\delta\varepsilon_{ij},\hat{\mathbf{W}}\rangle$ denotes the value of $\delta\varepsilon_{ij}$ at $\hat{\mathbf{W}}$. For any material obeying (1.7)–(1.9) one has

$$s^{ij}\delta\varepsilon_{ij} = C^{ijkl}\varepsilon_{kl}\delta\varepsilon_{ij} = \frac{1}{2}\delta(C^{ijkl}\varepsilon_{kl}\varepsilon_{ij}) = \frac{1}{2}\delta(s^{ij}\varepsilon_{ij}). \tag{1.17}$$

Under assumption (1.1),

$$s^{ij}\varepsilon_{ij} = s^{\alpha\beta}\varepsilon_{\alpha\beta} + 2s^{\alpha 3}\varepsilon_{\alpha 3}. \tag{1.18}$$

It then follows from (1.16)–(1.18) that (1.15) may be written

$$\delta\int_0^T[\mathcal{K}(t) - \mathcal{U}(t) + \mathcal{W}(t)]\,dt = 0, \tag{1.19}$$

where

$$\mathcal{K}(t) = \frac{1}{2}\int_\Omega \rho|\mathbf{W}_{,t}|^2 d\Omega \tag{1.20}$$

is the *kinetic energy* of the deformation,

$$\begin{aligned}\mathcal{U}(t) &= \frac{1}{2}\int_\Omega (s^{\alpha\beta}\varepsilon_{\alpha\beta} + 2s^{\alpha 3}\varepsilon_{\alpha 3})\,d\Omega \\ &= \frac{1}{2}\int_\Omega (s^{\alpha j}W_{j,\alpha} + \frac{1}{2}s^{\alpha\beta}\mathbf{W}_{,\alpha}\cdot\mathbf{W}_{,\beta})\,d\Omega\end{aligned} \tag{1.21}$$

is the *strain energy,* and

$$\mathcal{W}(t) = \int_\Omega \mathbf{F}\cdot\mathbf{W}\,d\Omega + \int_{\Gamma^D}\mathbf{f}\cdot\mathbf{W}\,d\Gamma \tag{1.22}$$

is the *work of external forces.* The variational formulation (1.19) with $\mathcal{K}$, $\mathcal{U}$ and $\mathcal{W}$ so defined, is Hamilton's principle for the membrane problem under consideration.

2. Systems of Interconnected Elastic Membranes

For $i = 1, \ldots, n$, let $\mathbf{a}_1^i, \mathbf{a}_2^i, \mathbf{a}_3^i$ be a right-handed orthonormal system in $\mathbb{R}^3$, let $\mathbf{p}_0^i$ be a fixed vector in $\mathbb{R}^3$, and Ω_i be a bounded, open, connected region in $\mathbb{R}^2$ with Lipschitz boundary consisting of a finite number of smooth simple curves. We set

$$\mathbf{p}_i(x_1, x_2) = \mathbf{p}_0^i + \sum_\alpha x_\alpha \mathbf{a}_\alpha^i, \tag{2.1}$$

$$\mathbf{r}_i(x_1, x_2, x_3) = \mathbf{p}_i(x_1, x_2) + x_3 \mathbf{a}_3^i,$$
$$\mathcal{P}_i = \{\mathbf{p}_i(x_1, x_2) | \, (x_1, x_2) \in \Omega_i\}.$$

The mapping $(x_1, x_2) \mapsto \mathbf{p}_i(x_1, x_2)$ is a homeomorphism of $\overline{\Omega}_i$ onto $\overline{\mathcal{P}}_i$ and of $\partial\Omega_i$ onto $\partial\mathcal{P}_i$, with

$$\mathbf{p}_i^{-1}(\mathbf{p}) = ((\mathbf{p} - \mathbf{p}_0^i) \cdot \mathbf{a}_1^i, (\mathbf{p} - \mathbf{p}_0^i) \cdot \mathbf{a}_2^i).$$

REMARK 2.1. As in the last section, Greek indices take values 1,2, but, unlike before, we *do not* assume summation convention and Roman indices are not necessarily restricted to set $\{1, 2, 3\}$. The range of such an index will be explicitly indicated whenever it is not clear fom context. We do, however, make the convention that

$$\sum_j := \sum_{j=1}^3 . \quad \square$$

The set $\mathcal{P}_i$ is considered as a reference cross-section of an infinite cylinder whose reference configuration is

$$\mathcal{B}_i = \{\mathbf{r}_i(x_1, x_2, x_3) \, | \, (x_1, x_2) \in \Omega_i, \; |x_3| < \infty\}.$$

We make the following hypotheses regarding the membranes $\mathcal{P}_i$.

$$\overline{\mathcal{P}}_i \cap \mathcal{P}_k = \emptyset, \;\; \forall i \neq k;$$
$$\cup_{i=1}^n \overline{\mathcal{P}}_i \;\; \text{is a connected set in } \mathbb{R}^3;$$
$$\overline{\mathcal{P}}_i \cap \overline{\mathcal{P}}_k \;\; \text{is either empty or is a finite union of components, } \forall i \neq k.$$

Each of the bodies $\mathcal{B}_i$ is assumed to satisfy the basic kinematic assumption (1.1): the position vector $\mathbf{R}^i(x_1, x_2, x_3)$ to the displaced particle originally at $\mathbf{r}_i(x_1, x_2, x_3)$ is given by

$$\mathbf{R}^i(x_1, x_2, x_3, t) = \mathbf{r}^i(x_1, x_2, x_3) + \mathbf{W}^i(x_1, x_2, t), \tag{2.2}$$
$$(x_1, x_2) \in \Omega_i, \; |x_3| < \infty, \; 1 = 1, \ldots, n.$$

2.1. Geometric Junction Conditions. We tentatively define a *joint*, or *junction*, of the multi-membrane system as a component (a maximal line segment; by definition, line segments are connected) which is a common edge of two or more reference cross-sections, that is, a maximal line segment where a $\overline{\mathcal{P}}_i$ and $\overline{\mathcal{P}}_k$ intersect for some $k > i$. However, this definition allows the possibility that the set of all such components does not form a mutually disjoint collection. If C_1 and C_2 are components such that $C_1 \cap C_2 \neq \emptyset$, we replace C_1 and C_2 by $C_1 \cap C_2$, $C_1 \setminus (C_1 \cap C_2)$ and $C_2 \setminus (C_1 \cap C_2)$, whose union is $C_1 \cup C_2$. Each set $C_\alpha \setminus (C_1 \cap C_2)$ is either a single line segment or the union of disjoint segments. In the latter case we consider $C_\alpha \setminus (C_1 \cap C_2)$ to consist of several separate components. In this manner one may eliminate the overlapping of components. Finally, we define the joints of the system to be the collection of disjoint, open line segments which are the *interiors* of the collection of non-overlapping components.

EXAMPLE 2.1. Let $\mathcal{P}_1, \mathcal{P}_2, \mathcal{P}_3$ be the planar regions in $\mathbb{R}^3$ defined by

$$\begin{aligned} \mathcal{P}_1 :&\ \ 0 < s_1, s_2 < 1,\ s_3 = 0, \\ \mathcal{P}_2 :&\ \ 0 < s_2, s_3 < 1,\ s_1 = 0, \\ \mathcal{P}_3 :&\ \ s_1 = s_3,\ 0 < s_1 < 1,\ 1/4 < s_2 < 3/4. \end{aligned}$$

To write these in the parametric form used above set

$$\begin{aligned} \Omega_1 = \Omega_2 &= \{(x_1, x_2) |\, 0 < x_1, x_2 < 1\}, \\ \Omega_3 &= \{(x_1, x_2) |\, 0 < x_1 < 1,\ 1/4 < x_2 < 3/4\}, \end{aligned}$$

$$\begin{aligned} \mathbf{a}_1^1 &= (1,0,0),\ \ \mathbf{a}_2^1 = (0,1,0), \\ \mathbf{a}_1^2 &= (0,1,0),\ \ \mathbf{a}_2^2 = (0,0,1), \\ \mathbf{a}_1^3 &= (1,0,1)/\sqrt{2},\ \ \mathbf{a}_2^3 = (0,1,0). \end{aligned}$$

The membranes intersect in components $\overline{\mathcal{P}}_i \cap \overline{\mathcal{P}}_k$ given by

$$\begin{aligned} C_1 := \overline{\mathcal{P}}_1 \cap \overline{\mathcal{P}}_2 :&\ \ 0 \le s_2 \le 1,\ s_1 = s_3 = 0, \\ C_2 := \overline{\mathcal{P}}_1 \cap \overline{\mathcal{P}}_3 :&\ \ 1/4 \le s_2 \le 3/4,\ s_1 = s_3 = 0, \\ C_3 := \overline{\mathcal{P}}_2 \cap \overline{\mathcal{P}}_3 &= \overline{\mathcal{P}}_1 \cap \overline{\mathcal{P}}_3, \end{aligned}$$

so that $C_2 = C_3 \subset C_1$. We replace C_1 and C_2 by $\tilde{C}_1$ and $\tilde{C}_2$, where

$$\begin{aligned} \tilde{C}_1 &= C_1 \cap C_2 = C_2, \\ \tilde{C}_2 = C_1 \setminus (C_1 \cap C_2) :&\ \ 0 \le s_2 < 1/4,\ 3/4 < s_2 \le 1,\ s_1 = s_3 = 0. \end{aligned}$$

Note that $C_2 \setminus (C_1 \cap C_2) = \emptyset$. Since $C_2 = C_3$, the same process eliminates C_3 as a joint. Finally, since $\tilde{C}_1$ contains two disjoint segments, the joints of the multi-link system are taken to be

$$J_1: \ 0 < s_2 < 1/4, \ s_1 = s_3 = 0,$$
$$J_2: \ 1/4 < s_2 < 3/4, \ s_1 = s_3 = 0,$$
$$J_3: \ 3/4 < s_2 < 1, \ s_1 = s_3 = 0.$$

$\overline{\mathcal{P}}_1$ and $\overline{\mathcal{P}}_2$ intersect along J_1 and J_3, while each pair among $\overline{\mathcal{P}}_1$, $\overline{\mathcal{P}}_2$ and $\overline{\mathcal{P}}_3$ intersects along J_2.

For any joint J of the multi-membrane system define

$$\mathcal{I}(J) = \{i \in [1, \dots, n] | \, J \subset \partial \mathcal{P}_i\}.$$

Thus $\mathcal{I}(J)$ is the index set of those $\mathcal{P}_i$ which share J. The deformation of the overall system is constrained by the following *geometric junction condition* at each joint J:

(2.3) $$\mathbf{W}^i(\mathbf{p}_i^{-1}(\mathbf{p}), t) = \mathbf{W}^j(\mathbf{p}_j^{-1}(\mathbf{p}), t), \ \ \mathbf{p} \in J, \ i, j \in \mathcal{I}(J).$$

Condition (2.3) simply requires that each membrane sharing J have the same displacement there.

2.2. Dynamic Conditions. The geometric joint conditions reflect the continuity of the overall structure. In addition, there are *dynamic junction conditions* which represent the balance of linear momentum at the joints. The simplest way to obtain these is to minimize the Hamiltonian associated with the structure. In carrying out the calculation, the variation must be taken with respect to displacements which satisfy the geometric boundary conditions and the geometric joint conditions. Of course, in addition to the dynamic junction conditions Hamilton's principle also yields the equations of motion and dynamic boundary conditions of the structure, as usual.

Set $\Lambda_i = \partial \mathcal{P}_i$. We assume that

(2.4) $$\Lambda_i = \overline{\Lambda}_i^C \cup \overline{\Lambda}_i^D \cup \overline{\Lambda}_i^J,$$

where Λ_i^C, Λ_i^D, Λ_i^J are disjoint and relatively open in Λ_i. This partition of Λ_i induces a similar partition of $\Gamma_i := \partial\Omega_i$ into

(2.5) $$\Gamma_i = \overline{\Gamma}_i^C \cup \overline{\Gamma}_i^D \cup \overline{\Gamma}_i^J,$$

where

$$\Lambda_i^C = \mathbf{p}_i(\Gamma_i^C), \ \ \Lambda_i^D = \mathbf{p}_i(\Gamma_i^D), \ \ \Lambda_i^J = \mathbf{p}_i(\Gamma_i^J).$$

The set Λ_i^C represents the part of Λ_i that is clamped, i.e.,

(2.6) $$\mathbf{W}^i = 0 \ \text{ on } \Gamma_i^C,$$

Λ_i^D is the part of Λ_i along which "dynamic" boundary conditions will hold, and Λ_i^J is that part of Λ_i such that

$$\Lambda_i^J \cap \Lambda_k \neq \emptyset \ \text{ for at least one } k \neq i.$$

Thus Λ_i^J consists of those joints of the multi-membrane system which belong to Λ_i. We label the entire collection of joints by $J_1, J_2, \dots, J_m$.

The total kinetic energy of the system is defined to be

$$\mathcal{K}(t) = \frac{1}{2}\sum_{i=1}^{n}\int_{\Omega_i} \rho_i |\mathbf{W}^i_{,t}|^2 d\Omega,$$

where $\rho_i(x_1, x_2)$ is the mass density per unit of reference area of $\mathcal{P}_i$. Let s_i^{jk} denote the stress tensor associated with the ith membrane. The total strain energy is defined to be

$$\mathcal{U}(t) = \frac{1}{2}\sum_{i=1}^{n}\int_{\Omega_i} (\sum_{\alpha,j} s_i^{\alpha j} W^i_{j,\alpha} + \frac{1}{2}\sum_{\alpha,\beta} s_i^{\alpha\beta}\mathbf{W}^i_{,\alpha}\cdot\mathbf{W}^i_{,\beta})\, d\Omega.$$

Let $\mathbf{F}^i$ denote an external body force acting on the ith membrane $\mathcal{P}_i$, $\mathbf{f}^i$ be a force acting along Λ_i^D and $\tilde{\mathbf{f}}^k$ be an applied force along the junction region J_k. The work of these external forces is defined to be

$$\begin{aligned}\mathcal{W}(t) &= \sum_{i=1}^{n}\left\{\int_{\mathcal{P}_i} \mathbf{F}^i\cdot\mathbf{W}^i\, d\Lambda_i + \int_{\Lambda_i^D} \mathbf{f}^i\cdot\mathbf{W}^i\, d\Lambda\right\} + \sum_{k=1}^{m}\int_{J_k} \tilde{\mathbf{f}}^k\cdot\mathbf{W}^{i_k}\, dJ_k\\ &= \sum_{i=1}^{n}\left\{\int_{\Omega_i} \mathbf{F}^i\cdot\mathbf{W}^i\, d\Omega_i + \int_{\Gamma_i^D} \mathbf{f}^i\cdot\mathbf{W}^i\, d\Gamma\right\} + \sum_{k=1}^{m}\int_{\Gamma_{i_k}(J_k)} \tilde{\mathbf{f}}^k\cdot\mathbf{W}^{i_k}\, d\Gamma,\end{aligned}$$

where

$$\Gamma_i(J_k) = \{(x_i, x_2) \in \Gamma_i^J \,|\, \mathbf{p}_i(x_1, x_2) \in J_k\} = \{\mathbf{p}_i^{-1}(\mathbf{p}) \,|\, \mathbf{p} \in J_k\},$$
$$i = 1, \dots, n;\ k = 1, \dots, m,$$

and where we may take

$$i_k = \min\{i \,|\, i \in \mathcal{I}(J_k)\}$$

in accordance with the geometric junction conditions (2.3). In the integrals over $\mathcal{P}_i$ and Λ_i^D, the integrands are evaluated at $\mathbf{p}_i^{-1}(\mathbf{p})$ with $\mathbf{p} \in \mathcal{P}_i$ and $\mathbf{p} \in \Lambda_i^D$, respectively, while the integrand in the integral over J_k is evaluated at $\mathbf{p}_{i_k}^{-1}(\mathbf{p})$, $\mathbf{p} \in J_k$. The dynamic equations of the multi-membrane system are *postulated* to be those associated with the Hamilton's principle

$$\delta\int_0^T [\mathcal{K}(t) - \mathcal{U}(t) + \mathcal{W}(t)]\, dt = 0, \tag{2.7}$$

in which the variation δ is taken with respect to displacements $(\mathbf{W}^1, \dots, \mathbf{W}^n)$ which respect the geometric boundary conditions (2.6) and geometric junction conditions (2.3) along each joint $J_1, \dots, J_m$. It is fairly obvious that we will obtain, in particular, the equations of motion (1.10) for each membrane of the system and the dynamic boundary conditions (1.14) along Γ_i^D. Thus we have for each $i = 1, \dots, n$ the equations of motion

$$(2.8)\quad \sum_{\alpha,j} s_{i,\alpha}^{\alpha j}\mathbf{a}_j^i + \sum_{\alpha,\beta}(s_i^{\alpha\beta}\mathbf{W}^i_{,\beta})_{,\alpha} + \mathbf{F}^i = \rho_i \mathbf{W}^i_{,tt}, \quad (x_1, x_2) \in \Omega_i, \ \ t > 0,$$

and the boundary conditions

$$(2.9)\qquad \sum_{\alpha,j} \nu_\alpha^i s_i^{\alpha j}\mathbf{a}_j^i + \sum_{\alpha,\beta} \nu_\alpha^i s_i^{\alpha\beta}\mathbf{W}^i_{,\beta} = \mathbf{f}^i \ \text{ on } \Gamma_i^D, \ \ t > 0.$$

There are, in addition, junction conditions given by the following variational equation:

$$(2.10)\quad \sum_{i=1}^n \int_{\Gamma_i^J}\left(\sum_{\alpha,j} \nu_\alpha^i s_i^{\alpha j}\mathbf{a}_j^i + \sum_{\alpha,\beta} \nu_\alpha^i s_i^{\alpha\beta}\mathbf{W}^i_{,\beta}\right) \cdot \hat{\mathbf{W}}^i \, d\Gamma = \sum_{k=1}^m \int_{J_k} \tilde{\mathbf{f}}^k \cdot \hat{\mathbf{W}}^{i_k} \, dJ_k$$

for all sufficiently smooth displacements $\hat{\mathbf{W}}^i$, $i = 1, \dots, n$, which satisfy the geometric boundary and junction conditions.

We have

$$\sum_{i=1}^n \int_{\Gamma_i^J} = \sum_{k=1}^m \sum_{i \in \mathcal{I}(J_k)} \int_{\Gamma_i(J_k)}.$$

The segment J_k may be parameterized as

$$J_k = \{\mathbf{q}_k^0 + s\boldsymbol{\tau}^{i_k} \,|\, a_k \le s \le b_k\},$$

where $\mathbf{q}_k^0 \in J_k$ is fixed but arbitrary. Since $J_k = \mathbf{p}_i(\Gamma_i(J_k))$, this parameterization of J_k induces a parameterization of the segment $\Gamma_i(J_k)$ given by given by

$$\Gamma_i(J_k) = \{\mathbf{x}_{ik}^0 + s(\tau_1, \tau_2) | \, a_k \le s \le b_k\},$$

where $\boldsymbol{\tau}^{i_k} = \tau_1 \mathbf{a}_1^{i_k} + \tau_2 \mathbf{a}_2^{i_k}$ and $\mathbf{x}_{ik}^0$ is some point of $\Gamma_i(J_k)$ depending on both i and k. Therefore

$$\int_{\Gamma_i(J_k)} \left(\sum_{\alpha,j} \nu^i_\alpha s^{\alpha j}_i \mathbf{a}^i_j + \sum_{\alpha,\beta} \nu^i_\alpha s^{\alpha\beta}_i \mathbf{W}^i_{,\beta} \right) \cdot \hat{\mathbf{W}}^i(\mathbf{x})\, d\Gamma$$

$$= \int_{a_k}^{b_k} \left(\sum_{\alpha,j} \nu^i_\alpha s^{\alpha j}_i \mathbf{a}^i_j + \sum_{\alpha,\beta} \nu^i_\alpha s^{\alpha\beta}_i \mathbf{W}^i_{,\beta} \right) \cdot \hat{\mathbf{W}}^i(\mathbf{x})\, ds$$

$$(\mathbf{x} = \mathbf{x}^0_{ik} + s(\tau_1, \tau_2))$$

$$= \int_{a_k}^{b_k} \left(\sum_{\alpha,j} \nu^i_\alpha s^{\alpha j}_i \mathbf{a}^i_j + \sum_{\alpha,\beta} \nu^i_\alpha s^{\alpha\beta}_i \mathbf{W}^i_{,\beta} \right) \cdot \hat{\mathbf{W}}^i(\mathbf{p}_i^{-1}(\mathbf{p}))\, ds$$

$$(\mathbf{p} = \mathbf{q}^0_k + s\boldsymbol{\tau}^{i_k})$$

$$= \int_{J_k} \left(\sum_{\alpha,j} \nu^i_\alpha s^{\alpha j}_i \mathbf{a}^i_j + \sum_{\alpha,\beta} \nu^i_\alpha s^{\alpha\beta}_i \mathbf{W}^i_{,\beta} \right) \cdot \hat{\mathbf{W}}^i(\mathbf{p}_i^{-1}(\mathbf{p}))\, dJ_k.$$

Thus (2.10) may be expressed as

$$\sum_{k=1}^m \sum_{i \in \mathcal{I}(J_k)} \int_{J_k} \left(\sum_{\alpha,j} \nu^i_\alpha s^{\alpha j}_i \mathbf{a}^i_j + \sum_{\alpha,\beta} \nu^i_\alpha s^{\alpha\beta}_i \mathbf{W}^i_{,\beta} \right) \cdot \hat{\mathbf{W}}^i(\mathbf{p}_i^{-1}(\mathbf{p}))\, dJ_k$$

$$= \sum_{k=1}^m \int_{J_k} \tilde{\mathbf{f}}^k \cdot \hat{\mathbf{W}}^{i_k}\, dJ_k.$$

Since the J_k's are mutually disjoint and the geometric junction constraints are local to each J_k individually, the last variational equation can hold if and only if

$$(2.11) \quad \sum_{i \in \mathcal{I}(J_k)} \int_{J_k} \left(\sum_{\alpha,j} \nu^i_\alpha s^{\alpha j}_i \mathbf{a}^i_j + \sum_{\alpha,\beta} \nu^i_\alpha s^{\alpha\beta}_i \mathbf{W}^i_{,\beta} \right) \cdot \hat{\mathbf{W}}^i(\mathbf{p}_i^{-1}(\mathbf{p}))\, dJ_k$$

$$= \int_{J_k} \tilde{\mathbf{f}}^k \cdot \hat{\mathbf{W}}^{i_k}\, dJ_k, \quad k = 1, \ldots, m.$$

From (2.3) we have

$$\hat{\mathbf{W}}^i(\mathbf{p}_i^{-1}(\mathbf{p})) = \hat{\mathbf{W}}^{i_k}(\mathbf{p}_{i_k}^{-1}(\mathbf{p})), \quad \forall i \in \mathcal{I}(J_k).$$

Since $\hat{\mathbf{W}}^{i_k}(\mathbf{p}_{i_k}^{-1}(\mathbf{p}))$ is arbitrary, it follows that

$$(2.12) \quad \sum_{i \in \mathcal{I}(J_k)} \left(\sum_{\alpha,j} \nu^i_\alpha s^{\alpha j}_i \mathbf{a}^i_j + \sum_{\alpha,\beta} \nu^i_\alpha s^{\alpha\beta}_i \mathbf{W}^i_{,\beta} \right) (\mathbf{p}_i^{-1}(\mathbf{p}), t)$$

$$= \tilde{\mathbf{f}}^k(\mathbf{p}_{i_k}^{-1}(\mathbf{p}), t), \quad \mathbf{p} \in J_k.$$

Equation (2.12) is a balance law for forces along the joint J_k and we refer to it as the *dynamic junction condition* there.

2.3. Linearization. The nonlinear equations of motion (2.8), dynamic boundary conditions (2.9) and dynamic junction conditions (2.12) are geometrically exact under the kinematic hypotheses imposed above (i.e., (1.1) and Hooke's Law). It is apparent that their linearizations about the equilibrium $\mathbf{W}^i = 0$, $i = 1, \ldots, n$, are the same as the linearizations of the equations

$$\sum_{\alpha,j} s_{i,\alpha}^{\alpha j}\mathbf{a}_j^i + \mathbf{F}^i = \rho_i \mathbf{W}^i_{,tt}, \quad (x_1, x_2) \in \Omega_i, \ t > 0; \tag{2.13}$$

$$\sum_{\alpha,j} \nu_\alpha^i s_i^{\alpha j}\mathbf{a}_j^i = \mathbf{f}^i \quad \text{on } \Gamma_i^D, \ t > 0; \tag{2.14}$$

$$\sum_{i\in\mathcal{I}(J_k)} \left(\sum_{\alpha,j} \nu_\alpha^i s_i^{\alpha j}\mathbf{a}_j^i \right)(\mathbf{p}_i^{-1}(\mathbf{p}), t) = \tilde{\mathbf{f}}^k(\mathbf{p}_{i_k}^{-1}(\mathbf{p}), t), \tag{2.15}$$

$$\mathbf{p} \in J_k, \ t > 0, \quad k = 1, \ldots, m.$$

The linearizations of (2.13)–(2.15) are obtained by replacing the nonlinear strain components $\varepsilon_{\alpha\beta}$ in the stress-strain law (1.7) by their linear approximations

$$\varepsilon_{\alpha\beta} = \frac{1}{2}(W_{\alpha,\beta} + W_{\beta,\alpha}). \tag{2.16}$$

Let us consider the case of an elastically isotropic body $\mathcal{B}_i$. Its stress-strain law is then

$$s_i^{jk} = 2\mu_i \varepsilon_{jk}^i + \lambda_i \sum_\ell \varepsilon_{\ell\ell}^i \delta_{jk},$$

where ε_{jk}^i are the linearized strains in $\mathcal{B}_i$ and λ_i and μ_i are its Lamé parameters. It follows from (2.16) that the linearized stress-displacement relation is

$$\begin{cases} s_i^{\alpha\beta} = \mu_i(W_{\alpha,\beta}^i + W_{\beta,\alpha}^i) + \lambda_i \sum_\gamma W_{\gamma,\gamma}^i \delta_{\alpha\beta}, \\ s_i^{\alpha 3} = \mu_i W_{3,\alpha}^i. \end{cases} \tag{2.17}$$

Substitution of (2.17) into (2.13) gives the linear equations of motion

$$\begin{cases} \sum_\alpha [(\lambda_i + \mu_i) W_{\alpha,\alpha\beta}^i + \mu_i W_{\beta,\alpha\alpha}^i] + F_\beta^i = \rho_i W_{\beta,tt}^i, \\ \mu_i \sum_\alpha W_{3,\alpha\alpha}^i + F_3^i = \rho_i W_{3,tt}^i \quad \text{in } \Omega, \ t > 0, \end{cases} \tag{2.18}$$

where $F_j^i = \mathbf{F}^i \cdot \mathbf{a}_j^i$. The boundary conditions (2.14) linearize to

$$\text{(2.19)} \qquad \begin{cases} \sum_\alpha [\mu_i \nu_\alpha^i (W_{\alpha,\beta}^i + W_{\beta,\alpha}^i) + \lambda_i W_{\alpha,\alpha}^i \nu_\beta^i] = f_\beta^i, \\ \mu_i \sum_\alpha \nu_\alpha^i W_{3,\alpha}^i = f_3^i \ \text{ on } \Gamma_i^D,\ t > 0, \end{cases}$$

where $f_j^i = \mathbf{f}^i \cdot \mathbf{a}_j^i$. The dynamic junction condition (2.15) may similarly be expressed as a linear equation in the displacements, using (2.17), but we shall not write it explicitly. The description of the linear system is completed by adjoining the geometric conditions

$$\text{(2.20)} \qquad \mathbf{W}^i = 0 \ \text{ on } \Gamma_i^C,\ t > 0,$$

$$\text{(2.21)} \quad \mathbf{W}^i(\mathbf{p}_i^{-1}(\mathbf{p}), t) = \mathbf{W}^j(\mathbf{p}_j^{-1}(\mathbf{p}), t), \qquad \mathbf{p} \in J_k;\ i, j \in \mathcal{I}(J_k);\ k = 1, \ldots, m,$$

and the initial conditions

$$\text{(2.22)} \qquad \mathbf{W}^i|_{t=0} = \mathbf{W}_0^i, \ \ \mathbf{W}_{,t}^i|_{t=0} = \mathbf{W}_1^i \ \text{ in } \Omega, , i = 1, \ldots, n.$$

Equations (2.18)–(2.21), and (2.15) with s_i^{jk} given by (2.17), describe the linearized motion of a system of interconnected isotropic elastic membranes. Let us note that the above system is also directly derivable from the variational principle

$$\delta \int_0^T [\mathcal{K}(t) - \mathcal{U}(t) + \mathcal{W}(t)]\, dt = 0,$$

where $\mathcal{K}$ and $\mathcal{W}$ are as above and

$$\mathcal{U}(t) = \frac{1}{2} \sum_{i=1}^n \int_{\Omega_i} \left(\sum_{\alpha,\beta} s_i^{\alpha\beta} W_{\beta,\alpha}^i + \mu_i |\nabla W_3^i|^2 \right) d\Omega,$$

with $\nabla W_3^i = \sum_\alpha W_{3,\alpha}^i \mathbf{a}_\alpha^i$ and $s_i^{\alpha\beta}$ given by (2.17).

2.4. Well-Posedness of the Linear Model. In this section it is proved that the linear model consisting of (2.13)–(2.15), (2.20)–(2.22) forms a well-posed problem in an appropriate function space. A linear strain tensor is assumed for each body $\mathcal{B}_i$, which is also supposed to be hookean with elasticity tensor C_i^{jklm} satisfying (1.8) and (1.9), but not necessarily isotropic. We do assume, however, that its elasticity tensor is *elliptic*: for every 3 by 3 symmetric tensor ε_{jk},

$$\text{(2.23)} \qquad \sum_{j,k,l,m} C_i^{jklm} \varepsilon_{jk} \varepsilon_{lm} \geq c_0 \sum_{j,k} \varepsilon_{jk}^2$$

for some $c_0 > 0$.

We begin by recalling the general *Korn's lemma* of 3-dimensional linear elasticity. Let $\mathcal{D}$ be a bounded, open connected set in $\mathbb{R}^3$ with Lipschitz boundary. We denote by $H^k(\mathcal{D})$ the standard Sobolev space of order k over $\mathcal{D}$. When $k = 0$ we write $L^2(\mathcal{D})$, as is customary, instead of $H^0(\mathcal{D})$. Let $\mathbf{a}_1, \mathbf{a}_2, \mathbf{a}_3$ be an orthonormal basis for $\mathbb{R}^3$ and define

$$V = \{\mathbf{W} = \sum_j W_j \mathbf{a}_j | \, W_j \in H^1(\mathcal{D})\},$$

$$H = \{\mathbf{W} = \sum_j W_j \mathbf{a}_j | \, W_j \in L^2(\mathcal{D})\},$$

with

$$\|\mathbf{W}\|_V = \left(\sum_j \|W_j\|^2_{H^1(\mathcal{D})}\right)^{1/2}, \quad \|\mathbf{W}\|_H = \left(\sum_j \|W_j\|^2_{L^2(\mathcal{D})}\right)^{1/2}.$$

For $\mathbf{W} \in V$, set

$$\varepsilon_{jk}(\mathbf{W}) = \frac{1}{2}(W_{j,k} + W_{k,j}), \quad e(\mathbf{W}) = \left(\sum_{j,k} \int_{\mathcal{D}} \varepsilon^2_{jk}(\mathbf{W}) \, d\mathcal{V}\right)^{1/2}.$$

With this notation, Korn's lemma may be stated as follows.

LEMMA 2.1. *There is a constant K depending only on $\mathcal{D}$ such that for all $\mathbf{W} \in V$,*

$$\|\mathbf{W}\|_V \le K(e(\mathbf{W}) + \|\mathbf{W}\|_H). \tag{2.24}$$

Since it is trivial that

$$e(\mathbf{W}) \le K\|\mathbf{W}\|_V,$$

Lemma 2.1 shows that the right hand side of (2.24) defines a norm on V which is *equivalent* to the $\|\cdot\|_V$ norm on this space. The inequality (2.24) is highly nontrivial. Many proofs of it have been given under various assumptions on the region $\mathcal{D}$; a relatively simple proof when $\mathcal{D}$ is a bounded Lipschitz domain may be found in [82].

REMARK 2.2. A well-known (and easy to prove) consequence of Lemma 2.1 and the compactness of the embedding $V \mapsto H$ is the following. Let $p(\cdot)$ be any continuous seminorm defined on V such that $e(\mathbf{W}) = 0$, $p(\mathbf{W}) = 0$ implies $\mathbf{W} = 0$. Then

$$\|\mathbf{W}\|_V \le K[e(\mathbf{W}) + p(\mathbf{W})], \quad \forall \mathbf{W} \in V. \tag{2.25}$$

Therefore, $e(\cdot)$ defines a *norm* on the closed subspace $\{\mathbf{W} \in V \mid p(\mathbf{W}) = 0\}$. One may choose, for example,

$$p(\mathbf{W}) = \left(\sum_j \|W_j\|^2_{H^{-r}((\partial\mathcal{D})_0)} \right)^{1/2},$$

where $r \geq -1/2$ and $(\partial\mathcal{D})_0$ is any nonempty, relatively open subset of $\partial\mathcal{D}$. In particular, one has

$$\|\mathbf{W}\|_V \leq Ke(\mathbf{W}), \quad \forall \mathbf{W} \in V \cap \{\mathbf{W} = 0 \text{ on } (\partial\mathcal{D})_0\}. \tag{2.26}$$ □

REMARK 2.3. Apropos to the membrane problem, let Ω be a bounded, open set in $\mathbb{R}^2$ with Lipschitz boundary and define

$$V = \{\mathbf{W} = \sum_j W_j \mathbf{a}_j \mid W_j \in H^1(\Omega)\},$$

$$H = \{\mathbf{W} = \sum_j W_j \mathbf{a}_j \mid W_j \in L^2(\Omega)\},$$

with

$$\|\mathbf{W}\|_V = \left(\sum_j \|W_j\|^2_{H^1(\Omega)} \right)^{1/2}, \quad \|\mathbf{W}\|_H = \left(\sum_j \|W_j\|^2_{L^2(\Omega)} \right)^{1/2}.$$

Then $\varepsilon_{33}(\mathbf{W}) = 0$ and $\varepsilon_{\alpha 3}(\mathbf{W}) = W_{3,\alpha}/2$. Define

$$e(\mathbf{W}) = \left[\int_\Omega \left(\sum_{\alpha,\beta} \varepsilon^2_{\alpha\beta}(\mathbf{W}) + \sum_\alpha \varepsilon^2_{\alpha 3}(\mathbf{W}) \right) d\Omega \right]^{1/2}.$$

These definitions of V, H and $e(\cdot)$ coincide with the previous ones if we take $\mathcal{D}$ to be the cylinder $\Omega \times (0,1)$ and consider only functions $\mathbf{W}$ which are independent of x_3. It is then obvious that (2.24) remains valid with V, H and $e(\cdot)$ so defined, as does Remark 2.2 with $(\partial\mathcal{D})_0$ replaced by Γ_0, where Γ_0 is any nonempty, relatively open subset of $\partial\Omega$ and K depends only on Ω. □

Now for $i = 1, \ldots, n$, let Ω_i and $\mathbf{a}^i_1, \mathbf{a}^i_2, \mathbf{a}^i_3$ be as defined at the beginning of this section, and set

$$V_i = \{\mathbf{W} = \sum_j W_j \mathbf{a}^i_j \mid W_j \in H^1(\Omega_i)\},$$

$$H_i = \{\mathbf{W} = \sum_j W_j \mathbf{a}^i_j \mid W_j \in L^2(\Omega_i)\},$$

$$V = \prod_{i=1}^n V_i, \quad H = \prod_{i=1}^n H_i.$$

An element $\mathbf{W} \in V$ is given by

$$\mathbf{W} = (\mathbf{W}^1, \dots, \mathbf{W}^n), \quad \mathbf{W}^i = \sum_j W_j^i \mathbf{a}_j^i \in V_i,$$

and similarly for H. The norms on H and V are respectively

$$\|\mathbf{W}\|_H = \left(\sum_{i=1}^n \int_{\Omega_i} \rho_i |\mathbf{W}^i|^2 d\Omega \right)^{1/2} \tag{2.27}$$

and

$$\|\mathbf{W}\|_V = \left(\sum_{i=1}^n \|\mathbf{W}^i\|_{V_i}^2 \right)^{1/2}, \tag{2.28}$$

where

$$\|\mathbf{W}^i\|_{V_i} = \left(\sum_j \|\mathbf{W}_j^i\|_{H^1(\Omega_i)}^2 \right)^{1/2}.$$

With these definitions, V is dense in H with compact embedding.

A continuous seminorm $e(\cdot)$ is defined on V by setting

$$e(\mathbf{W}) = \left(\sum_{i=1}^n e_i^2(\mathbf{W}^i) \right)^{1/2}, \tag{2.29}$$

where

$$\begin{aligned} e_i(\mathbf{W}^i) &= \left(\sum_{j,k} \int_{\Omega_i} \varepsilon_{jk}^2(\mathbf{W}^i)\, d\Omega \right)^{1/2} \\ &= \left[\int_{\Omega_i} \left(\sum_{\alpha,\beta} \varepsilon_{\alpha\beta}^2(\mathbf{W}^i) + \sum_\alpha \varepsilon_{\alpha 3}^2(\mathbf{W}^i) \right) d\Omega \right]^{1/2}. \end{aligned}$$

It follows immediately from Lemma 2.1 and Remark 2.3, applied in each region Ω_i, that (2.24) and (2.25) remain valid with these new definitions of V, H and $e(\cdot)$.

Let $\mathcal{V}$ denote the closed subspace of V consisting of functions $\mathbf{W} = (\mathbf{W}^1, \dots, \mathbf{W}^n)$ such that

$$\mathbf{W}^i \in V^i, \quad \mathbf{W}^i = 0 \text{ on } \Gamma_i^C, \tag{2.30}$$

$$\mathbf{W}^i(\mathbf{p}_i^{-1}(\mathbf{p})) = \mathbf{W}^j(\mathbf{p}_j^{-1}(\mathbf{p})), \quad \mathbf{p} \in J_k, \; i, j \in \mathcal{I}(J_k). \tag{2.31}$$

Then $\mathcal{V}$ is dense in H with compact embedding. The following lemma is the analog of (2.26) for systems of interconnected membranes.

LEMMA 2.2. *Assume that $\Gamma_i^C \neq \emptyset$ for at least one index i. Then there is a constant K depending only on $\Omega_1, \dots, \Omega_n$ such that*

$$\|\mathbf{W}\|_V \le Ke(\mathbf{W}), \quad \forall \mathbf{W} \in \mathcal{V}. \tag{2.32}$$

PROOF. Since the map $\mathbf{W} \mapsto \|\mathbf{W}^i\|_{H^{1/2}(\Gamma_i^C)}$ defines a continuous seminorm on V, it suffices to show that

$$\mathbf{W} \in V, \quad e(\mathbf{W}) = 0, \quad \mathbf{W}^i|_{\Gamma_i^C} = 0 \Longrightarrow \mathbf{W} = 0.$$

Without loss of generality, we may assume that $\Gamma_1^C \neq \emptyset$. Then since, in particular, $e_1(\mathbf{W}^1) = 0$, it follows from Remark 2.2 that $\mathbf{W}^1 = 0$. Suppose that $\mathcal{P}_2$ is joined to $\mathcal{P}_1$ along a joint J. Since $\mathbf{W}^1(\mathbf{p}_1^{-1}(\mathbf{p})) = 0$, $\mathbf{p} \in J$, the geometric junction condition (2.31) implies $\mathbf{W}^2(\mathbf{p}_2^{-1}(\mathbf{p})) = 0$, $\mathbf{p} \in J$, and then Remark 2.2 gives $\mathbf{W}^2 = 0$. Since the collection of plates forms a connected set, we may conclude in the same manner that $\mathbf{W}^i = 0$, $i = 1, \dots, n$. □

Having set up an appropriate function space framework, we may show that the problem (2.13)–(2.15), (2.20)–(2.22) is well-posed if the data $\mathbf{F}^i, \mathbf{f}^i$, $\mathbf{W}_0^i, \mathbf{W}_1^i$ are chosen in the proper spaces. Apart from the initial conditions (2.22), we adopt as the definition of a solution to our problem a solution of the variational equation

$$\begin{aligned}\sum_{i=1}^n \int_0^T \int_{\Omega_i} (\rho_i \mathbf{W}^i_{,tt} \cdot \hat{\mathbf{W}}^i + \sum_{\alpha,j} s_i^{\alpha j} \hat{W}^i_{j,\alpha})\, d\Omega dt \\ = \sum_{i=1}^n \left(\int_0^T \int_{\Omega_i} \mathbf{F}^i \cdot \hat{\mathbf{W}}^i d\Omega dt + \int_0^T \int_{\Gamma_i^D} \mathbf{f}^i \cdot \hat{\mathbf{W}}^i d\Gamma dt \right) \\ + \sum_{k=1}^m \int_0^T \int_{\Gamma_{i_k}(J_k)} \tilde{\mathbf{f}}^k \cdot \hat{\mathbf{W}}^{i_k}\, d\Gamma, \\ \forall \hat{\mathbf{W}} = (\hat{\mathbf{W}}^1, \dots, \hat{\mathbf{W}}^n) \in L^2(0,T;\mathcal{V}), \quad T > 0.\end{aligned} \tag{2.33}$$

Equivalently, for almost all $t > 0$,

$$\begin{aligned}\sum_{i=1}^n \int_{\Omega_i} (\rho_i \mathbf{W}^i_{,tt} \cdot \hat{\mathbf{W}}^i + \sum_{\alpha,j} s_i^{\alpha j} \hat{W}^i_{j,\alpha})\, d\Omega \\ = \sum_{i=1}^n \left(\int_{\Omega_i} \mathbf{F}^i \cdot \hat{\mathbf{W}}^i d\Omega + \int_{\Gamma_i^D} \mathbf{f}^i \cdot \hat{\mathbf{W}}^i d\Gamma \right) \\ + \sum_{k=1}^m \int_{\Gamma_{i_k}(J_k)} \tilde{\mathbf{f}}^k \cdot \hat{\mathbf{W}}^{i_k}\, d\Gamma, \quad \forall \hat{\mathbf{W}} = (\hat{\mathbf{W}}^1, \dots, \hat{\mathbf{W}}^n) \in \mathcal{V}.\end{aligned} \tag{2.34}$$

Equation (2.33) is obtained from (2.13)–(2.15), (2.20), (2.21) by multiplying (2.13) by $\hat{\mathbf{W}}^i$, integrating the product over $\Omega_i \times (0,T)$ and using integration by parts in the usual manner.

One may write

$$\int_{\Omega_i} s_i^{\alpha j} \hat{W}^i_{j,\alpha}\, d\Omega = \int_{\Omega_i} s_i^{\alpha j} \varepsilon_{\alpha j}(\hat{\mathbf{W}}^i)\, d\Omega.$$

The form

$$\sigma(\mathbf{W}, \hat{\mathbf{W}}) = \sum_{i=1}^{n} \int_{\Omega_i} \sum_{\alpha,j} s_i^{\alpha j} \varepsilon_{\alpha j}(\hat{\mathbf{W}}^i)\, d\Omega$$

is a symmetric, continuous bilinear form on V and, because of the ellipticity assumption (2.23) and Korn's lemma,

$$\text{(2.35)} \quad \begin{aligned} \sigma(\mathbf{W}, \mathbf{W}) &\geq c_0 \sum_{i=1}^{n} \int_{\Omega_i} \sum_{\alpha j} \varepsilon^2_{\alpha j}(\mathbf{W}^i)\, d\Omega \\ &\geq \frac{c_0}{2K^2} \|\mathbf{W}\|^2_V - c_0 \|\mathbf{W}\|^2_H, \quad \forall \mathbf{W} \in V. \end{aligned}$$

Therefore, $[\sigma(\mathbf{W}, \mathbf{W}) + c_0(\mathbf{W}, \mathbf{W})^2_H]^{1/2}$ defines a norm on $\mathcal{V}$ (and even on V) equivalent to that introduced above. We retopologize $\mathcal{V}$ with this new norm and denote by $\mathcal{V}'$ the dual space of $\mathcal{V}$. With H identified with its dual space we then have $\mathcal{V} \subset H \subset \mathcal{V}'$ with dense and compact embeddings. Let A denote the Riesz isomorphism of $\mathcal{V}$ onto $\mathcal{V}'$, so that

$$(\mathbf{W}, \hat{\mathbf{W}})_{\mathcal{V}} = \sigma(\mathbf{W}, \hat{\mathbf{W}}) + c_0(\mathbf{W}, \hat{\mathbf{W}})_H = \langle A\mathbf{W}, \hat{\mathbf{W}} \rangle_{\mathcal{V}},$$

where $\langle \mathbf{W}', \hat{\mathbf{W}} \rangle_{\mathcal{V}}$ denotes the pairing between an element $\mathbf{W}' \in \mathcal{V}'$ and $\hat{\mathbf{W}} \in \mathcal{V}$ in the $\mathcal{V}'$–$\mathcal{V}$ duality. If $\mathbf{W}' \in H$ one has $\langle \mathbf{W}', \hat{\mathbf{W}} \rangle_{\mathcal{V}} = (\mathbf{W}', \hat{\mathbf{W}})_H$, $\forall \hat{\mathbf{W}} \in \mathcal{V}$. With this notation, (2.34) may be written

$$\text{(2.36)} \quad \begin{aligned} &\langle \mathbf{W}_{,tt} + A\mathbf{W} - c_0\mathbf{W}, \hat{\mathbf{W}} \rangle_{\mathcal{V}} \\ &= \sum_{i=1}^{n} \left(\int_{\Omega_i} \mathbf{F}^i \cdot \hat{\mathbf{W}}^i d\Omega + \int_{\Gamma_i^D} \mathbf{f}^i \cdot \hat{\mathbf{W}}^i d\Gamma \right) \\ &\qquad + \sum_{k=1}^{m} \int_{\Gamma_{i_k}(J_k)} \tilde{\mathbf{f}}^k \cdot \mathbf{W}^{i_k}\, d\Gamma, \quad \forall \hat{\mathbf{W}} \in \mathcal{V}. \end{aligned}$$

The variational equation (2.36) falls within standard variational theory as found, for example, in [62, Chapter 3]. Let us first consider the case of homogeneous boundary and junction conditions.

THEOREM 2.1. *Assume that*

$$\mathbf{F}^i \in L^2(0,T; L^2(\Omega_i)), \ \mathbf{f}^i = 0, \ i = 1, \dots, n, \ \tilde{\mathbf{f}}^k = 0, \ k = 1, \dots, m,$$
$$\mathbf{W}_0 := (\mathbf{W}_0^1, \dots, \mathbf{W}_0^n) \in \mathcal{V}, \ \mathbf{W}_1 := (\mathbf{W}_1^1, \dots, \mathbf{W}_1^n) \in H.$$

Then there is a unique function

$$\mathbf{W} \in C([0,T];\mathcal{V}) \cap C^1([0,T];H) \cap C^2([0,T];\mathcal{V}')$$

which satisfies (2.36) *for almost all* $t \in [0,T]$ *and for which* $\mathbf{W}(0) = \mathbf{W}_0$, $\mathbf{W}_{,t}(0) = \mathbf{W}_1$.

The proof is just a matter of observing that

$$\sum_{i=1}^{n} \int_{\Omega_i} \mathbf{F}^i \cdot \hat{\mathbf{W}}^i d\Omega = (\mathbf{G}, \hat{\mathbf{W}})_H,$$

where

$$\mathbf{G} = (\rho_1^{-1}\mathbf{F}^1, \dots, \rho_n^{-1}\mathbf{F}^n) \in L^2(0,T;H),$$

and then applying standard theory. Note that it follows from (2.36) that $\mathbf{W}_{,tt} \in C([0,T];\mathcal{V}') \cap L^2(0,T;H)$.

Next, let us consider the case of inhomogeneous boundary conditions and assume that $\mathbf{F}^i = 0$. Introduce the space

$$D_A = \{\hat{\mathbf{W}} \in \mathcal{V} |\, A\hat{\mathbf{W}} \in H\}, \quad \|\hat{\mathbf{W}}\|_{D_A} = \|A\hat{\mathbf{W}}\|_H.$$

One has the dense and continuous embeddings $D_A \subset \mathcal{V} \subset H \subset \mathcal{V}' \subset D_A'$. Since

$$|\langle A\mathbf{W}, \hat{\mathbf{W}}\rangle_{\mathcal{V}}| = |(\mathbf{W}, A\hat{\mathbf{W}})_H| \le \|\mathbf{W}\|_H \|\hat{\mathbf{W}}\|_{D_A}, \quad \forall \mathbf{W} \in \mathcal{V},\ \hat{\mathbf{W}} \in D_A,$$

the operator may be extended to a mapping $H \mapsto D_A'$, also a homeomorphism, through the formula

$$\langle A\mathbf{W}, \hat{\mathbf{W}}\rangle_{D_A} = (\mathbf{W}, A\hat{\mathbf{W}})_H, \quad \forall \mathbf{W} \in H,\ \hat{\mathbf{W}} \in D_A.$$

The variational equation (2.36) may therefore be written in the weaker form

$$\langle \mathbf{W}_{,tt} + A\mathbf{W} - c_0\mathbf{W}, \hat{\mathbf{W}}\rangle_{D_A} = \sum_{i=1}^{n} \int_{\Gamma_i^D} \mathbf{f}^i \cdot \hat{\mathbf{W}}^i d\Gamma + \sum_{k=1}^{m} \int_{\Gamma_{i_k}(J_k)} \tilde{\mathbf{f}}^k \cdot \mathbf{W}^{i_k}\, d\Gamma, \quad \forall \hat{\mathbf{W}} \in D_A. \tag{2.37}$$

THEOREM 2.2. *Assume that* $\mathbf{F}^i = 0$,

$$\mathbf{f}^i \in L^2(0,T;L^2(\Gamma_i^D)), \quad i = 1,\dots,n,$$
$$\tilde{\mathbf{f}}^k \in L^2(0,T;L^2(\Gamma_{i_k}(J_k))), \quad k = 1,\dots,m,$$
$$\mathbf{W}_0 \in H, \quad \mathbf{W}_1 \in \mathcal{V}'.$$

Then there is a unique function

$$\mathbf{W} \in C([0,T];H) \cap C^1([0,T];\mathcal{V}') \cap C^2([0,T];D_A')$$

which satisfies (2.37) *for almost all* $t \in [0, T]$ *and for which* $\mathbf{W}(0) = \mathbf{W}_0$, $\mathbf{W}_{,t}(0) = \mathbf{W}_1$.

In fact, one has

$$\left|\sum_{i=1}^{n} \int_{\Gamma_i^D} \mathbf{f}^i \cdot \hat{\mathbf{W}}^i d\Gamma\right| + \left|\sum_{k=1}^{m} \int_{\Gamma_{i_k}(J_k)} \tilde{\mathbf{f}}^k \cdot \mathbf{W}^{i_k}\, d\Gamma\right|$$
$$\leq \|\hat{\mathbf{W}}\|_{\mathcal{V}} \left(\sum_{i=1}^{n} \|\mathbf{f}^i\|^2_{L^2(\Gamma_i^D)} + \sum_{k=1}^{m} \|\tilde{\mathbf{f}}^k\|^2_{L^2(\Gamma_{i_k}(J_k))}\right)^{1/2},$$

hence there exists an element $\mathbf{g} \in L^2(0, T; \mathcal{V}')$ such that

$$\sum_{i=1}^{n} \int_{\Gamma_i^D} \mathbf{f}^i \cdot \hat{\mathbf{W}}^i d\Gamma + \sum_{k=1}^{m} \int_{\Gamma_{i_k}(J_k)} \tilde{\mathbf{f}}^k \cdot \mathbf{W}^{i_k}\, d\Gamma = \langle \mathbf{g}, \hat{\mathbf{W}} \rangle_{\mathcal{V}}.$$

Standard variational theory then gives the conclusion of the Theorem.

REMARK 2.4. Consider the homogeneous problem $\mathbf{F}^i = 0$, $\mathbf{f}^i = 0$, $i = 1, \dots, n$, $\tilde{\mathbf{f}}^k = 0$, $k = 1, \dots, m$, and assume that the initial data satisfy

$$\mathbf{W}_0 \in D_A, \quad \mathbf{W}_1 \in \mathcal{V}.$$

It is standard that the solution $\mathbf{W}$ then has the regularity

$$\mathbf{W} \in C([0, T]; D_A) \cap C^1([0, T]; \mathcal{V}) \cap C^2([0, T]; H).$$

A fundamental question is whether or not $\mathbf{W}$ is then a *classical* solution. The answer amounts to determining the regularity of an element $\mathbf{U} = (\mathbf{U}^1, \dots, \mathbf{U}^n) \in D_A$. Such an element is characterized by the variational equation

$$\sigma(\mathbf{U}, \hat{\mathbf{U}}) = (\mathbf{F}, \hat{\mathbf{U}})_H, \quad \forall \hat{\mathbf{U}} \in \mathcal{V},$$

for some $\mathbf{F} \in H$, together with the geometric conditions

$$\mathbf{U}^i = 0 \text{ on } \Gamma_i^C, \; i = 1, \dots, n,$$
$$\mathbf{U}^i(\mathbf{p}_i^{-1}(\mathbf{p})) = \mathbf{U}^j(\mathbf{p}_j^{-1}(\mathbf{p})), \quad \mathbf{p} \in J_k, \; k = 1, \dots, m.$$

If we introduce the notation

$$V_i^s = \{\mathbf{U} = \sum_j U_j \mathbf{a}_j^i \,|\, U_j \in H^s(\Omega_i)\}, \quad V^s = \prod_{i=1}^{n} V_i^s, \tag{2.38}$$

the issue may be phrased as whether or not

$$D_A \subset V^s \cap \mathcal{V} \quad \text{for some } s > 3/2 \tag{2.39}$$

with continuous injection.

It is known that even for a single isotropic membrane the validity of (2.39) depends on the geometry of Ω and the particular boundary conditions. For example, (2.39) is true for any Lipschitz domain if $\overline{\Gamma}^C$ and $\overline{\Gamma}^D$ are disjoint. If $\overline{\Gamma}^C \cap \overline{\Gamma}^D \neq \emptyset$, (2.39) will hold if Γ^C and Γ^D meet in an angle smaller than π (measured in the interior of Ω). In this matter the reader is referred to Grisvard [30], [31], and Nicaise [78]. For systems of interconnected isotropic membranes, from the regularity results for a single membrane it is clear that there is H^s regularity ($s > 3/2$) in a neighborhood of each point of Γ_i^C, of Γ_i^D and of $\overline{\Gamma_i^C} \cap \overline{\Gamma_i^D}$ provided the angle condition just mentioned is satisfied. (Interior regularity is not at issue.) However, a precise description of the regularity possessed by $\mathbf{U}^i$ near Γ_i^J or, more specifically, its regularity at mixed corners (i.e., points of $\overline{\Gamma_i^J} \cap \overline{\Gamma_i^C}$ and $\overline{\Gamma_i^J} \cap \overline{\Gamma_i^D}$), remains unavailable. Let us mention, however, an important contribution of Nicaise [81] towards resolving this issue, wherin the precise singular behavior of solutions of the Laplace and biharmonic equations on networks at mixed corners is described. See also Nicaise [73], [74], where regularity of solutions (among other things) of "transmission problems" for general elliptic problems with general boundary and interface conditions is investigated. □

3. Controllability of Linked Isotropic Membranes

In this section we shall consider the question of exact controllability of solutions of the linearized equations of motion of the system of interconnected, elastically isotropic membranes which was derived in Section 2.3. We recall that such a system may be written (in the absence of distributed forces) as follows. The equations of motion are

$$\text{(3.1)} \qquad \begin{cases} \sum_\alpha s_{i,\alpha}^{\alpha\beta} = \rho_i W_{\beta,tt}^i, \\ \mu_i \Delta W_3^i = \rho_i W_{3,tt}^i, \quad \mathbf{x} \in \Omega_i,\ t > 0, \end{cases}$$

where $i = 1, \ldots, n$, $\Delta W_3^i = \sum_\alpha W_{3,\alpha\alpha}^i$ and where

$$s_i^{\alpha\beta} = \mu_i (W_{\alpha,\beta}^i + W_{\beta,\alpha}^i) + \lambda_i \Big(\sum_\gamma W_{\gamma,\gamma}^i\Big)\delta_{\alpha\beta}.$$

The boundary conditions are

$$\text{(3.2)} \qquad \begin{cases} \sum_\alpha \nu_\alpha^i s_i^{\alpha\beta} = f_\beta^i, \\ \sum_\alpha \nu_\alpha^i W_{3,\alpha}^i = f_3^i, \quad \mathbf{x} \in \Gamma_i^D,\ t > 0, \end{cases}$$

and

$$\mathbf{W}^i = 0, \;\; \mathbf{x} \in \Gamma_i^C, \; t > 0. \tag{3.3}$$

The junction conditions are

$$\mathbf{W}^i(\mathbf{p}_i^{-1}(\mathbf{p}), t) = \mathbf{W}^j(\mathbf{p}_j^{-1}(\mathbf{p}), t), \qquad \mathbf{p} \in J_k, \; i, j \in \mathcal{I}(J_k), \; k = 1, \dots, m, \tag{3.4}$$

and

$$\sum_{i \in \mathcal{I}(J_k)} \sum_\alpha \nu_\alpha^i \left(\sum_\beta s_i^{\alpha\beta} \mathbf{a}_\beta^i + \mu_i W_{3,\alpha} \mathbf{a}_3^i \right) (\mathbf{p}_i^{-1}(\mathbf{p}), t) = \tilde{\mathbf{f}}^k(\mathbf{p}_{i_k}^{-1}(\mathbf{p}), t), \;\; \mathbf{p} \in J_k, \; t > 0. \tag{3.5}$$

Given a suitable $T > 0$ and "arbitrary" initial data $(\mathbf{W}_0^i, \mathbf{W}_1^i)$ and final data $(\hat{\mathbf{W}}_0^i, \hat{\mathbf{W}}_1^i)$, the above system is *exactly controllable in time T* if there are *control functions* $\mathbf{f}^i = \sum_j f_j^i \mathbf{a}_j^i$, $\tilde{\mathbf{f}}^k$ such that the solution of (3.1)–(3.5) satisfying the initial conditions

$$\mathbf{W}^i|_{t=0} = \mathbf{W}_0^i, \;\; \mathbf{W}_{,t}^i|_{t=0} = \mathbf{W}_1^i, \;\; \mathbf{x} \in \Omega_i$$

achieves the state $(\hat{\mathbf{W}}_0^i, \hat{\mathbf{W}}_1^i)$ at time T:

$$\mathbf{W}^i|_{t=T} = \hat{\mathbf{W}}_0^i, \;\; \mathbf{W}_{,t}^i|_{t=T} = \hat{\mathbf{W}}_1^i, \;\; i = 1, \dots, n.$$

Of course, the data and controls must be chosen from appropriate function spaces for which the initial value problem is (at least) well- posed in some sense. Since the problem (3.1)–(3.5) is time reversible, one may assume that either the final state or the initial state vanishes. In the latter situation the exact controllability problem is called the *reachability problem.* Thus the reachability problem is that of showing that starting from the zero state, an "arbitrary state" $(\hat{\mathbf{W}}_0^i, \hat{\mathbf{W}}_1^i)$, $i = 1, \dots, n$, may be achieved in time T through an appropriate choice of control functions.

It is not to be expected that every system of interconnected membranes, no matter how configured, is exactly controllable. Indeed, our purpose is to determine those geometric properties of such a system which will guarantee its exact controllability. It will be fairly apparent that the sufficient conditions for exact controllability obtained below are (in all probability) rather far from necessary. However, it seems that methods other than those employed here will be required in order to further develop the theory.

Another point is that in order to obtain any exact controllability results at all we usually (but not always) find it necessary to employ controls not only along the exterior boundaries Λ_i^D but also in the junction regions Λ_i^J. It is not clear to what extent the latter requirement is due to the methods employed or whether there is something intrinsic to wave propagation in general systems of interconnected membranes which necessitates the use of junction based controls.

3.1. Observability Estimates for the Homogeneous Problem. In this section the *a priori* estimates needed to describe the set of reachable states will be derived. We begin by deriving certain basic identities. Let Ω and $\mathbf{a}_1, \mathbf{a}_2, \mathbf{a}_3$ be as defined in Section 1. We denote by $\mathbf{m}(x_1, x_2)$ the radial vector field

$$\mathbf{m}(x_1, x_2) = \sum_{\alpha}(x_\alpha - x_{0\alpha})\mathbf{a}_\alpha, \quad (x_1, x_2) \in \mathbb{R}^2,$$

where (x_{01}, x_{02}) is a fixed point in $\mathbb{R}^2$.

LEMMA 3.1. *Let ϕ be a sufficiently regular solution of the scalar equation*

$$\rho\phi_{,tt} - \mu\Delta\phi = 0 \quad \text{in } \Omega \times (0,T),$$

and set

$$X(t) = \rho \int_\Omega \phi_{,t}(\mathbf{m} \cdot \nabla\phi + \frac{1}{2}\phi)\, d\Omega,$$

where $\nabla\phi = \sum_\alpha \phi_{,\alpha}\mathbf{a}_\alpha$. Then

$$\begin{aligned}(3.6)\quad X(T) - X(0) &+ \frac{1}{2}\int_0^T \int_\Omega (\rho\phi_{,t}^2 + \mu|\nabla\phi|^2)\, d\Omega dt \\ &= \frac{1}{2}\int_0^T \int_\Gamma (\mathbf{m}\cdot\boldsymbol{\nu})\left[\rho\phi_{,t}^2 + \mu\left(\frac{\partial\phi}{\partial\boldsymbol{\nu}}\right)^2\right] d\Gamma dt \\ &\quad - \frac{\mu}{2}\int_0^T \int_\Gamma (\mathbf{m}\cdot\boldsymbol{\nu})\left(\frac{\partial\phi}{\partial\boldsymbol{\tau}}\right)^2 d\Gamma dt \\ &\quad + \mu\int_0^T \int_\Gamma \frac{\partial\phi}{\partial\boldsymbol{\nu}}\left[(\mathbf{m}\cdot\boldsymbol{\tau})\frac{\partial\phi}{\partial\boldsymbol{\tau}} + \frac{1}{2}\phi\right] d\Gamma dt.\end{aligned}$$

PROOF. The identity (3.6) follows from

$$0 = \int_0^T \int_\Omega (\mathbf{m}\cdot\nabla\phi + \frac{1}{2}\phi)(\rho\phi_{,tt} - \mu\Delta\phi)\, d\Omega dt$$

by appropriate integration by parts. It has been proved, either explicitly or implicitly, by many authors in various contexts (cf. [60, p. 143]). It will be valid for solutions having, for example, the regularity

$$\phi \in C([0,T]; H^s(\Omega)) \cap C^1([0,T]; H^{s-1}(\Omega))$$

for some $s > 3/2$. □

Let $\boldsymbol{\psi} : I\!R^2 \mapsto I\!R^2$ be a C^1 function. We introduce the notation

$$\boldsymbol{\psi}(x_1, x_2) = \sum_{\alpha} \psi_\alpha(x_1, x_2)\mathbf{a}_\alpha,$$
$$\varepsilon_{\alpha\beta}(\boldsymbol{\psi}) = \frac{1}{2}(\psi_{\alpha,\beta} + \psi_{\beta,\alpha}),$$
$$s^{\alpha\beta}(\boldsymbol{\psi}) = 2\mu\varepsilon_{\alpha\beta}(\boldsymbol{\psi}) + \lambda\delta_{\alpha\beta}\sum_{\gamma}\varepsilon_{\gamma\gamma}(\boldsymbol{\psi}),$$
$$S(\boldsymbol{\psi}) = (s^{\alpha\beta}(\boldsymbol{\psi})),$$

and we note the Green's formula

$$\int_\Omega (S(\boldsymbol{\psi}), \varepsilon(\boldsymbol{\phi}))\, d\Omega = -\int_\Omega \boldsymbol{\phi} \cdot \operatorname{div} S(\boldsymbol{\psi})\, d\Omega + \int_\Gamma \boldsymbol{\phi} \cdot (S(\boldsymbol{\psi})\boldsymbol{\nu})\, d\Gamma, \tag{3.7}$$

where $(S(\boldsymbol{\psi}), \varepsilon(\boldsymbol{\psi})) = \operatorname{trace}(S(\boldsymbol{\psi})\varepsilon(\boldsymbol{\phi}))$.

LEMMA 3.2. *Let* $\mathbf{h} = h_1\mathbf{a}_1 + h_2\mathbf{a}_2$ *be any* $W^{1,\infty}$ *vector field in* Ω *and* $\boldsymbol{\psi} = \psi_1\mathbf{a}_1 + \psi_2\mathbf{a}_2$, *where* $\psi_i \in H^s(\Omega)$ *for some* $s > 3/2$. *Then*

$$\begin{aligned}\int_\Omega (\nabla\boldsymbol{\psi}\mathbf{h}) \cdot \operatorname{div} S(\boldsymbol{\psi})\, d\Omega = &-\frac{1}{2}\int_\Gamma (\mathbf{h}\cdot\boldsymbol{\nu})(S(\boldsymbol{\psi}), \varepsilon(\boldsymbol{\psi}))\, d\Gamma \\ &+ \int_\Gamma (\nabla\boldsymbol{\psi}\mathbf{h}) \cdot S(\boldsymbol{\psi})\boldsymbol{\nu}\, d\Gamma - \int_\Omega Q(\nabla\boldsymbol{\psi})\, d\Omega,\end{aligned} \tag{3.8}$$

where $Q(\nabla\boldsymbol{\psi})$ *is a quadratic form in* $\psi_{i,j}$ *given by*

$$\begin{aligned}Q(\nabla\boldsymbol{\psi}) = \frac{1}{2}(h_{1,1} - h_{2,2})[(2\mu+\lambda)(\psi_{1,1}^2 - \psi_{2,2}^2) + \mu(\psi_{2,1}^2 - \psi_{1,2}^2)] \\ + (2\mu+\lambda)(h_{1,2}\psi_{2,1}\psi_{2,2} + h_{2,1}\psi_{1,1}\psi_{1,2}) \\ + \mu(h_{1,2}\psi_{1,1} + h_{2,1}\psi_{2,2})(\psi_{1,2} + \psi_{2,1}) \\ + \lambda(h_{1,2}\psi_{1,1}\psi_{2,1} + h_{2,1}\psi_{1,2}\psi_{2,2}).\end{aligned}$$

PROOF. From Green's formula (3.7) we have

$$\begin{aligned}\int_\Omega (\nabla\boldsymbol{\psi}\mathbf{h}) \cdot \operatorname{div} S(\boldsymbol{\psi})\, d\Omega = -\int_\Omega (S(\boldsymbol{\psi}), \varepsilon(\nabla\boldsymbol{\psi}\mathbf{h}))\, d\Gamma \\ + \int_\Gamma (\nabla\boldsymbol{\psi}\mathbf{h}) \cdot (S(\boldsymbol{\psi})\boldsymbol{\nu})\, d\Gamma,\end{aligned}$$

so to prove (3.8) we need to calculate $(S(\boldsymbol{\psi}), \varepsilon(\nabla\boldsymbol{\psi}\mathbf{h}))$. We have

$$\begin{aligned}(S(\boldsymbol{\psi}), \varepsilon(\nabla\boldsymbol{\psi}\mathbf{h})) = (2\mu+\lambda)[\psi_{1,1}(\nabla\psi_1\cdot\mathbf{h})_{,1} + \psi_{2,2}(\nabla\psi_2\cdot\mathbf{h})_{,2}] \\ + \lambda[\psi_{1,1}(\nabla\psi_2\cdot\mathbf{h})_{,2} + \psi_{2,2}(\nabla\psi_1\cdot\mathbf{h})_{,1}] \\ + \mu(\psi_{1,2} + \psi_{2,1})[(\nabla\psi_1\cdot\mathbf{h})_{,2} + (\nabla\psi_2\cdot\mathbf{h})_{,1}].\end{aligned}$$

A lengthy, but straightforward, calculation shows that the right side of the last equality may be written

$$\begin{aligned}
&\frac{1}{2}\operatorname{div}\{\mathbf{h}[(2\mu+\lambda)(\psi_{1,1}^2+\psi_{2,2}^2)+2\lambda\psi_{1,1}\psi_{2,2}+\mu(\psi_{1,2}+\psi_{2,1})^2]\}+Q(\nabla\boldsymbol{\psi})\\
&\qquad=\frac{1}{2}\operatorname{div}\{\mathbf{h}[(2\mu+\lambda)(\varepsilon_{11}^2+\varepsilon_{22}^2+2\lambda\varepsilon_{11}\varepsilon_{22}+4\mu\varepsilon_{12}^2]\}+Q(\nabla\boldsymbol{\psi})\\
&\qquad=\frac{1}{2}\operatorname{div}\{\mathbf{h}(S(\boldsymbol{\psi}),\varepsilon(\boldsymbol{\psi}))\}+Q(\nabla\boldsymbol{\psi}),
\end{aligned}$$

from which Lemma 3.2 follows. □

LEMMA 3.3. *Let $\boldsymbol{\psi}$ be a sufficiently regular solution of the system*

$$\rho\psi_{\beta,tt}-\sum_{\alpha}s_{,\alpha}^{\alpha\beta}(\boldsymbol{\psi})=0,\quad \beta=1,2,$$

and set

$$Y(t)=\rho\sum_{\beta}\int_{\Omega}\psi_{\beta,t}(\mathbf{m}\cdot\nabla\psi_{\beta}+\frac{1}{2}\psi_{\beta})\,d\Omega.$$

Then

$$\begin{aligned}
(3.9)\quad & Y(T)-Y(0)+\frac{1}{2}\int_0^T\int_{\Omega}\left[\sum_{\beta}\rho\psi_{\beta,t}^2+\sum_{\alpha,\beta}s^{\alpha\beta}(\boldsymbol{\psi})\varepsilon_{\alpha\beta}(\boldsymbol{\psi})\right]d\Omega dt\\
&=\frac{1}{2}\int_0^T\int_{\Gamma}(\mathbf{m}\cdot\boldsymbol{\nu})\left[\sum_{\beta}\rho\psi_{\beta,t}^2+\sum_{\alpha,\beta}\nu_{\alpha}s^{\alpha\beta}(\boldsymbol{\psi})\left(\frac{\partial\psi_{\beta}}{\partial\boldsymbol{\nu}}\right)\right]d\Gamma dt\\
&\quad-\frac{1}{2}\sum_{\alpha,\beta}\int_0^T\int_{\Gamma}(\mathbf{m}\cdot\boldsymbol{\nu})\tau_{\alpha}s^{\alpha\beta}(\boldsymbol{\psi})\left(\frac{\partial\psi_{\beta}}{\partial\boldsymbol{\tau}}\right)d\Gamma dt\\
&\quad+\sum_{\alpha,\beta}\int_0^T\int_{\Gamma}\nu_{\alpha}s^{\alpha\beta}(\boldsymbol{\psi})\left[(\mathbf{m}\cdot\boldsymbol{\tau})\frac{\partial\psi_{\beta}}{\partial\boldsymbol{\tau}}+\frac{1}{2}\psi_{\beta}\right]d\Gamma dt.
\end{aligned}$$

PROOF. A proof of Lemma 3.3 (in a somewhat different form) may be found in [45, Lemma 3.2]. It is obtained by carrying out appropriate integrations by parts in the identity

$$(3.10)\quad 0=\sum_{\beta}\int_0^T\int_{\Omega}[\mathbf{m}\cdot\nabla\psi_{\beta}+\frac{1}{2}\psi_{\beta}][\rho\psi_{\beta,tt}-\sum_{\alpha}s_{,\alpha}^{\alpha\beta}(\boldsymbol{\psi})]\,d\Omega dt.$$

For completeness, we shall carry out the calculations here. We write $s^{\alpha\beta}$ and $\varepsilon_{\alpha\beta}$ in place of $s^{\alpha\beta}(\boldsymbol{\psi})$ and $\varepsilon_{\alpha\beta}(\boldsymbol{\psi})$. One has

$$(3.11)\quad \sum_\beta \int_0^T \int_\Omega (\mathbf{m}\cdot\nabla\psi_\beta + \frac{1}{2}\psi_\beta)\rho\psi_{\beta,tt} d\Omega dt$$
$$= \rho \int_\Omega (\mathbf{m}\cdot\nabla\psi_\beta + \frac{1}{2}\psi_\beta)\psi_{\beta,t} d\Omega|_0^T$$
$$- \frac{\rho}{2}\int_0^T \int_\Omega \sum_\alpha \frac{\partial}{\partial x_\alpha}(m_\alpha \psi_{\beta,t}^2)\, d\Omega dt + \frac{\rho}{2}\int_0^T \int_\Omega \psi_{\beta,t}^2\, d\Omega dt$$
$$= \rho \int_\Omega (\mathbf{m}\cdot\nabla\psi_\beta + \frac{1}{2}\psi_\beta)\psi_{\beta,t} d\Omega|_0^T$$
$$+ \frac{\rho}{2}\int_0^T \int_\Omega \psi_{\beta,t}^2\, d\Omega dt - \frac{\rho}{2}\int_0^T \int_\Gamma (\mathbf{m}\cdot\boldsymbol{\nu})\psi_{\beta,t}^2\, d\Gamma dt,$$

where $m_\alpha = x_\alpha - x_{0\alpha}$. Also, with the help of Lemma 3.2 with $\mathbf{h} = \mathbf{m}$ we calculate

$$(3.12)\quad \sum_{\alpha,\beta} \int_\Omega (\mathbf{m}\cdot\nabla\psi_\beta + \frac{1}{2}\psi_\beta) s^{\alpha\beta}_{,\alpha}\, d\Omega = \int_\Omega (\nabla\boldsymbol{\psi}\mathbf{m} + \frac{1}{2}\boldsymbol{\psi})\cdot \operatorname{div} S(\boldsymbol{\psi})\, d\Omega$$
$$= -\frac{1}{2}\int_\Gamma (\mathbf{m}\cdot\boldsymbol{\nu}) s^{\alpha\beta}(\boldsymbol{\psi})\varepsilon_{\alpha\beta}(\boldsymbol{\psi})\, d\Gamma + \int_\Gamma (\nabla\boldsymbol{\psi}\mathbf{m})\cdot(S(\boldsymbol{\psi})\boldsymbol{\nu})\, d\Gamma$$
$$+ \frac{1}{2}\int_\Gamma \boldsymbol{\psi}\cdot(S(\boldsymbol{\psi})\boldsymbol{\nu})\, d\Gamma - \frac{1}{2}\int_\Omega s^{\alpha\beta}(\boldsymbol{\psi})\varepsilon_{\alpha\beta}(\boldsymbol{\psi})\, d\Omega.$$

On Γ one has

$$\psi_{\beta,\alpha} = \nu_\alpha \frac{\partial\psi_\beta}{\partial\boldsymbol{\nu}} + \tau_\alpha \frac{\partial\psi_\beta}{\partial\boldsymbol{\tau}},$$
$$\mathbf{m}\cdot\nabla\psi_\beta = (\mathbf{m}\cdot\boldsymbol{\nu})\frac{\partial\psi_\beta}{\partial\boldsymbol{\nu}} + (\mathbf{m}\cdot\boldsymbol{\tau})\frac{\partial\psi_\beta}{\partial\boldsymbol{\tau}},$$
$$\sum_{\alpha,\beta} s^{\alpha\beta}(\boldsymbol{\psi})\varepsilon_{\alpha\beta}(\boldsymbol{\psi}) = \sum_{\alpha,\beta}\left(\nu_\alpha \frac{\partial\psi_\beta}{\partial\boldsymbol{\nu}} + \tau_\alpha \frac{\partial\psi_\beta}{\partial\boldsymbol{\tau}}\right) s^{\alpha\beta}(\boldsymbol{\psi}).$$

Insertion of the last two expressions into (3.12) yields

$$(3.13)\quad \sum_{\alpha,\beta} \int_\Omega (\mathbf{m}\cdot\nabla\psi_\beta + \frac{1}{2}\psi_\beta) s^{\alpha\beta}_{,\alpha}\, d\Omega = \sum_{\alpha,\beta}\left\{\frac{1}{2}\int_\Gamma (\mathbf{m}\cdot\boldsymbol{\nu})\nu_\alpha s^{\alpha\beta}\frac{\partial\psi_\beta}{\partial\boldsymbol{\nu}}\, d\Gamma\right.$$
$$+ \int_\Gamma \nu_\alpha s^{\alpha\beta}[(\mathbf{m}\cdot\boldsymbol{\tau})\frac{\partial\psi_\beta}{\partial\boldsymbol{\tau}} + \frac{1}{2}\psi_\beta]\, d\Gamma$$
$$\left. - \frac{1}{2}\int_\Gamma (\mathbf{m}\cdot\boldsymbol{\nu})\tau_\alpha s^{\alpha\beta}\frac{\partial\psi_\beta}{\partial\boldsymbol{\tau}}\, d\Gamma - \frac{1}{2}\int_\Omega s^{\alpha\beta}\varepsilon_{\alpha\beta}\, d\Omega\right\}.$$

The identity (3.9) now follows from (3.10), (3.11) and (3.13). □

Let us now consider the homogeneous problem describing the motion of a system of interconnected membranes. Let the displacement vector of the i-th membrane be

$$\mathbf{\Psi}^i = \sum_j \Psi^i_j \mathbf{a}^i_j = \boldsymbol{\psi}^i + \Psi^i_3 \mathbf{a}^i_3, \quad \boldsymbol{\psi}^i = \sum_\alpha \Psi^i_\alpha \mathbf{a}^i_\alpha.$$

The system under consideration is then

$$(3.14) \qquad \begin{cases} \sum_\alpha s^{\alpha\beta}_{i,\alpha}(\boldsymbol{\psi}^i) = \rho_i \Psi^i_{\beta,tt}, \\ \mu_i \Delta \Psi^i_3 = \rho_i \Psi^i_{3,tt}, \quad \mathbf{x} \in \Omega_i, \ t > 0; \end{cases}$$

$$(3.15) \qquad \begin{cases} \sum_\alpha \nu^i_\alpha s^{\alpha\beta}_i(\boldsymbol{\psi}^i) = 0, \\ \boldsymbol{\nu}^i \cdot \nabla \Psi^i_3 = 0 \quad \mathbf{x} \in \Gamma^D_i, \ t > 0; \end{cases}$$

$$(3.16) \qquad \mathbf{\Psi}^i = 0, \quad \mathbf{x} \in \Gamma^C_i, \ t > 0;$$

$$(3.17) \quad \mathbf{\Psi}^i(\mathbf{p}_i^{-1}(\mathbf{p}), t) = \mathbf{\Psi}^j(\mathbf{p}_j^{-1}(\mathbf{p}), t), \qquad \mathbf{p} \in J_k, \ i, j \in \mathcal{I}(J_k), \ k = 1, \dots, m;$$

$$(3.18) \quad \sum_{i \in \mathcal{I}(J_k)} \sum_\alpha \nu^i_\alpha \left(\sum_\beta s^{\alpha\beta}_i(\boldsymbol{\psi}^i) \mathbf{a}^i_\beta + \mu_i \Psi_{3,\alpha} \mathbf{a}^i_3 \right) (\mathbf{p}_i^{-1}(\mathbf{p}), t) = 0, \qquad \mathbf{p} \in J_k, \ t > 0;$$

$$(3.19) \qquad \mathbf{\Psi}^i|_{t=0} = \mathbf{\Psi}^i_0, \quad \mathbf{\Psi}^i_{,t}|_{t=0} = \mathbf{\Psi}^i_1, \quad \mathbf{x} \in \Omega_i.$$

Let $E(t)$ denote the *total energy* associated with a solution of (3.14):

$$E(t) = \mathcal{K}(t) + \mathcal{U}(t)$$

where

$$\mathcal{K}(t) = \frac{1}{2} \sum_{i=1}^n \int_{\Omega_i} \rho_i |\mathbf{\Psi}^i_{,t}|^2 d\Omega,$$

$$\mathcal{U}(t) = \frac{1}{2} \sum_{i=1}^n \int_{\Omega_i} [\sum_{\alpha,\beta} s^{\alpha\beta}_i(\boldsymbol{\psi}^i) \varepsilon_{\alpha\beta}(\boldsymbol{\psi}^i) + \mu_i |\nabla \Psi^i_3|^2] \, d\Omega.$$

If $\mathbf{\Psi} = (\mathbf{\Psi}^1, \dots, \mathbf{\Psi}^n)$ satisfies (3.14)–(3.18), as a consequence of Hamilton's principle $E(t)$ is *independent of* t:

$$E(t) = E(0), \quad t \geq 0.$$

For $i = 1, \ldots, n$, set

$$\mathbf{m}^i(x_1, x_2) = \sum_\alpha (x_\alpha - x^i_{0\alpha})\mathbf{a}^i_\alpha, \quad (x_1, x_2) \in I\!R^2, \tag{3.20}$$

where (x^i_{01}, x^i_{02}) is a point in $I\!R^2$ which may depend on i.

LEMMA 3.4. *Let $\boldsymbol{\Psi} = (\boldsymbol{\Psi}^1, \ldots, \boldsymbol{\Psi}^n)$ be a sufficiently regular solution of* (3.14) *and set*

$$Z(t) = \sum_{i=1}^n \rho_i \int_{\Omega_i} \sum_j \Psi^i_{j,t}(\mathbf{m}^i \cdot \nabla \Psi^i_j + \Psi^i_j)\, d\Omega.$$

Then

$$\begin{aligned} Z(T) - Z(0) + \int_0^T E(t)\, dt = \frac{1}{2}\sum_{i=1}^n \Bigg\{ & \int_0^T \int_{\Gamma_i} (\mathbf{m}^i \cdot \boldsymbol{\nu}^i)\rho_i |\boldsymbol{\Psi}^i_{,t}|^2 d\Gamma dt \\ & + \int_0^T \int_{\Gamma_i} (\mathbf{m}^i \cdot \boldsymbol{\nu}^i) \left[\sum_{\alpha,\beta} \nu^i_\alpha s^{\alpha\beta}_i(\boldsymbol{\psi}^i) \frac{\partial \Psi^i_\beta}{\partial \nu^i} + \mu_i \left(\frac{\partial \Psi^i_3}{\partial \nu^i} \right)^2 \right] d\Gamma dt \\ & - \frac{1}{2} \int_0^T \int_{\Gamma_i} (\mathbf{m}^i \cdot \boldsymbol{\nu}^i) \left[\sum_{\alpha,\beta} \tau^i_\alpha s^{\alpha\beta}_i(\boldsymbol{\psi}^i) \frac{\partial \Psi^i_\beta}{\partial \tau^i} + \mu_i \left(\frac{\partial \Psi^i_3}{\partial \tau^i} \right)^2 \right] d\Gamma dt \\ & + \int_0^T \int_{\Gamma_i} (\mathbf{m}^i \cdot \boldsymbol{\tau}^i) \left[\sum_{\alpha,\beta} \nu^i_\alpha s^{\alpha\beta}_i(\boldsymbol{\psi}^i) \frac{\partial \Psi^i_\beta}{\partial \tau^i} + \mu_i \frac{\partial \Psi^i_3}{\partial \nu^i} \frac{\partial \Psi^i_3}{\partial \tau^i} \right] d\Gamma dt \\ & + \frac{1}{2} \int_0^T \int_{\Gamma_i} \left[\sum_{\alpha,\beta} \nu^i_\alpha s^{\alpha\beta}_i(\boldsymbol{\psi}^i) \Psi^i_\beta + \mu_i \Psi^i_3 \frac{\partial \Psi^i_3}{\partial \nu^i} \right] d\Gamma dt \Bigg\}. \end{aligned} \tag{3.21}$$

PROOF. We apply Lemmas 3.1 and 3.3 with $\Omega = \Omega_i$, $\mathbf{m} = \mathbf{m}^i$, $\phi = \Psi^i_3$, $\boldsymbol{\psi} = \boldsymbol{\psi}^i$, add the resulting identities (3.6) and (3.9) and then sum over i from 1 to n. $\square$

Now let $\boldsymbol{\Psi}$ be a sufficiently regular solution of (3.14)–(3.18). For example, it will suffice to have

$$\boldsymbol{\Psi} \in C([0,T]; V^s) \cap C^1([0,T]; V^{s-1})$$

for some $s > 3/2$ (see Remark 2.4). The identity (3.20) will be used to obtain the *a priori* estimates needed to describe the set of reachable states. We shall write $s^{\alpha\beta}_i$ and $\varepsilon^i_{\alpha\beta}$ as abbreviated notations for $s^{\alpha\beta}_i(\boldsymbol{\psi}^i)$ and $\varepsilon_{\alpha\beta}(\boldsymbol{\psi}^i)$, respectively. In the right side of (3.21) one may write

$$\sum_{i=1}^n \int_{\Gamma_i} = \sum_{i=1}^n \int_{\Gamma^C_i} + \sum_{i=1}^n \int_{\Gamma^D_i} + \sum_{i=1}^n \int_{\Gamma^J_i}.$$

We shall consider separately each of the sums on the right side of the last equality.

(a) Integrals over Γ_i^C. Since $\mathbf{\Psi} = 0$ on Γ_i^C, the sum of integrals over Γ_i^C on the right side of (3.21) reduces to

$$\frac{1}{2}\sum_{i=1}^{n}\int_{\Gamma_i^C}(\mathbf{m}^i\cdot\boldsymbol{\nu}^i)\left[\sum_{\alpha,\beta}\nu_\alpha^i s_i^{\alpha\beta}\frac{\partial\Psi_\beta^i}{\partial\boldsymbol{\nu}^i}+\mu_i\left(\frac{\partial\Psi_3^i}{\partial\boldsymbol{\nu}^i}\right)^2\right]d\Gamma.$$

The quantity in brackets is nonnegative since on Γ_i^C

$$\sum_{\alpha,\beta}\nu_\alpha^i s_i^{\alpha\beta}\frac{\partial\Psi_\beta^i}{\partial\boldsymbol{\nu}^i}=\sum_{\alpha,\beta}s_i^{\alpha\beta}\varepsilon_{\alpha\beta}^i\geq 0.$$

Let us assume that the multipliers $\mathbf{m}^i$ may be chosen so that

$$\mathbf{m}^i\cdot\boldsymbol{\nu}^i\leq 0\ \text{ on } \Gamma_i^C,\ i=1,\dots,n. \tag{3.22}$$

Then $\sum_{i=1}^n\int_{\Gamma_i^C}\leq 0$.

(b) *Integrals over* Γ_i^D. It follows from the boundary conditions (3.15) that the sum of integrals over Γ_i^D reduces to

$$\frac{1}{2}\sum_{i=1}^{n}\int_{\Gamma_i^D}(\mathbf{m}^i\cdot\boldsymbol{\nu}^i)\left[\rho_i|\mathbf{\Psi}_{,t}^i|^2-\sum_{\alpha,\beta}\tau_\alpha^i s_i^{\alpha\beta}\frac{\partial\Psi_\beta^i}{\partial\boldsymbol{\tau}^i}-\mu_i\left(\frac{\partial\Psi_3^i}{\partial\boldsymbol{\tau}^i}\right)^2\right]d\Gamma. \tag{3.23}$$

(c) Integrals over Γ_i^J. For any $\mathbf{f}\in L^1(\Gamma_i^J)$ we may write

$$\begin{aligned}\sum_{i=1}^{n}\int_{\Gamma_i^J}\mathbf{f}(\mathbf{x})\,d\Gamma&=\sum_{k=1}^{m}\sum_{i\in\mathcal{I}(J_k)}\int_{\Gamma_i(J_k)}\mathbf{f}(\mathbf{x})\,d\Gamma\\&=\sum_{k=1}^{m}\sum_{i\in\mathcal{I}(J_k)}\int_{J_k}\mathbf{f}(\mathbf{p}_i^{-1}(\mathbf{p}))\,dJ_k.\end{aligned}$$

Consider

$$\sum_{i\in\mathcal{I}(J_k)}\int_{J_k}(\mathbf{m}^i\cdot\boldsymbol{\nu}^i)\rho_i|\mathbf{\Psi}_{,t}^i|^2d\Gamma.$$

By utilizing the geometric junction conditions (3.17), this sum is seen to be the same as

$$\int_{J_k}|\mathbf{\Psi}_{,t}^{i_k}|^2\sum_{i\in\mathcal{I}(J_k)}(\mathbf{m}^i\cdot\boldsymbol{\nu}^i)\rho_i\,d\Gamma.$$

We assume that

$$\sum_{i\in\mathcal{I}(J_k)}\rho_i(\mathbf{m}^i\cdot\boldsymbol{\nu}^i)(\mathbf{p}_i^{-1}(\mathbf{p}))\leq 0,\ \ \forall\mathbf{p}\in J_k,\ k=1,\dots,m. \tag{3.24}$$

Then

$$\sum_{k=1}^{m}\sum_{i\in\mathcal{I}(J_k)}\int_{\Gamma_i(J_k)}(\mathbf{m}^i\cdot\boldsymbol{\nu}^i)\rho_i|\boldsymbol{\Psi}^i_{,t}|^2 d\Gamma\le 0. \tag{3.25}$$

Next, let us consider

$$\begin{aligned}\sum_{i\in\mathcal{I}(J_k)}\int_{J_k}&\left(\sum_{\alpha,\beta}\nu^i_\alpha s_i^{\alpha\beta}\Psi^i_\beta+\mu_i\Psi^i_3\frac{\partial\Psi^i_3}{\partial\nu^i}\right)dJ_k\\ &=\int_{J_k}\left[\sum_{i\in\mathcal{I}(J_k)}\left(\sum_{\alpha,\beta}\nu^i_\alpha s_i^{\alpha\beta}\mathbf{a}^i_\beta+\mu_i\frac{\partial\Psi^i_3}{\partial\nu^i}\mathbf{a}^i_3\right)\right]\cdot\boldsymbol{\Psi}^{i_k}dJ_k,\end{aligned}$$

in which we have again utilized (3.17). It follows from the dynamic junction conditions (3.18) that the last integral vanishes. Similarly, consider

$$\sum_{i\in\mathcal{I}(J_k)}\int_{J_k}(\mathbf{m}^i\cdot\boldsymbol{\tau}^i)\left(\sum_{\alpha,\beta}\nu^i_\alpha s_i^{\alpha\beta}\frac{\partial\Psi^i_\beta}{\partial\tau^i}+\mu_i\frac{\partial\Psi^i_3}{\partial\nu^i}\frac{\partial\Psi^i_3}{\partial\tau^i}\right)dJ_k. \tag{3.26}$$

Along J_k one has $\boldsymbol{\tau}^i=\pm\boldsymbol{\tau}^{i_k}$ which, along with (3.17), implies that

$$\frac{\partial\boldsymbol{\Psi}^i}{\partial\tau^i}(\mathbf{p}_i^{-1}(\mathbf{p}),t)=\pm\frac{\partial\boldsymbol{\Psi}^{i_k}}{\partial\tau^{i_k}}(\mathbf{p}_{i_k}^{-1}(\mathbf{p}),t),\quad \mathbf{p}\in J_k.$$

Therefore, (3.26) yields

$$\int_{J_k}\sum_{i\in\mathcal{I}(J_k)}(\mathbf{m}^i\cdot\boldsymbol{\tau}^{i_k})\left(\sum_{\alpha,\beta}\nu^i_\alpha s_i^{\alpha\beta}\mathbf{a}^i_\beta+\mu_i\frac{\partial\Psi^i_3}{\partial\nu^i}\mathbf{a}^i_3\right)\cdot\frac{\partial\boldsymbol{\Psi}^{i_k}}{\partial\tau^{i_k}}dJ_k. \tag{3.27}$$

Our next assumption on the multipliers is that

$$\mathbf{m}^i(\mathbf{p}_i^{-1}(\mathbf{p}))\cdot\boldsymbol{\tau}^{i_k}(\mathbf{p}_{i_k}^{-1}(\mathbf{p}))\ \text{ is independent of } i,\quad i\in\mathcal{I}(J_k). \tag{3.28}$$

It then follows from (3.18) that the integral (3.27) vanishes.

We summarize what we have obtained so far in the following lemma.

LEMMA 3.5. *Assume* (3.22), (3.24) *and* (3.28). *If* $\Psi = (\Psi^1, \dots, \Psi^n)$ *is a sufficiently regular solution of* (3.14)–(3.18), *then*

$$
\begin{aligned}
(3.29)\quad & Z(T) - Z(0) + TE(0) \\
&\le \frac{1}{2}\sum_{i=1}^{n}\int_0^T\int_{\Gamma_i^D}(\mathbf{m}^i\cdot\boldsymbol{\nu}^i)\left[\rho_i|\Psi^i_{,t}|^2 - \sum_{\alpha,\beta}\tau^i_\alpha s_i^{\alpha\beta}\frac{\partial\Psi^i_\beta}{\partial\tau^i} - \mu_i\left(\frac{\partial\Psi^i_3}{\partial\tau^i}\right)^2\right]d\Gamma dt \\
&+\frac{1}{2}\sum_{k=1}^{m}\sum_{i\in\mathcal{I}(J_k)}\int_0^T\int_{\Gamma_i(J_k)}(\mathbf{m}^i\cdot\boldsymbol{\nu}^i)\left\{\left[\sum_{\alpha,\beta}\nu^i_\alpha s_i^{\alpha\beta}\frac{\partial\Psi^i_\beta}{\partial\nu^i} + \mu_i\left(\frac{\partial\Psi^i_3}{\partial\nu^i}\right)^2\right]\right. \\
&\qquad\left. - \left[\sum_{\alpha,\beta}\tau^i_\alpha s_i^{\alpha\beta}\frac{\partial\Psi^i_\beta}{\partial\tau^i} + \mu_i\left(\frac{\partial\Psi^i_3}{\partial\tau^i}\right)^2\right]\right\}d\Gamma dt.
\end{aligned}
$$

REMARK 3.1. Note that

$$
(3.30)\qquad
\begin{aligned}
(\mathbf{m}^i\cdot\boldsymbol{\nu}^i)(\mathbf{p}_i^{-1}(\mathbf{p})) &= \sum_\alpha((\mathbf{p}-\mathbf{p}_0^i)\cdot\mathbf{a}^i_\alpha - x^i_{0\alpha})\nu^i_\alpha \\
&= (\mathbf{p}-\mathbf{p}_0^i-\mathbf{x}_0^i)\cdot\boldsymbol{\nu}^i,\quad \mathbf{p}\in\Lambda_i,
\end{aligned}
$$

where $\mathbf{x}_0^i = \sum_\alpha x^i_{0\alpha}\mathbf{a}^i_\alpha$. Similarly,

$$
(3.31)\qquad (\mathbf{m}^i\cdot\boldsymbol{\tau}^i)(\mathbf{p}_i^{-1}(\mathbf{p})) = (\mathbf{p}-\mathbf{p}_0^i-\mathbf{x}_0^i)\cdot\boldsymbol{\tau}^i,\quad \mathbf{p}\in\Lambda_i.
$$

Let $\mathbf{q}_0$ be any point of $I\!R^3$. It follows from (3.31) that condition (3.28) may be satisfied by choosing $\mathbf{x}_0^i$ to be the projection of $\mathbf{q}_0 - \mathbf{p}_0^i$ onto the plane containing $\mathcal{P}_i$, $\mathbf{i}\in\mathcal{I}(J_k)$, for then

$$
(3.32)\qquad
\begin{cases}
(\mathbf{m}^i\cdot\boldsymbol{\tau}^i)(\mathbf{p}_i^{-1}(\mathbf{p})) = (\mathbf{p}-\mathbf{q}_0)\cdot\boldsymbol{\tau}^i, \\
(\mathbf{m}^i\cdot\boldsymbol{\nu}^i)(\mathbf{p}_i^{-1}(\mathbf{p})) = (\mathbf{p}-\mathbf{q}_0)\cdot\boldsymbol{\nu}^i,\quad \mathbf{p}\in\Lambda_i,\ i\in\mathcal{I}(J_k),
\end{cases}
$$

hence

$$
\mathbf{m}^i(\mathbf{p}_i^{-1}(\mathbf{p}))\cdot\boldsymbol{\tau}^{i_k} = (\mathbf{p}-\mathbf{q}_0)\cdot\boldsymbol{\tau}^{i_k},\quad \mathbf{p}\in J_k,\ i\in\mathcal{I}(J_k),
$$

is independent of i. With this choice of the points $\mathbf{x}_0^i$, condition (3.24) takes the form

$$
(3.33)\qquad \sum_{i\in\mathcal{I}(J_k)}\rho_i(\mathbf{p}-\mathbf{q}_0)\cdot\boldsymbol{\nu}^i \le 0,\quad \mathbf{p}\in J_k.
$$

Note that $(\mathbf{p}-\mathbf{q}_0)\cdot\boldsymbol{\nu}^i$ is either strictly positive, strictly negative or identically zero on J_k since $\boldsymbol{\nu}^i$ is constant there. It is clear that (3.32) will hold if at most one of the $\mathbf{m}^i\cdot\boldsymbol{\nu}^i$, say $\mathbf{m}^q\cdot\boldsymbol{\nu}^q$, is positive, if $\mathbf{m}^i\cdot\boldsymbol{\nu}^i \le 0$ for $i\ne q$ and if $\rho_q(\mathbf{m}^q\cdot\boldsymbol{\nu}^q)$ is sufficiently small relative to $\rho_i|\mathbf{m}^i\cdot\boldsymbol{\nu}^i|$ for those $i\ne q$ for which $\mathbf{m}^i\cdot\boldsymbol{\nu}^i$ is not identically zero. □

We are going to write the integral over $\Gamma_i(J_k)$ in (3.29) in terms of normal and tangential derivatives. On Γ_i one may write

$$s_i^{\alpha\beta} = \mu_i(\Psi^i_{\alpha,\beta} + \Psi^i_{\beta,\alpha}) + \lambda_i(\sum_\gamma \Psi^i_{\gamma,\gamma})\delta_{\alpha\beta}$$
$$= \mu_i \left(\nu^i_\beta \frac{\partial \Psi^i_\alpha}{\partial \nu^i} + \nu^i_\alpha \frac{\partial \Psi^i_\beta}{\partial \nu^i} + \tau^i_\beta \frac{\partial \Psi^i_\alpha}{\partial \tau^i} + \tau^i_\alpha \frac{\partial \Psi^i_\beta}{\partial \tau^i} \right)$$
$$+ \lambda_i \delta_{\alpha\beta} \sum_\gamma \left(\nu^i_\gamma \frac{\partial \Psi^i_\gamma}{\partial \nu^i} + \tau^i_\gamma \frac{\partial \Psi^i_\gamma}{\partial \tau^i} \right),$$

from which follows

$$(3.34)\quad S_i(\boldsymbol{\psi}^i)\boldsymbol{\nu}^i := \sum_{\alpha,\beta} \nu^i_\alpha s_i^{\alpha\beta} \mathbf{a}^i_\beta = \mu_i \left(\boldsymbol{\tau}^i \cdot \frac{\partial \boldsymbol{\psi}^i}{\partial \boldsymbol{\nu}^i} + \boldsymbol{\nu}^i \cdot \frac{\partial \boldsymbol{\psi}^i}{\partial \boldsymbol{\tau}^i} \right) \boldsymbol{\tau}^i$$
$$+ \left[(2\mu_i + \lambda_i) \left(\boldsymbol{\nu}^i \cdot \frac{\partial \boldsymbol{\psi}^i}{\partial \boldsymbol{\nu}^i} \right) + \lambda_i \left(\boldsymbol{\tau}^i \cdot \frac{\partial \boldsymbol{\psi}^i}{\partial \boldsymbol{\tau}^i} \right) \right] \boldsymbol{\nu}^i,$$

$$(3.35)\quad S_i(\boldsymbol{\psi}^i)\boldsymbol{\tau}^i := \sum_{\alpha,\beta} \tau^i_\alpha s_i^{\alpha\beta} \mathbf{a}^i_\beta$$
$$= \left[(2\mu_i + \lambda_i) \left(\boldsymbol{\tau}^i \cdot \frac{\partial \boldsymbol{\psi}^i}{\partial \boldsymbol{\tau}^i} \right) + \lambda_i \left(\boldsymbol{\nu}^i \cdot \frac{\partial \boldsymbol{\psi}^i}{\partial \boldsymbol{\nu}^i} \right) \right] \boldsymbol{\tau}^i$$
$$+ \mu_i \left(\boldsymbol{\tau}^i \cdot \frac{\partial \boldsymbol{\psi}^i}{\partial \boldsymbol{\nu}^i} + \boldsymbol{\nu}^i \cdot \frac{\partial \boldsymbol{\psi}^i}{\partial \boldsymbol{\tau}^i} \right) \boldsymbol{\nu}^i.$$

It may then be verified that

$$\sum_{\alpha,\beta} \nu^i_\alpha s_i^{\alpha\beta} \frac{\partial \Psi^i_\beta}{\partial \nu^i} - \sum_{\alpha,\beta} \tau^i_\alpha s_i^{\alpha\beta} \frac{\partial \Psi^i_\beta}{\partial \tau^i}$$
$$= \mu_i \left[\left(\boldsymbol{\tau}^i \cdot \frac{\partial \boldsymbol{\psi}^i}{\partial \boldsymbol{\nu}^i} \right)^2 - \left(\boldsymbol{\nu}^i \cdot \frac{\partial \boldsymbol{\psi}^i}{\partial \boldsymbol{\tau}^i} \right)^2 \right]$$
$$+ (2\mu_i + \lambda_i) \left[\left(\boldsymbol{\nu}^i \cdot \frac{\partial \boldsymbol{\psi}^i}{\partial \boldsymbol{\nu}^i} \right)^2 - \left(\boldsymbol{\tau}^i \cdot \frac{\partial \boldsymbol{\psi}^i}{\partial \boldsymbol{\tau}^i} \right)^2 \right].$$

and therefore

$$
\begin{aligned}
(3.36)\quad \sum_{i\in\mathcal{I}(J_k)} \int_{\Gamma_i(J_k)} (\mathbf{m}^i\cdot\boldsymbol{\nu}^i) &\left\{ \left[\sum_{\alpha,\beta} \nu^i_\alpha s_i^{\alpha\beta}\frac{\partial \Psi^i_\beta}{\partial \boldsymbol{\nu}^i} + \mu_i\left(\frac{\partial \Psi^i_3}{\partial \boldsymbol{\nu}^i}\right)^2 \right] \right. \\
&\left. - \left[\sum_{\alpha,\beta} \tau^i_\alpha s_i^{\alpha\beta}\frac{\partial \Psi^i_\beta}{\partial \boldsymbol{\tau}^i} + \mu_i\left(\frac{\partial \Psi^i_3}{\partial \boldsymbol{\tau}^i}\right)^2 \right] \right\} d\Gamma \\
= \sum_{i\in\mathcal{I}(J_k)} \int_{\Gamma_i(J_k)} (\mathbf{m}^i\cdot\boldsymbol{\nu}^i) &\left\{ \mu_i \left[\left(\boldsymbol{\tau}^i\cdot\frac{\partial \boldsymbol{\psi}^i}{\partial \boldsymbol{\nu}^i}\right)^2 - \left(\boldsymbol{\nu}^i\cdot\frac{\partial \boldsymbol{\psi}^i}{\partial \boldsymbol{\tau}^i}\right)^2 \right] \right. \\
&+ (2\mu_i+\lambda_i)\left[\left(\boldsymbol{\nu}^i\cdot\frac{\partial \boldsymbol{\psi}^i}{\partial \boldsymbol{\nu}^i}\right)^2 - \left(\boldsymbol{\tau}^i\cdot\frac{\partial \boldsymbol{\psi}^i}{\partial \boldsymbol{\tau}^i}\right)^2 \right] \\
&\left. + \mu_i\left(\frac{\partial \Psi^i_3}{\partial \boldsymbol{\nu}^i}\right)^2 - \mu_i\left(\frac{\partial \Psi^i_3}{\partial \boldsymbol{\tau}^i}\right)^2 \right\} d\Gamma.
\end{aligned}
$$

The next thing to be proved is that, under some additional conditions on the multipliers $\mathbf{m}^i$ and parameters μ_i, λ_i, one may, with the aid of the dynamic junction conditions, eliminate the normal derivatives from the right side of (3.36). This is important in order to establish controllability results which utilize only physically meaningful controls (forces and moments) in the junction regions.

To this end, let us denote by $S_{\boldsymbol{\nu}^i}(\boldsymbol{\psi}^i)$ and $S_{\boldsymbol{\tau}^i}(\boldsymbol{\psi}^i)$ the normal and tangential components, respectively, of $S_i(\boldsymbol{\psi}^i)\boldsymbol{\nu}^i$ in (3.34):

$$
S_i(\boldsymbol{\psi}^i)\boldsymbol{\nu}^i = S_{\boldsymbol{\tau}^i}(\boldsymbol{\psi}^i)\boldsymbol{\tau}^i + S_{\boldsymbol{\nu}^i}(\boldsymbol{\psi}^i)\boldsymbol{\nu}^i,
$$

where

$$
\begin{aligned}
S_{\boldsymbol{\tau}^i}(\boldsymbol{\psi}^i) &= \mu_i\left(\boldsymbol{\tau}^i\cdot\frac{\partial \boldsymbol{\psi}^i}{\partial \boldsymbol{\nu}^i} + \boldsymbol{\nu}^i\cdot\frac{\partial \boldsymbol{\psi}^i}{\partial \boldsymbol{\tau}^i}\right), \\
S_{\boldsymbol{\nu}^i}(\boldsymbol{\psi}^i) &= (2\mu_i+\lambda_i)\left(\boldsymbol{\nu}^i\cdot\frac{\partial \boldsymbol{\psi}^i}{\partial \boldsymbol{\nu}^i}\right) + \lambda_i\left(\boldsymbol{\tau}^i\cdot\frac{\partial \boldsymbol{\psi}^i}{\partial \boldsymbol{\tau}^i}\right).
\end{aligned}
$$

Then

$$
S_i(\boldsymbol{\psi}^i)\boldsymbol{\tau}^i = S_{\boldsymbol{\tau}^i}(\boldsymbol{\psi}^i)\boldsymbol{\nu}^i + \left[(2\mu_i+\lambda_i)\left(\boldsymbol{\tau}^i\cdot\frac{\partial \boldsymbol{\psi}^i}{\partial \boldsymbol{\tau}^i}\right) + \lambda_i\left(\boldsymbol{\nu}^i\cdot\frac{\partial \boldsymbol{\psi}^i}{\partial \boldsymbol{\nu}^i}\right)\right]\boldsymbol{\tau}^i.
$$

The right side of (3.36) may be written (suppressing the argument ψ^i in $S_{\tau^i}(\psi^i)$ and $S_{\nu^i}(\psi^i)$)

$$(3.37)\quad \int_{\Gamma_i(J_k)} (\mathbf{m}^i\cdot\boldsymbol{\nu}^i)\left\{ S_{\tau^i}\left(\frac{1}{\mu_i}S_{\tau^i} - 2\boldsymbol{\nu}^i\cdot\frac{\partial\psi^i}{\partial\tau^i}\right)\right.$$
$$+\frac{1}{2\mu_i+\lambda_i}\left(S_{\nu^i}-\lambda_i\left(\boldsymbol{\tau}^i\cdot\frac{\partial\psi^i}{\partial\tau^i}\right)\right)^2 - (2\mu_i+\lambda_i)\left(\boldsymbol{\tau}^i\cdot\frac{\partial\psi^i}{\partial\tau^i}\right)^2$$
$$\left.+\mu_i\left(\frac{\partial\Psi_3^i}{\partial\nu^i}\right)^2 - \mu_i\left(\frac{\partial\Psi_3^i}{\partial\tau^i}\right)^2\right\}d\Gamma$$
$$=\int_{\Gamma_i(J_k)} (\mathbf{m}^i\cdot\boldsymbol{\nu}^i)\left\{\frac{1}{\mu_i}S^2_{\tau^i} + \frac{1}{2\mu_i+\lambda_i}S^2_{\nu^i} + \mu_i\left(\frac{\partial\Psi_3^i}{\partial\nu^i}\right)^2\right.$$
$$-2S_{\tau^i}\left(\boldsymbol{\nu}^i\cdot\frac{\partial\psi^i}{\partial\tau^i}\right) - \frac{2\lambda_i}{2\mu_i+\lambda_i}S_{\nu^i}\left(\boldsymbol{\tau}^i\cdot\frac{\partial\psi^i}{\partial\tau^i}\right)$$
$$\left.+\frac{\lambda_i^2-(2\mu_i+\lambda_i)^2}{2\mu_i+\lambda_i}\left(\boldsymbol{\tau}^i\cdot\frac{\partial\psi^i}{\partial\tau^i}\right)^2 - \mu_i\left(\frac{\partial\Psi_3^i}{\partial\tau^i}\right)^2\right\}d\Gamma.$$

Let $0<\delta<1$ and set

$$\delta_i = \delta\,\mathrm{sgn}\,(\mathbf{m}^i\cdot\boldsymbol{\nu}^i).$$

(Note that $\mathbf{m}^i\cdot\boldsymbol{\nu}^i$ must be of constant sign on $\Gamma_i(J_k)$.) The integral on the right side of (3.37) does not exceed

$$\int_{\Gamma_i(J_k)} (\mathbf{m}^i\cdot\boldsymbol{\nu}^i)\left\{(1+\delta_i)\left[\frac{1}{\mu_i}S^2_{\tau^i} + \frac{1}{2\mu_i+\lambda_i}S^2_{\nu^i} + \frac{1}{\mu_i}\left(\mu_i\frac{\partial\Psi_3^i}{\partial\nu^i}\right)^2\right]\right.$$
$$+\frac{1}{\delta_i}\left[\mu_i\left(\boldsymbol{\nu}^i\cdot\frac{\partial\psi^i}{\partial\tau^i}\right)^2 - \delta_i\mu_i\left(\frac{\partial\Psi_3^i}{\partial\tau^i}\right)^2\right.$$
$$\left.\left.+\frac{(1+\delta_i)\lambda_i^2-\delta_i(2\mu_i+\lambda_i)^2}{2\mu_i+\lambda_i}\left(\boldsymbol{\tau}^i\cdot\frac{\partial\psi^i}{\partial\tau^i}\right)^2\right]\right\}d\Gamma$$
$$\le (1+\delta_i)\int_{\Gamma_i(J_k)} (\mathbf{m}^i\cdot\boldsymbol{\nu}^i)\left[\frac{1}{\mu_i}S^2_{\tau^i} + \frac{1}{2\mu_i+\lambda_i}S^2_{\nu^i} + \frac{1}{\mu_i}\left(\mu_i\frac{\partial\Psi_3^i}{\partial\nu^i}\right)^2\right]d\Gamma$$
$$+\tilde{C}_i\int_{\Gamma_i(J_k)}\left|\frac{\partial\boldsymbol{\Psi}^i}{\partial\tau^i}\right|^2 d\Gamma,$$

where $\tilde{C}_i$ is a constant depending on the set of elastic parameters μ_i, λ_i and on

$$\max_{\mathbf{p}\in J_k} |(\mathbf{m}^i \cdot \boldsymbol{\nu}^i)(\mathbf{p}_i^{-1}(\mathbf{p}))|.$$

The dynamic junction condition (3.18) may be written

$$\sum_{i\in\mathcal{I}(J_k)} \left(S_{\boldsymbol{\nu}^i}\boldsymbol{\nu}^i + S_{\boldsymbol{\tau}^i}\boldsymbol{\tau}^i + \mu_i \frac{\partial \Psi_3^i}{\partial \boldsymbol{\nu}^i}\mathbf{a}_3^i \right) = 0. \tag{3.38}$$

We define, for $i \in \mathcal{I}(J_k)$,

$$\sigma_i(J_k) = \begin{cases} 1 & \text{if } \boldsymbol{\tau}^i = \boldsymbol{\tau}^{i_k} \\ -1 & \text{if } \boldsymbol{\tau}^i = -\boldsymbol{\tau}^{i_k}. \end{cases}$$

Equation (3.38) may be decoupled into

$$\sum_{i\in\mathcal{I}(J_k)} \sigma_i(J_k) S_{\boldsymbol{\tau}^i} = 0 \tag{3.39}$$

and

$$\sum_{i\in\mathcal{I}(J_k)} \left(S_{\boldsymbol{\nu}^i}\boldsymbol{\nu}^i + \mu_i \frac{\partial \Psi_3^i}{\partial \boldsymbol{\nu}^i}\mathbf{a}_3^i \right) = 0. \tag{3.40}$$

Our final conditions on the multipliers are as follows:

$$\sum_{i\in\mathcal{I}(J_k)} a_i = 0, \quad \sum_{i\in\mathcal{I}(J_k)} a_i^2 = 1 \Longrightarrow \sum_{i\in\mathcal{I}(J_k)} (\mathbf{m}^i \cdot \boldsymbol{\nu}^i)\frac{a_i^2}{\mu_i} < 0, \tag{3.41}$$

and

$$\sum_{i\in\mathcal{I}(J_k)} (b_i\boldsymbol{\nu}^i + c_i\mathbf{a}_3^i) = 0, \quad \sum_{i\in\mathcal{I}(J_k)} (b_i^2 + c_i^2) = 1 \Longrightarrow \tag{3.42}$$

$$\sum_{i\in\mathcal{I}(J_k)} (\mathbf{m}^i \cdot \boldsymbol{\nu}^i)\left(\frac{b_i^2}{2\mu_i + \lambda_i} + \frac{c_i^2}{\mu_i} \right) < 0.$$

When (3.41) and (3.42) are satisfied it follows that for δ sufficiently small,

$$\sum_{i\in\mathcal{I}(J_k)} \sigma_i(J_k) S_{\boldsymbol{\tau}^i} = 0 \Longrightarrow \sum_{i\in\mathcal{I}(J_k)} (1+\delta_i)(\mathbf{m}^i \cdot \boldsymbol{\nu}^i)\frac{S_{\boldsymbol{\tau}^i}^2}{\mu_i} \le 0,$$

and

$$\sum_{i\in\mathcal{I}(J_k)} \left(S_{\boldsymbol{\nu}^i}\boldsymbol{\nu}^i + \mu_i \frac{\partial \Psi_3^i}{\partial \boldsymbol{\nu}^i}\mathbf{a}_3^i \right) = 0 \Longrightarrow$$

$$\sum_{i\in\mathcal{I}(J_k)} (1+\delta_i)(\mathbf{m}^i \cdot \boldsymbol{\nu}^i)\left[\frac{S_{\boldsymbol{\nu}^i}^2}{2\mu_i + \lambda_i} + \frac{1}{\mu_i}\left(\mu_i \frac{\partial \Psi_3^i}{\partial \boldsymbol{\nu}^i} \right)^2 \right] \le 0.$$

Therefore, we have the estimate

$$\sum_{i\in\mathcal{I}(J_k)}\int_{\Gamma_i(J_k)}(\mathbf{m}^i\cdot\boldsymbol{\nu}^i)\Bigg\{\Bigg[\sum_{\alpha,\beta}\nu^i_\alpha s_i^{\alpha\beta}\frac{\partial\Psi^i_\beta}{\partial\nu^i}+\mu_i\left(\frac{\partial\Psi^i_3}{\partial\nu^i}\right)^2\Bigg]$$

$$-\Bigg[\sum_{\alpha,\beta}\tau^i_\alpha s_i^{\alpha\beta}\frac{\partial\Psi^i_\beta}{\partial\tau^i}+\mu_i\left(\frac{\partial\Psi^i_3}{\partial\tau^i}\right)^2\Bigg]\Bigg\}d\Gamma$$

$$\le\sum_{i\in\mathcal{I}(J_k)}\tilde{C}_i\int_{\Gamma_i(J_k)}\left|\frac{\partial\boldsymbol{\Psi}^i}{\partial\tau^i}\right|^2d\Gamma=C_k\int_{\Gamma_{i_k}(J_k)}\left|\frac{\partial\boldsymbol{\Psi}^{i_k}}{\partial\tau^{i_k}}\right|^2d\Gamma,$$

where $C_k=\sum_{i\in\mathcal{I}(J_k)}\tilde{C}_i$. As a consequence of this inequality and Lemma 3.5, we obtain the following result.

THEOREM 3.1. *Assume* (3.22), (3.24), (3.28), (3.41) *and* (3.42). *If* $\boldsymbol{\Psi}=(\boldsymbol{\Psi}^1,\dots,\boldsymbol{\Psi}^n)$ *is a sufficiently regular solution of* (3.14)–(3.18), *then*

$$(3.43)\quad Z(T)-Z(0)+TE(0)$$

$$\le\frac{1}{2}\sum_{i=1}^n\int_0^T\int_{\Gamma_i^D}(\mathbf{m}^i\cdot\boldsymbol{\nu}^i)\Bigg[\rho_i|\boldsymbol{\Psi}^i_{,t}|^2-\sum_{\alpha,\beta}\tau^i_\alpha s_i^{\alpha\beta}\frac{\partial\Psi^i_\beta}{\partial\tau^i}-\mu_i\left(\frac{\partial\Psi^i_3}{\partial\tau^i}\right)^2\Bigg]d\Gamma dt$$

$$+C\sum_{k=1}^m\int_0^T\int_{\Gamma_{i_k}(J_k)}\left|\frac{\partial\boldsymbol{\Psi}^{i_k}}{\partial\tau^{i_k}}\right|^2d\Gamma dt.$$

Let us define

$$\Gamma_i^{D^+}=\{\mathbf{x}\in\Gamma_i^D\,|\,(m^i\cdot\boldsymbol{\nu}^i)(\mathbf{x})\ge0\},\quad\Gamma_i^{D^-}=\Gamma_i^D\setminus\Gamma_i^{D^+}.$$

COROLLARY 3.1. *Assume* (3.22), (3.24), (3.28), (3.41) *and* (3.42). *Suppose further that either*

(i) $\Gamma_i^C\ne\emptyset$ *for at least one index* i*; or*

(ii) $\Gamma_i^{D^+}\ne\emptyset$ *for at least one index* i.

If $\boldsymbol{\Psi}=(\boldsymbol{\Psi}^1,\dots,\boldsymbol{\Psi}^n)$ *is a sufficiently regular solution of* (3.14)–(3.18), *then there is a* $T_0>0$ *such that for* $T>T_0$ *there is a constant* C_T *such that*

$$(3.44)\quad\|(\boldsymbol{\Psi}(0),\boldsymbol{\Psi}_{,t}(0))\|^2_{\mathcal{V}\times H}\le C_T\Bigg\{\sum_{i=1}^n\Bigg[\int_0^T\int_{\Gamma_i^{D^+}}(|\boldsymbol{\Psi}^i_{,t}|^2+|\boldsymbol{\Psi}^i|^2)\,d\Gamma dt$$

$$+\int_0^T\int_{\Gamma_i^{D^-}}\left|\frac{\partial\boldsymbol{\Psi}^i}{\partial\tau^i}\right|^2d\Gamma dt\Bigg]+\sum_{k=1}^m\int_0^T\int_{\Gamma_{i_k}(J_k)}\left|\frac{\partial\boldsymbol{\Psi}^{i_k}}{\partial\tau^{i_k}}\right|^2d\Gamma dt\Bigg\}.$$

PROOF. We write

$$\sum_{\alpha,\beta} \tau^i_\alpha s^{\alpha\beta}_i \frac{\partial \Psi^i_\beta}{\partial \tau^i} = (S_i(\psi^i)\tau^i) \cdot \frac{\partial \psi^i}{\partial \tau^i},$$

(see (3.35), (3.35)). On Γ^D_i, $S_i(\psi^i)\nu^i = 0$ hence, from (3.34),

$$\tau^i \cdot \frac{\partial \psi^i}{\partial \nu^i} = -\nu^i \cdot \frac{\partial \psi^i}{\partial \tau^i},$$

$$\nu^i \cdot \frac{\partial \psi^i}{\partial \nu^i} = -\frac{\lambda_i}{2\mu_i + \lambda_i} \tau^i \cdot \frac{\partial \psi^i}{\partial \tau^i}.$$

Therefore, on Γ^D_i we have

$$S_i(\psi^i)\tau^i = \frac{(2\mu_i + \lambda_i)^2 - \lambda_i^2}{2\mu_i + \lambda_i} \left(\tau^i \cdot \frac{\partial \psi^i}{\partial \tau^i} \right) \tau^i$$

so that

$$0 \le (S_i(\psi^i)\tau^i) \cdot \frac{\partial \psi^i}{\partial \tau^i} = \frac{(2\mu_i + \lambda_i)^2 - \lambda_i^2}{2\mu_i + \lambda_i} \left(\tau^i \cdot \frac{\partial \psi^i}{\partial \tau^i} \right)^2 \le C \left| \frac{\partial \Psi^i}{\partial \tau^i} \right|^2.$$

The right side of (3.44) is therefore an upper bound for the right side of (3.43). If $\Gamma^C_i \neq \emptyset$ for at least one index i, it follows easily from Lemma 2.2 that

$$|Z(t)| \le C_0 E(t) = C_0 E(0) \sim \|(\Psi(0), \Psi_{,t}(0))\|^2_{\mathcal{V}\times H}, \quad 0 \le t \le T,$$

for an appropriate constant C_0. We therefore obtain (3.44) with $T_0 = 2C_0$. On the other hand, if $\Gamma^C_i = \emptyset$ for all i but $\Gamma^{D+}_i \neq \emptyset$ for some i we have (cf. Remark 2.2)

$$|Z(t)| \le C_0 \left(E(0) + \sum_{i=1}^n \int_{\Gamma^{D+}_i} |\Psi^i(t)|^2 \, d\Gamma \right).$$

A calculation of Komornik [43] gives

$$\sup_{[0,T]} \int_{\Gamma^{D+}_i} |\Psi^i(t)|^2 \, d\Gamma \le (1 + 1/T) \int_0^T \int_{\Gamma^{D+}_i} |\Psi^i(t)|^2 \, d\Gamma dt$$
$$+ \int_0^T \int_{\Gamma^{D+}_i} |\Psi^i_{,t}(t)|^2 \, d\Gamma dt.$$

The estimate (3.44) then follows from the last two inequalities since

$$E(0) + \sum_{i=1}^n \int_{\Gamma^{D+}_i} |\Psi^i(0)|^2 \, d\Gamma \sim \|(\Psi(0), \Psi_{,t}(0))\|^2_{\mathcal{V}\times H}.$$

□

REMARK 3.2. The inequalities in (3.41) and (3.42) will, of course, hold if $\mathbf{m}^i \cdot \boldsymbol{\nu}^i \leq 0$ for all $i \in \mathcal{I}(J_k)$. A less trivial sufficient condition for their validity is that, at each joint J_k, at most one of the $\mathbf{m}^i \cdot \boldsymbol{\nu}^i$, say $\mathbf{m}^q \cdot \boldsymbol{\nu}^q$, is positive, $\mathbf{m}^i \cdot \boldsymbol{\nu}^i < 0$ if $i \neq q$, $i \in \mathcal{I}(J_k)$, and $(\mathbf{m}^q \cdot \boldsymbol{\nu}^q)/\mu_q$ and $(\mathbf{m}^q \cdot \boldsymbol{\nu}^q)/\lambda_q$ are sufficiently small relative to $|\mathbf{m}^i \cdot \boldsymbol{\nu}^i|/\mu_i$ and $|\mathbf{m}^i \cdot \boldsymbol{\nu}^i|/\lambda_i$, respectively, for $i \neq q$ and $i \in \mathcal{I}(J_k)$. Suppose, for example, that $\mathcal{I}(J_k) = [1, 2, \ldots, q]$ and set $m_i = (\mathbf{m}^i \cdot \boldsymbol{\nu}^i)/\mu_i$. By utilizing the constraint in (3.41) we may write

$$\sum_{i=1}^{q} m_i a_i^2 = m_q \left(\sum_{i=1}^{q-1} a_i \right)^2 + \sum_{i=1}^{q-1} m_i a_i^2,$$

which is negative if $m_q \geq 0$, $m_i < 0$ for $i = 1, \ldots, q-1$ and m_q is sufficiently small with respect to $|m_1|, \ldots, |m_{q-1}|$. A more precise condition is that the determinants

$$D_j = \begin{vmatrix} m_q + m_1 & m_q & \cdots & m_q \\ m_q & m_q + m_2 & \cdots & m_q \\ \vdots & \vdots & \ddots & \vdots \\ m_q & m_q & \cdots & m_q + m_j \end{vmatrix}$$

satisfy

$$(-1)^j D_j > 0 \quad \text{for } j = 1, \ldots, q-1. \tag{3.45}$$

Similarly, set $\bar{m}_i = (\mathbf{m}^i \cdot \boldsymbol{\nu}^i)/(2\mu_i + \lambda_i)$ and write, utilizing the constraint in (3.42),

$$\sum_{i=1}^{q} (\bar{m}_i b_i^2 + m_i c_i^2) = \bar{m}_q \left[\sum_{i=1}^{q-1} [b_i(\boldsymbol{\nu}^i \cdot \boldsymbol{\nu}^q) + c_i(\mathbf{a}_3^i \cdot \boldsymbol{\nu}^q)] \right]^2$$
$$+ m_q \left[\sum_{i=1}^{q-1} [b_i(\boldsymbol{\nu}^i \cdot \mathbf{a}_3^q) + c_i(\mathbf{a}_3^i \cdot \mathbf{a}_3^q)] \right]^2 + \sum_{i=1}^{q-1} (\bar{m}_i \mathbf{b}_i^2 + m_i \mathbf{c}_i^2),$$

which is negative if $\bar{m}_i$, m_i are negative for $i = 1, \ldots, q-1$, and if $\bar{m}_q$, m_q are positive but sufficiently small with respect to the values of $|\bar{m}_i|$, $|m_i|$, respectively, for $i = 1, \ldots, q-1$. Again, one may write a more precise condition analogous to (3.45), in which one should note that $\mathbf{a}_3^i \cdot \mathbf{a}_3^q = -\boldsymbol{\nu}^i \cdot \boldsymbol{\nu}^q$ and $\mathbf{a}_3^i \cdot \boldsymbol{\nu}^q = \boldsymbol{\nu}^i \cdot \mathbf{a}_3^q$. □

3.2. A Priori Estimates for Serially Connected Membranes. Consider n interconnected membranes with joints $J_1, \ldots, J_m$. The configuration is called *serially connected* if

$$\begin{cases} \mathbf{a}_3^i = \pm \mathbf{a}_3^j, & \forall i, j = 1, \ldots, n, \\ m = n - 1. \end{cases}$$

By reindexing the membranes and joints, if necessary, we may assume that

$$\begin{cases} \overline{\mathcal{P}}_i \cap \overline{\mathcal{P}}_{i+1} = J_i, & i = 1, \dots, n-1, \\ \overline{\mathcal{P}}_i \cap \overline{\mathcal{P}}_j = \emptyset, & |i-j| > 1. \end{cases}$$

Thus $\mathcal{I}(J_k) = [k, k+1]$. Further, without loss of generality, we may suppose that $\mathbf{a}_3^i = \mathbf{a}_3^j = (0,0,1)$ (hence

$$\boldsymbol{\nu}^i = -\boldsymbol{\nu}^{i+1}, \quad \boldsymbol{\tau}^i = -\boldsymbol{\tau}^{i+1} \text{ on } J_i)$$

and that

$$\mathcal{P}_i = \Omega_i, \quad \Lambda_i^C = \Gamma_i^C, \quad \Lambda_i^D = \Gamma_i^D,$$

$$\begin{cases} \Gamma_k(J_k) = \Gamma_{k+1}(J_k) = J_k, & k = 1, \dots, n-1, \\ \Gamma_i(J_k) = \emptyset, & |i-k| > 1. \end{cases}$$

If $\mathcal{P}_i$ is serially connected to $\mathcal{P}_{i+1}$, the geometric and dynamic conditions along J_i reduce to, respectively,

$$\boldsymbol{\psi}^i = \boldsymbol{\psi}^{i+1}, \quad \Psi_3^i = \Psi_3^{i+1} \text{ along } J_i, \tag{3.46}$$

and

$$\begin{cases} S_{\boldsymbol{\tau}^i}(\boldsymbol{\psi}^i) = S_{\boldsymbol{\tau}^{i+1}}(\boldsymbol{\psi}^{i+1}), \\ S_{\boldsymbol{\nu}^i}(\boldsymbol{\psi}^i) = S_{\boldsymbol{\nu}^{i+1}}(\boldsymbol{\psi}^{i+1}), \\ \mu_i \dfrac{\partial \Psi_3^i}{\partial \boldsymbol{\nu}^i} = -\mu_{i+1} \dfrac{\partial \Psi_3^{i+1}}{\partial \boldsymbol{\nu}^{i+1}} \text{ along } J_i, \end{cases} \tag{3.47}$$

so that the transverse motion is not coupled to the in-plane motion, as is to be expected.

When membranes are serially connected, it is possible to eliminate the integrals over J_k in the estimate (3.29), provided the parameters ρ_i, μ_i, λ_i satisfy certain monotonicity conditions. To prove this statement, we choose as multipliers

$$\mathbf{m}^i(\mathbf{x}) = \mathbf{x} - \mathbf{x}_0, \quad \mathbf{x} \in \mathbb{R}^2, \quad i = 1, \dots, n,$$

where $\mathbf{x}_0 \in \mathbb{R}^2$ *does not depend on* i, and we write $\mathbf{m} = \mathbf{m}^1$.

THEOREM 3.2. *Assume that* $\mathbf{x}_0 \in \mathbb{R}^2$ *may be chosen so that*

$$(\mathbf{x} - \mathbf{x}_0) \cdot \boldsymbol{\nu}^i \le 0 \quad on\ \Gamma_i^C,\ i = 1, \dots, n. \tag{3.48}$$

Suppose further that for the given choice of $\mathbf{x}_0$ *and each* $i = 1, \dots, n-1$,

$$\rho_i \le \rho_{i+1}, \quad \mu_i \ge \mu_{i+1}, \quad \lambda_i \ge \lambda_{i+1} \tag{3.49}$$

if $\mathbf{m} \cdot \boldsymbol{\nu}^i > 0$ *on* J_i;

$$\rho_i \ge \rho_{i+1}, \quad \mu_i \le \mu_{i+1}, \quad \lambda_i \le \lambda_{i+1} \tag{3.50}$$

if $\mathbf{m} \cdot \boldsymbol{\nu}^i < 0$ *on* J_i. *If* $\boldsymbol{\Psi} = (\boldsymbol{\Psi}^1, \dots, \boldsymbol{\Psi}^n)$ *is a sufficiently regular solution of* (3.14)–(3.16), (3.46), (3.47), *then*

$$(3.51) \quad Z(T) - Z(0) + TE(0)$$
$$\leq \frac{1}{2} \sum_{i=1}^{n} \int_0^T \int_{\Gamma_i^D} (\mathbf{m} \cdot \boldsymbol{\nu}^i) \left[\rho_i |\boldsymbol{\Psi}^i_{,t}|^2 - \sum_{\alpha,\beta} \tau^i_\alpha s_i^{\alpha\beta} \frac{\partial \Psi^i_\beta}{\partial \tau^i} - \mu_i \left(\frac{\partial \Psi^i_3}{\partial \tau^i} \right)^2 \right] d\Gamma dt.$$

PROOF. Since $\mathbf{m}^i$ is independent of i, clearly (3.28) will be satisfied. Suppose that $\mathbf{m} \cdot \boldsymbol{\nu}^i > 0$ on J_i. Condition (3.24) will then hold if and only if $\rho_i \leq \rho_{i+1}$, since $\boldsymbol{\nu}^i = -\boldsymbol{\nu}^{i+1}$ on J_i. The sum of integrals over $\Gamma_i(J_k)$ in (3.29) may, in the present context, be written (see (3.37))

(3.52)
$$\int_{J_k} \mathbf{m} \cdot \boldsymbol{\nu}^k \Bigg\{ \left[\frac{S^2_{\tau^k}}{\mu_k} - \frac{S^2_{\tau^{k+1}}}{\mu_{k+1}} \right] + \left[\mu_k \left(\frac{\partial \Psi^k_3}{\partial \nu^k} \right)^2 - \mu_{k+1} \left(\frac{\partial \Psi^{k+1}_3}{\partial \nu^{k+1}} \right)^2 \right]$$
$$- \left[\mu_k \left(\frac{\partial \Psi^k_3}{\partial \tau^k} \right)^2 - \mu_{k+1} \left(\frac{\partial \Psi^{k+1}_3}{\partial \tau^{k+1}} \right)^2 \right] + \left[\frac{S^2_{\nu^k}}{2\mu_k + \lambda_k} - \frac{S^2_{\nu^{k+1}}}{2\mu_{k+1} + \lambda_{k+1}} \right]$$
$$- 2 \left[S_{\tau^k} \left(\boldsymbol{\nu}^k \cdot \frac{\partial \boldsymbol{\psi}^k}{\partial \tau^k} \right) - S_{\tau^{k+1}} \left(\boldsymbol{\nu}^{k+1} \cdot \frac{\partial \boldsymbol{\psi}^{k+1}}{\partial \tau^{k+1}} \right) \right]$$
$$- 2 \left[\frac{\lambda_k}{2\mu_k + \lambda_k} S_{\nu^k} \left(\boldsymbol{\tau}^k \cdot \frac{\partial \boldsymbol{\psi}^k}{\partial \tau^k} \right) - \frac{\lambda_{k+1}}{2\mu_{k+1} + \lambda_{k+1}} S_{\nu^{k+1}} \left(\boldsymbol{\tau}^{k+1} \cdot \frac{\partial \boldsymbol{\psi}^{k+1}}{\partial \tau^{k+1}} \right) \right]$$
$$- 4 \Bigg[\frac{\mu_k(\mu_k + \lambda_k)}{2\mu_k + \lambda_k} \left(\boldsymbol{\tau}^k \cdot \frac{\partial \boldsymbol{\psi}^k}{\partial \tau^k} \right)^2$$
$$- \frac{\mu_{k+1}(\mu_{k+1} + \lambda_{k+1})}{2\mu_{k+1} + \lambda_{k+1}} \left(\boldsymbol{\tau}^{k+1} \cdot \frac{\partial \boldsymbol{\psi}^{k+1}}{\partial \tau^{k+1}} \right)^2 \Bigg] \Bigg\} d\Gamma.$$

By invoking (3.46) and (3.47), the first three brackets in (3.52) are seen to be nonpositive exactly when $\mu_k \geq \mu_{k+1}$, and the fifth bracket vanishes. The rest may be written

$$\int_{J_k} \mathbf{m} \cdot \boldsymbol{\nu}^k \Bigg\{ \left(\frac{1}{2\mu_k + \lambda_k} - \frac{1}{2\mu_{k+1} + \lambda_{k+1}} \right) S^2_{\nu^k}$$
$$- 2 \left(\frac{\lambda_k}{2\mu_k + \lambda_k} - \frac{\lambda_{k+1}}{2\mu_{k+1} + \lambda_{k+1}} \right) S_{\nu^k} \left(\boldsymbol{\tau}^k \cdot \frac{\partial \boldsymbol{\psi}^k}{\partial \tau^k} \right)$$
$$- 4 \left[\frac{\mu_k(\mu_k + \lambda_k)}{2\mu_k + \lambda_k} - \frac{\mu_{k+1}(\mu_{k+1} + \lambda_{k+1})}{2\mu_{k+1} + \lambda_{k+1}} \right] \left(\boldsymbol{\tau}^k \cdot \frac{\partial \boldsymbol{\psi}^k}{\partial \tau^k} \right)^2 \Bigg\} d\Gamma.$$

When $\mu_k \geq \mu_{k+1}$, it is easily checked that the quadratic form inside the braces is nonpositive if and only if

$$2\mu_k + \lambda_k \geq 2\mu_{k+1} + \lambda_{k+1} \text{ and } |\lambda_k - \lambda_{k+1}| \leq (2\mu_k + \lambda_k) - (2\mu_{k+1} + \lambda_{k+1}),$$

both of which will hold if $\lambda_k \geq \lambda_{k+1}$.

The proof is similar when (3.50) is satisfied. □

REMARK 3.3. Recall that $\mathbf{m} \cdot \boldsymbol{\nu}^i$ must be of constant sign on J_i. If, for a particular index i, $\mathbf{m} \cdot \boldsymbol{\nu}^i = 0$ on J_i, no relation between the material parameters of $\mathcal{P}_i$ and $\mathcal{P}_{i+1}$ need be assumed. Also, since the choice of an $\mathbf{x}_0$ satisfying (3.48) is not unique, for certain configurations it may be possible to choose distinct values of $\mathbf{x}_0$ such that for certain indices i, $\mathbf{m} \cdot \boldsymbol{\nu}^i > 0$ on J_i for the first choice of $\mathbf{x}_0$ while, for the second, $\mathbf{m} \cdot \boldsymbol{\nu}^i < 0$ on J_i. In such situations either (3.49) or (3.50) is sufficient for the conclusion of the theorem. These observations suggest that the above restrictions on the material parameters are due to the method of proof employed and are not intrinsic to the validity of the theorem. □

REMARK 3.4. The same proof shows that Theorem 3.2 is valid also in the more general situation where the J_i's are any Lipschitz continuous curves. Of course, in this situation $\mathbf{m} \cdot \boldsymbol{\nu}^i$ need not be of constant sign on J_i so that it is necessary to assume *both* $\mathbf{m} \cdot \boldsymbol{\nu}^i \geq 0$ on J_i and (3.49), or $\mathbf{m} \cdot \boldsymbol{\nu}^i \leq 0$ on J_i and (3.50). □

3.3. A Priori Estimates for Single Jointed Membrane Systems. We consider a membrane system containing a single junction region J, i.e., $m = 1$. In this case it is possible to prove, at least for some specific configurations, a stronger result than Theorem 3.1.

We first choose the multipliers $\mathbf{m}^i$ as in Remark 3.1, that is,

$$\begin{cases} (\mathbf{m}^i \cdot \boldsymbol{\nu}^i)(\mathbf{p}_i^{-1}(\mathbf{p})) = (\mathbf{p} - \mathbf{q}_0) \cdot \boldsymbol{\nu}^i, \\ (\mathbf{m}^i \cdot \boldsymbol{\tau}^i)(\mathbf{p}_i^{-1}(\mathbf{p})) = (\mathbf{p} - \mathbf{q}_0) \cdot \boldsymbol{\tau}^i, \ \mathbf{p} \in \Lambda_i, \end{cases} \tag{3.53}$$

where $\mathbf{q}_0$ is to be determined. Then we have the following result.

THEOREM 3.3. *Assume that* $m = 1$ *and that* $\mathbf{q}_0$ *may be chosen such that*

$$(\mathbf{p} - \mathbf{q}_0)\boldsymbol{\nu}^i \leq 0, \ \ \mathbf{p} \in \Lambda_i^C, \ i = 1, \ldots, n, \tag{3.54}$$

$$(\mathbf{p} - \mathbf{q}_0)\boldsymbol{\nu}^i = 0, \ \ \mathbf{p} \in J, \ i = 1, \ldots, n. \tag{3.55}$$

If $\Psi = (\Psi^1, \dots, \Psi^n)$ *is a sufficiently regular solution of* (3.14)–(3.18), *then*

$$(3.56)\quad Z(T) - Z(0) + TE(0)$$

$$\leq \frac{1}{2}\sum_{i=1}^{n}\int_0^T\int_{\Gamma_i^D}(\mathbf{m}^i\cdot\boldsymbol{\nu}^i)\left[\rho_i|\Psi^i_{,t}|^2 - \sum_{\alpha,\beta}\tau^i_\alpha s_i^{\alpha\beta}\frac{\partial\Psi^i_\beta}{\partial\tau^i} - \mu_i\left(\frac{\partial\Psi^i_3}{\partial\tau^i}\right)^2\right]d\Gamma dt$$

PROOF. Condition (3.55) means that $\mathbf{q}_0$ may be chosen along the straight line containing J. When it is satisfied, so are (3.24) (3.41) and (3.42) ((3.28) holds because of our choice of multipliers) and all integrals over $\Gamma_i(J_k)$ in (3.43) vanish. □

Unfortunately, in many cases of interest, it will not be possible to choose $\mathbf{q}_0$ according to (3.55) while still maintaining the inequalities (3.54). We shall consider one such situation and show that $\mathbf{q}_0$ may be chosen in such a way that (3.24), (3.41) and (3.42) are satisfied without any restrictions on the parameters ρ_i, μ_i λ_i.

To this end, we consider a folded membrane, i.e., $n = 2$, $m = 1$, we assume that $\Lambda_2^C = \emptyset$ and that $\mathbf{q}_0$ is such that

$$(3.57)\qquad (\mathbf{p} - \mathbf{q}_0)\cdot\boldsymbol{\nu}^1 \leq 0,\ \ p \in \Lambda_1^C.$$

Condition (3.57) will continue to hold if $\mathbf{q}_0$ is replaced by $\mathbf{q}_0 + r\mathbf{a}_3^1$, where r is any real number. Let

$$\delta_r^i(\mathbf{p}) = \frac{(\mathbf{p} - \mathbf{q}_0 - r\mathbf{a}_3^1)\cdot\boldsymbol{\nu}^i}{\|\mathbf{p} - (\mathbf{q}_0 + r\mathbf{a}_3^1)\|}, i = 1, 2,$$

which is, of course, nothing but the cosine of the angle between $\mathbf{p}-(\mathbf{q}_0+r\mathbf{a}_3^1)$ and the normal vector $\boldsymbol{\nu}^i$ at $\mathbf{p}$. Then

$$\lim_{r\to\pm\infty}\delta_r^1(\mathbf{p}) = 0,\ \ \mathbf{p}\in J,$$

and, since the membranes are nonserial,

$$\lim_{r\to\pm\infty}\delta_r^2(\mathbf{p}) \neq 0,\ \ \mathbf{p}\in J,$$

with the convergence uniform on J. By properly choosing the sign of r, we may assure that

$$\lim_r \delta_r^2(\mathbf{p}) < 0.$$

It then follows that

$$\sum_{i=1}^{2}\rho_i(\mathbf{p} - \mathbf{q}_0 - r\mathbf{a}_3^1)\cdot\boldsymbol{\nu}^i = [\rho_1\delta_r^1(\mathbf{p}) + \rho_2\delta_r^2(\mathbf{p})]\|\mathbf{p} - \mathbf{q}_0 - r\mathbf{a}_3^1\| \leq 0,\ \ \mathbf{p}\in J,$$

if r is sufficiently large and of proper sign. Similarly

$$\sum_{i=1}^{2} \frac{a_i^2}{\mu_i}(\mathbf{p} - \mathbf{q}_0 - r\mathbf{a}_3^1) \cdot \boldsymbol{\nu}^i = \|\mathbf{p} - \mathbf{q}_0 - r\mathbf{a}_3^1\| \left(\frac{\delta_r^1}{\mu_1} + \frac{\delta_r^2}{\mu_2} \right) \mathbf{a}_1^2 \leq 0, \quad \mathbf{p} \in J,$$

and likewise for (3.42). We therefore have the following result.

THEOREM 3.4. *Assume that* $n = 2$, $m = 1$, *that* (3.57) *holds and that* $\Gamma_2^C = \emptyset$. *If* $\boldsymbol{\Psi} = (\boldsymbol{\Psi}^1, \boldsymbol{\Psi}^2)$ *is a sufficiently regular solution of* (3.14)–(3.18), *then the estimate* (3.43) *is valid for* $\boldsymbol{\Psi}$.

REMARK 3.5. A completely analogous result is valid in the situation where $n \geq 2$ membranes are connected along a single joint, provided that $\Gamma_i^C = \emptyset$ for $i \geq 2$ and the configuration is such that

$$\lim_{r} \delta_r^i(\mathbf{p}) < 0, \quad \mathbf{p} \in J, \ i = 2, \ldots, n,$$

for a suitable choice of the sign of r. □

3.4. The Reachable States. The *a priori* estimate of Corollary 3.1 will be utilized in this section to determine the reachable states of the system (3.1)–(3.5). Since this estimate is proved only for "sufficiently regular" solutions of (3.14)–(3.18), the following *regularity hypothesis* will be imposed (see Remark 2.4):

$$D_A \subset V^s \cap \mathcal{V} \tag{3.58}$$

with continuous injection for some $s > 3/2$.

This regularity assumption imposes a further constraint on the configuration of the system of membranes, but we do not know how to quantify this constraint in terms of simple geometric properties of the configuration.

To describe the set of reachable states, we apply the *Hilbert Uniqueness Method* introduced by J.-L. Lions in [61]. Let $(\boldsymbol{\Psi}_0, \boldsymbol{\Psi}_1) \in D_A$ and let $\boldsymbol{\Psi}$ be the solution of (3.14)–(3.18) with *final data* $(\boldsymbol{\Psi}_0, \boldsymbol{\Psi}_1)$ at time T:

$$\boldsymbol{\Psi}|_{t=T} = \boldsymbol{\Psi}_0, \quad \boldsymbol{\Psi}_{,t}|_{t=T} = \boldsymbol{\Psi}_1. \tag{3.59}$$

We introduce a seminorm on $D_A \times \mathcal{V}$ by defining

$$\|(\boldsymbol{\Psi}_0, \boldsymbol{\Psi}_1)\|_F = \sum_{i=1}^{n} \left\{ \int_0^T \int_{\Gamma_i^{D+}} (|\boldsymbol{\Psi}_{,t}^i|^2 + |\boldsymbol{\Psi}^i|^2) \, d\Gamma dt \right.$$
$$\left. + \int_0^T \int_{\Gamma_i^{D-}} \left| \frac{\partial \boldsymbol{\Psi}^i}{\partial \tau^i} \right|^2 d\Gamma dt \right\} + \sum_{k=1}^{m} \int_0^T \int_{\Gamma_{i_k}(J_k)} \left| \frac{\partial \boldsymbol{\Psi}^{i_k}}{\partial \tau^{i_k}} \right|^2 d\Gamma dt.$$

According to Corollary 3.1 (with $E(0)$ replaced by $E(T)$, which is permissible since the problem for $\boldsymbol{\Psi}$ is time reversible), $\|\cdot\|_F$ defines a *norm* if $T > T_0$ and if (3.24), (3.28), (3.41) and (3.42) are satisfied. Let F denote the Hilbert space which is the completion of $D_A \times \mathcal{V}$ with respect to $\|\cdot\|_F$. Corollary 3.1 gives

$$F \subset \mathcal{V} \times H \tag{3.60}$$

with dense and continuous injection. If $(\Psi_0, \Psi_1) \in F$ then, by definition of the space F, the corresponding solution Ψ of (3.14)–(3.18) satisfies

$$\begin{aligned}
\Psi^i_{,t}\big|_{\Sigma_i^{D+}} &\in L^2(0,T;\mathcal{L}^2(\Gamma_i^{D^+})),\\
\frac{\partial \Psi^i}{\partial \tau^i}\bigg|_{\Sigma_i^{D-}} &\in L^2(0,T;\mathcal{L}^2(\Gamma_i^{D^-})),\\
\frac{\partial \Psi^{i_k}}{\partial \tau^{i_k}}\bigg|_{\Sigma_{i_k}(J_k)} &\in L^2(0,T;\mathcal{L}^2(\Gamma_{i_k}(J_k))).
\end{aligned}$$

In the above

$$\Sigma_i^{D\pm} = \Gamma_i^{D\pm} \times (0,T),\ \ \Sigma_{i_k}(J_k) = \Gamma_{i_k}(J_k) \times (0,T),$$

$$\mathcal{L}^2(\Gamma_i^{D\pm}) = \{\sum_j f_j \mathbf{a}_j^i | \, f_j \in L^2(\Gamma_i),\ \operatorname{supp} f_j \subset \Gamma_i^{D\pm}\},$$

and similarly for $\mathcal{L}^2(\Gamma_{i_k}(J_k))$.

Let $\mathbf{W} = (\mathbf{W}^1, \dots, \mathbf{W}^n)$ be the solution of (3.1)–(3.5) on $(0,T)$ with zero initial data

$$\mathbf{W}|_{t=0} = \mathbf{W}_{,t}|_{t=0} = 0, \tag{3.61}$$

where $T > T_0$, and let Ψ be the solution of (3.14)–(3.18), (3.59) with $(\Psi_0, \Psi_1) \in D_A \times \mathcal{V}$. Form the expression

$$\begin{aligned}
0 = \sum_{i=1}^n \int_0^T \int_{\Omega_i} \{\sum_\beta [\sum_\alpha s^{\alpha\beta}_{i,\alpha}(\mathbf{w}^i) - \rho_i W^i_{\beta,tt}\psi^i_\beta]\\
+ (\mu_i \Delta W^i_3 - \rho_i W^i_{3,tt})\psi^i_3\}\, d\Omega dt,
\end{aligned}$$

where $\mathbf{w}^i = \sum_\alpha W^i_\alpha \mathbf{a}^i_\alpha$. We formally integrate by parts to transfer the differentiations from $\mathbf{W}$ to Ψ and obtain

$$\begin{aligned}
\sum_{i=1}^n \int_{\Omega_i} \rho_i[\Psi^i_0 \cdot \mathbf{W}^i_{,t}(T) - \Psi^i_1 \cdot \mathbf{W}^i(T)]\, d\Omega\\
= \sum_{i=1}^n \int_0^T \int_{\Gamma_i^D} \mathbf{f}^i \cdot \Psi^i \, d\Gamma dt + \sum_{k=1}^m \int_0^T \int_{\Gamma_{i_k}} \tilde{\mathbf{f}}^k \cdot \Psi^{i_k}\, d\Gamma dt.
\end{aligned}$$

This last equality may be written, in view of (3.60), as

$$(3.62)\quad \langle(\mathbf{W}_{,t}(T), -\mathbf{W}(T)), (\mathbf{\Psi}_0, \mathbf{\Psi}_1)\rangle_F = \sum_{i=1}^{n} \int_0^T \int_{\Gamma_i^D} \mathbf{f}^i \cdot \mathbf{\Psi}^i \, d\Gamma dt + \sum_{k=1}^{m} \int_0^T \int_{\Gamma_{i_k}} \tilde{\mathbf{f}}^k \cdot \mathbf{\Psi}^{i_k} \, d\Gamma dt, \quad \forall (\mathbf{\Psi}_0, \mathbf{\Psi}_1) \in F,$$

where $\langle g', g\rangle_F$ denotes the duality pairing between an element $g \in F$ and $g' \in F'$, the dual space of F.

In (3.62), let us choose

$$(3.63)\qquad \mathbf{f}^i = -\left[\left(\frac{\partial}{\partial t}\right)^2 \mathbf{\Psi}^i - \mathbf{\Psi}^i\right]\Bigg|_{\Sigma_i^{D+}} - \left(\frac{\partial}{\partial \tau^i}\right)^2 \mathbf{\Psi}^i \Bigg|_{\Sigma_i^{D-}},$$

where the "derivatives" on the right side of (3.63) are defined as follows:

$$\left\langle -\left(\frac{\partial}{\partial t}\right)^2 \mathbf{\Psi}^i\Bigg|_{\Sigma_i^{D+}}, \hat{\mathbf{\Psi}}\right\rangle = \int_0^T \int_{\Gamma_i^{D+}} \mathbf{\Psi}^i_{,t} \cdot \hat{\mathbf{\Psi}}_{,t} \, d\Gamma dt, \qquad \forall \hat{\mathbf{\Psi}} \in H^1(0,T; \mathcal{L}^2(\Gamma_i^{D^+})),$$

$$\left\langle -\left(\frac{\partial}{\partial \tau^i}\right)^2 \mathbf{\Psi}^i\Bigg|_{\Sigma_i^{D-}}, \hat{\mathbf{\Psi}}\right\rangle = \int_0^T \int_{\Gamma_i^{D-}} \frac{\partial \mathbf{\Psi}^i}{\partial \tau^i} \cdot \frac{\partial \hat{\mathbf{\Psi}}}{\partial \tau^i} \, d\Gamma dt, \qquad \forall \hat{\mathbf{\Psi}} \in L^2(0,T; \mathcal{H}^1(\Gamma_i^{D^-})),$$

where

$$\mathcal{H}^1(\Gamma_i^{D^-}) = \{\hat{\mathbf{\Psi}} \in \mathcal{L}^2(\Gamma_i^{D^-}) | \frac{\partial \hat{\mathbf{\Psi}}}{\partial \tau^i} \in \mathcal{L}^2(\Gamma_i^{D^-})\}.$$

Thus

$$(3.64)\qquad \left(\frac{\partial}{\partial \tau^i}\right)^2 \mathbf{\Psi}^i \Bigg|_{\Sigma_i^{D-}} \in L^2(0,T; (\mathcal{H}^1(\Gamma_i^{D^-}))')$$

and

$$(3.65)\qquad \left(\frac{\partial}{\partial t}\right)^2 \mathbf{\Psi}^i \Bigg|_{\Sigma_i^{D+}} \in (H^1(0,T; \mathcal{L}^2(\Gamma_i^{D^+})))'.$$

REMARK 3.6. Note that the quantity on the left side of (3.65) *is not* the t derivative of $\mathbf{\Psi}^i_{,t}$ in the sense of distributions. Also, one has $(\mathcal{H}^1(\Gamma_i^{D^-}))' = \mathcal{H}^{-1}(\Gamma_i^{D^-})$ if $\Gamma_i^C \neq \emptyset$ and, in this case, the left side of (3.64) is the tangential derivative of $(\partial \mathbf{\Psi}^i / \partial \tau^i)|_{\Sigma_i^{D-}}$ in the sense of distributions. □

Similarly, we select

$$\text{(3.66)}\qquad \tilde{\mathbf{f}}^k = -\left(\frac{\partial}{\partial\tau^{i_k}}\right)^2 \mathbf{\Psi}^{i_k}\Big|_{\Sigma_{i_k}(J_k)} \in L^2(0,T;(\mathcal{H}^1(\Gamma_{i_k}(J_k)))').$$

With these choices of $\mathbf{f}^i$, $\tilde{\mathbf{f}}^k$, $(\mathbf{W}_{,t}(T), -\mathbf{W}(T))$ is a linear function of $(\mathbf{\Psi}_0, \mathbf{\Psi}_1)$:

$$(\mathbf{W}_{,t}(T), -\mathbf{W}(T)) = L(\mathbf{\Psi}_0, \mathbf{\Psi}_1)$$

for some linear operator L and (3.62) becomes

$$\begin{aligned}\langle L(\mathbf{\Psi}_0,\mathbf{\Psi}_1),(\mathbf{\Psi}_0,\mathbf{\Psi}_1)\rangle_F &= \sum_{i=1}^{n}\left\{\int_0^T\int_{\Gamma_i^{D+}}(|\mathbf{\Psi}^i_{,t}|^2+|\mathbf{\Psi}^i|^2)\,d\Gamma dt\right.\\ &\quad\left.+\int_0^T\int_{\Gamma_i^{D-}}\left|\frac{\partial\mathbf{\Psi}^i}{\partial\tau^i}\right|^2 d\Gamma dt\right\}+\sum_{k=1}^{m}\int_0^T\int_{\Gamma_{i_k}(J_k)}\left|\frac{\partial\mathbf{\Psi}^{i_k}}{\partial\tau^{i_k}}\right|^2 d\Gamma dt\\ &= \|(\mathbf{\Psi}_0,\mathbf{\Psi}_1)\|_F^2.\end{aligned}$$

Therefore, L is just the Riesz isomorphism of F onto F'. It follows that the set of states $(\mathbf{W}(T), \mathbf{W}_{,t}(T))$ which are reachable in time $T > T_0$, starting from the zero state, is characterized by $(\mathbf{W}_{,t}(T), -\mathbf{W}(T)) \in F'$ if one utilizes controls

$$\text{(3.67)}\qquad \mathbf{f}^i \in (H^1(0,T;\mathcal{L}^2(\Gamma_i^{D^+})))' \bigoplus L^2(0,T;(\mathcal{H}^1(\Gamma_i^{D^-}))')$$

along Σ_i^D, and

$$\text{(3.68)}\qquad \tilde{\mathbf{f}}^k \in L^2(0,T;(\mathcal{H}^1(\Gamma_{i_k}(J_k)))')$$

along $\Sigma_{i_k}(J_k)$. Since $F \subset \mathcal{V} \times H$ one has $\mathcal{V}' \times H \subset F'$. We have therefore obtained the following result.

THEOREM 3.5. *Let the hypotheses of Corollary* 3.1 *hold and let* $T > T_0$. *Assume that* $\mathbf{W}_0 \in H$, $\mathbf{W}_1 \in \mathcal{V}'$ *and that* (3.58) *holds. Then there are controls* $\mathbf{f}^i$, $\tilde{\mathbf{f}}^k$ *satisfying* (3.67), (3.68) *such that the solution of* (3.1)–(3.5), (3.61) *satisfies*

$$\mathbf{W}|_{t=T} = \mathbf{W}_0, \quad \mathbf{W}_{,t}|_{t=T} = \mathbf{W}_1.$$

REMARK 3.7. When $\mathbf{f}^i$, $\tilde{\mathbf{f}}^k$ satisfy (3.67) and (3.68), respectively, it may be proved that the problem (3.1)–(3.5), (3.61) has a unique solution, defined by transposition. This solution exists only in a very weak sense and satisfies (3.62) by virtue of the sense in which solution is defined. The reader is referred to [44] for a discussion of this point in a general context. □

3.4.1. *Serially Connected Membranes.* It follows from Theorem 3.2 that for n serially connected membranes the following estimate is valid.

COROLLARY 3.2. *Suppose that the hypotheses of Theorem 3.2 are satisfied. If* $\mathbf{\Psi} = (\mathbf{\Psi}^1, \dots, \mathbf{\Psi}^n)$ *is a sufficiently regular solution of* (3.14)–(3.16), (3.46), (3.47) *on* $(0,T)$ *with* $(\mathbf{\Psi}(T), \mathbf{\Psi}_{,t}(T)) = (\mathbf{\Psi}_0, \mathbf{\Psi}_1)$, *there is a* $T_0 > 0$ *such that for* $T > T_0$,

$$\|(\mathbf{\Psi}_0, \mathbf{\Psi}_1)\|^2_{\mathcal{V}\times H} \le C_T \sum_{i=1}^n \left\{ \int_0^T \int_{\Gamma_i^{D+}} (|\mathbf{\Psi}^i_{,t}|^2 + |\mathbf{\Psi}^i|^2)\, d\Gamma dt + \int_0^T \int_{\Gamma_i^{D-}} \left| \frac{\partial \mathbf{\Psi}^i}{\partial \tau^i} \right|^2 d\Gamma dt \right\}. \tag{3.69}$$

As a consequence of Corollary 3.2 the following controllability result which, unlike Theorem 3.5, does not involve controls in the joints, is obtained.

THEOREM 3.6. *Let the hypotheses of Theorem 3.2 hold, assume that* (3.58) *holds and let* $T > T_0$. *Then*

$$H \times \mathcal{V}' \subset \{(\mathbf{W}(T), \mathbf{W}_{,t}(T)) |\, \mathbf{f}^i \ \text{satisfies (3.67) and}\ \tilde{\mathbf{f}}^k = 0\}, \tag{3.70}$$

where $\mathbf{W}$ *is the solution of* (3.1)–(3.5), (3.61) *corresponding to* $\mathbf{f}^i$ *and* $\tilde{\mathbf{f}}^k = 0$. *Further, if*

$$\mathbf{m} \cdot \boldsymbol{\nu}^i \ge 0 \ \text{ on } \Gamma_i^D, \tag{3.71}$$

(that is, if $\Gamma_i^{D^-} = \emptyset$*) then*

$$\mathcal{V} \times H \subset \{(\mathbf{W}(T), \mathbf{W}_{,t}(T)) |\, \mathbf{f}^i \in \mathcal{L}^2(\Sigma_i^D) \ \text{and}\ \tilde{\mathbf{f}}^k = 0\}. \tag{3.72}$$

PROOF. The inclusion (3.70) follows from Corollary 3.2 in the same way that Theorem 3.5 was deduced from Corollary 3.1. If (3.71) is satisfied, the estimate (3.69) reduces to

$$\|(\mathbf{\Psi}_0, \mathbf{\Psi}_1)\|^2_{\mathcal{V}\times H} \le C_T \sum_{i=1}^n \int_0^T \int_{\Gamma_i^D} (|\mathbf{\Psi}^i_{,t}|^2 + |\mathbf{\Psi}^i|^2)\, d\Gamma dt.$$

Then, by the process of "weakening the norm" (see, e.g., [60, p. 200]), one may obtain

$$\|(\mathbf{\Psi}_0, \mathbf{\Psi}_1)\|^2_{H\times\mathcal{V}'} \le C_T \sum_{i=1}^n \int_0^T \int_{\Gamma_i^D} |\mathbf{\Psi}^i|^2\, d\Gamma dt. \tag{3.73}$$

As above, for $T > T_0$ one may introduce a norm $\|\cdot\|_G$ through

$$\|(\mathbf{\Psi}_0, \mathbf{\Psi}_1)\|^2_G = \sum_{i=1}^n \int_0^T \int_{\Gamma_i^D} |\mathbf{\Psi}^i|^2\, d\Gamma dt$$

and a space G which is the completion of $D_A \times \mathcal{V}$ with respect to this norm. The set of reachable states $(\mathbf{W}(T), \mathbf{W}_{,t}(T))$, using controls $\mathbf{f}^i \in L^2(\Sigma_i^D)$, $\tilde{\mathbf{f}}^k = 0$, is characterized by $(\mathbf{W}_{,t}(T), -\mathbf{W}(T)) \in G'$. The estimate (3.73) shows that G is continuously and densely embedded in $H \times \mathcal{V}'$, hence $H \times \mathcal{V} \subset G'$. □

3.4.2. *Membrane Transmission Problems.* Let Ω be any bounded, connected open set in $\mathbb{R}^2$ with a Lipschitz continuous boundary consisting of a finite number of smooth simple arcs, and let Ω_1 be a simply connected open set in $\mathbb{R}^2$ with smooth boundary such that $\overline{\Omega}_1 \subset \Omega$. Set

$$\Omega_2 = \Omega \setminus \Omega_1, \quad \Gamma = \partial\Omega, \quad \Gamma_1 = \partial\Omega_1, \quad \Gamma_2 = \partial\Omega_2 = \Gamma \cup \Gamma_2.$$

We imagine that Ω represents the equilibrium position of a membrane composed of two parts, Ω_1 and Ω_2, each of which is homogeneous and elastically isotropic but with possibly differing material properties. Let $\boldsymbol{\nu}^i$ denote the exterior unit normal to Γ_i, so that $\boldsymbol{\nu}^1 = -\boldsymbol{\nu}^2$ on Γ_1, and let $\mathbf{i}, \mathbf{j}, \mathbf{k}$ be the natural basis for $\mathbb{R}^3$. If

$$\mathbf{W}^i = W_1^i \mathbf{i} + W_2^i \mathbf{j} + W_3^i \mathbf{k} := \mathbf{w}^i + W_3^i \mathbf{k}$$

is the displacement vector of Ω_i, for small displacements the motion is described by

$$\begin{cases} \sum_\alpha s_{i,\alpha}^{\alpha\beta}(\mathbf{w}^i) = \rho_i W_{\beta,tt}^i, \\ \mu_i \Delta W_3^i = \rho_i W_{3,tt}^i, \quad \mathbf{x} \in \Omega_i,\ t > 0; \end{cases} \tag{3.74}$$

$$\mathbf{w}^2 = 0, \quad W_3^2 = 0, \quad \mathbf{x} \in \Gamma^C,\ t > 0; \tag{3.75}$$

$$\begin{cases} \sum_\alpha \nu_\alpha^2 s_2^{\alpha\beta}(\mathbf{w}^2) = f_\beta, \\ \mu_2 \dfrac{\partial W_3^2}{\partial \boldsymbol{\nu}^2} = f_3, \quad \mathbf{x} \in \Gamma^D,\ t > 0, \end{cases} \tag{3.76}$$

where $\Gamma = \overline{\Gamma}^C \cup \overline{\Gamma}^D$ as above and $\mathbf{f} = f_1\mathbf{i} + f_2\mathbf{j} + f_3\mathbf{k}$ is an applied force on Γ^D;

$$\begin{cases} S_{\boldsymbol{\tau}^1}(\mathbf{w}^1) = S_{\boldsymbol{\tau}^2}(\mathbf{w}^2), \\ S_{\boldsymbol{\nu}^1}(\mathbf{w}^1) = S_{\boldsymbol{\nu}^2}(\mathbf{w}^2), \\ \mu_1 \dfrac{\partial W_3^1}{\partial \boldsymbol{\nu}^1} = -\mu_2 \dfrac{\partial W_3^2}{\partial \boldsymbol{\nu}^2}, \quad x \in \Gamma_1,\ t > 0, \end{cases} \tag{3.77}$$

where $\boldsymbol{\tau}^i$ is the unit tangent vector to Γ_i, positively oriented with respect to $\mathbf{k}$ (hence $\boldsymbol{\tau}^1 = -\boldsymbol{\tau}^2$ on Γ_i). The initial conditions are

$$\mathbf{W}^i|_{t=0} = \mathbf{W}_{,t}^i|_{t=0} = 0, \quad \mathbf{x} \in \Omega_i.$$

The problem (3.74)–(3.77) is known in the literature as a *problem of transmission*. It obviously falls within the framework of serially connected membranes discussed above in which Γ_1 plays the role of the junction region (cf. Remark 3.4). However, the present problem is somewhat simpler in that there is no difficulty concerning regularity of the solution since Γ_1 is a smooth curve entirely contained within Ω; one has in this case

$$W_j^1 \in C([0,T]; H^2(\Omega_1)), \;\; W_j^2 \in C([0,T]; H^s(\Omega_2)), \;\; j = 1,2,3,$$

for some $s > 3/2$ provided $\overline{\Gamma}^C$ and $\overline{\Gamma}^D$ meet in a strictly convex angle. (The assumption that Γ_1 is a smooth curve may be weakened to allow for certain piecewise smooth curves; see Nicaise [73], [74].) The techniques above therefore yield, for example, the following result.

THEOREM 3.7. *Suppose there is a point* $\mathbf{x}_0 \in \Omega_1$ *such that*

$$\begin{aligned}(\mathbf{x} - \mathbf{x}_0) \cdot \boldsymbol{\nu}^1 &\geq 0 \;\; \text{on } \Gamma_1,\\ (\mathbf{x} - \mathbf{x}_0) \cdot \boldsymbol{\nu}^2 &\geq 0 \;\; \text{on } \Gamma^D,\\ (\mathbf{x} - \mathbf{x}_0) \cdot \boldsymbol{\nu}^2 &\leq 0 \;\; \text{on } \Gamma^C.\end{aligned}$$

Assume further that

$$\rho_1 \leq \rho_2, \;\; \mu_1 \geq \mu_2, \;\; \lambda_1 \geq \lambda_2.$$

Then for $T > T_0$,

$$\mathcal{V} \times H \subset \{(\mathbf{W}(T), \mathbf{W}_{,t}(T)) \,|\, \mathbf{f} \in \mathcal{L}^2(\Sigma^D)\}.$$

REMARK 3.8. Since there is no coupling between the pair $\mathbf{w}^1, \mathbf{w}^2$ and the pair W_3^1, W_3^2 in the system (3.74)– (3.77), the last result is equivalent to two separate reachability results, one for the problem

$$\begin{aligned}\mu_i \Delta W_3^i &= \rho_i W_{3,tt}^i, \;\; \mathbf{x} \in \Omega_i, \; t > 0,\\ W_3^2 &= 0, \;\; \mathbf{x} \in \Gamma^C, \; t > 0,\\ \mu_2 \frac{\partial W_3^2}{\partial \boldsymbol{\nu}^2} &= f_3, \;\; \mathbf{x} \in \Gamma^D, \; t > 0,\end{aligned}$$

$$\begin{cases} W_3^1 = W_3^2, \\ \mu_1 \dfrac{\partial W_3^1}{\partial \boldsymbol{\nu}^1} = -\mu_2 \dfrac{\partial W_3^2}{\partial \boldsymbol{\nu}^2}, \;\; \mathbf{x} \in \Gamma_1, \; t > 0, \end{cases}$$

and the other for an analogous transmission problem involving the equations of plane elasticity. The former controllability result has been obtained by Lions [60, Chapter VI] in the case where the control is in the Dirichlet, rather than Neumann, boundary condition (i.e., $W_3^2 = f_3$ on Γ^D). In this case, geometric restrictions on Γ are unnecessary.

CHAPTER VII

Modeling and Controllability of Systems of Linked Plates

The main purpose of this Chapter is to derive the dynamic equations of motion of a system of interconnected thin, elastic plates. In particular, we seek to determine the transmission conditions which describe how disturbances in one member of the system are propagated into the others. The approach taken is similar to that of Chapter VI. We first derive the equations of motion and boundary conditions of a general nonlinear plate, and corresponding Hamilton's principle, under a specific kinematic hypothesis on the deformation process and assuming a linear stress-strain relation. Linearizing this system around the trivial equilibrium leads to a system known in the literature as the Reissner, or Reissner-Mindlin, plate model for the transverse displacement and shear angles, and the usual equations of plane elasticity for in-plane deformation.

The dynamical system describing the transient behavior of a collection of interconnected plates is derived by minimizing an appropriate Lagrangian over a class of deformations which satisfies certain geometric constraints, in particular, in the junction regions where two or more plates are joined together. The form of these constraints at a joint depends on the nature of the joint (e.g., rigid, hinged, etc.) and each choice leads to particular balance laws for the transmission of linear and angular momentum through the joint. In the resulting dynamical system there are five unknown functions associated with each plate of the system: three components of the displacement vector and two shear angles. All displacements and shear angles are coupled through the geometric and dynamic junction conditions. The shear angles and transverse displacement of each plate are further coupled in the equations of motion through a coupling parameter called the shear modulus of the plate. In Section 6, we consider the limit of the dynamical system (in the case of semi-rigid joints) as all shear moduli tend to infinity (this corresponds to the absence of transverse shearing) and obtain a limit system which models the transient motion of a collection of interconnected Kirchhoff plates with semi-rigid joints.

Exact controllability of a system of linked Reissner plates is considered in Section 7.6, but only in the case of serially connected plates.

1. Modeling of Dynamic Nonlinear Elastic Plates

In this section the dynamic equations and edge conditions which describe the motion of a class of nonlinear elastic thin plates will be derived. Although the models which are presented are known in the literature, their derivations are included here for the sake of completeness and because they may not be familiar to the reader. Our presentation is adapted from Wempner [108, Chapter 7]. Summation convention will be assumed throughout this section and the next.

Let Ω is a bounded, open, connected set in $\mathbb{R}^2$ with Lipschitz continuous boundary Γ consisting of a finite number of smooth, simple curves, and let $\mathbf{a}_1, \mathbf{a}_2, \mathbf{a}_3$ be a fixed, right-handed orthonormal basis for $\mathbb{R}^3$. We set

$$Q = \{(x_1, x_2, x_3) |\, (x_1, x_2) \in \overline{\Omega},\ \ -h_1(x_1, x_2) \leq x_3 \leq h_2(x_1, x_2)\},$$

where h_1, h_2 are positive, differentiable functions on $\overline{\Omega}$. The reference configuration of the plate is given by

$$\mathcal{B} = \{\mathbf{p}_0 + x_\alpha \mathbf{a}_\alpha + x_3 \mathbf{a}_3 := \mathbf{p}(x_1, x_2) + x_3 \mathbf{a}_3 |\, (x_1, x_2, x_3) \in Q\},$$

where $\mathbf{p}_0$ is a fixed vector in $\mathbb{R}^3$ and $\mathbf{p}(x_1, x_2) = \mathbf{p}_0 + x_\alpha \mathbf{a}_\alpha$ The planar region

$$\mathcal{P} = \{\mathbf{p}(x_1, x_2) |\, (x_1, x_2) \in \Omega\}$$

is called the *reference surface.* Note that the plate may have variable thickness and is not necessarily symmetric with respect to its reference surface.

Basic kinematic hypothesis. The position vector at time t to the displaced location of the particle situated at $\mathbf{p}(x_1, x_2) + x_3 \mathbf{a}_3$ in the reference configuration is

$$\mathbf{R}(x_1, x_2, x_3, t) = \mathbf{P}(x_1, x_2, t) + x_3 \mathbf{A}_3(x_1, x_2, t), \tag{1.1}$$

where $\mathbf{A}_3 \neq 0$.

Assumption (1.1) means that the x_3 line remains straight in the deformed plate. Set

$$\mathbf{G}_i = \mathbf{R}_{,i}, \quad \mathbf{A}_\alpha = \mathbf{G}_\alpha|_{x_3=0} = \mathbf{P}_{,\alpha},$$

where a subscript after a comma means differentiation with respect to the indicated spatial variable. Then $\mathbf{A}_1, \mathbf{A}_2$ are vectors which are tangent to the deformed reference surface. From (1.1) one has

$$\mathbf{G}_\alpha = \mathbf{A}_\alpha + x_3 \mathbf{A}_{3,\alpha}, \quad \mathbf{G}_3 = \mathbf{A}_3. \tag{1.2}$$

It follows from (1.2) and (1.2) that the components ε_{ij} of the strain tensor are given by

$$(1.3)\qquad \begin{cases} \varepsilon_{\alpha\beta} = \frac{1}{2}[\mathbf{A}_\alpha \cdot \mathbf{A}_\beta + x_3(\mathbf{A}_\alpha \cdot \mathbf{A}_{3,\beta} + \mathbf{A}_\beta \cdot \mathbf{A}_{3,\alpha}) \\ \qquad\qquad + x_3^2 \mathbf{A}_{3,\alpha} \cdot \mathbf{A}_{3,\beta} - \delta_{\alpha\beta}], \\ \varepsilon_{\alpha 3} = \frac{1}{2}(\mathbf{A}_\alpha \cdot \mathbf{A}_3 + x_3 \mathbf{A}_3 \cdot \mathbf{A}_{3,\alpha}), \\ \varepsilon_{33} = \frac{1}{2}(\mathbf{A}_3 \cdot \mathbf{A}_3 - 1). \end{cases}$$

We introduce

$$(1.4)\qquad \mathbf{W} = \mathbf{P} - \mathbf{p}, \quad \mathbf{U} = \mathbf{A}_3 - \mathbf{a}_3.$$

The quantity $\mathbf{W}(x_1, x_2, t)$ is the displacement of the point $\mathbf{p}(x_1, x_2)$ in the reference surface and $\mathbf{U}$ represents the "rotation" of the $\mathbf{a}_3$ line (this will be made precise below). Set

$$(1.5)\qquad \begin{aligned} \mathbf{r}(x_1, x_2, x_3, t) &= \mathbf{W}(x_1, x_2, t) + x_3 \mathbf{U}(x_1, x_2, t) \\ &:= W_i(x_1, x_2, t)\mathbf{a}_i + x_3 U_i(x_1, x_2, t)\mathbf{a}_i, \end{aligned}$$

which is the *displacement* of the point $\mathbf{p}(x_1, x_2) + x_3\mathbf{a}_3$. The strains may then be written in terms of the displacement $\mathbf{W}$ and rotation $\mathbf{U}$ by noting that $\mathbf{A}_3 = \mathbf{U} + \mathbf{a}_3$ and $\mathbf{A}_\alpha = \mathbf{W}_{,\alpha} + \mathbf{a}_\alpha$.

1.1. Equations of Motion. Let s^{ij} denote the components of the second Piola-Kirchhoff stress tensor and $\mathbf{t}^i$ be the stress vectors, so that

$$(1.6)\qquad \mathbf{t}^i = s^{ij}\mathbf{G}_j.$$

If $\bar{\mathbf{F}}$ denotes the body force per unit of undeformed volume, the three-dimensional conditions for the balance of forces is

$$(1.7)\qquad \mathbf{t}^i_{,i} + \bar{\mathbf{F}} = \rho \mathbf{R}_{,tt},$$

where ρ is the mass of the plate per unit of reference volume. If (1.7) is integrated in x_3 through the thickness of the plate the result is

$$(1.8)\qquad \int_{-h_1}^{h_2} \mathbf{t}^\alpha_{,\alpha} dx_3 + \mathbf{t}^3|_{x_3=-h_1}^{x_3=h_2} + \int_{-h_1}^{h_2} \bar{\mathbf{F}}\, dx_3 = \int_{-h_1}^{h_2} \rho \mathbf{R}_{,tt} dx_3.$$

Let

$$(1.9)\qquad \mathbf{N}^i = \int_{-h_1}^{h_2} \mathbf{t}^i dx_3 = \int_{-h_1}^{h_2} s^{ij}\mathbf{G}_j dx_3.$$

$\mathbf{N}^i$ is the resultant stress per unit of reference area. One has

$$\mathbf{N}^\alpha_{,\alpha} = \int_{-h_1}^{h_2} \mathbf{t}^\alpha_{,\alpha} dx_3 + h_{2,\alpha}\mathbf{t}^\alpha|_{x_3=h_2} + h_{1,\alpha}\mathbf{t}^\alpha|_{x_3=-h_1}$$

and therefore (1.8) may be written

$$(1.10)\quad \mathbf{N}^{\alpha}_{,\alpha} + [\mathbf{t}^3 - h_{2,\alpha}\mathbf{t}^{\alpha}]_{x_3=h_2} - [\mathbf{t}^3 + h_{1,\alpha}\mathbf{t}^{\alpha}]_{x_3=-h_1} + \int_{-h_1}^{h_2} \bar{\mathbf{F}}\,dx_3 = \int_{-h_1}^{h_2} \rho(\mathbf{W}_{,tt} + x_3\mathbf{U}_{,tt})\,dx_3.$$

Since the mapping $(x_1, x_2, x_3) \mapsto x_i\mathbf{a}_i$ is orthogonal with determinant unity, the vector

$$\mathbf{n}^+ = (-h_{2,\alpha}\mathbf{a}_\alpha + \mathbf{a}_3)/\sqrt{1 + |\nabla h_2|^2}$$

(resp., $\mathbf{n}^- = -(h_{1,\alpha}\mathbf{a}_\alpha + \mathbf{a}_3)/\sqrt{1+|\nabla h_1|^2}$), where $\nabla h_\mu = h_{\mu,\alpha}\mathbf{a}_\alpha$, is a unit normal vector to the upper surface $\mathbf{p}(x_1, x_2) + h_2(x_1, x_2)\mathbf{a}_3$ (resp., to the lower surface $\mathbf{p}(x_1, x_2) - h_1(x_1, x_2)\mathbf{a}_3$) of the plate and points towards the exterior of the body. Therefore

$$[\mathbf{t}^3 - h_{2,\alpha}\mathbf{t}^{\alpha}]_{x_3=h_2} = \sqrt{1 + |\nabla h_2|^2}\,\mathbf{t}^+,$$
$$-[\mathbf{t}^3 + h_{1,\alpha}\mathbf{t}^{\alpha}]_{x_3=-h_1} = \sqrt{1 + |\nabla h_1|^2}\,\mathbf{t}^-,$$

where $\mathbf{t}^{\pm} = \mathbf{n}^{\pm}_i\mathbf{t}^i$ denote the applied tractions on the upper and lower surfaces, respectively. Set

$$\mathbf{F} = \sqrt{1 + |\nabla h_2|^2}\,\mathbf{t}^+ + \sqrt{1 + |\nabla h_1|^2}\,\mathbf{t}^- + \int_{-h_1}^{h_2} \bar{\mathbf{F}}\,dx_3,$$

the resultant body force per unit of reference area, and

$$m(x_1, x_2) = \int_{-h_1}^{h_2} \rho(x_1, x_2, x_3)\,dx_3, \quad m_\rho(x_1, x_2) = \int_{-h_1}^{h_2} x_3\rho(x_1, x_2, x_3)\,dx_3.$$

With this notation, equation (1.10) becomes

$$(1.11)\qquad \mathbf{N}^{\alpha}_{,\alpha} + \mathbf{F} = m\mathbf{W}_{,tt} + m_\rho\mathbf{U}_{,tt}.$$

From (1.2) and (1.4) we have

$$\mathbf{G}_\alpha = (\mathbf{W}_{,\alpha} + \mathbf{a}_\alpha) + x_3\mathbf{U}_\alpha, \quad \mathbf{G}_3 = \mathbf{U} + \mathbf{a}_3$$

and therefore

$$\mathbf{N}^i = \int_{-h_1}^{h_2} (s^{i\beta}\mathbf{G}_\beta + s^{i3}\mathbf{G}_3)\,dx_3 = N^{i\beta}(\mathbf{W}_{,\beta} + \mathbf{a}_\beta) + M^{i\beta}\mathbf{U}_{,\beta} + N^{i3}(\mathbf{U} + \mathbf{a}_3),$$

where

$$(1.12)\qquad N^{ij} = \int_{-h_1}^{h_2} s^{ij}\,dx_3, \quad M^{ij} = \int_{-h_1}^{h_2} x_3 s^{ij}\,dx_3$$

are called the *generalized stresses* and *generalized moments*, respectively. Equation (1.11) may therefore be written

$$(1.13)\quad \frac{\partial}{\partial x_\alpha}[N^{\alpha\beta}(\mathbf{W}_{,\beta}+\mathbf{a}_\beta)+M^{\alpha\beta}\mathbf{U}_{,\beta}+N^{\alpha 3}(\mathbf{U}+\mathbf{a}_3)]+\mathbf{F} = m\mathbf{W}_{,tt}+m_\rho\mathbf{U}_{,tt}.$$

Next, multiply (1.7) by x_3 and integrate through the plate thickness. One obtains

$$(1.14)\quad \int_{-h_1}^{h_2} x_3\mathbf{t}^\alpha_{,\alpha}dx_3 - \int_{-h_1}^{h_2}\mathbf{t}^3dx_3 + x_3\mathbf{t}^3|_{x_3=-h_1}^{x_3=h_2} + \int_{-h_1}^{h_2} x_3\bar{\mathbf{F}}\,dx_3 = m_\rho\mathbf{W}_{,tt}+I_\rho\mathbf{U}_{,tt}.$$

where

$$I_\rho(x_1,x_2)=\int_{-h_1}^{h_2} x_3^2\rho(x_1,x_2,x_3)\,dx_3.$$

The resultant stress couple per unit of reference area is

$$\mathbf{M}^\alpha=\int_{-h_1}^{h_2} x_3\mathbf{t}^\alpha dx_3=\int_{-h_1}^{h_2} x_3 s^{\alpha j}\mathbf{G}_j dx_3 = M^{\alpha\beta}(\mathbf{W}_{,\beta}+\mathbf{a}_\beta)+I^{\alpha\beta}\mathbf{U}_{,\beta}+M^{\alpha 3}(\mathbf{U}+\mathbf{a}_3),$$

where

$$I^{\alpha\beta}=\int_{-h_1}^{h_2} x_3^2 s^{\alpha\beta}dx_3.$$

Equation (1.14) may be written

$$(1.15)\quad \mathbf{M}^\alpha_{,\alpha}-\mathbf{N}^3+\mathbf{C}=m_\rho\mathbf{W}_{,tt}+I_\rho\mathbf{U}_{,tt},$$

where

$$\mathbf{C}=h_2[\mathbf{t}^3-h_{2,\alpha}\mathbf{t}^\alpha]_{x_3=h_2}+h_1[\mathbf{t}^3+h_{1,\alpha}\mathbf{t}^\alpha]_{x_3=-h_1}+\int_{-h_1}^{h_2} x_3\bar{\mathbf{F}}\,dx_3 = h_2\sqrt{1+|\nabla h_2|^2}\,\mathbf{t}^+ - h_1\sqrt{1+|\nabla h_1|^2}\,\mathbf{t}^- + \int_{-h_1}^{h_2} x_3\bar{\mathbf{F}}\,dx_3$$

is the resultant applied couple per unit of reference area and

$$I_\rho(x_1,x_2)=\int_{-h_1}^{h_2} x_3^2\,\rho(x_1,x_2,x_3)\,dx_3.$$

When written in terms of generalized stresses and generalized moments, (1.15) becomes

$$(1.16)\quad \frac{\partial}{\partial x_\alpha}[M^{\alpha\beta}(\mathbf{W}_{,\beta}+\mathbf{a}_\beta)+I^{\alpha\beta}\mathbf{U}_{,\beta}+M^{\alpha 3}(\mathbf{U}+\mathbf{a}_3)]$$
$$-[N^{3\beta}(\mathbf{W}_{,\beta}+\mathbf{a}_\beta)+M^{3\beta}\mathbf{U}_{,\beta}+N^{33}(\mathbf{U}+\mathbf{a}_3)]+\mathbf{C}$$
$$=m_\rho\mathbf{W}_{,tt}+I_\rho\mathbf{U}_{,tt}.$$

The equations of motion are (1.13) and (1.16). They comprise six scalar equations for the 21 scalar functions $W_i, U_i, N^{ij}, M^{ij}, I^{\alpha\beta}$ (taking into account the symmetry of s^{ij}).

1.2. Edge Conditions. Let $\partial\mathcal{P}$ denote the two-dimensional boundary of the reference surface $\mathcal{P}$ and suppose that

$$(1.17)\qquad \partial\mathcal{P}=\overline{\Lambda}^C\cup\overline{\Lambda}^D,$$

where Λ^C, Λ^D are mutually disjoint, relatively open subsets of $\partial\mathcal{P}$ and $\Lambda^D\neq\emptyset$. (As in Section 1.2, the superscripts C and D are used to indicate "clamped" and "dynamic," respectively.) The partition (1.17) induces a similar partition of $\Gamma:=\partial\Omega$ into

$$(1.18)\qquad \Gamma=\overline{\Gamma}^C\cup\overline{\Gamma}^D$$

through the mapping

$$\Gamma\mapsto\partial\mathcal{P}:(x_1,x_2)\mapsto\mathbf{p}(x_1,x_2).$$

Let (ν_1,ν_2) denote the unit exterior normal vector to Γ and $(-\nu_2,\nu_1)$ be the unit tangent vector to Γ which is positively oriented with respect to the vector $(0,0,1)$. Let $\bar{\mathbf{f}}$ denote a given distribution of forces along $\Lambda^D+x_3\mathbf{a}_3$, $-h_1\le x_3\le h_2$.The three-dimensional condition for the balance of forces along the edge of the plate corresponding to Λ^D is

$$(1.19)\qquad \nu_\alpha\mathbf{t}^\alpha=\bar{\mathbf{f}}.$$

Proceeding as above, we impose the conditions

$$\int_{-h_1}^{h_2}\nu_\alpha\mathbf{t}^\alpha dx_3=\int_{-h_1}^{h_2}\bar{\mathbf{f}}\,dx_3:=\mathbf{f},$$
$$\int_{-h_1}^{h_2}x_3\nu_\alpha\mathbf{t}^\alpha dx_3=\int_{-h_1}^{h_2}x_3\bar{\mathbf{f}}\,dx_3:=\mathbf{c}.$$

One thereby obtains the edge conditions

$$(1.20)\qquad \nu_\alpha\mathbf{N}^\alpha=\mathbf{f},\ \ \nu_\alpha\mathbf{M}^\alpha=\mathbf{c}\ \text{ on }\Gamma^D,\ t>0$$

or, in expanded form,

$$\text{(1.21)}\qquad \begin{cases} \nu_\alpha[N^{\alpha\beta}(\mathbf{W}_{,\beta}+\mathbf{a}_\beta)+M^{\alpha\beta}\mathbf{U}_{,\beta}+N^{\alpha 3}(\mathbf{U}+\mathbf{a}_3)]=\mathbf{f},\\ \nu_\alpha[M^{\alpha\beta}(\mathbf{W}_{,\beta}+\mathbf{a}_\beta)+I^{\alpha\beta}\mathbf{U}_{,\beta}+M^{\alpha 3}(\mathbf{U}+\mathbf{a}_3)]=\mathbf{c}. \end{cases}$$

1.3. Hamilton's Principle. Let $\hat{\mathbf{r}}$ be a sufficiently smooth function defined on $\overline{Q}\times[0,T]$ such that $\hat{\mathbf{r}}|_{t=0}=\hat{\mathbf{r}}|_{t=T}=0$ and assume that $\hat{\mathbf{r}}$ satisfies the same "geometric" conditions as the actual displacement $\mathbf{r}=\mathbf{W}+x_3\mathbf{U}$ of the plate. For definiteness, let us suppose that the geometric condition is

$$\text{(1.22)}\qquad \mathbf{r}=0,\ \ (x_1,x_2)\in\Gamma^C,\ -h_1(x_1,x_2)<x_3<h_2(x_1,x_2).$$

From (1.7)

$$\int_0^T\int_Q(\mathbf{t}^i_{,i}+\bar{\mathbf{F}})\cdot\hat{\mathbf{r}}\,dQdt=\int_0^T\int_Q\rho\mathbf{r}_{,tt}\cdot\hat{\mathbf{r}}\,dQdt,$$

This equation may be written

$$\int_0^T\int_{\partial Q}n_i\mathbf{t}^i\cdot\hat{\mathbf{r}}_i dSdt+\int_0^T\int_Q(-\mathbf{t}^i\cdot\hat{\mathbf{r}}_{,i}+\bar{\mathbf{F}}\cdot\hat{\mathbf{r}})\,dQdt$$
$$=-\int_0^T\int_Q\rho\mathbf{r}_{,t}\cdot\hat{\mathbf{r}}_{,t}dQdt,$$

where $n_i\mathbf{a}_i$ is the exterior unit normal to $\partial\mathcal{B}$ (so that (n_1,n_2,n_3) is the unit exterior normal to ∂Q). If in this variational equation we consider only test functions $\hat{\mathbf{r}}=\hat{\mathbf{W}}+x_3\hat{\mathbf{U}}$ of the form (1.5), it may be written

$$\text{(1.23)}\qquad \int_0^T\int_\Omega(\mathbf{F}\cdot\hat{\mathbf{W}}+\mathbf{C}\cdot\hat{\mathbf{U}})\,d\Omega dt+\int_0^T\int_{\Gamma^D}(\mathbf{f}\cdot\hat{\mathbf{W}}+\mathbf{c}\cdot\hat{\mathbf{U}})\,d\Gamma dt$$
$$-\int_0^T\int_\Omega(\mathbf{N}^\alpha\cdot\hat{\mathbf{W}}_{,\alpha}+\mathbf{M}^\alpha\cdot\hat{\mathbf{U}}_{,\alpha}+\mathbf{N}^3\cdot\hat{\mathbf{U}})\,d\Omega dt$$
$$=-\int_0^T\int_\Omega[(m\mathbf{W}_{,t}+m_\rho\mathbf{U}_{,t})\cdot\hat{\mathbf{W}}_{,t}+(m_\rho\mathbf{W}_{,t}+I_\rho\mathbf{U}_{,t})\cdot\hat{\mathbf{U}}_{,t}]\,d\Omega dt.$$

Conversely, if the variational equation (1.23) holds for all test functions $\hat{\mathbf{r}}=\hat{\mathbf{W}}+x_3\hat{\mathbf{U}}$ as described above, then $\mathbf{W},\mathbf{U}$ and the resultant stresses and moments satisfy the two-dimensional plate equations and edge conditions (1.11), (1.15) and (1.20). The variational formulation (1.23) of the boundary value problem is the principle of virtual work for the problem under consideration.

For a given admissible displacement $\mathbf{r}$ satisfying (1.22), the associated *kinetic energy* is

$$\begin{aligned}\mathcal{K}(t) &= \frac{1}{2}\int_Q \rho|\mathbf{r}_{,t}|^2 dQ \\ &= \frac{1}{2}\int_\Omega (m|\mathbf{W}_{,t}|^2 + 2m_\rho \mathbf{W}_{,t}\cdot\mathbf{U}_{,t} + I_\rho|\mathbf{U}_{,t}|^2)\, d\Omega,\end{aligned} \tag{1.24}$$

and the associated *work* done by the external forces is

$$\begin{aligned}\mathcal{W}(t) &= \int_Q \bar{\mathbf{F}}\cdot\mathbf{r}dQ + \int_{\partial Q} n_i\mathbf{t}^i\cdot\mathbf{r}\, dS \\ &= \int_\Omega (\mathbf{F}\cdot\mathbf{W} + \mathbf{C}\cdot\mathbf{U})\, d\Omega + \int_{\Gamma^D} (\mathbf{f}\cdot\mathbf{W} + \mathbf{c}\cdot\mathbf{U})\, d\Gamma.\end{aligned} \tag{1.25}$$

By utilizing the notation (1.24), (1.25) and recalling (1.2) and the symmetry of s^{ij}, the principle of virtual work may be written

$$\delta\int_0^T [\mathcal{K}(t) + \mathcal{W}(t)]\, dt - \int_0^T\int_Q s^{ij}\delta\varepsilon_{ij} d\Omega dx_3 dt = 0, \tag{1.26}$$

where δ represents the Fréchet derivative with respect to an admissible displacement $\mathbf{r}$, i.e., a sufficiently smooth displacement of the form (1.5) defined on $\overline{Q}\times[0,T]$ which matches given displacements of the body at $t=0$ and $t=T$ and satisfies the geometric boundary condition (1.22).

If it is assumed that the material in question is hookean, ie., satisfies (1.8), with elasticity tensor C^{ijkl} satisfying (1.8) and (1.9), it follows from (1.17) that (1.26) may be written

$$\delta\int_0^T [\mathcal{K}(t) + \mathcal{W}(t) - \mathcal{U}(t)]dt = 0, \tag{1.27}$$

where $\mathcal{U}$ is the *strain energy functional* given by

$$\mathcal{U}(t) = \frac{1}{2}\int_Q s^{ij}\varepsilon_{ij} dQ. \tag{1.28}$$

Thus, the actual evolution of deformations is a stationary state of the *Lagrangian*

$$\mathcal{L} = \int_0^T [\mathcal{K}(t) + \mathcal{W}(t) - \mathcal{U}(t)]dt.$$

This is Hamilton's principle for our plate problem. Since $\mathbf{W}$ and $\mathbf{U}$ are independent quantities, (1.27) is equivalent to the two equations

$$\delta_{\mathbf{W}}\mathcal{L} = 0 \text{ and } \delta_{\mathbf{U}}\mathcal{L} = 0,$$

where the subscript denotes differentiation with respect to the indicated variable. The equation $\delta_{\mathbf{W}}\mathcal{L} = 0$ yields (1.11) and the first of the edge conditions in (1.20), while the second gives (1.15) and the remaining edge condition. Because of assumption (1.8) the resultants $N^{ij}, M^{ij}, I^{\alpha\beta}$ are now expressed in terms of the strains ε_{ij} and various resultants in x_3 of the coefficients of elasticity C^{ijkl}. Consequently, under assumption (1.8) equations (1.13) and (1.16) comprise six equations for the six scalar components of the displacements $\mathbf{W}$ and $\mathbf{U}$.

REMARK 1.1. One may, of course, consider geometric constraints other than, or in addition to, (1.22). for example, suppose that

$$\partial\mathcal{P} = \overline{\Lambda}^C \cup \overline{\Lambda}^D \cup \overline{\Lambda}^S,$$

where Λ^C, Λ^D are as before and Λ^S corresponds to a *simply supported* part of $\partial\mathcal{P}$, that is,

$$\mathbf{W} = 0 \quad \text{on } \Gamma^S.$$

Here Γ^S is the part of $\partial\Omega$ corresponding to Λ^S. Application of Hamilton's principle leads to the dynamic condition

$$\nu_\alpha \mathbf{M}_\alpha = \mathbf{c} \quad \text{on } \Gamma^S.$$

The complete set on edge conditions in this situation is therefore (1.20),

$$\mathbf{W} = 0, \;\; \mathbf{U} = 0 \;\; \text{on } \Gamma^C,$$
$$\mathbf{W} = 0, \;\; \nu_\alpha \mathbf{M}_\alpha = \mathbf{c} \;\; \text{on } \Gamma^S. \quad \Box$$

1.4. Additional Kinematic and Material Assumptions. In this subsection the additional kinematic requirement

$$|\mathbf{A}_3| = 1 \;\; \text{and that } \mathbf{A}_3 \cdot \mathbf{a}_3 > 0. \tag{1.29}$$

is imposed. The first condition in (1.29) is equivalent to the vanishing of the extensional strain ε_{33} and the second asserts that the x_3-line rotates through an angle of less than 90 degrees. The two conditions together imply that

$$U_3 = -1 + \sqrt{1 - |U_\alpha \mathbf{a}_\alpha|} := -1 + \sqrt{1 - |\mathbf{u}|^2} \tag{1.30}$$

where $\mathbf{u} = U_\alpha \mathbf{a}_\alpha$. Therefore

$$\mathbf{U} = \mathbf{u} + (-1 + \sqrt{1 - |\mathbf{u}|^2})\,\mathbf{a}_3. \tag{1.31}$$

The kinetic energy is written

$$\mathcal{K}(t) = \frac{1}{2}\int_\Omega [m|\mathbf{W}_{,t}|^2 + +2m_\rho(\mathbf{w}_{,t} \cdot \mathbf{u}_{,t} + W_{3,t}U_{3,t}) + I_\rho(|\mathbf{u}_{,t}|^2 + U_{3,t}^2)]\, d\Omega,$$

where $\mathbf{w} = W_\alpha \mathbf{a}_\alpha$ and U_3 given by (1.30). The work functional is expressed as

$$\mathcal{W}(t) = \int_\Omega [\mathbf{F} \cdot \mathbf{W} + (C_\mu \mathbf{a}_\mu) \cdot \mathbf{u} + C_3 U_3] \, d\Omega + \int_{\Gamma^D} [\mathbf{f} \cdot \mathbf{W} + (c_\mu \mathbf{a}_\mu) \cdot \mathbf{u} + c_3 U_3] \, d\Gamma$$

where

$$\mathbf{C} = C_i \mathbf{a}_i, \quad \mathbf{c} = c_i \mathbf{a}_i.$$

We assume the geometric constraint (1.22). The necessary conditions that $\mathcal{L}$ be stationary are

$$\delta_{\mathbf{W}} \mathcal{L} = 0, \quad \delta_{\mathbf{u}} \mathcal{L} = 0,$$

which comprise five equations for the five unknowns W_i, U_μ.

The variational equation $\delta_{\mathbf{W}} \mathcal{L} = 0$ again leads to (1.13) and (1.21) in which $\mathbf{U}$ is to be replaced in terms of $\mathbf{u}$ by means of (1.31). We obtain, therefore

$$\begin{aligned} \frac{\partial}{\partial x_\alpha} \{ N^{\alpha\beta} (\mathbf{W}_{,\beta} + \mathbf{a}_\beta) + M^{\alpha\beta} [\mathbf{u}_{,\beta} + (\sqrt{1 - |\mathbf{u}|^2})_{,\beta} \mathbf{a}_3] \\ + N^{\alpha 3} [\mathbf{u} + (\sqrt{1 - |\mathbf{u}|^2}) \mathbf{a}_3] \} + \mathbf{F} \\ = m \mathbf{W}_{,tt} + m_\rho [\mathbf{u}_{,tt} + (\sqrt{1 - |\mathbf{u}|^2})_{,tt} \mathbf{a}_3], \end{aligned} \tag{1.32}$$

$$\begin{aligned} \nu_\alpha \{ N^{\alpha\beta} (\mathbf{W}_{,\beta} + \mathbf{a}_\beta) + M^{\alpha\beta} [\mathbf{u}_{,\beta} + (\sqrt{1 - |\mathbf{u}|^2})_{,\beta} \mathbf{a}_3] \\ + N^{\alpha 3} [\mathbf{u} + (\sqrt{1 - |\mathbf{u}|^2}) \mathbf{a}_3] \} = \mathbf{f} \quad \text{on } \Gamma^D. \end{aligned} \tag{1.33}$$

In terms of components, (1.32) is expressed as

$$\left\{ \begin{aligned} \frac{\partial}{\partial x_\alpha} [N^{\alpha\beta} (W_{\mu,\beta} + \delta_{\mu\beta}) + M^{\alpha\beta} U_{\mu,\beta} + N^{\alpha 3} U_\mu] + F_\mu \\ = m W_{\mu,tt} + m_\rho U_{\mu,tt}, \\ \frac{\partial}{\partial x_\alpha} [N^{\alpha\beta} W_{3,\beta} + M^{\alpha\beta} (\sqrt{1 - |\mathbf{u}|^2})_{,\beta} + N^{\alpha 3} (\sqrt{1 - |\mathbf{u}|^2})] + F_3 \\ = m W_{3,tt} + m_\rho (\sqrt{1 - |\mathbf{u}|^2})_{,tt}, \end{aligned} \right. \tag{1.34}$$

where $\mathbf{F} = F_i \mathbf{a}_i$, while the edge conditions (1.33) are

$$\left\{ \begin{aligned} & \nu_\alpha [N^{\alpha\beta} (W_{\mu,\beta} + \delta_{\alpha\beta}) + M^{\alpha\beta} U_{\mu,\beta} + N^{\alpha 3} U_\mu] = f_\mu, \\ & \nu_\alpha [N^{\alpha\beta} (W_{3,\beta} + M^{\alpha\beta} (\sqrt{1 - |\mathbf{u}|^2})_{,\beta} \\ & \qquad + N^{\alpha 3} (\sqrt{1 - |\mathbf{u}|^2})] = f_3 \quad \text{on } \Gamma^D, \end{aligned} \right. \tag{1.35}$$

where $\mathbf{f} = f_i \mathbf{a}_i$.

Let us next calculate $\delta_{\mathbf{u}}\mathcal{L}$. One has for the first variation of kinetic energy

$$\langle \delta_{\mathbf{u}}\mathcal{K}, \hat{\mathbf{u}}\rangle = \int_\Omega [(m_\rho \mathbf{w}_{,t} + I_\rho \mathbf{u}_{,t})\cdot \hat{\mathbf{u}}_{,t} + (m_\rho W_{3,t} + I_\rho U_{3,t})\langle \delta_{\mathbf{u}} U_{3,t}, \hat{\mathbf{u}}\rangle]\, d\Omega,$$

where

$$U_{3,t} = -\frac{\mathbf{u}\cdot\mathbf{u}_{,t}}{(1-|\mathbf{u}|^2)^{1/2}},$$

$$\langle \delta_{\mathbf{u}} U_{3,t}, \hat{\mathbf{u}}\rangle = -\frac{\mathbf{u}_{,t}\cdot\hat{\mathbf{u}} + \mathbf{u}\cdot\hat{\mathbf{u}}_{,t}}{(1-|\mathbf{u}|^2)^{1/2}} - \frac{(\mathbf{u}\cdot\mathbf{u}_{,t})(\mathbf{u}\cdot\hat{\mathbf{u}})}{(1-|\mathbf{u}|^2)^{3/2}}.$$

Therefore

(1.36)
$$\langle \delta_{\mathbf{u}}\mathcal{K}, \hat{\mathbf{u}}\rangle = \int_\Omega \Bigg[(m_\rho \mathbf{w}_{,t} + I_\rho \mathbf{u}_{,t})\cdot \hat{\mathbf{u}}_{,t} + \left(I_\rho \frac{\mathbf{u}\cdot\mathbf{u}_{,t}}{(1-|\mathbf{u}|^2)^{1/2}} - m_\rho W_{3,t}\right) \times \left(\frac{\mathbf{u}_{,t}\cdot\hat{\mathbf{u}} + \mathbf{u}\cdot\hat{\mathbf{u}}_{,t}}{(1-|\mathbf{u}|^2)^{1/2}} + \frac{(\mathbf{u}\cdot\mathbf{u}_{,t})(\mathbf{u}\cdot\hat{\mathbf{u}})}{(1-|\mathbf{u}|^2)^{3/2}}\right)\Bigg]\, d\Omega.$$

Similarly,

(1.37)
$$\begin{aligned}\langle \delta_{\mathbf{u}}\mathcal{W}, \hat{\mathbf{u}}\rangle &= \int_\Omega [(C_\mu \mathbf{a}_\mu)\cdot\hat{\mathbf{u}} + C_3\langle \delta_{\mathbf{u}} U_3, \hat{\mathbf{u}}\rangle]\, d\Omega \\ &\quad + \int_{\Gamma^D} [(c_\mu \mathbf{a}_\mu)\cdot\hat{\mathbf{u}} + c_3\langle \delta_{\mathbf{u}} U_3, \hat{\mathbf{u}}\rangle]\, d\Gamma \\ &= \int_\Omega \left[(C_\mu \mathbf{a}_\mu)\cdot\hat{\mathbf{u}} - C_3\frac{\mathbf{u}\cdot\hat{\mathbf{u}}}{(1-|\mathbf{u}|^2)^{1/2}}\right] d\Omega \\ &\quad + \int_{\Gamma^D} \left[(c_\mu \mathbf{a}_\mu)\cdot\hat{\mathbf{u}} - c_3\frac{\mathbf{u}\cdot\hat{\mathbf{u}}}{(1-|\mathbf{u}|^2)^{1/2}}\right] d\Gamma.\end{aligned}$$

The variation $\delta_{\mathbf{u}}\mathcal{U}$ of the strain energy is given by

$$\delta_{\mathbf{u}}\mathcal{U} = \int_Q (s^{\alpha\beta}\delta_{\mathbf{u}}\varepsilon_{\alpha\beta} + 2s^{3\beta}\delta_{\mathbf{u}}\varepsilon_{3\beta})\, dQ.$$

One has

$$\delta_{\mathbf{u}}\varepsilon_{\alpha\beta} = \frac{1}{2}(\mathbf{G}_\alpha\cdot\delta_{\mathbf{u}}\mathbf{G}_\beta + \mathbf{G}_\beta\cdot\delta_{\mathbf{u}}\mathbf{G}_\alpha),$$
$$\delta_{\mathbf{u}}\varepsilon_{3\beta} = \frac{1}{2}(\mathbf{G}_3\cdot\delta_{\mathbf{u}}\mathbf{G}_\beta + \mathbf{G}_\beta\cdot\delta_{\mathbf{u}}\mathbf{G}_3),$$

where

$$\mathbf{G}_\beta = \mathbf{R}_{,\beta} = \mathbf{W}_{,\beta} + \mathbf{a}_\beta + x_3\mathbf{U}_{,\beta},$$
$$\mathbf{G}_3 = \mathbf{A}_3 = \mathbf{U} + \mathbf{a}_3 = \mathbf{u} + (U_3+1)\mathbf{a}_3.$$

Consequently,

$$(1.38)\quad \langle \delta_{\mathbf{u}}\mathcal{U}, \hat{\mathbf{u}}\rangle = \int_Q [s^{\alpha\beta}\mathbf{G}_\beta \cdot \langle \delta_{\mathbf{u}}\mathbf{G}_\alpha, \hat{\mathbf{u}}\rangle + s^{3\beta}\mathbf{G}_\beta \cdot \langle \delta_{\mathbf{u}}\mathbf{G}_3, \hat{\mathbf{u}}\rangle + s^{3\beta}\mathbf{G}_3 \cdot \langle \delta_{\mathbf{u}}\mathbf{G}_\alpha, \hat{\mathbf{u}}\rangle]\, dQ.$$

A straightforward calculation gives

$$\langle \delta_{\mathbf{u}}\mathbf{G}_3, \hat{\mathbf{u}}\rangle = \hat{\mathbf{u}} - \frac{\mathbf{u}\cdot\hat{\mathbf{u}}}{(1-|\mathbf{u}|^2)^{1/2}}\mathbf{a}_3,$$

$$\begin{aligned}\langle \delta_{\mathbf{u}}\mathbf{G}_\beta, \hat{\mathbf{u}}\rangle &= x_3(\hat{\mathbf{u}}_{,\beta} + \langle \delta_{\mathbf{u}}U_{3,\beta}, \hat{\mathbf{u}}\rangle)\\ &= x_3\left\{\hat{\mathbf{u}}_{,\beta} - \left[\frac{\mathbf{u}_{,\beta}\cdot\hat{\mathbf{u}} + \mathbf{u}\cdot\hat{\mathbf{u}}_{,\beta}}{(1-|\mathbf{u}|^2)^{1/2}} + \frac{(\mathbf{u}\cdot\mathbf{u}_{,\beta})(\mathbf{u}\cdot\hat{\mathbf{u}})}{(1-|\mathbf{u}|^2)^{3/2}}\right]\mathbf{a}_3\right\},\end{aligned}$$

hence

$$\begin{aligned}\mathbf{G}_\beta \cdot \langle \delta_{\mathbf{u}}\mathbf{G}_\alpha, \hat{\mathbf{u}}\rangle =& x_3(\mathbf{W}_{,\beta} + \mathbf{a}_\beta + x_3\mathbf{U}_{,\beta})\\ &\cdot\left\{\hat{\mathbf{u}}_{,\alpha} - \left[\frac{\mathbf{u}_{,\alpha}\cdot\hat{\mathbf{u}} + \mathbf{u}\cdot\hat{\mathbf{u}}_{,\alpha}}{(1-|\mathbf{u}|^2)^{1/2}} + \frac{(\mathbf{u}\cdot\mathbf{u}_{,\alpha})(\mathbf{u}\cdot\hat{\mathbf{u}})}{(1-|\mathbf{u}|^2)^{3/2}}\right]\mathbf{a}_3\right\}\\ =& x_3(\mathbf{w}_{,\beta} + \mathbf{a}_\beta + x_3\mathbf{u}_{,\beta})\cdot\hat{\mathbf{u}}_{,\alpha}\\ &- x_3(W_{3,\beta} + x_3U_{3,\beta})\left[\frac{\mathbf{u}_{,\alpha}\cdot\hat{\mathbf{u}} + \mathbf{u}\cdot\hat{\mathbf{u}}_{,\alpha}}{(1-|\mathbf{u}|^2)^{1/2}}\right.\\ &\left. + \frac{(\mathbf{u}\cdot\mathbf{u}_{,\alpha})(\mathbf{u}\cdot\hat{\mathbf{u}})}{(1-|\mathbf{u}|^2)^{3/2}}\right]\end{aligned}$$

Therefore

$$(1.39)\quad \int_{-h_1}^{h_2} s^{\alpha\beta}\mathbf{G}_\beta\cdot\langle \delta_{\mathbf{u}}\mathbf{G}_\alpha, \hat{\mathbf{u}}\rangle\, dx_3 = [M^{\alpha\beta}(\mathbf{w}_\beta + \mathbf{a}_\beta) + I^{\alpha\beta}\mathbf{u}_\beta]\cdot\hat{\mathbf{u}}_{,\alpha} - (M^{\alpha\beta}W_{3,\beta} + I^{\alpha\beta}U_{3,\beta})\left[\frac{\mathbf{u}_{,\alpha}\cdot\hat{\mathbf{u}} + \mathbf{u}\cdot\hat{\mathbf{u}}_{,\alpha}}{(1-|\mathbf{u}|^2)^{1/2}} + \frac{(\mathbf{u}\cdot\mathbf{u}_{,\alpha})(\mathbf{u}\cdot\hat{\mathbf{u}})}{(1-|\mathbf{u}|^2)^{3/2}}\right].$$

Similarly,

$$(1.40)\quad \int_{-h_1}^{h_2} s^{3\beta}\mathbf{G}_\beta\cdot\langle \delta_{\mathbf{u}}\mathbf{G}_3, \hat{\mathbf{u}}\rangle\, dx_3 = [N^{3\beta}(\mathbf{w}_\beta + \mathbf{a}_\beta) + M^{3\beta}\mathbf{u}_\beta]\cdot\hat{\mathbf{u}}_{,\alpha} - (N^{3\beta}W_{3,\beta} + M^{3\beta}U_{3,\beta})\frac{\mathbf{u}\cdot\hat{\mathbf{u}}}{(1-|\mathbf{u}|^2)^{1/2}},$$

and

$$(1.41)\quad \int_{-h_1}^{h_2} s^{3\beta}\mathbf{G}_3\cdot\langle \delta_{\mathbf{u}}\mathbf{G}_\beta, \hat{\mathbf{u}}\rangle\, dx_3 = M^{3\beta}\mathbf{u}\cdot\hat{\mathbf{u}}_{,\beta} - M^{3\beta}(U_3+1)\left[\frac{\mathbf{u}_{,\alpha}\cdot\hat{\mathbf{u}} + \mathbf{u}\cdot\hat{\mathbf{u}}_{,\alpha}}{(1-|\mathbf{u}|^2)^{1/2}} + \frac{(\mathbf{u}\cdot\mathbf{u}_{,\alpha})(\mathbf{u}\cdot\hat{\mathbf{u}})}{(1-|\mathbf{u}|^2)^{3/2}}\right].$$

It follows from (1.36)–(1.41) that the variational equation $\delta_{\mathbf{u}}\mathcal{L} = 0$ may be written

$$\begin{aligned}0 = \int_0^T \int_\Omega \{[(m_\rho \mathbf{w}_{,t} + I_\rho \mathbf{u}_{,t}) \cdot \hat{\mathbf{u}}_{,t} + \mathcal{N}_1(W_3, \mathbf{u})) \cdot \hat{\mathbf{u}}_{,t} + \mathcal{N}_2(W_3, \mathbf{u}) \cdot \hat{\mathbf{u}}]\\ - [M^{\alpha\beta}(\mathbf{w}_{,\beta} + \mathbf{a}_\beta) + I^{\alpha\beta}\mathbf{u}_{,\beta} + M^{3\alpha}\mathbf{u} - \mathcal{M}_\alpha(\mathbf{W}, \mathbf{u})] \cdot \hat{\mathbf{u}}_{,\alpha}\\ - [N^{3\beta}(\mathbf{w}_{,\beta} + \mathbf{a}_\beta) + M^{3\beta}\mathbf{u}_{,\beta} - \mathcal{N}_3(\mathbf{W}, \mathbf{u})] \cdot \hat{\mathbf{u}}\} d\Omega dt\\ + \int_0^T \int_\Omega (C_\mu \mathbf{a}_\mu - C_3 \mathcal{N}_4(\mathbf{u})) \cdot \hat{\mathbf{u}}\, d\Omega dt\\ + \int_0^T \int_{\Gamma^D} (c_\mu \mathbf{a}_\mu - c_3 \mathcal{N}_4(\mathbf{u})) \cdot \hat{\mathbf{u}}\, d\Omega dt,\end{aligned}$$

where $\mathcal{N}_1, \dots, \mathcal{N}_4, \mathcal{M}_\alpha$ are nonlinear partial differential operators defined as follows:

$$\mathcal{N}_1(W_3, \mathbf{u},) = \left(I_\rho \frac{\mathbf{u} \cdot \mathbf{u}_{,t}}{(1 - |\mathbf{u}|^2)^{1/2}} - m_\rho W_{3,t}\right) \frac{\mathbf{u}}{(1 - |\mathbf{u}|^2)^{1/2}},$$

$$\begin{aligned}\mathcal{N}_2(W_3, \mathbf{u}) = \left(I_\rho \frac{\mathbf{u} \cdot \mathbf{u}_{,t}}{(1 - |\mathbf{u}|^2)^{1/2}} - m_\rho W_{3,t}\right)\\ \times \left(\frac{\mathbf{u}_{,t}}{(1 - |\mathbf{u}|^2)^{1/2}} + \frac{(\mathbf{u} \cdot \mathbf{u}_{,t})\mathbf{u}}{(1 - |\mathbf{u}|^2)^{3/2}}\right),\end{aligned}$$

$$\begin{aligned}\mathcal{N}_3(\mathbf{W}, \mathbf{u}) = [M^{\alpha\beta}W_{3,\beta} + I^{\alpha\beta}U_{3,\beta} + M^{3\alpha}(U_3 + 1)] \Bigg[\frac{\mathbf{u}_{,\alpha}}{(1 - |\mathbf{u}|^2)^{1/2}}\\ + \frac{\mathbf{u} \cdot \mathbf{u}_{,\alpha}}{(1 - |\mathbf{u}|^2)^{3/2}}\mathbf{u}\Bigg] + (N^{3\beta}W_{3,\beta} + M^{3\beta}U_{3,\beta})\frac{\mathbf{u}}{(1 - |\mathbf{u}|^2)^{1/2}},\end{aligned}$$

$$\mathcal{N}_4(\mathbf{u}) = \frac{\mathbf{u}}{1 - |\mathbf{u}|^2},$$

$$\mathcal{M}_\alpha(\mathbf{W}, \mathbf{u}) = [M^{\alpha\beta}W_{3,\beta} + I^{\alpha\beta}U_{3,\beta} + M^{3\alpha}(U_3 + 1)]\frac{\mathbf{u}}{(1 - |\mathbf{u}|^2)^{1/2}}.$$

In all of these expressions, U_3 is given by (1.30). The corresponding equations of motion and edge conditions are, respectively,

$$\begin{aligned}(1.42)\quad \frac{\partial}{\partial x_\alpha}[M^{\alpha\beta}(\mathbf{w}_{,\beta} + \mathbf{a}_\beta) + I^{\alpha\beta}\mathbf{u}_{,\beta} + M^{3\alpha}\mathbf{u} - \mathcal{M}_\alpha(\mathbf{W}, \mathbf{u})]\\ - [N^{3\beta}(\mathbf{w}_{,\beta} + \mathbf{a}_\beta) + M^{3\beta}\mathbf{u}_{,\beta} - \mathcal{N}_3(\mathbf{W}, \mathbf{u})] + C_\mu \mathbf{a}_\mu - C_3 \mathcal{N}_4(\mathbf{u})\\ = \frac{\partial}{\partial t}[m_\rho \mathbf{w}_{,t} + I_\rho \mathbf{u}_{,t} + \mathcal{N}_1(W_3, \mathbf{u})] - \mathcal{N}_2(W_3, \mathbf{u}),\end{aligned}$$

and

$$(1.43)\quad \nu_\alpha[M^{\alpha\beta}(\mathbf{w}_{,\beta}+\mathbf{a}_\beta)+I^{\alpha\beta}\mathbf{u}_{,\beta}+M^{3\alpha}\mathbf{u}-\mathcal{M}_\alpha(\mathbf{u})] = c_\mu\mathbf{a}_\mu - c_3\mathcal{N}_4(\mathbf{u}) \text{ on } \Gamma^D.$$

The complete set of partial differential equations and edge conditions for the functions $\mathbf{W}$ and $\mathbf{u}$ are equations (1.32) and (1.42), edge conditions (1.33) and (1.43), and the geometric boundary conditions

$$(1.44)\qquad \mathbf{W}=0,\quad \mathbf{u}=0 \text{ on } \Gamma^C.$$

REMARK 1.2. In addition to (1.29), a kinematic assumption that is standard in thin plate theory is the *Kirchhoff hypothesis* that $\mathbf{A}_3$ be normal the the deformed reference surface. Then

$$\mathbf{A}_3=\frac{\mathbf{A}_1\times\mathbf{A}_2}{|\mathbf{A}_1\times\mathbf{A}_2|}=\frac{(\mathbf{W}_{,1}+\mathbf{a}_1)\times(\mathbf{W}_{,2}+\mathbf{a}_2)}{|(\mathbf{W}_{,1}+\mathbf{a}_1)\times(\mathbf{W}_{,2}+\mathbf{a}_2)|},$$

so that $\mathbf{U}=\mathbf{A}_3-\mathbf{a}_3$ is expressed in terms of $\mathbf{W}_{,\alpha}$. The kinetic energy therefore involves the derivatives $\mathbf{W}_{,t}$ and $\mathbf{W}_{,t\alpha}$ while the strain energy contains spatial derivatives of $\mathbf{W}$ up to second order. The geometric boundary condition $\mathbf{u}=0$ in (1.44) becomes

$$W_{2,1}W_{3,2}-W_{2,2}W_{3,1}=W_{3,1},\quad W_{1,2}W_{3,1}-W_{1,1}W_{3,2}=W_{3,2} \text{ on } \Gamma^C.$$

When Hamilton's principle is applied in this context, the only variation to be taken is with respect to $\mathbf{W}$. The result is a nonlinear system of three scalar equations for the components of $\mathbf{W}$ which involves spatial derivatives of $\mathbf{W}$ up to fourth order and of $\mathbf{W}_{,t}$ up to second order, together with appropriate edge conditions on Γ^D. When $h_1=h_2\equiv$ const., the linearization of this system in the case of an homogeneous, elastically isotropic material is the *Kirchhoff plate system;* another specialization of it is the *von Kármán system* (see, e.g., [51, Chapter I] and Section 2.5 below.) □

1.5. Rotations Associated with Plate Deformation. The kinematic assumption for this section is (1.1). Recall that the vectors $\mathbf{A}_\alpha$ are tangent to the deformed reference surface. Let us set

$$(1.45)\qquad \mathbf{t}_\alpha=\frac{\mathbf{A}_\alpha}{|\mathbf{A}_\alpha|}=\frac{\mathbf{a}_\alpha+\mathbf{W}_{i,\alpha}\mathbf{a}_i}{|\mathbf{a}_\alpha+\mathbf{W}_{i,\alpha}\mathbf{a}_i|},\quad \mathbf{n}=\frac{\mathbf{A}_1\times\mathbf{A}_2}{|\mathbf{A}_1\times\mathbf{A}_2|},$$

so that $\mathbf{t}_\alpha$ is a unit tangent vector and $\mathbf{n}$ a unit normal vector to the deformed reference surface. Introduce a right-handed orthonormal basis $\mathbf{A}'_i$ such that

$$(1.46)\qquad \mathbf{A}'_3=\mathbf{n},\quad \mathbf{A}'_1\cdot\mathbf{A}_2=\mathbf{A}'_2\cdot\mathbf{A}_1,\quad \mathbf{A}'_1\cdot\mathbf{A}_1>0.$$

These conditions uniquely determine the $\mathbf{A}'_i$ basis.

The $\mathbf{A}'_i$ basis may be obtained from the $\mathbf{a}_i$ basis through an orthogonal transformation. This transformation may be considered to consist of three separate rotations: a rotation through an angle ψ_1 around the $\mathbf{a}_1$-axis, followed by a rotation through an angle ψ_2 around the rotated $\mathbf{a}_2$-axis, followed by a rotation through an angle ψ_3 around the twice rotated $\mathbf{a}_3$-axis. The angles ψ_1, ψ_2 are associated with plate bending, while ψ_3 is associated with twisting of the reference surface. We choose the positive direction of rotation to be counterclockwise. One then has

$$\tan\psi_1 = -\frac{\mathbf{n}\cdot\mathbf{a}_2}{\mathbf{n}\cdot\mathbf{a}_3} = \frac{W_{3,2}(W_{1,1}+1) - W_{1,2}W_{3,1}}{(W_{1,1}+1)(W_{2,2}+1) - W_{1,2}W_{2,1}}, \tag{1.47}$$

$$\tan\psi_2 = -\frac{\mathbf{n}\cdot\mathbf{a}_1}{\mathbf{n}\cdot\mathbf{a}_3} = \frac{W_{3,1}(W_{2,2}+1) - W_{2,1}W_{3,2}}{(W_{1,1}+1)(W_{2,2}+1) - W_{1,2}W_{2,1}}. \tag{1.48}$$

In order to express ψ_3 in terms of displacements, we note that the initial rotation through ψ_1 around the $\mathbf{a}_1$-axis maps $\mathbf{a}_2$ to the vector $\mathbf{a}_2'$, where

$$\mathbf{a}_2' = (\cos\psi_1)\mathbf{a}_2 + (\sin\psi_1)\mathbf{a}_3, \tag{1.49}$$

while the rotation through ψ_2 about $\mathbf{a}_2'$ maps $\mathbf{a}_1$ to $\mathbf{a}_1'$, where

$$\mathbf{a}_1' = (\cos\psi_2)\mathbf{a}_1 + (\sin\psi_1\sin\psi_2)\mathbf{a}_2 - (\cos\psi_1\sin\psi_2)\mathbf{a}_3. \tag{1.50}$$

The image of $\mathbf{a}_3$ after these two rotations is $\mathbf{A}_3'$, and $\mathbf{a}_1', \mathbf{a}_2'$ are in the plane spanned by $\mathbf{t}_1, \mathbf{t}_2$. The final rotation through ψ_3 maps the orthonormal vectors $\mathbf{a}_1', \mathbf{a}_2'$ onto $\mathbf{A}_1', \mathbf{A}_2'$, respectively. A little geometry reveals that ψ_3 is the average of two rotations θ_1 and θ_2 about $\mathbf{A}_3'$, the first mapping $\mathbf{a}_1'$ to $\mathbf{t}_1$ and the second mapping $\mathbf{a}_2'$ to $\mathbf{t}_2$. Consequently,

$$\sin^2\psi_3 = \frac{1}{2}(1 - \cos\theta_1\cos\theta_2 + \sin\theta_1\sin\theta_2),$$

where

$$\cos\theta_\alpha = \mathbf{a}_\alpha'\cdot\mathbf{t}_\alpha, \quad \sin\theta_1 = \pm\mathbf{a}_2'\cdot\mathbf{t}_1, \quad \sin\theta_2 = \mp\mathbf{a}_1'\cdot\mathbf{t}_2.$$

Thus

$$\sin^2\psi_3 = \frac{1}{2}[1 - (\mathbf{a}_1'\cdot\mathbf{t}_1)(\mathbf{a}_2'\cdot\mathbf{t}_2) - (\mathbf{a}_1'\cdot\mathbf{t}_2)(\mathbf{a}_2'\cdot\mathbf{t}_1)]. \tag{1.51}$$

In view of (1.45), (1.49) and (1.50), (1.51) expresses ψ_3 in terms of the displacements, but the relationship is obviously very complicated in general.

We next consider the rotation angles ϕ_1 around the $\mathbf{a}_1$-axis and ϕ_2 around the $\mathbf{a}_2$-axis needed to map $\mathbf{a}_3$ to a unit vector along $\mathbf{A}_3$. Since $\mathbf{A}_3 = U_\alpha\mathbf{a}_\alpha + (U_3+1)\mathbf{a}_3$, one has

$$\tan\phi_1 = -\frac{U_2}{1+U_3}, \quad \tan\phi_2 = \frac{U_1}{1+U_3}. \tag{1.52}$$

The differences $\phi_\alpha - \psi_\alpha$ are the *transverse shear angles*.

We now define the *vector rotation angle* associated with the deformation of the plate as

$$\mathbf{\Phi}(x_1, x_2, t) = \phi_\alpha(x_1, x_2, t)\mathbf{a}_\alpha + \psi_3(x_1, x_2, t)\mathbf{a}_3. \tag{1.53}$$

The part $\phi_\alpha \mathbf{a}_\alpha$ measures rotations due to bending and transverse shearing in the plate, while $\psi_3 \mathbf{a}_3$ is a measure of rotation due to in-plane twisting of the plate. Equations (1.45), (1.47), (1.48), (1.51)–(1.53) give the (nonlinear) relationship between the displacements $\mathbf{W}, \mathbf{U}$ and the vector rotation angle $\mathbf{\Phi}$. Let us note that when (x_1, x_2) is a point on $\partial\Omega$ we may express $\phi_\alpha \mathbf{a}_\alpha$ in terms of the normal vector $\boldsymbol{\nu} = \nu_1 \mathbf{a}_1 + \nu_2 \mathbf{a}_2$ and the tangent vector $\boldsymbol{\tau} = -\nu_2 \mathbf{a}_1 + \nu_1 \mathbf{a}_2$, in which case (1.53) takes the form

$$\mathbf{\Phi} = (\boldsymbol{\nu} \cdot \boldsymbol{\phi})\boldsymbol{\nu} + (\boldsymbol{\tau} \cdot \boldsymbol{\phi})\boldsymbol{\tau} + \psi_3 \mathbf{a}_3, \tag{1.54}$$

where $\boldsymbol{\phi} = \phi_\alpha \mathbf{a}_\alpha$.

2. Linearization

The kinematic conditions (1.1) and (1.29) are assumed. It is further assumed that

$$h_1(x_1, x_2) = h_2(x_1, x_2) \equiv \frac{h}{2} = \text{const.},$$

and that

$$\rho(x_1, x_2, x_3) = \rho(x_1, x_2, -x_3), \quad |x_3| \le h/2.$$

It follows, in particular, that $m_\rho = 0$.

2.1. Linearization of Equations of Motion. We are going to linearize (1.32) and (1.42) around the equilibrium $W_i = U_\alpha = 0$. It is apparent from the structure of these equations that the linearization of this system about zero is the same as the linearization of the system

$$\begin{cases} N^{\alpha j}_{,\alpha} + F_j = m W_{j,tt}, \\ M^{\alpha\beta}_{,\alpha} - N^{3\beta} + C_\beta = I_\rho U_{\beta,tt}, \end{cases} \tag{2.1}$$

since all of the omitted terms are at least of second order in the displacements.

In order to calculate the generalized stresses $N^{\alpha j}$ and generalized moments $M^{\alpha\beta}$ in terms of the strains, a specific stress-strain relation will be assumed. We set, following Reissner [88],

$$\begin{cases} \varepsilon_{11} = \dfrac{1}{E}(s^{11} - \sigma s^{22}) - \dfrac{\bar\sigma}{\bar E} s^{33}, \\ \varepsilon_{22} = \dfrac{1}{E}(s^{22} - \sigma s^{11}) - \dfrac{\bar\sigma}{\bar E} s^{33}, \\ \varepsilon_{33} = \dfrac{1}{\bar E}(s^{33} - \bar\sigma s^{\alpha\alpha}), \\ \varepsilon_{12} = \dfrac{1+\sigma}{E} s^{12}, \quad \varepsilon_{13} = \dfrac{1}{2G} s^{13}, \quad \varepsilon_{23} = \dfrac{1}{2G} s^{23}. \end{cases} \tag{2.2}$$

The quantities $E, \bar E, \sigma, \bar\sigma, G$ are positive constants which satisfy

$$0 < \sigma < 1/2, \quad \bar E(1-\sigma) - E\bar\sigma > 0, \quad \bar E(1-\sigma) - 2E\bar\sigma^2 > 0.$$

When $E = \bar{E}$ and $\sigma = \bar{\sigma}$, (2.2) are the strain-stress relations of an isotropic material. Otherwise, they indicate that the material is isotropic in directions parallel to the reference plane but have different material properties in the transverse direction.

To calculate the generalized stresses and moments in terms of displacements it is necessary to solve (2.2) for s^{ij} in terms of ε_{kl}. We introduce

$$\hat{\mu} = \frac{\bar{E}[\bar{E}(1-\sigma) - E\bar{\sigma}]}{2[\bar{E}(1-\sigma) - 2E\bar{\sigma}^2]}, \quad \bar{\lambda} = \frac{E\bar{E}\bar{\sigma}}{\bar{E}(1-\sigma) - 2E\bar{\sigma}^2},$$

and further define η and $\bar{\mu}$ through the equations

$$\frac{E}{1-\sigma^2}\left[1 + \frac{\bar{\lambda}\bar{\sigma}(1+\sigma)}{\bar{E}}\right] = 2\bar{\mu} + \bar{\lambda},$$
$$\frac{E}{1-\sigma^2}\left[\sigma + \frac{\bar{\lambda}\bar{\sigma}(1+\sigma)}{\bar{E}}\right] = 2\eta + \bar{\lambda}.$$

Note that $\bar{\mu} > \eta$ since $\sigma < 1$. With these definitions it may be verified that

$$\begin{cases} s^{11} = 2\bar{\mu}\varepsilon_{11} + 2\eta\varepsilon_{22} + \bar{\lambda}\varepsilon_{kk}, \\ s^{22} = 2\eta\varepsilon_{11} + 2\bar{\mu}\varepsilon_{22} + \bar{\lambda}\varepsilon_{kk}, \\ s^{33} = 2\hat{\mu}\varepsilon_{33} + \bar{\lambda}\varepsilon_{kk}, \\ s^{12} = 2(\bar{\mu} - \eta)\varepsilon_{12}, \quad s^{13} = 2G\varepsilon_{13}, \quad s^{23} = 2G\varepsilon_{23}. \end{cases}$$

Under the further assumption $\varepsilon_{33} = 0$ which is in effect, these reduce to

$$\text{(2.3)} \qquad \begin{cases} s^{11} = (2\mu + \lambda)\varepsilon_{11} + \lambda\varepsilon_{22}, \\ s^{22} = \lambda\varepsilon_{11} + (2\mu + \lambda)\varepsilon_{22}, \\ s^{33} = \bar{\lambda}\varepsilon_{\beta\beta}, \\ s^{12} = 2\mu\varepsilon_{12}, \quad s^{13} = 2G\varepsilon_{13}, \quad s^{23} = 2G\varepsilon_{23}, \end{cases}$$

where

$$\mu = \bar{\mu} - \eta, \quad \lambda = \bar{\lambda} + 2\eta.$$

REMARK 2.1. When $E = \bar{E}$ and $\sigma = \bar{\sigma}$, then $\eta = 0$, $\hat{\mu} = \bar{\mu} = \mu$, $\bar{\lambda} = \lambda$ and λ, μ reduce to the Lamé parameters

$$\lambda = \frac{E\sigma}{(1+\sigma)(1-2\sigma)}, \quad \mu = \frac{E}{2(1+\sigma)}. \quad \square$$

From (1.3) one has

$$\text{(2.4)} \qquad \int_{-h/2}^{h/2} \varepsilon_{\alpha\beta} dx_3 = \frac{h}{2}(W_{\alpha,\beta} + W_{\beta,\alpha} + W_{i,\alpha}W_{i,\beta}) + \frac{h^3}{24}U_{i,\alpha}U_{i,\beta},$$

$$\text{(2.5)} \qquad \int_{-h/2}^{h/2} \varepsilon_{\alpha 3} dx_3 = \frac{h}{2}[(U_\alpha + W_{3,\alpha}) + U_i W_{i,\alpha}],$$

where U_3 is given by (1.30). It follows from (2.3)–(2.5) that the generalized stresses are

$$N^{11} = h(2\mu+\lambda)(W_{1,1} + \frac{1}{2}W_{i,1}W_{i,1} + \frac{h^2}{24}U_{i,1}U_{i,1}) + h\lambda(W_{2,2} + \frac{1}{2}W_{i,2}W_{i,2} + \frac{h^2}{24}U_{i,2}U_{i,2}), \tag{2.6}$$

$$N^{22} = h(2\mu+\lambda)(W_{2,2} + \frac{1}{2}W_{i,2}W_{i,2} + \frac{h^2}{24}U_{i,2}U_{i,2}) + h\lambda(W_{1,1} + \frac{1}{2}W_{i,1}W_{i,1} + \frac{h^2}{24}U_{i,1}U_{i,1}), \tag{2.7}$$

$$N^{12} = h\mu(W_{1,2} + W_{2,1} + W_{i,2}W_{i,2} + \frac{h^2}{24}U_{i,1}U_{i,2}), \tag{2.8}$$

$$N^{\alpha 3} = Gh(U_\alpha + W_{3,\alpha} + U_i W_{i,\alpha}). \tag{2.9}$$

Similarly, the generalized moments $M^{\alpha\beta}$ are given by

$$M^{11} = \frac{h^3}{12}[(2\mu+\lambda)(U_{1,1} + U_{i,1}W_{i,1}) + \lambda(U_{2,2} + U_{i,2}W_{i,2})], \tag{2.10}$$

$$M^{22} = \frac{h^3}{12}[(2\mu+\lambda)(U_{2,2} + U_{i,2}W_{i,2}) + \lambda(U_{1,1} + U_{i,1}W_{i,1})], \tag{2.11}$$

$$M^{12} = \frac{h^3}{12}\mu(U_{1,2} + U_{2,1} + U_{i,2}W_{i,1} + U_{i,1}W_{i,2}). \tag{2.12}$$

The *linearized* versions of (2.6)–(2.12) are

$$\begin{cases} N^{11} = h[(2\mu+\lambda)W_{1,1} + \lambda W_{2,2}], \\ N^{22} = h[(2\mu+\lambda)W_{2,2} + \lambda W_{1,1}], \\ N^{12} = h\mu(W_{1,2} + W_{2,1}), \\ N^{\alpha 3} = Gh(U_\alpha + W_{3,\alpha}), \end{cases} \tag{2.13}$$

$$\begin{cases} M^{11} = \dfrac{h^3}{12}[(2\mu+\lambda)U_{1,1} + \lambda U_{2,2}], \\ M^{22} = \dfrac{h^3}{12}[(2\mu+\lambda)U_{2,2} + \lambda U_{1,1}], \\ M^{12} = \dfrac{h^3}{12}\mu(U_{1,2} + U_{2,1}). \end{cases} \tag{2.14}$$

The linearized equations of motion are therefore (2.1) with $N^{\alpha j}$ and $M^{\alpha\beta}$ given by (2.13) and (2.14), respectively. This system is sometimes referred to as the *Reissner* model, or the *Reissner-Mindlin* model. Let us rewrite it in a different form. Let $\mathcal{S}$ denote the set of 2×2 symmetric tensors. We define a mapping $C : \mathcal{S} \mapsto \mathcal{S}$ by

$$C[\varepsilon] = h[\lambda(\varepsilon_{\alpha\alpha})\mathcal{I} + 2\mu\varepsilon], \quad \varepsilon \in \mathcal{S}, \tag{2.15}$$

where $\mathcal{I}$ is the identity in $\mathcal{S}$. Note that

$$C^{\alpha\beta}[\varepsilon]\hat{\varepsilon}_{\alpha\beta} = h[\lambda\varepsilon_{\alpha\alpha}\hat{\varepsilon}_{\beta\beta} + 2\mu\varepsilon_{\alpha\beta}\hat{\varepsilon}_{\alpha\beta}], \quad \forall\varepsilon, \hat{\varepsilon} \in \mathcal{S}, \tag{2.16}$$

where $C[\varepsilon] := (C^{\alpha\beta}[\varepsilon])$. For any differentiable function $\mathbf{w} = w_\alpha \mathbf{a}_\alpha : \overline{\Omega} \mapsto \mathbb{R}^2$ define $\varepsilon(\mathbf{w}) \in \mathcal{S}$ by

$$\varepsilon_{\alpha\beta}(\mathbf{w}) = \frac{1}{2}(w_{\alpha,\beta} + w_{\beta,\alpha}), \quad \varepsilon(\mathbf{w}) = (\varepsilon_{\alpha\beta}(\mathbf{w})). \tag{2.17}$$

Then we see that

$$N^{\alpha\beta} = C^{\alpha\beta}[\varepsilon(\mathbf{w})], \quad M^{\alpha\beta} = C^{\alpha\beta}[\varepsilon(\mathbf{u})].$$

Thus the Reissner model may be expressed as

$$\frac{\partial}{\partial x_\alpha} C^{\alpha\beta}[\varepsilon(\mathbf{w})] + F_\beta = mW_{\beta,tt}, \tag{2.18}$$

$$\begin{cases} Gh\dfrac{\partial}{\partial x_\alpha}(U_\alpha + W_{3,\alpha}) + F_3 = mW_{3,tt}, \\ \dfrac{h^2}{12}\dfrac{\partial}{\partial x_\alpha} C^{\alpha\beta}[\varepsilon(\mathbf{u})] - Gh(U_\beta + W_{3,\beta}) + C_\beta = I_\rho U_{\beta,tt}. \end{cases} \tag{2.19}$$

2.2. Linearization of Edge Conditions. The same argument as in the last subsection shows that the linearization of the edge conditions (1.33), (1.43) leads to

$$\nu_\alpha N^{\alpha j} = f_j, \quad \nu_\alpha M^{\alpha\beta} = c_\beta \ \text{ on } \Gamma^D,$$

where $N^{\alpha j}$ and $M^{\alpha\beta}$ are given by (2.13) and (2.14), respectively. In terms of the matrix C these boundary conditions read

$$\nu_\alpha C^{\alpha\beta}[\varepsilon(\mathbf{w})] = f_\beta \ \text{ on } \Gamma^D, \tag{2.20}$$

$$\begin{cases} Gh\nu_\alpha(U_\alpha + W_{3,\alpha}) = f_3, \\ \dfrac{h^2}{12}\nu_\alpha C^{\alpha\beta}[\varepsilon(\mathbf{u})] = c_\beta \ \text{ on } \Gamma^D. \end{cases} \tag{2.21}$$

Of course, one retains the geometric boundary condition

$$W_i = 0, \quad U_\alpha = 0 \ \text{ on } \Gamma^C. \tag{2.22}$$

It is seen that $\mathbf{w} = W_\alpha \mathbf{a}_\alpha$ is not coupled to $W_3, \mathbf{u} = U_\alpha \mathbf{a}_\alpha$ in either the equations of motion or the boundary conditions.

2.3. Hamilton's Principle for the Reissner Model. It is easily verified that the system (2.18)–(2.21) is equivalent to the variational principle

$$\delta \int_0^T [\mathcal{K}(t) + \mathcal{W}(t) - \mathcal{U}(t)]\, dt = 0, \tag{2.23}$$

where

$$\mathcal{K}(t) = \frac{1}{2} \int_\Omega [m(|\mathbf{w}_{,t}|^2 + W_{3,t}^2) + I_\rho |\mathbf{u}_{,t}|^2]\, d\Omega, \tag{2.24}$$

$$\mathcal{W}(t) = \int_\Omega [F_j W_j + C_\beta U_\beta]\, d\Omega + \int_{\Gamma_i} [f_j W_j + c_\beta U_\beta]\, d\Gamma, \tag{2.25}$$

$$\mathcal{U}(t) = \frac{1}{2} \int_\Omega \{C^{\alpha\beta}[\varepsilon(\mathbf{w})]\varepsilon_{\alpha\beta}(\mathbf{w}) + \frac{h^2}{12} C^{\alpha\beta}[\varepsilon(\mathbf{u})]\varepsilon_{\alpha\beta}(\mathbf{u}) + Gh|\mathbf{u} + \nabla W_3|^2\} d\Omega, \tag{2.26}$$

where $\nabla W_3 = W_{3,\alpha}\mathbf{a}_\alpha$. In (2.23) the variation is taken with respect to sufficiently smooth functions which satisfy the geometric boundary condition (2.22).

2.4. Linearization of the Vector Rotation Angle. We seek the linearization of the vector rotation angle (1.53). It follows immediately from (1.52) that ϕ_α linearizes to

$$\phi_1 = -U_2, \quad \phi_2 = U_1. \tag{2.27}$$

To linearize (1.51) we retain up to quadratic terms on each side of that equation. The approximations of $\mathbf{a}_1'$ and $\mathbf{a}_2'$ are

$$\mathbf{a}_1' = [1 - (1/2)\psi_2^2]\mathbf{a}_1 + \psi_1\psi_2\mathbf{a}_2 - \psi_2\mathbf{a}_3,$$
$$\mathbf{a}_2' = [1 - (1/2)\psi_1^2]\mathbf{a}_2 + \psi_1\mathbf{a}_3.$$

We also have

$$|\mathbf{a}_1 + W_{i,1}\mathbf{a}_i|^{-1} = 1 - W_{1,1} - \frac{1}{2}|\mathbf{W}_{,1}|^2 + \frac{3}{2}W_{1,1}^2 + \cdots,$$
$$|\mathbf{a}_2 + W_{i,2}\mathbf{a}_i|^{-1} = 1 - W_{2,2} - \frac{1}{2}|\mathbf{W}_{,2}|^2 + \frac{3}{2}W_{2,2}^2 + \cdots,$$

and therefore

$$\mathbf{a}_1' \cdot \mathbf{t}_1 = \frac{\mathbf{a}_1' \cdot (\mathbf{a}_1 + W_{i,1}\mathbf{a}_i)}{|\mathbf{a}_1 + W_{i,1}\mathbf{a}_i|}$$
$$= 1 - \frac{1}{2}|\mathbf{W}_{,1}|^2 + \frac{1}{2}W_{1,1}^2 - \frac{1}{2}\psi_2^2 - \psi_2 W_{3,1} + \cdots,$$

$$\mathbf{a}_2' \cdot \mathbf{t}_2 = 1 - \frac{1}{2}|\mathbf{W}_{,2}|^2 + \frac{1}{2}W_{2,2}^2 - \frac{1}{2}\psi_1^2 + \psi_1 W_{3,2} + \cdots,$$
$$\mathbf{a}_1' \cdot \mathbf{t}_2 = W_{1,2} - W_{1,2}W_{2,2} + \psi_1\psi_2 - \psi_2 W_{3,2} + \cdots,$$
$$\mathbf{a}_2' \cdot \mathbf{t}_1 = W_{1,2} - W_{2,1}W_{1,1} + \psi_1 W_{3,1} + \cdots.$$

It follows that

$$(\mathbf{a}_1' \cdot \mathbf{t}_1)(\mathbf{a}_2' \cdot \mathbf{t}_2) + (\mathbf{a}_1' \cdot \mathbf{t}_2)(\mathbf{a}_2' \cdot \mathbf{t}_1) = 1 + W_{1,2}W_{2,1}$$
$$- \frac{1}{2}(W_{2,1}^2 + W_{1,2}^2 + W_{3,1}^2 + W_{3,2}^2 + \psi_1^2 + \psi_2^2) + \psi_1 W_{3,2} - \psi_2 W_{3,1} + \cdots.$$

From (1.47) and (1.48) we have

$$\psi_1 = W_{3,2} + \cdots, \quad \psi_2 = -W_{3,1} + \cdots,$$

so that

$$(\mathbf{a}_1' \cdot \mathbf{t}_1)(\mathbf{a}_2' \cdot \mathbf{t}_2) + (\mathbf{a}_1' \cdot \mathbf{t}_2)(\mathbf{a}_2' \cdot \mathbf{t}_1) = 1 - \frac{1}{2}(W_{1,2} - W_{2,1})^2 + \cdots$$

and therefore from (1.51)

$$\psi_3^2 \approx \frac{1}{4}(W_{1,2} - W_{2,1})^2 = \frac{1}{4}|\operatorname{curl}\mathbf{w}|^2,$$

where $\mathbf{w} = W_\alpha \mathbf{a}_\alpha$. Since $\operatorname{curl}\mathbf{w} = (W_{2,1} - W_{1,2})\mathbf{a}_3$ corresponds to counterclockwise rotation about $\mathbf{a}_3$ we take

$$\psi_3 = \frac{1}{2}(W_{2,1} - W_{1,2}). \tag{2.28}$$

It follows from (1.53)–(2.28) that the linearization of the vector rotation angle is

$$\mathbf{\Phi} = -U_2\mathbf{a}_1 + U_1\mathbf{a}_2 + \frac{1}{2}(W_{2,1} - W_{1,2})\mathbf{a}_3. \tag{2.29}$$

At a point of $\partial\Omega$,

$$W_{2,1} - W_{1,2} = \boldsymbol{\tau} \cdot \frac{\partial \mathbf{w}}{\partial \boldsymbol{\nu}} - \boldsymbol{\nu} \cdot \frac{\partial \mathbf{w}}{\partial \boldsymbol{\tau}}$$

and (2.29) may then be written

$$\mathbf{\Phi} = -(\boldsymbol{\tau} \cdot \mathbf{u})\boldsymbol{\nu} + (\boldsymbol{\nu} \cdot \mathbf{u})\boldsymbol{\tau} + \frac{1}{2}\left(\boldsymbol{\tau} \cdot \frac{\partial \mathbf{w}}{\partial \boldsymbol{\nu}} - \boldsymbol{\nu} \cdot \frac{\partial \mathbf{w}}{\partial \boldsymbol{\tau}}\right)\mathbf{a}_3, \tag{2.30}$$

where $\mathbf{w} = W_\alpha \mathbf{a}_\alpha$ and $\mathbf{u} = U_\alpha \mathbf{a}_\alpha$.

2.5. The Kirchhoff Plate Model. The Kirchhoff model corresponds to the absence of transverse shearing. This may be expressed by saying that $\mathbf{A}_3$ is a unit vector which is normal the the deformed reference plane or, what is the same thing, that $\phi_i = \psi_i$. In the linear situation, this means that

$$U_1 = \psi_2 = -W_{3,1}, \quad U_2 = -\psi_1 = -W_{3,2}. \tag{2.31}$$

To obtain the Kirchhoff model in a formal way, we eliminate $U_\alpha + W_{3,\alpha}$ between the equations in (2.19). This leads to the equation

$$DU_{\alpha,\alpha\beta\beta} + C_{\alpha,\alpha} + F_3 - mW_{3,tt} = I_\rho U_{\alpha,\alpha tt},$$

where $D = (2\mu + \lambda)h^3/12$. If (2.31) is used to substitute for U_α in terms of $W_{3,\alpha}$ the result is the *Kirchhoff plate equation*

$$mW_{3,tt} - I_\rho W_{3,\alpha\alpha tt} + DW_{3,\alpha\alpha\beta\beta} = F_3 + C_{\alpha,\alpha}. \tag{2.32}$$

The geometric boundary condition for W_3 on Γ^C follows from (2.22) and (2.31):

$$W_3 = 0, \quad \frac{\partial W_3}{\partial \boldsymbol{\nu}} = 0 \text{ on } \Gamma^C. \tag{2.33}$$

To obtain an edge condition for W_3 on Γ^D we substitute (2.31) into the last two equations in (2.21), multiply the first of these equations by ν_1, the second by ν_2 and add the products. The result may be written

$$-D[W_{3,\alpha\alpha} + (1-\beta)B_1(W_3)] = \boldsymbol{\nu}\cdot\mathbf{c} \text{ on } \Gamma^D, \tag{2.34}$$

where $\beta = \lambda/(2\mu+\lambda)$ satisfies $0 < \beta < 1$ and where

$$B_1(W_3) = 2\nu_1\nu_2 W_{3,12} - \nu_1^2 W_{3,22} - \nu_2^2 W_{3,11}.$$

This boundary condition is a balance law for the bending moment about the tangent vector $\boldsymbol{\tau}$. There is a second edge condition for W_3 along Γ^D, representing the balance of shear forces, which cannot be deduced from purely formal calculations on (2.20) and (2.21), using (2.31). It is

$$\begin{aligned} -D\left[\frac{\partial}{\partial \boldsymbol{\nu}} W_{3,\alpha\alpha} + (1-\beta)\frac{\partial}{\partial \boldsymbol{\tau}} B_2(W_3)\right] + I_\rho \frac{\partial}{\partial \boldsymbol{\nu}} W_{3,tt} \\ = f_3 - \frac{\partial}{\partial \boldsymbol{\tau}}(\boldsymbol{\tau}\cdot\mathbf{c}) - \boldsymbol{\nu}\cdot\mathbf{C} \text{ on } \Gamma^D, \end{aligned} \tag{2.35}$$

where

$$B_2(W_3) = \nu_1\nu_2(W_{3,22} - W_{3,11}) + (\nu_1^2 - \nu_2^2)W_{3,12}.$$

Starting from the Reissner system, a rigorous derivation of the system (2.32)–(2.35) may be given by passing to the limit as $G \to \infty$; see [51, Chapter II] and Section 6 below.

3. Systems of Linked Reissner Plates

For $i = 1, \ldots, n$ let $\mathbf{a}_1^i, \mathbf{a}_2^i, \mathbf{a}_3^i$ be a right- handed orthonormal system in $I\!R^3$, let $\mathbf{p}_0^i$ be a fixed vector in $I\!R^3$, and Ω_i be a bounded, open, connected region in $I\!R^2$ with Lipschitz boundary. Summation convention is no longer assumed but, as above, Greek indices take values 1,2 and $\sum_j := \sum_{j=1}^3$. We set

$$\mathbf{p}_i(x_1, x_2) = \mathbf{p}_0^i + \sum_\alpha x_\alpha \mathbf{a}_\alpha^i,$$

$$\mathcal{P}_i = \{\mathbf{p}_i(x_1, x_2) | \, (x_1, x_2) \in \Omega_i\}.$$

The mapping $(x_1, x_2) \mapsto \mathbf{p}_i(x_1, x_2)$ is a homeomorphism of $\overline{\Omega}_i$ onto $\overline{\mathcal{P}}_i$ and of $\partial\Omega$ onto $\partial\mathcal{P}$, with

$$\mathbf{p}_i^{-1}(\mathbf{p}) = ((\mathbf{p} - \mathbf{p}_0^i) \cdot \mathbf{a}_1^i, (\mathbf{p} - \mathbf{p}_0^i) \cdot \mathbf{a}_2^i).$$

The set $\mathcal{P}_i$ is considered as the reference surface of a thin plate whose reference configuration is

$$\mathcal{B}_i = \{\mathbf{p}_i(x_1, x_2) + x_3 \mathbf{a}_3^i \, | \, (x_1, x_2) \in \Omega_i, \ |x_3| < h/2\},$$

where h is a positive constant not depending on i. We make the following assumptions.

$$\overline{\mathcal{P}}_i \cap \mathcal{P}_k = \emptyset, \ \ \forall i \neq k;$$

$$\cup_{i=1}^n \overline{\mathcal{P}}_i \ \text{ is a connected set in } I\!R^3;$$

$$\overline{\mathcal{P}}_i \cap \overline{\mathcal{P}}_k \ \text{ is either empty or is a finite union of components, } \ \forall i \neq k$$

Each plate is assumed to satisfy the basic kinematic assumption (1.1): the position vector $\mathbf{R}^i(x_1, x_2, x_3)$ to the displaced particle originally at $\mathbf{p}_i(x_1, x_2) + x_3 \mathbf{a}_3^i$ is given by

$$\text{(3.1)} \quad \mathbf{R}^i(x_1, x_2, x_3) = \mathbf{P}^i(x_1, x_2) + x_3 \mathbf{A}_3^i(x_1, x_2), \qquad (x_1, x_2) \in \Omega_i, \ |x_3| < h/2, \ 1 = 1, \ldots, n,$$

where $\mathbf{A}_3^i \neq 0$.

As before, we set $\mathbf{A}_\alpha^i = \mathbf{R}_{,\alpha}^i|_{x_3=0}$,

$$\mathbf{P}^i - \mathbf{p}_i := \mathbf{W}^i = \sum_j W_j^i \mathbf{a}_j^i,$$

$$\mathbf{A}_3^i - \mathbf{a}_3^i := \mathbf{U}^i = \sum_j U_j^i \mathbf{a}_j^i,$$

$$\mathbf{w}^i = \sum_\alpha W_\alpha^i \mathbf{a}_\alpha^i, \ \ \mathbf{u}^i = \sum_\alpha U_\alpha^i \mathbf{a}_\alpha^i.$$

3.1. Geometric Junction Conditions. The *joints*, or *junction regions* of the multi-plate system are defined *exactly* as in Section 2.1. Thus each joint is an open line segment that is a common edge of two or more reference surfaces. As in Section 2.1, it may be assumed without loss of generality that the joints are mutually disjoint. The index set $\mathcal{E}(J)$ of a joint J is

$$\mathcal{I}(J) = \{i \in [1,\dots,n] | \, J \subset \partial\mathcal{P}_i\}.$$

Thus $\mathcal{I}(J)$ is the index set of those $\mathcal{P}_i$ which share J. We shall distinguish several types of joints by considering various geometric constraints on joint deformation.

(I) J is a *connected joint* if

$$\mathbf{W}^i(\mathbf{p}_i^{-1}(\mathbf{p})) = \mathbf{W}^j(\mathbf{p}_j^{-1}(\mathbf{p})), \quad \forall \mathbf{p} \in J, \ \forall i,j \in \mathcal{I}(J). \tag{3.2}$$

Condition (3.2) simply requires continuity of displacements along the joints.

Let $\boldsymbol{\tau}^i$ denote the unit tangent vector to $\partial\mathcal{P}_i$ that is positively oriented with respect to $\mathbf{a}_3^i$.

(II) J is a *hinged joint* if it is a connected joint and if

$$\left(\frac{\partial}{\partial\boldsymbol{\tau}^i}\right)^2 \mathbf{W}^i(\mathbf{p}_i^{-1}(\mathbf{p})) = 0, \quad \forall \mathbf{p} \in J, \ \forall i \in \mathcal{I}(J). \tag{3.3}$$

Condition (3.3) means that J remains straight under deformation.

Let $\boldsymbol{\Phi}^i$ denote the vector rotation angle associated with the deformation of $\mathcal{P}_i$.

(III) J is a *semi-rigid joint* if J is a connected joint, if

$$\boldsymbol{\Phi}^i(\mathbf{p}_i^{-1}(\mathbf{p})) = \boldsymbol{\Phi}^j(\mathbf{p}_j^{-1}(\mathbf{p})), \quad \forall \mathbf{p} \in J, \ \forall i,j \in \mathcal{I}(J), \tag{3.4}$$

and if

$$(\mathbf{A}_1^i \cdot \mathbf{A}_2^i)(\mathbf{p}_i^{-1}(\mathbf{p})) = 0, \quad \forall \mathbf{p} \in J, \ \forall i \in \mathcal{I}(J). \tag{3.5}$$

Conditions (3.4) and (3.5) require continuity of vector rotation angles and vanishing of the in-plane twisting strains, respectively, along the joint, but allow for bending along the length of the joint and rotation of the entire multi-link system around the joint.

(IV) J is a *rigid joint* if J is a semi-rigid joint and if

$$\left(\frac{\partial}{\partial\boldsymbol{\tau}^i}\right)^2 \mathbf{W}^i(\mathbf{p}_i^{-1}(\mathbf{p})) = 0, \quad \forall \mathbf{p} \in J, \ \forall i \in \mathcal{I}(J). \tag{3.6}$$

Thus a rigid joint is a semi-rigid joint which remains straight under deformation.

It is obvious that other *ad hoc*, but physically reasonable, geometric constraints on the deformation of J could be imposed; for example, one might require that J neither bend nor stretch under deformation. The latter constraint is formulated as

$$\frac{\partial}{\partial \boldsymbol{\tau}^i}(\boldsymbol{\tau}^i \cdot \mathbf{w}^i)(\mathbf{p}_i^{-1}(\mathbf{p})) = 0, \quad \forall \mathbf{p} \in J,\ \forall i \in \mathcal{I}(J).$$

3.2. Linearization of the Geometric Joint Conditions. All of the geometric joint conditions are linear in the displacements with the exception of those appearing in the definition of a semi-rigid joint. Let us examine those conditions in the linear context of the last section. Let $\boldsymbol{\nu}^i = \sum_\alpha \nu^i_\alpha \mathbf{a}^i_\alpha$ denote the unit exterior normal vector to $\partial\mathcal{P}_i$. Then $\boldsymbol{\tau}^i = -\nu^i_2 \mathbf{a}^i_1 + \nu^i_1 \mathbf{a}^i_2$ is the unit tangent vector to $\partial\mathcal{P}_i$ that is positively oriented with respect to a^i_3. It follows from (2.30) that (3.4) then takes the form

$$\begin{aligned}(3.7)\quad & -(\boldsymbol{\tau}^i \cdot \mathbf{u}^i)\boldsymbol{\nu}^i + (\boldsymbol{\nu}^i \cdot \mathbf{u}^i)\boldsymbol{\tau}^i + \frac{1}{2}\left(\boldsymbol{\tau}^i \cdot \frac{\partial \mathbf{w}^i}{\partial \boldsymbol{\nu}^i} - \boldsymbol{\nu}^i \cdot \frac{\partial \mathbf{w}^i}{\partial \boldsymbol{\tau}^i}\right)\mathbf{a}^i_3 \\ &= -(\boldsymbol{\tau}^j \cdot \mathbf{u}^j)\boldsymbol{\nu}^j + (\boldsymbol{\nu}^j \cdot \mathbf{u}^j)\boldsymbol{\tau}^j + \frac{1}{2}\left(\boldsymbol{\tau}^j \cdot \frac{\partial \mathbf{w}^j}{\partial \boldsymbol{\nu}^j} - \boldsymbol{\nu}^j \cdot \frac{\partial \mathbf{w}^j}{\partial \boldsymbol{\tau}^j}\right)\mathbf{a}^j_3\end{aligned}$$

while (3.5) becomes

$$\mathbf{a}^i_1 \cdot \mathbf{w}^i_{,2} + \mathbf{a}^i_2 \cdot \mathbf{w}^i_{,1} = 0.$$

The latter equation is the same as

$$\boldsymbol{\tau}^i \cdot \frac{\partial \mathbf{w}^i}{\partial \boldsymbol{\nu}^i} + \boldsymbol{\nu}^i \cdot \frac{\partial \mathbf{w}^i}{\partial \boldsymbol{\tau}^i} = 0$$

and therefore (3.7) may be written

$$\begin{aligned}(3.8)\quad & -(\boldsymbol{\tau}^i \cdot \mathbf{u}^i)\boldsymbol{\nu}^i + (\boldsymbol{\nu}^i \cdot \mathbf{u}^i)\boldsymbol{\tau}^i - \left(\boldsymbol{\nu}^i \cdot \frac{\partial \mathbf{w}^i}{\partial \boldsymbol{\tau}^i}\right)\mathbf{a}^i_3 \\ &= -(\boldsymbol{\tau}^j \cdot \mathbf{u}^j)\boldsymbol{\nu}^j + (\boldsymbol{\nu}^j \cdot \mathbf{u}^j)\boldsymbol{\tau}^j - \left(\boldsymbol{\nu}^j \cdot \frac{\partial \mathbf{w}^j}{\partial \boldsymbol{\tau}^j}\right)\mathbf{a}^j_3.\end{aligned}$$

Along J one has $\boldsymbol{\tau}^i = \pm\boldsymbol{\tau}^j$ for every $i, k \in \mathcal{I}(J)$. We define

$$(3.9)\qquad \sigma_{ij}(J) = \begin{cases} 1 & \text{if } \boldsymbol{\tau}^i = \boldsymbol{\tau}^j \text{ on } J \\ -1 & \text{if } \boldsymbol{\tau}^i = -\boldsymbol{\tau}^j \text{ on } J. \end{cases}$$

It follows that (3.8) is equivalent to

$$(3.10)\qquad (\boldsymbol{\nu}^i \cdot \mathbf{u}^i) = \sigma_{ij}(\boldsymbol{\nu}^j \cdot \mathbf{u}^j),$$

$$(3.11)\quad (\boldsymbol{\tau}^i \cdot \mathbf{u}^i)\boldsymbol{\nu}^i + \left(\boldsymbol{\nu}^i \cdot \frac{\partial \mathbf{w}^i}{\partial \boldsymbol{\tau}^i}\right)\mathbf{a}^i_3 = (\boldsymbol{\tau}^j \cdot \mathbf{u}^j)\boldsymbol{\nu}^j + \left(\boldsymbol{\nu}^j \cdot \frac{\partial \mathbf{w}^j}{\partial \boldsymbol{\tau}^j}\right)\mathbf{a}^j_3$$

for all $i, k \in \mathcal{I}(J)$. However, as will be proved in a moment, the continuity condition (3.2) implies that

$$-\frac{\partial W_3^i}{\partial \tau^i}\nu^i + \left(\nu^i \cdot \frac{\partial \mathbf{w}^i}{\partial \tau^i}\right)\mathbf{a}_3^i = -\frac{\partial W_3^j}{\partial \tau^j}\nu^j + \left(\nu^j \cdot \frac{\partial \mathbf{w}^j}{\partial \tau^j}\right)\mathbf{a}_3^j, \tag{3.12}$$

and therefore (3.11) reduces to

$$\tau^i \cdot (\mathbf{u}^i + \nabla W_3^i)\nu^i = \tau^j \cdot (\mathbf{u}^j + \nabla W_3^j)\nu^j, \quad \forall i, k \in \mathcal{I}(J), \tag{3.13}$$

where $\nabla W_3^i = \sum_\alpha W_{3,\alpha}^i \mathbf{a}_\alpha^i$. If the cardinality ℓ of $\mathcal{I}(J)$ is greater than two, or if $\mathcal{I}(J) = [i, j]$ and $\nu^i \neq -\nu^j$, then (3.13) implies that

$$\tau^i \cdot (\mathbf{u}^i + \nabla W_3^i) = 0.$$

On the other hand, if $\mathcal{I}(J) = [i, j]$ and $\nu^i = -\nu^j$, then $\mathbf{a}_3^i = -\sigma_{ij}\mathbf{a}_3^j$ so the continuity condition $\mathbf{W}^i = \mathbf{W}^j$ implies $W_3^i = -\sigma_{ij} W_3^j$ and so $\tau^i \cdot \nabla W_3^i = -\tau^j \cdot \nabla W_3^j$. Therefore (3.13) simplifies to $\tau^i \cdot \mathbf{u}^i = \tau^j \cdot \mathbf{u}^j$ which, when combined with (3.10), implies that $\mathbf{u}^i = -\sigma_{ij}\mathbf{u}^j$. Thus with (3.2), conditions (3.4) and (3.5) linearize to

$$\begin{cases} \mathbf{u}^i = -\sigma_{ij}\mathbf{u}^j \ \text{ if } \mathcal{I}(J) = [i, j] \text{ and } \nu^i = -\nu^j, \\ (\nu^i \cdot \mathbf{u}^i) = \sigma_{ij}(\nu^j \cdot \mathbf{u}^j), \quad \tau^i \cdot (\mathbf{u}^i + \nabla W_3^i) = 0 \\ \qquad\qquad\qquad\qquad \text{otherwise, } \forall i, j \in \mathcal{I}(J). \end{cases} \tag{3.14}$$

The conditions $\mathcal{I}(J) = [i, j]$, $\nu^i = -\nu^j$, mean that only two plates are connected along J and that their reference surfaces are coplanar. In this case we say that the two plates are *serially connected* along J.

REMARK 3.1. In the absence of transverse shearing, the condition $\tau^i \cdot (\mathbf{u}^i + \nabla W_3^i) = 0$ is automatically satisfied since $\mathbf{u}^i = -\nabla W_3^i$ in this case, while (3.10) then becomes

$$\frac{\partial W_3^i}{\partial \nu^i}(\mathbf{p}_i^{-1}(\mathbf{p})) = \sigma_{ij}\frac{\partial W_3^j}{\partial \nu^j}(\mathbf{p}_j^{-1}(\mathbf{p})), \quad \forall \mathbf{p} \in J. \tag{3.15}$$

It will be shown in section 6 that (3.15) is the linearization of the nonlinear constraint

$$\mathbf{n}^i(\mathbf{p}_i^{-1}(\mathbf{p})) \cdot \mathbf{n}^j(\mathbf{p}_j^{-1}(\mathbf{p})) = \mathbf{a}_3^i \cdot \mathbf{a}_3^j, \quad \forall \mathbf{p} \in J, \tag{3.16}$$

where $\mathbf{n}_i$ denotes the unit normal vector to the deformed reference surface $\mathcal{P}_i$. Equation (3.16) states the geometric constraint that the angle between the normals along J to any two plates meeting there is the same after deformation as before. This is why such a joint is called semi-rigid. □

Let us now verify (3.12). From the continuity condition (3.2) we obtain

$$\frac{\partial \mathbf{w}^{\imath}}{\partial \boldsymbol{\tau}^{\imath}} + \frac{\partial W_3^{\imath}}{\partial \boldsymbol{\tau}^{\imath}} \mathbf{a}_3^{\imath} = \sigma_{\imath\jmath} \left(\frac{\partial \mathbf{w}^{\jmath}}{\partial \boldsymbol{\tau}^{\jmath}} + \frac{\partial W_3^{\jmath}}{\partial \boldsymbol{\tau}^{\jmath}} \mathbf{a}_3^{\jmath} \right), \tag{3.17}$$

so that

$$\boldsymbol{\nu}^{\imath} \cdot \frac{\partial \mathbf{w}^{\imath}}{\partial \boldsymbol{\tau}^{\imath}} = \sigma_{\imath\jmath} \left(\boldsymbol{\nu}^{\imath} \cdot \frac{\partial \mathbf{w}^{\jmath}}{\partial \boldsymbol{\tau}^{\jmath}} + \frac{\partial W_3^{\jmath}}{\partial \boldsymbol{\tau}^{\jmath}} (\boldsymbol{\nu}^{\imath} \cdot \mathbf{a}_3^{\jmath}) \right). \tag{3.18}$$

Since

$$\frac{\partial \mathbf{w}^{\jmath}}{\partial \boldsymbol{\tau}^{\jmath}} = \left(\boldsymbol{\nu}^{\jmath} \cdot \frac{\partial \mathbf{w}^{\jmath}}{\partial \boldsymbol{\tau}^{\jmath}} \right) \boldsymbol{\nu}^{\jmath} + \left(\boldsymbol{\tau}^{\jmath} \cdot \frac{\partial \mathbf{w}^{\jmath}}{\partial \boldsymbol{\tau}^{\jmath}} \right) \boldsymbol{\tau}^{\jmath},$$

it follows from (3.17) that

$$\boldsymbol{\nu}^{\imath} \cdot \frac{\partial \mathbf{w}^{\imath}}{\partial \boldsymbol{\tau}^{\imath}} = \sigma_{\imath\jmath} \left[\left(\boldsymbol{\nu}^{\jmath} \cdot \frac{\partial \mathbf{w}^{\jmath}}{\partial \boldsymbol{\tau}^{\jmath}} \right) \boldsymbol{\nu}^{\imath} \cdot \boldsymbol{\nu}^{\jmath} + \frac{\partial W_3^{\jmath}}{\partial \boldsymbol{\tau}^{\jmath}} (\boldsymbol{\nu}^{\imath} \cdot \mathbf{a}_3^{\jmath}) \right].$$

It also follows from (3.17) that

$$\frac{\partial W_3^{\imath}}{\partial \boldsymbol{\tau}^{\imath}} = \sigma_{\imath\jmath} \left[\left(\boldsymbol{\nu}^{\jmath} \cdot \frac{\partial \mathbf{w}^{\jmath}}{\partial \boldsymbol{\tau}^{\jmath}} \right) \mathbf{a}_3^{\imath} \cdot \boldsymbol{\nu}^{\jmath} + \frac{\partial W_3^{\jmath}}{\partial \boldsymbol{\tau}^{\jmath}} (\mathbf{a}_3^{\imath} \cdot \mathbf{a}_3^{\jmath}) \right],$$

hence

$$\begin{aligned} -\frac{\partial W_3^{\imath}}{\partial \boldsymbol{\tau}^{\imath}} \boldsymbol{\nu}^{\imath} + \left(\boldsymbol{\nu}^{\imath} \cdot \frac{\partial \mathbf{w}^{\imath}}{\partial \boldsymbol{\tau}^{\imath}} \right) \mathbf{a}_3^{\imath} &\\ = \sigma_{\imath\jmath} \Bigg[\left(\boldsymbol{\nu}^{\jmath} \cdot \frac{\partial \mathbf{w}^{\jmath}}{\partial \boldsymbol{\tau}^{\jmath}} \right) &((\boldsymbol{\nu}^{\imath} \cdot \boldsymbol{\nu}^{\jmath}) \mathbf{a}_3^{\imath} - (\mathbf{a}_3^{\imath} \cdot \boldsymbol{\nu}^{\jmath}) \boldsymbol{\nu}^{\imath}) \\ &- \frac{\partial W_3^{\jmath}}{\partial \boldsymbol{\tau}^{\jmath}} ((\mathbf{a}_3^{\imath} \cdot \mathbf{a}_3^{\jmath}) \boldsymbol{\nu}^{\imath} - (\boldsymbol{\nu}^{\imath} \cdot \mathbf{a}_3^{\jmath}) \mathbf{a}_3^{\imath}) \Bigg]. \end{aligned} \tag{3.19}$$

Since $\mathbf{a}_3^{\imath} \cdot \mathbf{a}_3^{\jmath} = \sigma_{\imath\jmath} (\boldsymbol{\nu}^{\imath} \cdot \boldsymbol{\nu}^{\jmath})$ and $\mathbf{a}_3^{\imath} \cdot \boldsymbol{\nu}^{\jmath} = -\sigma_{\imath\jmath} (\mathbf{a}_3^{\jmath} \cdot \boldsymbol{\nu}^{\imath})$, (3.12) follows from (3.19).

3.3. Dynamic Joint Conditions. We wish to derive the balance laws of linear momentum and angular momentum at a joint. As usual, this is done through a variational formulation of the boundary value problem associated with the structure in which the variation is taken with respect to test functions which satisfy, in particular, the geometric joint conditions.

In what follows attention will be restricted to *linear* dynamics and constraints. Set $\Lambda_\imath = \partial \mathcal{P}_\imath$. We assume that

$$\Lambda_\imath = \overline{\Lambda}_\imath^C \cup \overline{\Lambda}_\imath^D \cup \overline{\Lambda}_\imath^J, \tag{3.20}$$

where Λ_i^C, Λ_i^D, Λ_i^J are disjoint and relatively open in Λ_i. We then have the associated partition

$$\Gamma_i = \overline{\Gamma}_i^C \cup \overline{\Gamma}_i^D \cup \overline{\Gamma}_i^J, \tag{3.21}$$

where

$$\Lambda_i^C = \mathbf{p}_i(\Gamma_i^C), \ \Lambda_i^D = \mathbf{p}_i(\Gamma_i^D), \ \Lambda_i^J = \mathbf{p}_i(\Gamma_i^J).$$

The set Λ_i^C represents the part of Λ_i that is clamped, i.e.,

$$\mathbf{w}^i = 0, \ \mathbf{u}^i = 0, \ W_3^i = 0 \ \text{ on } \Gamma_i^C, \tag{3.22}$$

Λ_i^D is the part of Λ_i along which "dynamic" boundary conditions hold, and Λ_i^J consists of those joints of the multi-link system which belong to Λ_i. We label the entire collection of joints by $J_1, J_2, \ldots, J_m$. Obviously, geometric constraints other than, or in addition to, (3.22) could also be considered (cf. Remark 1.1).

Let $\lambda_i, \mu_i, G_i, m_i, I_{\rho_i}$ be the elastic parameters associated with the plate whose reference surface is $\mathcal{P}_i$ and set

$$C_i[\varepsilon] = h[\lambda_i(\varepsilon_{11} + \varepsilon_{22})\mathcal{I} + 2\mu_i\varepsilon], \ \forall \varepsilon \in \mathcal{S}.$$

For $\varepsilon, \hat{\varepsilon} \in \mathcal{S}$ we write $(\varepsilon, \hat{\varepsilon}) = \sum_{\alpha,\beta} \varepsilon_{\alpha\beta}\hat{\varepsilon}_{\alpha\beta}$. The dynamic equations of the multi-link system of plates are *postulated* to be those associated with the Hamilton's principle for a system of linked Reissner plates (cf. Section 2.3):

$$\delta \int_0^T [\mathcal{K}(t) - \mathcal{U}(t) + \mathcal{W}(t)]\, dt = 0, \tag{3.23}$$

where

$$\mathcal{K}(t) = \frac{1}{2}\sum_{i=1}^n \int_{\Omega_i} [m_i(|\mathbf{w}^i_{,t}|^2 + |W^i_{3,t}|^2) + I_{\rho_i}|\mathbf{u}^i_{,t}|^2]\, d\Omega,$$

$$\mathcal{U}(t) = \frac{1}{2}\sum_{i=1}^n \int_{\Omega_i} \{(C_i[\varepsilon(\mathbf{w}^i)], \varepsilon(\mathbf{w}^i)) + \frac{h^2}{12}(C_i[\varepsilon(\mathbf{u}^i)], \varepsilon(\mathbf{u}^i)) + G_i h|\mathbf{u}^i + \nabla W_3^i|^2\} d\Omega,$$

$$\mathcal{W}(t) = \sum_{i=1}^n \left\{ \int_{\Omega_i} (\mathbf{F}_0^i \cdot \mathbf{w}^i + F_3^i W_3^i + \mathbf{C}^i \cdot \mathbf{u}^i)\, d\Omega + \int_{\Gamma_i^D} (\mathbf{f}_0^i \cdot \mathbf{w}^i + f_3^i W_3^i + \mathbf{c}^i \cdot \mathbf{u}^i)\, d\Gamma \right\}.$$

Here $\mathbf{F}_0^i = \sum_\alpha F_\alpha^i \mathbf{a}_\alpha^i$ denotes the resultant in-plane component of applied force in the ith plate, $\mathbf{C}^i = \sum_\alpha C_\alpha^i \mathbf{a}_\alpha^i$ is the resultant in-plane component of the applied couple, and similarly for $\mathbf{f}_0^i$ and $\mathbf{c}^i$. The variation in (3.23) must be taken with respect to displacements which respect the geometric junction conditions imposed along $J_1, \dots, J_m$, as well as the clamped conditions (3.22).

As a result of the calculation (3.23) we obtain, in particular, the Reissner plate equations (2.18), (2.19) for each plate in the multi-link system, and the boundary conditions (2.20), (2.21) along each Γ_i^D. We write the equations of motion in the following manner:

$$(3.24) \qquad \begin{cases} m_i \mathbf{w}^i_{,tt} - \operatorname{div} C_i[\varepsilon(\mathbf{w}^i)] = \mathbf{F}^i_0, \\ m_i W^i_{3,tt} - G_i h \operatorname{div}(\mathbf{u}^i + \nabla W^i_3) = F^i_3, \\ I_{\rho_i} \mathbf{u}^i_{,tt} - \dfrac{h^2}{12} \operatorname{div} C_i[\varepsilon(\mathbf{u}^i)] + G_i h(\mathbf{u}^i + \nabla W^i_3) = \mathbf{C}_i, \end{cases}$$

where

$$\operatorname{div} C_i[\varepsilon(\mathbf{w})] = \sum_\alpha \frac{\partial}{\partial x_\alpha} C_i^{\alpha\beta}[\varepsilon(\mathbf{w})]\mathbf{a}^i_\beta.$$

The boundary conditions along Γ_i^D may be written

$$(3.25) \qquad \begin{cases} C_i[\varepsilon(\mathbf{w}^i)]\boldsymbol{\nu}^i = \mathbf{f}^i_0, \\ G_i h \boldsymbol{\nu}^i \cdot (\mathbf{u}^i + \nabla W^i_3) = f^i_3, \\ \dfrac{h^2}{12} C_i[\varepsilon(\mathbf{u}^i)]\boldsymbol{\nu}^i = \mathbf{c}^i \quad \text{on } \Gamma_i^D, \end{cases}$$

where

$$C_i[\varepsilon(\mathbf{w}^i)]\boldsymbol{\nu}^i = \sum_\alpha \nu^i_\alpha C_i^{\alpha\beta}[\varepsilon(\mathbf{w}^i)]\mathbf{a}^i_\beta.$$

In addition to equations (3.24), (3.25) and the clamped conditions (3.22), the following variational junction condition must be satisfied:

$$(3.26) \quad \sum_{i=1}^n \int_{\Gamma_i^J} \Big\{ \hat{\mathbf{w}}^i \cdot (C_i[\varepsilon(\mathbf{w}^i)]\boldsymbol{\nu}^i) + G_i h \hat{W}^i_3 \boldsymbol{\nu}^i \cdot (\mathbf{u}^i + \nabla W^i_3) \\ + \frac{h^2}{12} \hat{\mathbf{u}}^i \cdot (C_i[\varepsilon(\mathbf{u}^i)]\boldsymbol{\nu}^i) \Big\} \, d\Gamma = 0$$

for all displacements $\hat{\mathbf{w}}^i + \hat{W}^i_3 \mathbf{a}^i_3$ and "rotations" $\mathbf{u}^i$ which satisfy the geometric junction conditions.By employing the notation of Section 2.2, (3.26) may be expressed as

$$\sum_{k=1}^m \sum_{i \in \mathcal{I}(J_k)} \int_{J_k} \Big(\{C_i[\varepsilon(\mathbf{w}^i)]\boldsymbol{\nu}^i + G_i h \boldsymbol{\nu}^i \cdot (\mathbf{u}^i + \nabla W^i_3)\mathbf{a}^i_3\} \cdot \hat{\mathbf{W}}^i(\mathbf{p}_i^{-1}(\mathbf{p})) \\ + \frac{h^2}{12} \hat{\mathbf{u}}^i \cdot (C_i[\varepsilon(\mathbf{u}^i)]\boldsymbol{\nu}^i)(\mathbf{p}_i^{-1}(\mathbf{p})) \Big) \, dJ_k = 0,$$

which can hold if and only if

$$\sum_{i\in\mathcal{I}(J_k)}\int_{J_k}\Big(\{C_i[\varepsilon(\mathbf{w}^i)]\boldsymbol{\nu}^i+G_ih\boldsymbol{\nu}^i\cdot(\mathbf{u}^i+\nabla W_3^i)\mathbf{a}_3^i\}\cdot\hat{\mathbf{W}}^i(\mathbf{p}_i^{-1}(\mathbf{p})) + \frac{h^2}{12}\hat{\mathbf{u}}^i\cdot(C_i[\varepsilon(\mathbf{u}^i)]\boldsymbol{\nu}^i)(\mathbf{p}_i^{-1}(\mathbf{p}))\Big)\,dJ_k=0,\quad k=1,\dots,m. \tag{3.27}$$

The dynamic junction conditions may be read off from (3.27).

3.3.1. *Dynamic conditions at a connected joint.* If J_k is a connected joint, then

$$\hat{\mathbf{W}}^i(\mathbf{p}_i^{-1}(\mathbf{p}))=\hat{\mathbf{W}}^{i_k}(\mathbf{p}_{i_k}^{-1}(\mathbf{p}))\ \ \forall i\in\mathcal{I}(J_k).$$

Since $\hat{\mathbf{W}}^{i_k}(\mathbf{p}_{i_k}^{-1}(\mathbf{p}))$ is arbitrary, as is $\hat{\mathbf{u}}^i$ $(i\in\mathcal{I}(J_k))$ one obtains from (3.27)

$$\sum_{i\in\mathcal{I}(J_k)}\{C_i[\varepsilon(\mathbf{w}^i)]\boldsymbol{\nu}^i+G_ih\boldsymbol{\nu}^i\cdot(\mathbf{u}^i+\nabla W_3^i)\mathbf{a}_3^i\}(\mathbf{p}_i^{-1}(\mathbf{p}))=0,\quad \forall\mathbf{p}\in J_k, \tag{3.28}$$

$$(C_i[\varepsilon(\mathbf{u}^i)]\boldsymbol{\nu}^i)(\mathbf{p}_i^{-1}(\mathbf{p}))=0,\quad \forall i\in\mathcal{I}(J_k),\ \forall\mathbf{p}\in J_k. \tag{3.29}$$

Equation (3.28) is a balance law for forces along the joint while (3.29) states that the bending moment of each plate sharing the joint J_k must vanish there.

3.3.2. *Dynamic conditions at a hinged joint.* In this case, the variational equation (3.27) must hold for arbitrary (sufficiently regular) $\hat{\mathbf{u}}^i$ and for $\hat{\mathbf{W}}^i$ satisfying (3.2) and (3.3). One therefore obtains once again the condition (3.29) for the vanishing of bending moments. In addition,

$$\sum_{i\in\mathcal{I}(J_k)}\int_{J_k}\{C_i[\varepsilon(\mathbf{w}^i)]\boldsymbol{\nu}^i+G_ih\boldsymbol{\nu}^i\cdot(\mathbf{u}^i+\nabla W_3^i)\mathbf{a}_3^i\}\cdot\hat{\mathbf{W}}^{i_k}(\mathbf{p}_{i_k}^{-1}(\mathbf{p}))\,dJ_k=0 \tag{3.30}$$

for all functions $\hat{\mathbf{W}}^{i_k}$ of the form

$$\hat{\mathbf{W}}^{i_k}(\mathbf{p}_{i_k}^{-1}(\mathbf{p}))=\mathbf{C}_k^0+s\mathbf{C}_k^1,\quad(\mathbf{p}=\mathbf{q}_k^0+s\boldsymbol{\tau}^{i_k}),$$

where $\mathbf{C}_k^0,\mathbf{C}_k^1$ are fixed but arbitrary vectors in $\mathbb{R}^3$. It is then a consequence of (3.30) that the following *non local* balance laws must hold along J_k:

$$\sum_{i\in\mathcal{I}(J_k)}\int_{a_k}^{b_k}\{C_i[\varepsilon(\mathbf{w}^i)]\boldsymbol{\nu}^i+G_ih\boldsymbol{\nu}^i\cdot(\mathbf{u}^i+\nabla W_3^i)\mathbf{a}_3^i\}ds=0, \tag{3.31}$$

$$\sum_{i\in\mathcal{I}(J_k)}\int_{a_k}^{b_k}s\{C_i[\varepsilon(\mathbf{w}^i)]\boldsymbol{\nu}^i+G_ih\boldsymbol{\nu}^i\cdot(\mathbf{u}^i+\nabla W_3^i)\mathbf{a}_3^i\}ds=0, \tag{3.32}$$

where

$$J_k = \{\mathbf{q}_k^0 + s\boldsymbol{\tau}^{i_k} \,|\, a_k \le s \le b_k\}.$$

The first is a balance law for the average of forces along J_k and the second is a balance law for the angular momentum with respect to the point $\mathbf{q}_k^0$.

3.3.3. *Dynamic conditions at a semi-rigid joint.* We have once again

$$\text{(3.33)} \quad \sum_{i\in\mathcal{I}(J_k)} \int_{J_k} \Big(\{C_i[\varepsilon(\mathbf{w}^i)]\boldsymbol{\nu}^i + G_i h \boldsymbol{\nu}^i \cdot (\mathbf{u}^i + \nabla W_3^i)\mathbf{a}_3^i\} \cdot \hat{\mathbf{W}}^i + \frac{h^2}{12}\hat{\mathbf{u}}^i \cdot (C_i[\varepsilon(\mathbf{u}^i)]\boldsymbol{\nu}^i) \Big) (\mathbf{p}_i^{-1}(\mathbf{p}))\, dJ_k = 0$$

for all sufficiently regular test functions $\hat{\mathbf{W}}, \hat{\mathbf{u}}$ which satisfy (3.2), (3.10) and (3.14). We resolve $C_i[\varepsilon(\mathbf{u}^i)]\boldsymbol{\nu}^i$ into its normal and tangential components:

$$C_i[\varepsilon(\mathbf{u}^i)]\boldsymbol{\nu}^i = C_{\boldsymbol{\nu}^i}(\mathbf{u}^i)\boldsymbol{\nu}^i + C_{\boldsymbol{\tau}^i}(\mathbf{u}^i)\boldsymbol{\tau}^i,$$
$$C_{\boldsymbol{\nu}^i}(\mathbf{u}^i) = \boldsymbol{\nu}^i \cdot C_i[\varepsilon(\mathbf{u}^i)]\boldsymbol{\nu}^i, \quad C_{\boldsymbol{\tau}^i}(\mathbf{u}^i) = \boldsymbol{\tau}^i \cdot C_i[\varepsilon(\mathbf{u}^i)]\boldsymbol{\nu}^i.$$

Equation (3.33) may be written in terms of these components as

$$\text{(3.34)} \quad \sum_{i\in\mathcal{I}(J_k)} \int_{J_k} \Big(\{C_i[\varepsilon(\mathbf{w}^i)]\boldsymbol{\nu}^i + G_i h \boldsymbol{\nu}^i \cdot (\mathbf{u}^i + \nabla W_3^i)\mathbf{a}_3^i\} \cdot \hat{\mathbf{W}}^i + \frac{h^2}{12}\Big[(\boldsymbol{\nu}^i \cdot \hat{\mathbf{u}}^i)C_{\boldsymbol{\nu}^i}(\mathbf{u}^i) + \Big(\boldsymbol{\tau}^i \cdot \hat{\mathbf{u}}^i + \frac{\partial \hat{W}_3^i}{\partial \boldsymbol{\tau}^i}\Big) C_{\boldsymbol{\tau}^i}(\mathbf{u}^i) + \hat{W}_3^i \frac{\partial}{\partial \boldsymbol{\tau}^i} C_{\boldsymbol{\tau}^i}(\mathbf{u}^i)\Big]\Big)\, dJ_k = 0.$$

We have integrated by parts along J_k to obtain the last integral, utilizing test function $\hat{W}_3^i$ which vanish at the ends of J_k. By utilizing the geometric constraint (3.2) we may conclude from (3.34) that

$$\text{(3.35)} \quad \sum_{i\in\mathcal{I}(J_k)} \{C_i[\varepsilon(\mathbf{w}^i)]\boldsymbol{\nu}^i + G_i h \boldsymbol{\nu}^i \cdot (\mathbf{u}^i + \nabla W_3^i)\mathbf{a}_3^i\} = -\frac{h^2}{12} \sum_{i\in\mathcal{I}(J_k)} \frac{\partial}{\partial \boldsymbol{\tau}^i} C_{\boldsymbol{\tau}^i}(\mathbf{u}^i)\mathbf{a}_3^i,$$

and

$$\text{(3.36)} \quad \sum_{i\in\mathcal{I}(J_k)} \int_{J_k} \Big[(\boldsymbol{\nu}^i \cdot \hat{\mathbf{u}}^i)C_{\boldsymbol{\nu}^i}(\mathbf{u}^i) + \Big(\boldsymbol{\tau}^i \cdot \hat{\mathbf{u}}^i + \frac{\partial \hat{W}_3^i}{\partial \boldsymbol{\tau}^i}\Big) C_{\boldsymbol{\tau}^i}(\mathbf{u}^i)\Big]\, dJ_k = 0.$$

The ith term on the right side of (3.35) represents the force exerted on the joint due to twisting of the ith plate about $\boldsymbol{\nu}^i$ In addition, by utilizing

(3.10) and (3.14) in (3.36), we obtain the following conditions on balance of moments, where $\sigma_i(J_k) := \sigma_{ii_k}$:

$$\sum_{i\in\mathcal{I}(J_k)} \sigma_i(J_k) C_{\nu^i}(\mathbf{u}^i) = 0, \tag{3.37}$$

$$C_{\tau^i}(\mathbf{u}^i) = C_{\tau^j}(\mathbf{u}^j) \ \text{ if } \mathcal{I}(J_k) = [i,j] \text{ and } \boldsymbol{\nu}^i = -\boldsymbol{\nu}^j. \tag{3.38}$$

Let us note that when (3.38) holds, ie., when two plates are serially connected, the right side of (3.35) vanishes since $\boldsymbol{\tau}^i = \sigma_{ij}\boldsymbol{\tau}^j$ and $\mathbf{a}_3^i = -\sigma_{ij}\mathbf{a}_3^j$ in this situation, so (3.35) reduces to

$$\begin{aligned} G_i h\boldsymbol{\nu}^i \cdot (\mathbf{u}^i + \nabla W_3^i) &= \sigma_{ij} G_j h \boldsymbol{\nu}^j \cdot (\mathbf{u}^j + \nabla W_3^j), \\ C_i[\varepsilon(\mathbf{w}^i)]\boldsymbol{\nu}^i &= -C_j[\varepsilon(\mathbf{w}^j)]\boldsymbol{\nu}^j \ \text{ if } \mathcal{I}(J_k) = [i,j] \text{ and } \boldsymbol{\nu}^i = -\boldsymbol{\nu}^j. \end{aligned} \tag{3.39}$$

Furthermore, at a serial joint, (3.37) reads

$$C_{\nu^i}(\mathbf{u}^i) = \sigma_{ij} C_{\nu^j}(\mathbf{u}^j)$$

which, when combined with (3.38), yields

$$C_i[\varepsilon(\mathbf{u}^i)]\boldsymbol{\nu}^i = \sigma_{ij} C_j[\varepsilon(\mathbf{u}^j)]\boldsymbol{\nu}^j, \quad \text{if } \mathcal{I}(J_k) = [i,j] \text{ and } \boldsymbol{\nu}^i = -\boldsymbol{\nu}^j. \tag{3.40}$$

Therefore, the dynamic junction conditions at a semi-rigid serial joint are (3.39) and (3.40); at a semi-rigid non serial joint they are (3.35) and (3.37). As is to be expected, at a serial joint there is no coupling in either the geometric or dynamic junction conditions between the in-plane displacements $\mathbf{w}^i, \mathbf{w}^j$ and the components $W_3^i, W_3^j, \mathbf{u}^i, \mathbf{u}^j$ related to bending.

3.3.4. *Dynamic conditions at a rigid joint.* We write (3.33) as

$$\begin{aligned} \sum_{i\in\mathcal{I}(J_k)} \int_{J_k} \Bigg(&\{C_i[\varepsilon(\mathbf{w}^i)]\boldsymbol{\nu}^i + G_i h \boldsymbol{\nu}^i \cdot (\mathbf{u}^i + \nabla W_3^i)\mathbf{a}_3^i\} \cdot \hat{\mathbf{W}}^i \\ &+ \frac{h^2}{12}\Bigg[(\boldsymbol{\nu}^i \cdot \hat{\mathbf{u}}^i) C_{\nu^i}(\mathbf{u}^i) + \left(\boldsymbol{\tau}^i \cdot \hat{\mathbf{u}}^i + \frac{\partial \hat{W}_3^i}{\partial \tau^i}\right) C_{\tau^i}(\mathbf{u}^i) \\ &\qquad - \frac{\partial \hat{W}_3^i}{\partial \tau^i} C_{\tau^i}(\mathbf{u}^i)\Bigg]\Bigg)\, dJ_k = 0 \end{aligned} \tag{3.41}$$

for sufficiently regular test functions satisfying (3.2), (3.3), (3.10) and (3.14). Unlike the case of a semi-rigid joint, it is not permissible to utilize test functions such that $\hat{W}_3^i = 0$ at the ends of J_k, since $\hat{\mathbf{W}}^i$ must be linear along the joint. From (3.2) and (3.30),

$$\hat{\mathbf{W}}^i(\mathbf{p}_i^{-1}(\mathbf{p})) = \mathbf{C}_k^0 + s\mathbf{C}_k^1, \quad (\mathbf{p} = \mathbf{q}_k^0 + s\boldsymbol{\tau}^{i_k}),$$

hence $\hat{W}_3^i = (\mathbf{C}_k^0 + s\mathbf{C}_k^1) \cdot \mathbf{a}_3^i$ and

$$\frac{\partial \hat{W}_3^i}{\partial \tau^i}(\mathbf{p}_i^{-1}(\mathbf{p})) = \sigma_i(J_k)\mathbf{C}_k^1 \cdot \mathbf{a}_3^i$$

along J_k. We then deduce from (3.41) that (3.37), (3.38) must hold and that

$$\sum_{i\in\mathcal{I}(J_k)} \int_{a_k}^{b_k} \Big(\{C_i[\varepsilon(\mathbf{w}^i)]\boldsymbol{\nu}^i + G_i h\boldsymbol{\nu}^i \cdot (\mathbf{u}^i + \nabla W_3^i)\mathbf{a}_3^i\} \cdot (\mathbf{C}_k^0 + s\mathbf{C}_k^1) - \frac{h^2}{12}\sigma_i(J_k)(\mathbf{C}_k^1 \cdot \mathbf{a}_3^i)C_{\tau^i}(\mathbf{u}^i)\Big)\, ds = 0$$

for arbitrary vectors $\mathbf{C}_k^0, \mathbf{C}_k^1$. By setting $\mathbf{C}_k^1 = 0$ the balance of average forces law (3.31) is obtained. In addition, we deduce the following balance law for angular momentum along J_k:

$$\sum_{i\in\mathcal{I}(J_k)} \int_{a_k}^{b_k} \Big(s\{C_i[\varepsilon(\mathbf{w}^i)]\boldsymbol{\nu}^i + G_i h\boldsymbol{\nu}^i \cdot (\mathbf{u}^i + \nabla W_3^i)\mathbf{a}_3^i\} - \frac{h^2}{12}\sigma_i(J_k)C_{\tau^i}(\mathbf{u}^i)\mathbf{a}_3^i\Big)\, ds = 0.$$

3.4. Junctions With Masses and Applied Forces. In the derivation of the dynamic junction conditions it is tacitly assumed that the junction regions are massless and that no external forces act there. Suppose now that there is a linear distribution of point masses along the joint J_k, and let $\tilde{m}_k$ denote the linear mass density along J_k. Furthermore, let $\tilde{\mathbf{f}}^k$ and $\tilde{\mathbf{c}}^k$ denote a distribution of forces and couples, respectively, applied along J_k. One may imagine that $\tilde{\mathbf{f}}^k$ and $\tilde{\mathbf{c}}^k$ are applied along J_k to the edge of a particular plate and that the masses are likewise attached to that plate. These forces and couples, and the energy due to the mass distribution, are then transmitted to other plates of the system through the connection conditions along J_k. For definiteness, we suppose that $\tilde{\mathbf{f}}^k$, $\tilde{\mathbf{c}}^k$ and the mass distribution are associated with the plate of index $i_k = \min\{i|\, i \in \mathcal{I}(J_k)\}$. (In fact, as explained above $\tilde{\mathbf{f}}^k$ and $\tilde{\mathbf{c}}^k$ are resultants in z of a distribution of forces and couples applied over the full edge $\Sigma_k := J_k \times [-h/2, h/2]\mathbf{a}_3^{i_k}$, and $\tilde{m}_k$ is the z-resultant of a planar density of point masses distributed over Σ_k.) The effect of the distribution of masses and applied forces and couples is to modify the kinetic energy $\mathcal{K}$ of the system and the work $\mathcal{W}$ done on the system. The new expressions are

$$\mathcal{K}(t) = \frac{1}{2}\sum_{i=1}^{n} \int_{\Omega_i} [m_i|\mathbf{W}_{,t}^i|^2 + I_{\rho_i}|\mathbf{u}_{,t}^i|^2]\, d\Omega + \frac{1}{2}\sum_{k=1}^{m} \int_{\Gamma_{i_k}(J_k)} (\tilde{m}_k|\mathbf{W}_{,t}^{i_k}|^2 + I_{\tilde{\rho}_k}|\mathbf{u}_{,t}^{i_k}|^2)\, d\Gamma,$$

where $I_{\tilde{\rho}_k}$ is the moment of inertia of J_k, and

$$\mathcal{W}(t) = \sum_{i=1}^{n} \left\{ \int_{\Omega_i} (\mathbf{F}^i \cdot \mathbf{W}^i + \mathbf{C}^i \cdot \mathbf{u}^i)\, d\Omega + \int_{\Gamma_i^D} (\mathbf{f}^i \cdot \mathbf{W}^i + \mathbf{c}^i \cdot \mathbf{u}^i)\, d\Gamma \right\} + \sum_{k=1}^{m} \int_{\Gamma_{i_k}(J_k)} (\tilde{\mathbf{f}}^k \cdot \mathbf{W}^{i_k} + \tilde{\mathbf{c}}^k \cdot \mathbf{u}^{i_k})\, d\Gamma.$$

One may now proceed to derive the dynamic conditions via Hamilton's principle. The details of the new calculation, which is similar to that carried out in the last subsection, will not be provided. The only change is in the variational junction condition (3.27) which, in the present situation, is

$$\text{(3.42)} \quad \sum_{i\in\mathcal{I}(J_k)} \int_{J_k} \Big(\{C_i[\varepsilon(\mathbf{w}^i)]\boldsymbol{\nu}^i + G_i h \boldsymbol{\nu}^i \cdot (\mathbf{u}^i + \nabla W_3^i)\mathbf{a}_3^i\} \cdot \hat{\mathbf{W}}^i(\mathbf{p}_i^{-1}(\mathbf{p})) + \frac{h^2}{12}\hat{\mathbf{u}}^i \cdot (C_i[\varepsilon(\mathbf{u}^i)]\boldsymbol{\nu}^i)(\mathbf{p}_i^{-1}(\mathbf{p})) \Big)\, dJ_k + \int_{J_k} [(\tilde{m}_k \mathbf{W}^{i_k}_{,tt} - \tilde{\mathbf{f}}^k) \cdot \hat{W}^{i_k} + (I_{\tilde{\rho}_k}\mathbf{u}^{i_k}_{,tt} - \tilde{\mathbf{c}}^k) \cdot \hat{\mathbf{u}}^{i_k}]\, dJ_k = 0, \quad k = 1, \dots, m.$$

The dynamic junction conditions may be derived from (3.42)

3.4.1. *Dynamic conditions at a connected joint.* By use of the geometric joint condition at a connected joint one immediately obtains from (3.42) the following conditions along J_k.

$$\text{(3.43)} \quad \sum_{i\in\mathcal{I}(J_k)} \{C_i[\varepsilon(\mathbf{w}^i)]\boldsymbol{\nu}^i + G_i h \boldsymbol{\nu}^i \cdot (\mathbf{u}^i + \nabla W_3^i)\mathbf{a}_3^i\} = \tilde{\mathbf{f}}^k - \tilde{m}_k \mathbf{W}^{i_k}_{,tt},$$

$$\text{(3.44)} \quad \begin{cases} \dfrac{h^2}{12} C_{i_k}[\varepsilon(\mathbf{u}^{i_k})]\boldsymbol{\nu}^{i_k} = \tilde{\mathbf{c}}^k - I_{\tilde{\rho}_k}\mathbf{u}^{i_k}_{,tt}, \\ C_i[\varepsilon(\mathbf{u}^i)]\boldsymbol{\nu}^i = 0, \quad i \in \mathcal{I}(J_k),\ i \neq i_k. \end{cases}$$

Conditions (3.44) indicate that at a connected joint neither the rotational energy nor the applied couple are transmitted to the connecting plates of the system.

3.4.2. *Dynamic conditions at a hinged joint.* One obtains, once again, (3.44) for the balance of bending moments, while (3.43) is replaced by the non local balance laws

$$(3.45) \quad \sum_{i\in\mathcal{I}(J_k)} \int_{a_k}^{b_k} \{C_i[\varepsilon(\mathbf{w}^i)]\boldsymbol{\nu}^i + G_i h\boldsymbol{\nu}^i \cdot (\mathbf{u}^i + \nabla W_3^i)\mathbf{a}_3^i\} ds = \int_{a_k}^{b_k} (\tilde{\mathbf{f}}^k - \tilde{m}_k \mathbf{W}_{,tt}^{i_k})\, ds$$

and

$$(3.46) \quad \sum_{i\in\mathcal{I}(J_k)} \int_{a_k}^{b_k} s\{C_i[\varepsilon(\mathbf{w}^i)]\boldsymbol{\nu}^i + \boldsymbol{\nu}^i \cdot (\mathbf{u}^i + \nabla W_3^i)\mathbf{a}_3^i\} ds = \int_{a_k}^{b_k} s(\tilde{\mathbf{f}}^k - \tilde{m}_k \mathbf{W}_{,tt}^{i_k})\, ds.$$

3.4.3. *Dynamic conditions at a semi-rigid joint.* The balance law for linear momentum now takes the form

$$(3.47) \quad \sum_{i\in\mathcal{I}(J_k)} \{C_i[\varepsilon(\mathbf{w}^i)]\boldsymbol{\nu}^i + G_i h\boldsymbol{\nu}^i \cdot (\mathbf{u}^i + \nabla W_3^i)\mathbf{a}_3^i\} = -\frac{h^2}{12} \sum_{i\in\mathcal{I}(J_k)} \frac{\partial}{\partial \boldsymbol{\tau}^i} C_{\boldsymbol{\tau}^i}(\mathbf{u}^i)\mathbf{a}_3^i + \tilde{\mathbf{f}}^k - \tilde{m}_k \mathbf{W}_{,tt}^{i_k} + \frac{\partial}{\partial \boldsymbol{\tau}^{i_k}} \left[\boldsymbol{\tau}^{i_k} \cdot (\tilde{\mathbf{c}}^k - I_{\tilde{\rho}_k} \mathbf{u}_{,tt}^{i_k})\right] \mathbf{a}_3^{i_k},$$

while the balance law for angular momentum is

$$(3.48) \quad \sum_{i\in\mathcal{I}(J_k)} \sigma_i(J_k) C_{\boldsymbol{\nu}^i}(\mathbf{u}^i) = \boldsymbol{\nu}^{i_k} \cdot (\tilde{\mathbf{c}}^k - I_{\tilde{\rho}_k} \mathbf{u}_{,tt}^{i_k}).$$

3.4.4. *Dynamic conditions at a rigid joint.* The balance law (3.48) is retained, but (3.47) is replaced by (3.45) together with

$$(3.49) \quad \sum_{i\in\mathcal{I}(J_k)} \int_{a_k}^{b_k} \Big(s\{C_i[\varepsilon(\mathbf{w}^i)]\boldsymbol{\nu}^i + G_i h\boldsymbol{\nu}^i \cdot (\mathbf{u}^i + \nabla W_3^i)\mathbf{a}_3^i\} - \frac{h^2}{12}\sigma_i(J_k) C_{\boldsymbol{\tau}^i}(\mathbf{u}^i)\mathbf{a}_3^i\Big)\, ds = \int_{a_k}^{b_k} s[(\tilde{m}_k \mathbf{W}_{,tt}^{i_k} - \tilde{\mathbf{f}}^k) + \boldsymbol{\tau}^{i_k} \cdot (I_{\tilde{\rho}_k} \mathbf{u}_{,tt}^{i_k} - \tilde{\mathbf{c}}^k)\mathbf{a}_3^{i_k}]\, ds.$$

4. Well-posedness of Systems of Linked Reissner Plates

In this section it will be demonstrated that the mathematical model of a system of linked Reissner plates derived in the last section comprises a well-posed mathematical problem in a function space appropriate to the model.

Let Ω, $\mathbf{a}_1, \mathbf{a}_2, \mathbf{a}_3$ be as defined at the beginning of this chapter. For any unit vector $\mathbf{a} \in \mathbb{R}^3$ and $s \geq 0$ let $H^s(\Omega; [\mathbf{a}])$ denote the Hilbert space consisting of functions $f(\cdot)\mathbf{a}$, where $f \in H^s(\Omega)$, the standard (real) Sobolev space of order s. Set

$$H = \bigoplus_{j=1}^{3} H^0(\Omega; [\mathbf{a}_j]) \bigotimes \bigoplus_{\alpha=1}^{2} H^0(\Omega; [\mathbf{a}_\alpha]),$$

$$V = \bigoplus_{j=1}^{3} H^1(\Omega; [\mathbf{a}_j]) \bigotimes \bigoplus_{\alpha=1}^{2} H^1(\Omega; [\mathbf{a}_\alpha]).$$

A function $\mathbf{R}$ in H (resp., in V) is given by the pair $\mathbf{R} = (\mathbf{W}, \mathbf{u})$, where $\mathbf{W}(\cdot) = \sum_j W_j(\cdot)\mathbf{a}_j$, $\mathbf{u}(\cdot) = \sum_\alpha U_\alpha(\cdot)\mathbf{a}_\alpha$ with $W_j, U_\alpha \in H^0(\Omega)$ (resp., $W_j, U_\alpha \in H^1(\Omega)$). The space V is dense in H with compact injection.

Let us further introduce a continuous seminorm on V, denoted by $\sigma(\cdot)$ and defined by

$$\sigma(\mathbf{R}) = \Big(\int_\Omega \{(C[\varepsilon(\mathbf{w})], \varepsilon(\mathbf{w})) + \frac{h^2}{12}(C[\varepsilon(\mathbf{u})], \varepsilon(\mathbf{u})) + Gh|\mathbf{u} + \nabla W_3|^2\} d\Omega\Big)^{1/2},$$

where $C^{\alpha\beta}[\varepsilon(\cdot)]$ and $\varepsilon_{\alpha\beta}(\cdot)$ are defined in (2.15) and (2.17) and $\mathbf{w} = \sum_\alpha W_\alpha \mathbf{a}_\alpha$. It follows from Korn's lemma (see Section 2.4) that there is a constant $K > 0$ depending only on Ω, h and the elastic parameters such that

$$\|\mathbf{R}\|_V \leq K[\sigma(\mathbf{R}) + \|\mathbf{R}\|_H], \quad \forall \mathbf{R} \in V. \tag{4.1}$$

One may replace $\|\mathbf{R}\|_H$ on the right side of (4.1) by any continuous seminorm $p(\cdot)$ defined on V such that $\sigma(\mathbf{R}) = 0$, $p(\mathbf{R}) = 0$ implies $\mathbf{R} = 0$ (cf. Remark 2.2). In particular,

$$\|\mathbf{R}\|_V \leq K\sigma(\mathbf{R}), \quad \forall \mathbf{R} \in V \text{ such that } \mathbf{R} = 0 \text{ on } \Gamma_0, \tag{4.2}$$

where Γ_0 is any nonempty, relatively open subset of $\partial\Omega$.

4.1. Function Spaces for Linked Plates. For $i = 1, \dots, n$, let Ω_i, $\mathbf{a}_1^i, \mathbf{a}_2^i, \mathbf{a}_3^i$ be as defined in Section 3. We form the spaces H and V above in which Ω is replaced by Ω_i and $\mathbf{a}_j$ by $\mathbf{a}_j^i$, and denote these spaces by H_i and V_i, respectively. We then set

$$H = \prod_{i=1}^{n} H_i, \quad V = \prod_{i=1}^{n} V_i.$$

If $\mathbf{R} \in H$ then $\mathbf{R} = (\mathbf{R}^1, \mathbf{R}^2, \dots, \mathbf{R}^n)$, where

$$\mathbf{R}^i = (\mathbf{W}^i, \mathbf{u}^i), \quad \mathbf{W}^i = \sum_j W_j^i \mathbf{a}_j^i, \quad \mathbf{u}^i = \sum_\alpha U_\alpha^i \mathbf{a}_\alpha^i. \tag{4.3}$$

The norms on H and V are defined to be, respectively,

$$\|\mathbf{R}\|_H = \left[\sum_{i=1}^{n} \int_{\Omega_i} (m_i |\mathbf{W}^i|^2 + I_{\rho_i} |\mathbf{u}^i|^2) \, d\Omega \right]^{1/2}$$

and

$$\|\mathbf{R}\|_V = \left(\sum_{i=1}^{n} \|\mathbf{R}^i\|_{V_i}^2 \right)^{1/2}.$$

A continuous seminorm $\sigma(\cdot)$ is defined on V by setting

$$\sigma(\mathbf{R}) = \left[\sum_{i=1}^{n} \sigma_i^2(\mathbf{R}^i) \right]^{1/2},$$

where

$$\sigma_i(\mathbf{R}^i) = \Big(\int_{\Omega_i} \{ (C_i[\varepsilon(\mathbf{w}^i)], \varepsilon(\mathbf{w}^i)) + \frac{h^2}{12} (C_i[\varepsilon(\mathbf{u}^i)], \varepsilon(\mathbf{u}^i)) + G_i h |\mathbf{u}^i + \nabla W_3^i|^2 \} d\Omega \Big)^{1/2},$$

It follows immediately from (4.1), applied in each region Ω_i using the spaces V_i, H_i and the seminorm σ_i, that

$$\|\mathbf{R}\|_V \le K[\sigma(\mathbf{R}) + \|\mathbf{R}\|_H], \quad \forall \mathbf{R} \in V.$$

Let $\mathcal{V}$ denote the *closed* subspace of V consisting of functions $\mathbf{R} = (\mathbf{R}^1, \dots, \mathbf{R}^n)$ such that

$$\mathbf{R}^i \in V_i, \tag{4.4}$$

$$\mathbf{R}^i = 0 \text{ on } \Gamma_i^C, \tag{4.5}$$

$$\mathbf{R} \text{ satisfies the geometric junction conditions.} \tag{4.6}$$

For instance, if J_k is a connected joint, (4.6) requires that

$$\mathbf{W}^i(\mathbf{p}_i^{-1}(\mathbf{p})) = \mathbf{W}^j(\mathbf{p}_j^{-1}(\mathbf{p})), \quad \forall i, j \in \mathcal{I}(J_k),$$

while if J_k is hinged it is also required that

$$\left(\frac{\partial}{\partial \tau^i}\right)^2 \mathbf{W}^i(\mathbf{x}) = 0, \ \forall \mathbf{x} \in \Gamma_i(J_k).$$

In each of these two cases there are no constraints on the $\mathbf{u}^i$ components along J_k. Such constraints would be present, however, if J_k is semi-rigid or rigid, according to the definitions of the last section.

The following lemma is the analog of (4.2) for systems of linked plates.

LEMMA 4.1. *Assume that $\Gamma_i^C \neq \emptyset$ for at least one index i and that each joint is either semi-rigid or rigid. Then*

$$\sigma(\mathbf{R}) \geq k\|\mathbf{R}\|_V, \ \forall \mathbf{R} \in \mathcal{V}, \tag{4.7}$$

for some $k > 0$.

PROOF. Without loss of generality, we may assume that $\Gamma_1^C \neq \emptyset$ and that each joint is semi-rigid. Since the map $V \mapsto \mathbb{R}^+ : \mathbf{R} \mapsto \|\mathbf{R}^1\|_{L^2(\Gamma_1^C)}$ defines a continuous seminorm on V, it is sufficient to prove that

$$\sigma(\mathbf{R}) = 0, \ \mathbf{R}^1|_{\Gamma_1^C} = 0 \Longrightarrow \mathbf{R} = 0. \tag{4.8}$$

Since $\sigma(\mathbf{R}) = 0$ implies $\sigma_i(\mathbf{R}^i) = 0$, $i = 1, \dots, n$, it follows from (4.8) and (4.2) that $\mathbf{R}^1 = 0$. Suppose that $\mathcal{P}_2$ is connected to $\mathcal{P}_1$ along a joint J. According to the definition of a semi-rigid joint we have, for all $\mathbf{p} \in J$,

$$\mathbf{W}^1(\mathbf{p}_1^{-1}(\mathbf{p})) = \mathbf{W}^2(\mathbf{p}_2^{-1}(\mathbf{p})), \tag{4.9}$$

$$(\boldsymbol{\nu}^1 \cdot \mathbf{u}^1)(\mathbf{p}_1^{-1}(\mathbf{p})) = (\boldsymbol{\nu}^2 \cdot \mathbf{u}^2)(\mathbf{p}_2^{-1}(\mathbf{p})), \tag{4.10}$$

$$[\boldsymbol{\tau}^1 \cdot (\mathbf{u}^1 + \nabla W_3^1)\boldsymbol{\nu}^1](\mathbf{p}_1^{-1}(\mathbf{p})) = [\boldsymbol{\tau}^2 \cdot (\mathbf{u}^2 + \nabla W_3^2)\boldsymbol{\nu}^2](\mathbf{p}_2^{-1}(\mathbf{p})). \tag{4.11}$$

Since $\mathbf{R}^1 = 0$, it follows from (4.9) and (4.10) that

$$W^2|_{\Gamma_2(J)} = 0 \ \ \boldsymbol{\nu}^2 \cdot \mathbf{u}^2 = 0, \tag{4.12}$$

so that, in particular,

$$\left.\frac{\partial W_3^2}{\partial \tau^2}\right|_{\Gamma_2(J)} = 0. \tag{4.13}$$

Then (4.11) gives

$$\boldsymbol{\tau}^2 \cdot \mathbf{u}^2 = 0. \tag{4.14}$$

Therefore $\mathbf{R}^2 = 0$, and so forth. □

4.2. Existence and Uniqueness of Solutions. For reference, let us write down the complete system under consideration.

Equations of motion. For $i = 1, \dots, n$,

$$\begin{cases} m_i \mathbf{w}^i_{,tt} - \operatorname{div} C_i[\varepsilon(\mathbf{w}^i)] = \mathbf{F}^i_0, \\ m_i W^i_{3,tt} - G_i h \operatorname{div}(\mathbf{u}^i + \nabla W^i_3) = F^i_3, \\ I_{\rho_i} \mathbf{u}^i_{,tt} - \dfrac{h^2}{12} \operatorname{div} C_i[\varepsilon(\mathbf{u}^i)] + G_i h(\mathbf{u}^i + \nabla W^i_3) = \mathbf{C}_i. \end{cases} \tag{4.15}$$

Boundary conditions along Γ^C_i, $i = 1, \dots, n$:

$$\mathbf{R}^i = (\mathbf{W}^i, \mathbf{u}^i) = 0. \tag{4.16}$$

Boundary conditions along Γ^D_i, $i = 1, \dots, n$:

$$\begin{cases} C_i[\varepsilon(\mathbf{w}^i)]\boldsymbol{\nu}^i = \mathbf{f}^i_0, \\ G_i h \boldsymbol{\nu}^i \cdot (\mathbf{u}^i + \nabla W^i_3) = f^i_3, \\ \dfrac{h^2}{12} C_i[\varepsilon(\mathbf{u}^i)]\boldsymbol{\nu}^i = \mathbf{c}^i. \end{cases} \tag{4.17}$$

In addition, the geometric and associated dynamic junction conditions must hold along the joints $J_1, \dots, J_m$. If, for instance, J_k is a semi-rigid, non serial joint, the geometric conditions along J_k are

$$\begin{cases} \mathbf{W}^i(\mathbf{p}_i^{-1}(\mathbf{p})) = \mathbf{W}^j(\mathbf{p}_j^{-1}(\mathbf{p})), \\ (\boldsymbol{\nu}^i \cdot \mathbf{u}^i)(\mathbf{p}_i^{-1}(\mathbf{p})) = (\boldsymbol{\nu}^j \cdot \mathbf{u}^j)(\mathbf{p}_j^{-1}(\mathbf{p})), \\ \boldsymbol{\tau}^i \cdot (\mathbf{u}^i + \nabla W^i_3)(\mathbf{p}_i^{-1}(\mathbf{p})) = 0, \quad \forall i, j \in \mathcal{I}(J_k), \ \mathbf{p} \in J_k, \end{cases} \tag{4.18}$$

while the dynamic conditions along J_k are

$$\sum_{i \in \mathcal{I}(J_k)} \{C_i[\varepsilon(\mathbf{w}^i)]\boldsymbol{\nu}^i + G_i h \boldsymbol{\nu}^i \cdot (\mathbf{u}^i + \nabla W^i_3)\mathbf{a}^i_3\} = -\frac{h^2}{12} \sum_{i \in \mathcal{I}(J_k)} \frac{\partial}{\partial \boldsymbol{\tau}^i} C_{\boldsymbol{\tau}^i}(\mathbf{u}^i)\mathbf{a}^i_3, \tag{4.19}$$

and

$$\sum_{i \in \mathcal{I}(J_k)} \sigma_i(J_k) C_{\boldsymbol{\nu}^i}(\mathbf{u}^i) = 0. \tag{4.20}$$

We identify the Hilbert space H with its dual space. Since $\mathcal{V}$ is dense in H with continuous injection, we may write $\mathcal{V} \subset H \subset \mathcal{V}'$, where $\mathcal{V}'$ denotes the dual space of $\mathcal{V}$. The duality between elements $\mathbf{R} \in \mathcal{V}$ and $\mathbf{R}' \in \mathcal{V}'$ is denoted by $\langle \mathbf{R}', \mathbf{R} \rangle_{\mathcal{V}}$. If $\mathbf{R}' \in H$, one has $\langle \mathbf{R}', \mathbf{R} \rangle_{\mathcal{V}} = (\mathbf{R}', \mathbf{R})_H$, $\forall \mathbf{R} \in \mathcal{V}$.

Let $\sigma(\mathbf{R}, \hat{\mathbf{R}})$ denote the symmetric bilinear form associated with the seminorm e, that is

$$\sigma(\mathbf{R}, \hat{\mathbf{R}}) = \sum_{i=1}^{n} \sigma_i(\mathbf{R}^i, \hat{\mathbf{R}}^i),$$

where

$$\sigma_i(\mathbf{R}^i, \hat{\mathbf{R}}^i) = \int_{\Omega_i} \{(C_i[\varepsilon(\mathbf{w}^i)], \varepsilon(\hat{\mathbf{w}}^i)) + \frac{h^2}{12}(C_i[\varepsilon(\mathbf{u}^i)], \varepsilon(\hat{\mathbf{u}}^i)) + G_i h(\mathbf{u}^i + \nabla W_3^i) \cdot (\hat{\mathbf{u}}^j + \nabla \hat{W}_3^j)\} d\Omega, \quad \forall \mathbf{R}, \hat{\mathbf{R}} \in V_i.$$

Regardless of what the specific geometric and dynamic junction conditions happen to be, a solution of the system of linked plates is, *by definition*, a solution of the variational equation

$$\text{(4.21)} \quad \int_0^T [\langle \mathbf{R}_{,tt}, \hat{\mathbf{R}} \rangle + \sigma(\mathbf{R}, \hat{\mathbf{R}})]\, dt = \sum_{i=1}^{n} \int_0^T \int_{\Omega_i} (\mathbf{F}^i \cdot \hat{\mathbf{W}}^i + \mathbf{C}^i \cdot \hat{\mathbf{u}}^i)\, d\Omega dt + \sum_{i=1}^{n} \int_0^T \int_{\Gamma_i^D} (\mathbf{f}^i \cdot \hat{\mathbf{W}}^i + \mathbf{c}^i \cdot \hat{\mathbf{u}}^i)\, d\Gamma dt, \quad \forall \hat{\mathbf{R}} \in L^2(0,T; \mathcal{V}),\ \forall T > 0,$$

or, equivalently, for almost all $t > 0$,

$$\text{(4.22)} \quad \langle \mathbf{R}_{,tt}, \hat{\mathbf{R}} \rangle_{\mathcal{V}} + \sigma(\mathbf{R}, \hat{\mathbf{R}}) = \sum_{i=1}^{n} \int_{\Omega_i} (\mathbf{F}^i \cdot \hat{\mathbf{W}}^i + \mathbf{C}^i \cdot \hat{\mathbf{u}}^i)\, d\Omega + \sum_{i=1}^{n} \int_{\Gamma_i^D} (\mathbf{f}^i \cdot \hat{\mathbf{W}}^i + \mathbf{c}^i \cdot \hat{\mathbf{u}}^i)\, d\Gamma, \quad \forall \hat{\mathbf{R}} \in \mathcal{V}.$$

Equation (4.21) is exactly the variational equation associated with the Hamilton's principle which was used to derive the system in the first place. As is usual, the geometric boundary condition (4.16) and the geometric junction conditions are built into the space $\mathcal{V}$ while, at least formally, all of the dynamic requirements are satisfied as a consequence of the variational equation (4.22). We have not as yet specified the regularity a solution is required to possess; of course, this will depend on the assumptions placed on the data $\mathbf{F}^i, \mathbf{C}^i, \mathbf{f}^i, \mathbf{c}^i$ and on the initial data

$$\text{(4.23)} \quad \mathbf{R}|_{t=0} = \mathbf{R}_0, \quad \mathbf{R}_{,t}|_{t=0} = \mathbf{R}_1.$$

Let us first consider the case of homogeneous boundary data, $\mathbf{f}^i = 0$, $\mathbf{c}^i = 0$, and assume of the rest of the data that

$$\text{(4.24)} \quad (\mathbf{F}^i, \mathbf{C}^i) \in L^2(0,T; H_i), \quad \mathbf{R}_0 \in \mathcal{V}, \ \mathbf{R}_1 \in H.$$

The right side of (4.21) may be rewritten

$$\sum_{i=1}^{n}\int_0^T\int_{\Omega_i}(\mathbf{F}^i\cdot\hat{\mathbf{W}}^i+\mathbf{C}^i\cdot\hat{\mathbf{u}}^i)\,d\Omega dt=\int_0^T(\mathbf{G},\mathbf{R})_H dt,\quad \forall\hat{\mathbf{R}}\in H,$$

where

$$\mathbf{G}=(\mathbf{G}^1,\dots,\mathbf{G}^n)\in L^2(0,T;H),\quad \mathbf{G}^i=(m_i^{-1}\mathbf{F}^i,I_{\rho_i}^{-1}\mathbf{C}^i),$$

and therefore (4.22) may be written

$$\langle\mathbf{R}_{,tt},\hat{\mathbf{R}}\rangle_{\mathcal{V}}+\sigma(\mathbf{R},\hat{\mathbf{R}})=(\mathbf{G},\hat{\mathbf{R}})_H,\quad\forall\hat{\mathbf{R}}\in\mathcal{V}. \tag{4.25}$$

The bilinear form $\sigma(\cdot,\cdot)$ is nonnegative on V. If this form is strictly coercive on $\mathcal{V}$, the problem (4.25), (4.23) falls within the framework of standard variational theory as may be found, for example, in Lions-Magenes [62, Chapter 3]. Otherwise, the change of variable $\mathbf{R}=e^{\delta t}\tilde{\mathbf{R}}$ ($\delta>0$) leads to the equation

$$\langle\tilde{\mathbf{R}}_{,tt},\hat{\mathbf{R}}\rangle+\delta(\tilde{\mathbf{R}}_{,t},\hat{\mathbf{R}})_H+\sigma(\mathbf{R},\hat{\mathbf{R}})+\delta^2(\mathbf{R},\hat{\mathbf{R}})_H=(\mathbf{H},\hat{\mathbf{R}})_H,\quad\forall\hat{\mathbf{R}}\in\mathcal{V},$$

where $\mathbf{H}(t)=e^{-\delta t}\mathbf{G}(t)$ and where, according to (4.1), the form $\sigma(\cdot,\cdot)+\delta^2(\cdot,\cdot)_H$ is strictly coercive on V. In either case, standard variational theory gives the following result.

THEOREM 4.1. *Under assumptions* (4.24), *the problem* (4.25), (4.23) *has a unique solution* $\mathbf{R}$ *with regularity*

$$\mathbf{R}\in C([0,T];\mathcal{V})\cap C^1([0,T];H). \tag{4.26}$$

It follows from (4.25) and (4.26) that

$$\mathbf{R}_{,tt}\in C([0,T];\mathcal{V}')\cap L^2(0,T;H).$$

Now consider the case when the boundary conditions are non homogeneous. Without loss of generality we may assume that $\mathbf{F}^i=0$ and $\mathbf{C}^i=0$. Write

$$\mathbf{f}^i=\sum_j f_j^i\mathbf{a}_j^1,\quad \mathbf{c}^i=\sum_\alpha c_\alpha^i\mathbf{a}_\alpha^1,$$

and let us assume that

$$f_j^i,c_\alpha^i\in L^2(0,T;L^2(\Gamma_i^D)),\quad \mathbf{R}_0\in H,\ \mathbf{R}_1\in\mathcal{V}'. \tag{4.27}$$

Then

$$\left|\int_{\Gamma_i^D}(\mathbf{f}^i\cdot\hat{\mathbf{W}}^i+\mathbf{c}^i\cdot\hat{\mathbf{u}}^i)\,d\Gamma\right|\le C(\mathbf{f}^i,\mathbf{c}^i)\|\hat{\mathbf{R}}\|_{V_i}$$

for some constant $C(\mathbf{f}^i, \mathbf{c}^i)$. It follows that there exists an element $\mathbf{g} \in L^2(0,T;\mathcal{V}')$ such that

$$\sum_{i=1}^{n} \int_{\Gamma_i^D} (\mathbf{f}^i \cdot \hat{\mathbf{W}}^i + \mathbf{c}^i \cdot \hat{\mathbf{u}}^i)\, d\Gamma = \langle \mathbf{g}, \hat{\mathbf{R}} \rangle_{\mathcal{V}}, \quad \forall \hat{\mathbf{R}} \in \mathcal{V},$$

so that (4.22) may be written

$$\langle \mathbf{R}_{,tt}, \hat{\mathbf{R}} \rangle + \sigma(\mathbf{R}, \hat{\mathbf{R}}) = \langle \mathbf{g}, \hat{\mathbf{R}} \rangle_{\mathcal{V}}, \quad \forall \hat{\mathbf{R}} \in \mathcal{V}. \tag{4.28}$$

Once again, standard variational theory may be applied in order to deduce the following result.

THEOREM 4.2. *Under assumptions* (4.27), *the problem* (4.28), (4.23) *has a unique solution* $\mathbf{R}$ *with regularity*

$$\mathbf{R} \in C([0,T];H) \cap C^1([0,T];\mathcal{V}'). \tag{4.29}$$

In the context of the last theorem, equation (4.28) must be understood in $D_A - D_A'$ duality, as in Section 2.4.

REMARK 4.1. As in the case of systems of linked membranes, the issue of *regularity of the solution* remains open, in particular, under what conditions on the data will the system admit a *classical* solution? This question is open even for the pure initial value problem (i.e., $\mathbf{F}^i = \mathbf{f}^i = 0$, $\mathbf{C}^i = \mathbf{c}^i = 0$). The main question, of course, is regularity of the solution in a neighborhood of a junction region (cf. Remark 2.4).

5. Controllability of Linked Reissner Plates

It is possible to prove exact controllability results for interconnected Reissner plates similar to those obtained in Chapter VI for interconnected membranes. However, derivation of the necessary observability estimates is even more computationally intensive than before and in the end the results could only be considered as *formal*, depending as they necessarily will on the assumption that solutions of the homogeneous system have the necessary regularity to justify the computations. The desired regularity properties will depend on the geometry of the configuration of plates in an unknown (to us) manner and it seems difficult to resolve this problem. For this reason, we shall consider the exact controllability question only for *transmission problems* for thin plates, where regularity of solutions can be made precise.

Let us consider a bounded, open, connected set Ω in $\mathbb{R}^2$ with Lipschitz boundary consisting of a finite number of smooth simple arcs, a bounded, simply connected open set Ω_1 with smooth boundary such that $\overline{\Omega}_1 \subset \Omega$, and set

$$\Omega_2 = \Omega \setminus \Omega_1, \quad \Gamma = \partial\Omega, \quad \Gamma_1 = \partial\Omega_1, \Gamma_2 = \partial\Omega_2 = \Gamma \cup \Gamma_1.$$

We consider the following system of dynamic equations, transmission conditions and boundary conditions (where $\dot{} = \partial/\partial t$):

$$\text{(5.1)} \qquad \begin{cases} I_{\rho_i}\ddot{\mathbf{u}}^i - \operatorname{div} C_i[\varepsilon(\mathbf{u}^i)] + G_i h(\nabla W^i + \mathbf{u}^i) = 0, \\ m_i \ddot{W}^i - G_i h \operatorname{div}(\nabla W^i + \mathbf{u}^i) = 0 \quad \text{in } \Omega_i,\ t > 0,\ \ i = 1,2; \end{cases}$$

$$\text{(5.2)} \qquad \begin{cases} W^1 = W^2, \ \ \mathbf{u}^1 = \mathbf{u}^2, \\ G_1 h(\nabla W^1 + \mathbf{u}^1)\cdot \boldsymbol{\nu}^1 + G_2 h(\nabla W^2 + \mathbf{u}^2)\cdot \boldsymbol{\nu}^2 = 0, \\ C_1[\varepsilon(\mathbf{u}^1)]\boldsymbol{\nu}^1 + C_2[\varepsilon(\mathbf{u}^2)]\boldsymbol{\nu}^2 = 0 \quad \text{on } \Gamma_1,\ t > 0; \end{cases}$$

$$\text{(5.3)} \qquad W^2 = 0, \ \ \mathbf{u}^2 = 0 \quad \text{on } \Gamma^C,\ t > 0;$$

$$\text{(5.4)} \qquad \begin{cases} C_2[\varepsilon(\mathbf{u}^2)]\boldsymbol{\nu}^2 = \mathbf{c}, \\ G_2 h(\nabla W^2 + \mathbf{u}^2)\cdot \boldsymbol{\nu}^2 = f \quad \text{on } \Gamma^D,\ t > 0. \end{cases}$$

Initial conditions are

$$\text{(5.5)} \qquad W^i|_{t=0} = \dot{W}^i|_{t=0} = 0, \ \ \mathbf{u}^i|_{t=0} = \dot{\mathbf{u}}^i|_{t=0} = 0 \quad \text{in } \Omega_i,\ i = 1,2.$$

Equations (5.1)–(5.4) are those governing the transverse displacements W^i and the rotations $\mathbf{u}^i = (U_1^i, U_2^i)$ of two serial Reissner plates connected by a semi-rigid joint Γ_1 (see (3.14), (3.39) and (3.40)) except that we have written C_i instead of $(h^2/12)C_i$. In (5.4), f and $\mathbf{c} = (c_1, c_2)$ are the control inputs. Of course, there are other equations and transmission conditions for the in-plane displacements, but we have already studied exact controllability of this system in Section 3.4.2, Chapter VI.

5.1. Controllability in Transmission Problems for Thin Plates.

5.1.1. *Observability Estimates and Consequences.* Let $\mathbf{i}, \mathbf{j}, \mathbf{k}$ denote the natural basis for $\mathbb{R}^3$ and set

$$\mathcal{H}^s(\Omega_i) = H^s(\Omega_i; [\mathbf{i}]) \bigoplus H^s(\Omega_i; [\mathbf{j}]) \bigoplus H^s(\Omega_i; [\mathbf{k}]),$$
$$H_i = \mathcal{H}^0(\Omega_i), \ \ V_i = \mathcal{H}^1(\Omega_i), \ \ H = H_1 \times H_2, \ \ V = V_1 \times V_2.$$

If $\boldsymbol{\Phi}$ belongs to H or V, then $\boldsymbol{\Phi} = (\boldsymbol{\Phi}^1, \boldsymbol{\Phi}^2)$ where

$$\boldsymbol{\Phi}^i = \boldsymbol{\varphi}^i + Z^i\mathbf{k}, \ \ \boldsymbol{\varphi}^i = \sum_\alpha \varphi_\alpha^i \mathbf{a}_\alpha.$$

As a norm on H we take

$$\|\boldsymbol{\Phi}\|_H = \left[\sum_{i=1}^{2} \int_{\Omega_i} (m_i |Z^i|^2 + I_{\rho_i} |\boldsymbol{\varphi}^i|^2)\, d\Omega\right]^{1/2}.$$

Let

$$\mathcal{V} = \{\boldsymbol{\Phi} \in V : \boldsymbol{\Phi}^2|_{\Gamma^C} = 0, \ \ \boldsymbol{\Phi}^1|_{\Gamma_1} = \boldsymbol{\Phi}^2|_{\Gamma_1}\},$$

and define

$$\|\boldsymbol{\Phi}\|_{\mathcal{V}} = \left[\sum_{i=1}^{2} \int_{\Omega_i} [(C_i[\varepsilon(\boldsymbol{\varphi}^i)], \varepsilon(\boldsymbol{\varphi}^i)) + G_i h|\nabla Z^i + \boldsymbol{\varphi}^i|^2]\, d\Omega\right]^{1/2}.$$

Then $\|\cdot\|_{\mathcal{V}}$ is a semi-norm on $\mathcal{V}$ which is a norm equivalent to the $\|\cdot\|_V$ norm on $\mathcal{V}$ if $\Gamma^C \neq \emptyset$. To simplify the discussion, such shall be assumed to be the case.

We also consider the homogeneous problem

$$\text{(5.6)} \qquad \begin{cases} I_{\rho_i}\ddot{\boldsymbol{\varphi}}^i - \operatorname{div} C_i[\varepsilon(\boldsymbol{\varphi}^i)] + G_i h(\nabla Z^i + \boldsymbol{\varphi}^i) = 0, \\ m_i \ddot{Z}^i - G_i h \operatorname{div}(\nabla Z^i + \boldsymbol{\varphi}^i) = 0 \quad \text{in } \Omega_i,\ t > 0, \ \ i = 1, 2; \end{cases}$$

$$\text{(5.7)} \qquad \begin{cases} Z^1 = Z^2, \ \ \boldsymbol{\varphi}^1 = \boldsymbol{\varphi}^2, \\ G_1 h(\nabla Z^1 + \boldsymbol{\varphi}^1)\cdot \boldsymbol{\nu}^1 + G_2 h(\nabla Z^2 + \boldsymbol{\varphi}^2)\cdot \boldsymbol{\nu}^2 = 0, \\ C_1[\varepsilon(\boldsymbol{\varphi}^1)]\boldsymbol{\nu}^1 + C_2[\varepsilon(\boldsymbol{\varphi}^2)]\boldsymbol{\nu}^2 = 0 \quad \text{on } \Gamma_1,\ t > 0; \end{cases}$$

$$\text{(5.8)} \qquad Z^2 = 0, \ \ \boldsymbol{\varphi}^2 = 0 \quad \text{on } \Gamma^C,\ t > 0;$$

$$\text{(5.9)} \qquad \begin{cases} C_2[\varepsilon(\boldsymbol{\varphi}^2)]\boldsymbol{\nu}^2 = 0, \\ G_2 h(\nabla Z^2 + \boldsymbol{\varphi}^2)\cdot \boldsymbol{\nu}^2 = 0 \quad \text{on } \Gamma^D,\ t > 0; \end{cases}$$

$$\text{(5.10)} \qquad \begin{cases} \boldsymbol{\varphi}^i_{t=0} = \boldsymbol{\varphi}^i_0, \ \ \dot{\boldsymbol{\varphi}}^i{}_{t=0} = \boldsymbol{\varphi}^i_1, \\ Z^i_{t=0} = Z^i_0, \ \ \dot{Z}^i_{t=0} = Z^i_1 \quad \text{in } \Omega_i, \ \ i = 1, 2. \end{cases}$$

We shall write

$$\boldsymbol{\Phi}_\alpha = (\boldsymbol{\Phi}^1_\alpha, \boldsymbol{\Phi}^2_\alpha), \ \ \boldsymbol{\Phi}^i_\alpha = \boldsymbol{\varphi}^i_\alpha + Z^i_\alpha \mathbf{k}, \ \ \alpha = 0, 1.$$

Let $\mathcal{V}'$ denote the dual space of $\mathcal{V}$ with respect to H and A denote the Riesz isomorphism of $\mathcal{V}$ onto $\mathcal{V}'$. The weak formulation of (5.6)–(5.9) is

$$\sum_{i=1}^{2} \int_{\Omega_i} \{m_i \ddot{Z}^i \hat{Z}^i + I_{\rho_i}\ddot{\boldsymbol{\varphi}}^i \hat{\boldsymbol{\varphi}}^i + G_i h(\nabla Z^i + \boldsymbol{\varphi}^i)\cdot(\nabla \hat{Z}^i + \hat{\boldsymbol{\varphi}}^i)$$
$$+ (C_i[\varepsilon(\boldsymbol{\varphi}^i)], \varepsilon(\hat{\boldsymbol{\varphi}}^i))\}d\Omega = 0, \ \ \forall \hat{\boldsymbol{\Phi}} = (\hat{\boldsymbol{\Phi}}^1, \hat{\boldsymbol{\Phi}}^2) \in \mathcal{V},$$

where $\hat{\boldsymbol{\Phi}}^i = \hat{\varphi}^i + \hat{Z}^i\mathbf{k}$. Equivalently,

$$(5.11) \qquad \begin{cases} \ddot{\boldsymbol{\Phi}} + A\boldsymbol{\Phi} = 0 \text{ in } \mathcal{V}', \\ \boldsymbol{\Phi}|_{t=0} = \boldsymbol{\Phi}_0, \quad \dot{\boldsymbol{\Phi}}|_{t=0} = \boldsymbol{\Phi}_1. \end{cases}$$

Standard theory gives: if $(\boldsymbol{\Phi}_0, \boldsymbol{\Phi}_1) \in \mathcal{V} \times H$ (resp., $D_A \times \mathcal{V}$), (5.11) has a unique solution with

$$\boldsymbol{\Phi} \in C([0,T]; \mathcal{V}) \times C^1([0,T]; H) \times C^2([0,T]; \mathcal{V}')$$

(resp.,

$$\boldsymbol{\Phi} \in C([0,T]; D_A) \times C^1([0,T]; \mathcal{V}) \times C^2([0,T]; H)).$$

Let $(\boldsymbol{\Psi}^1, \boldsymbol{\Psi}^2) \in D_A$. Since Γ_1 is smooth and strictly contained in Ω, one may use standard techniques of elliptic regularity to show that $\boldsymbol{\Phi}^1 \in \mathcal{H}^2(\Omega_1)$. From some results of Nicaise [78] one also has $\boldsymbol{\Phi}^2 \in \mathcal{H}^s(\Omega_2)$ for some $s \in (1,2)$ which depends on the angle of intersection of Γ^C and Γ^D, where $s > 3/2$ if this angle is strictly smaller than π (measured in the interior of Ω). In this case, $\boldsymbol{\Phi}$ will satisfy the transmission conditions in the sense of traces.

The observability estimate to be proved is as follows.

PROPOSITION 5.1. *Assume that Γ^C and Γ^D meet in a strictly convex angle and that there is a point $\mathbf{x}_0 \in \Omega_1$ such that*

$$(5.12) \qquad \begin{cases} (x - x_0) \cdot \boldsymbol{\nu}^1 \geq 0 & \textit{on } \Gamma_1, \\ (x - x_0) \cdot \boldsymbol{\nu}^2 \geq 0 & \textit{on } \Gamma^D, \\ (x - x_0) \cdot \boldsymbol{\nu}^2 \leq 0 & \textit{on } \Gamma^C. \end{cases}$$

Suppose further that

$$\rho_1 \leq \rho_2, \ G_1 \geq G_2, \ \mu_1 \geq \mu_2, \ \lambda_1 \geq \lambda_2.$$

Let $(\boldsymbol{\Phi}_0, \boldsymbol{\Phi}_1) \in \mathcal{V} \times H$ and $\boldsymbol{\Phi}$ be the solution of (5.6)–(5.10). *Then there is a $T_0 > 0$ such that for $T > T_0$,*

$$\|(\boldsymbol{\Phi}_0, \boldsymbol{\Phi}_1)\|^2_{H \times \mathcal{V}'} \leq C_T \int_0^T \int_{\Gamma^D} |\boldsymbol{\Phi}|^2 d\Gamma dt.$$

From this proposition one may immediately deduce a reachability theorem for the problem (5.1)–(5.5). Write

$$\mathbf{R} = (\mathbf{R}^1, \mathbf{R}^2), \quad \mathbf{R}^i = \mathbf{u}^i + W^i\mathbf{k}, \quad \mathbf{f} = \mathbf{c} + f\mathbf{k}.$$

If $\mathbf{f} \in \mathcal{L}^2(\Sigma^D)$, where $\Sigma^D = \Gamma^D \times (0,T)$, problem (5.1)–(5.5) has a unique solution $\mathbf{R}$ with

$$\mathbf{R} \in C([0,T]; H) \times C^1([0,T]; \mathcal{V}') \times C^2([0,T]; (D_A)')$$

(see Theorem 4.2). Let

$$\mathcal{R}_T = \{(\mathbf{R}(T), \dot{\mathbf{R}}(T)) \,|\, \mathbf{f} \in \mathcal{L}^2(\Sigma^D)\}.$$

THEOREM 5.1. *Under the assumptions of Proposition* 5.1, *for* $T > T_0$ *one has* $\mathcal{V} \times H \subset \mathcal{R}_T$.

REMARK 5.1. A similar result has been proved in [46] in the case where the controls act through the Dirichlet, rather than Neumann, boundary conditions on Γ^D, i.e., $\mathbf{u}^2 = \mathbf{c}$, $W^2 = f$. In that situation no restriction on the geometry of Γ is needed and one obtains the sharper result $H \times \mathcal{V}' = \mathcal{R}_T$. □

PROOF OF PROPOSITION 5.1. The proof is based on the following identities, in which we have written

$$Q_i = \Omega_i \times (0,T), \quad \Sigma_i = \Gamma_i \times (0,T).$$

LEMMA 5.1. *Let* $\mathbf{\Phi} = (\mathbf{\Phi}^1, \mathbf{\Phi}^2)$ *be a sufficiently regular solution of* (5.6) *and define*

$$P_i(t) = m_i \int_{\Omega_i} \dot{Z}^i(\mathbf{m}\cdot\nabla Z^i + \delta Z^i)\, d\Omega,$$
$$Q_i(t) = I_{\rho_i} \int_{\Omega_i} \dot{\varphi}^i \cdot (\nabla \varphi^i \mathbf{m} + \delta \varphi^i)\, d\Omega,$$

where $\mathbf{m}(\mathbf{x}) = \mathbf{x} - \mathbf{x}_0$ *and* $\delta \in I\!R$. *Then*

$$(5.13)\quad P_i(T) - P_i(0) + (1-\delta)\int_{Q_i} m_i|\dot{Z}^i|^2 dQ$$
$$- \int_{Q_i} G_i h(\mathbf{m}\cdot\nabla Z^i + \delta Z^i)\operatorname{div}(\nabla Z^i + \varphi^i)\, dQ = \frac{m_i}{2}\int_{\Sigma_i} (\mathbf{m}\cdot\boldsymbol{\nu}^i)|\dot{Z}^i|^2 d\Sigma;$$

$$(5.14)\quad Q_i(T) - Q_i(0) + \int_{Q_i} \{(1-\delta) I_{\rho_i}|\dot{\varphi}^i|^2 + \delta(C_i[\varepsilon(\varphi^i)], \varepsilon(\varphi^i))\} dQ$$
$$+ \int_{Q_i} G_i h(\nabla\varphi^i\mathbf{m} + \delta\varphi^i)\cdot(\nabla Z^i + \varphi^i)\, dQ$$
$$= \frac{1}{2}\int_{\Sigma_i} (\mathbf{m}\cdot\boldsymbol{\nu}^i)\left\{ I_{\rho_i}|\dot{\varphi}^i|^2 + (C_i[\varepsilon(\varphi^i)]\boldsymbol{\nu}^i)\cdot\frac{\partial\varphi^i}{\partial\boldsymbol{\nu}^i} - (C_i[\varepsilon(\varphi^i)]\boldsymbol{\tau}^i)\cdot\frac{\partial\varphi^i}{\partial\boldsymbol{\tau}^i}\right\} d\Sigma$$
$$+ \int_{\Sigma_i} (C_i[\varepsilon(\varphi^i)]\boldsymbol{\nu}^i)\cdot\left[(\mathbf{m}\cdot\boldsymbol{\tau}^i)\frac{\partial\varphi^i}{\partial\boldsymbol{\tau}^i} + \delta\varphi^i\right] d\Sigma.$$

PROOF. The relation (5.13) follows from

$$0 = \int_{Q_i} (\mathbf{m}\cdot\nabla Z^i + \delta Z^i)[m_i \ddot{Z}^i - G_i h \operatorname{div}(\nabla Z^i + \varphi^i)]\, dQ$$

upon noting that

$$\int_{Q_i} (\mathbf{m} \cdot \nabla Z^i + \delta Z^i) m_i \ddot{Z}^i \, dQ = P_i(t)\big|_0^T$$
$$+ (1-\delta) \int_{Q_i} m_i |\dot{Z}^i|^2 dQ - \frac{m_i}{2} \int_{\Sigma_i} (\mathbf{m} \cdot \boldsymbol{\nu}^i) |\dot{Z}^i|^2 d\Sigma.$$

To prove (5.14) we form

$$0 = \int_{Q_i} (\nabla \boldsymbol{\varphi}^i \mathbf{m} + \delta \boldsymbol{\varphi}^i) \cdot \{ I_{\rho_i} \ddot{\boldsymbol{\varphi}}^i - \operatorname{div} C_i[\varepsilon(\boldsymbol{\varphi}^i)] + G_i h (\nabla Z^i + \boldsymbol{\varphi}^i) \} \, dQ.$$

Consider

$$\int_{Q_i} (\nabla \boldsymbol{\varphi}^i \mathbf{m} + \delta \boldsymbol{\varphi}^i) \cdot \{ I_{\rho_i} \ddot{\boldsymbol{\varphi}}^i - \operatorname{div} C_i[\varepsilon(\boldsymbol{\varphi}^i)] \} \, dQ$$
$$= \sum_\beta \int_{Q_i} (\mathbf{m} \cdot \nabla \varphi^i_\beta + \delta \varphi^i_\beta) \{ I_{\rho_i} \ddot{\varphi}^i_\beta - \sum_\alpha C^{\alpha\beta}_{i,\alpha} [\varepsilon(\boldsymbol{\varphi}^i)] \} \, dQ,$$

where $C_i^{\alpha\beta}$ are the components of the matrix C_i. According to (2.16) we have

$$\sum_{\alpha,\beta} C_i^{\alpha\beta}[\varepsilon] \varepsilon_{\alpha\beta} = \frac{h^3}{12} [\lambda_i (\sum_\sigma \varepsilon_{\sigma\sigma})^2 + 2\mu_i \sum_{\alpha,\beta} \varepsilon^2_{\alpha\beta}]$$

Keeping this last equality in mind, one may proceed in a manner *identical* to the proof of Lemma 3.3, Chapter VI, to show that

$$\sum_\beta \int_{Q_i} (\mathbf{m} \cdot \nabla \varphi^i_\beta + \delta \varphi^i_\beta) \{ I_{\rho_i} \ddot{\varphi}^i_\beta - \sum_\alpha C^{\alpha\beta}_{i,\alpha} [\varepsilon(\boldsymbol{\varphi}^i)] \} \, dQ$$
$$= Q_i(t) - Q_i(0) + \int_{Q_i} \{ (1-\delta) I_{\rho_i} |\dot{\boldsymbol{\varphi}}^i|^2 + \delta \sum_{\alpha,\beta} C_i^{\alpha\beta} [\varepsilon(\boldsymbol{\varphi}^i)] \varepsilon_{\alpha\beta}(\boldsymbol{\varphi}^i) \} \, dQ$$
$$- \frac{1}{2} \int_{\Sigma_i} (\mathbf{m} \cdot \boldsymbol{\nu}^i) \left\{ I_{\rho_i} |\dot{\boldsymbol{\varphi}}^i|^2 + \sum_{\alpha,\beta} C_i^{\alpha\beta} [\varepsilon(\boldsymbol{\varphi}^i)] \left[\nu^i_\alpha \frac{\partial \varphi^i_\beta}{\partial \nu^i} - \tau^i_\alpha \frac{\partial \varphi^i_\beta}{\partial \tau^i} \right] \right\} d\Sigma$$
$$- \int_{\Sigma_i} \sum_{\alpha,\beta} \nu^i_\alpha C_i^{\alpha\beta} [\varepsilon(\boldsymbol{\varphi}^i)] \left[(\mathbf{m} \cdot \boldsymbol{\tau}^i) \frac{\partial \varphi^i_\beta}{\partial \tau^i} + \delta \varphi^i_\beta \right] d\Sigma.$$

When written in vector form, it is seen that the integrals over Σ_i in the last relation are exactly those on the right side of (5.14), while the integral over Q_i on the right is the same as the first integral on the left side of (5.14). □

Let us replace δ by $(1-\delta)$ in (5.14) and then add the result to (5.13). We obtain

$$
\begin{aligned}
(5.15)\quad R_\imath(T) - R_\imath(0) &+ \int_{Q_\imath} \{(1-\delta)m_\imath|\dot{Z}^\imath|^2 + \delta I_{\rho_\imath}|\dot{\varphi}^\imath|^2 \\
&+ (1-\delta)(C_\imath[\varepsilon(\varphi^\imath)], \varepsilon(\varphi^\imath))\}\, dQ \\
&+ \int_{Q_\imath} G_\imath h\{(\nabla\varphi^\imath \mathbf{m} + (1-\delta)\varphi^\imath)\cdot(\nabla Z^\imath + \varphi^\imath) \\
&- (\mathbf{m}\cdot\nabla Z^\imath + \delta Z^\imath)\operatorname{div}(\nabla Z^\imath + \varphi^\imath)\}\, dQ \\
&= \frac{1}{2}\int_{\Sigma_\imath} (\mathbf{m}\cdot\boldsymbol{\nu}^\imath)\left\{m_\imath|\dot{Z}^\imath|^2 + I_{\rho_\imath}|\dot{\varphi}^\imath|^2\right. \\
&+ \left.(C_\imath[\varepsilon(\varphi^\imath)]\boldsymbol{\nu}^\imath)\cdot\frac{\partial\varphi^\imath}{\partial\boldsymbol{\nu}^\imath} - (C_\imath[\varepsilon(\varphi^\imath)]\boldsymbol{\tau}^\imath)\cdot\frac{\partial\varphi^\imath}{\partial\boldsymbol{\tau}^\imath}\right\} d\Sigma \\
&+ \int_{\Sigma_\imath} (C_\imath[\varepsilon(\varphi^\imath)]\boldsymbol{\nu}^\imath)\cdot\left[(\mathbf{m}\cdot\boldsymbol{\tau}^\imath)\frac{\partial\varphi^\imath}{\partial\boldsymbol{\tau}^\imath} + (1-\delta)\varphi^\imath\right] d\Sigma,
\end{aligned}
$$

where $R_\imath(t) = P_\imath(t) + Q_\imath(t)$. By an elementary calculation we obtain

$$
\begin{aligned}
(5.16)\quad &\int_{\Omega_\imath} [(\nabla\varphi^\imath \mathbf{m} + (1-\delta)\varphi^\imath)\cdot(\nabla Z^\imath + \varphi^\imath) \\
&\qquad - (\mathbf{m}\cdot\nabla Z^\imath + \delta Z^\imath)\operatorname{div}(\nabla Z^\imath + \varphi^\imath)\}\, d\Omega \\
&= \delta\int_{\Omega_\imath} (|\nabla Z^\imath|^2 - |\varphi^\imath|^2)\, d\Omega - \frac{1}{2}\int_{\Gamma_\imath} (\mathbf{m}\cdot\boldsymbol{\nu}^\imath)\left\{\left(\frac{\partial Z^\imath}{\partial\boldsymbol{\nu}^\imath} + \boldsymbol{\nu}^\imath\cdot\varphi^\imath\right)^2\right. \\
&\quad \left. - \left(\frac{\partial Z^\imath}{\partial\boldsymbol{\tau}^\imath} + \boldsymbol{\tau}^\imath\cdot\varphi^\imath\right)^2 - 2\left(\frac{\partial Z^\imath}{\partial\boldsymbol{\nu}^\imath} + \boldsymbol{\nu}^\imath\cdot\varphi^\imath\right)(\boldsymbol{\nu}^\imath\cdot\varphi^\imath)\right\} d\Gamma \\
&\quad - \int_{\Gamma_\imath} \left(\frac{\partial Z^\imath}{\partial\boldsymbol{\nu}^\imath} + \boldsymbol{\nu}^\imath\cdot\varphi^\imath\right)\left[(\mathbf{m}\cdot\boldsymbol{\tau}^\imath)\frac{\partial Z^\imath}{\partial\boldsymbol{\tau}^\imath} + \delta Z^\imath\right] d\Gamma.
\end{aligned}
$$

Use of (5.16) in (5.15) gives the identity

$$
\begin{aligned}
(5.17)\quad & R_\imath(T) - R_\imath(0) + \int_{Q_\imath} \{(1-\delta)m_\imath|\dot{Z}^\imath|^2 + \delta I_{\rho_\imath}|\dot{\varphi}^\imath|^2 \\
&+ (1-\delta)(C_\imath[\varepsilon(\varphi^\imath)], \varepsilon(\varphi^\imath)) + \delta G_\imath h(|\nabla Z^\imath|^2 - |\varphi^\imath|^2)\}\,dQ \\
&= \frac{1}{2}\int_{\Sigma_\imath} (\mathbf{m}\cdot\nu^\imath)\{m_\imath|\dot{Z}^\imath|^2 d + I_{\rho_\imath}|\dot{\varphi}^\imath|^2\}\,d\Sigma \\
&+ \frac{1}{2}\int_{\Sigma_\imath} (\mathbf{m}\cdot\nu^\imath)\left\{(C_\imath[\varepsilon(\varphi^\imath)]\nu^\imath)\cdot\frac{\partial\varphi^\imath}{\partial\nu^\imath} - (C_\imath[\varepsilon(\varphi^\imath)]\tau^\imath)\cdot\frac{\partial\varphi^\imath}{\partial\tau^\imath}\right. \\
&+ G_\imath h\left(\frac{\partial Z^\imath}{\partial\nu^\imath} + \nu^\imath\cdot\varphi^\imath\right)^2 - G_\imath h\left(\frac{\partial Z^\imath}{\partial\tau^\imath} + \tau^\imath\cdot\varphi^\imath\right)^2 \\
&\left. - 2G_\imath h\left(\frac{\partial Z^\imath}{\partial\nu^\imath} + \nu^\imath\cdot\varphi^\imath\right)(\nu^\imath\cdot\varphi^\imath)\right\}d\Sigma \\
&+ \int_{\Sigma_\imath}\left\{(C_\imath[\varepsilon(\varphi^\imath)]\nu^\imath)\cdot\left[(\mathbf{m}\cdot\tau^\imath)\frac{\partial\varphi^\imath}{\partial\tau^\imath} + (1-\delta)\varphi^\imath\right]d\Sigma\right. \\
&\left. + G_\imath h\left(\frac{\partial Z^\imath}{\partial\nu^\imath} + \nu^\imath\cdot\varphi^\imath\right)\left[(\mathbf{m}\cdot\tau^\imath)\frac{\partial Z^\imath}{\partial\tau^\imath} + \delta Z^\imath\right]\right\}d\Sigma.
\end{aligned}
$$

Lemma 5.2. *Let $\Phi = (\Phi^1, \Phi^2)$ be a sufficiently smooth solution of* (5.6)–(5.9). *Then*

$$
\begin{aligned}
(5.18)\quad & R(t) - R(0) + \sum_{\imath=1}^{2}\int_{Q_\imath} \{(1-\delta)m_\imath|\dot{Z}^\imath|^2 + \delta I_{\rho_\imath}|\dot{\varphi}^\imath|^2 \\
&+ (1-\delta)(C_\imath[\varepsilon(\varphi^\imath)], \varepsilon(\varphi^\imath)) + \delta G_\imath h(|\nabla Z^\imath|^2 - |\varphi^\imath|^2)\}\,dQ \\
&= \mathcal{J}(\Sigma^D) + \mathcal{J}(\Sigma^C) + \mathcal{J}(\Sigma_1),
\end{aligned}
$$

where $R(t) = R_1(t) + R_2(t)$ and

$$
\begin{aligned}
\mathcal{J}(\Sigma^D) &= \frac{1}{2}\int_{\Sigma^D} (\mathbf{m}\cdot\nu^2)\left\{\rho_2|\dot{Z}^2|^2 + I_{\rho_2}|\dot{\varphi}^2|^2\right\}d\Sigma \\
&- \frac{1}{2}\int_{\Sigma^D} (\mathbf{m}\cdot\nu^2)\left\{(C_2[\varepsilon(\varphi^2)]\tau^2)\cdot\frac{\partial\varphi^2}{\partial\tau^2} + G_2 h\left(\frac{\partial Z^2}{\partial\tau^2} + \tau^2\cdot\varphi^2\right)^2\right\}d\Sigma,
\end{aligned}
$$

$$
\begin{aligned}
\mathcal{J}(\Sigma^C) = \frac{1}{2}\int_{\Sigma^C} (\mathbf{m}\cdot\nu^2)&\left\{(C_2[\varepsilon(\varphi^2)]\nu^2)\cdot\frac{\partial\varphi^2}{\partial\nu^2}\right. \\
&\left. + G_2 h\left(\frac{\partial Z^2}{\partial\nu^2} + \nu^2\cdot\varphi^2\right)^2\right\}d\Sigma,
\end{aligned}
$$

$$\mathcal{J}(\Sigma_1) = \frac{1}{2}\int_{\Sigma_1} (\mathbf{m}\cdot\boldsymbol{\nu}^1)\Big\{(\rho_1-\rho_2)|\dot{Z}^1|^2 + (I_{\rho_1}-I_{\rho_2})|\dot{\boldsymbol{\varphi}}^1|^2$$
$$+ G_1 h(1-G_1G_2^{-1})\left(\frac{\partial Z^1}{\partial \boldsymbol{\nu}^1} + \boldsymbol{\nu}^1\cdot\boldsymbol{\varphi}^1\right)^2$$
$$- h(G_1-G_2)\left(\frac{\partial Z^1}{\partial \boldsymbol{\tau}^1} + \boldsymbol{\tau}^1\cdot\boldsymbol{\varphi}^1\right)^2\Big\}\, d\Sigma$$
$$+\frac{1}{2}\int_{\Sigma_1} (\mathbf{m}\cdot\boldsymbol{\nu}^1)\Big\{(C_1[\varepsilon(\boldsymbol{\varphi}^1)]\boldsymbol{\nu}^1)\cdot\frac{\partial \boldsymbol{\varphi}^1}{\partial \boldsymbol{\nu}^1} - (C_2[\varepsilon(\boldsymbol{\varphi}^2)]\boldsymbol{\nu}^2)\cdot\frac{\partial \boldsymbol{\varphi}^2}{\partial \boldsymbol{\nu}^2}$$
$$- (C_1[\varepsilon(\boldsymbol{\varphi}^1)]\boldsymbol{\tau}^1)\cdot\frac{\partial \boldsymbol{\varphi}^1}{\partial \boldsymbol{\tau}^1} + (C_2[\varepsilon(\boldsymbol{\varphi}^2)]\boldsymbol{\tau}^2)\cdot\frac{\partial \boldsymbol{\varphi}^2}{\partial \boldsymbol{\tau}^2}\Big\}\, d\Sigma.$$

PROOF. We sum (5.17) for $i = 1, 2$ and note that

$$\int_{\Gamma_2} = \int_{\Gamma^D} + \int_{\Gamma^C} + \int_{\Gamma_1}.$$

By using the boundary conditions (5.8) and (5.9), it is seen that the integrals over Σ^D and Σ^C reduce to $\mathcal{J}(\Sigma^D)$ and $\mathcal{J}(\Sigma^C)$, respectively. On Γ_1 we have $\boldsymbol{\nu}^2 = -\boldsymbol{\nu}^1$, $\boldsymbol{\tau}^2 = -\boldsymbol{\tau}^1$. Therefore

$$\sum_{i=1}^{2}\int_{\Sigma_1} (\mathbf{m}\cdot\boldsymbol{\nu}^i)(m_i|\dot{Z}^i|^2 + I_{\rho_i}|\dot{\boldsymbol{\varphi}}^i|^2)\, d\Sigma$$
$$= \int_{\Sigma_1} (\mathbf{m}\cdot\boldsymbol{\nu}^1)\left[(\rho_1-\rho_2)|\dot{Z}^1|^2 + (I_{\rho_1}-I_{\rho_2})|\dot{\boldsymbol{\varphi}}^1|^2\right] d\Sigma,$$

since $\dot{Z}^1 = \dot{Z}^2$ and $\dot{\varphi}^1 = \dot{\varphi}^2$ on Σ_1. Also, in view of the transmission conditions (5.7),

$$\sum_{i=1}^{2}\int_{\Sigma_i} (\mathbf{m}\cdot\boldsymbol{\nu}^i)\Big\{(C_i[\varepsilon(\boldsymbol{\varphi}^i)]\boldsymbol{\nu}^i)\cdot\frac{\partial \boldsymbol{\varphi}^i}{\partial \boldsymbol{\nu}^i} - (C_i[\varepsilon(\boldsymbol{\varphi}^i)]\boldsymbol{\tau}^i)\cdot\frac{\partial \boldsymbol{\varphi}^i}{\partial \boldsymbol{\tau}^i}$$
$$+ G_i h\left(\frac{\partial Z^i}{\partial \boldsymbol{\nu}^i} + \boldsymbol{\nu}^i\cdot\boldsymbol{\varphi}^i\right)^2\Big\}\, d\Sigma$$
$$= \int_{\Sigma_1} (\mathbf{m}\cdot\boldsymbol{\nu}^1)\Big\{(C_1[\varepsilon(\boldsymbol{\varphi}^1)]\boldsymbol{\nu}^1)\cdot\frac{\partial \boldsymbol{\varphi}^1}{\partial \boldsymbol{\nu}^1} - (C_2[\varepsilon(\boldsymbol{\varphi}^2)]\boldsymbol{\nu}^2)\cdot\frac{\partial \boldsymbol{\varphi}^2}{\partial \boldsymbol{\nu}^2}$$
$$- (C_1[\varepsilon(\boldsymbol{\varphi}^1)]\boldsymbol{\tau}^1)\cdot\frac{\partial \boldsymbol{\varphi}^1}{\partial \boldsymbol{\tau}^1} + (C_2[\varepsilon(\boldsymbol{\varphi}^2)]\boldsymbol{\tau}^2)\cdot\frac{\partial \boldsymbol{\varphi}^2}{\partial \boldsymbol{\tau}^2}$$
$$+ G_1 h(1-G_1G_2^{-1})\left(\frac{\partial Z^1}{\partial \boldsymbol{\nu}^1} + \boldsymbol{\nu}^1\cdot\boldsymbol{\varphi}^1\right)^2\Big\}\, d\Sigma.$$

By noting that

$$\frac{\partial \varphi^2}{\partial \boldsymbol{\tau}^2} = -\frac{\partial \varphi^1}{\partial \boldsymbol{\tau}^1}, \quad \frac{\partial Z^2}{\partial \boldsymbol{\tau}^2} = -\frac{\partial Z^1}{\partial \boldsymbol{\tau}^1} \quad \text{on } \Sigma_1,$$

one finds that

$$\sum_{i=1}^{2} \int_{\Sigma_i} (\mathbf{m}\cdot\boldsymbol{\nu}^i) \left\{ G_i h \left(\frac{\partial Z^i}{\partial \boldsymbol{\tau}^i} + \boldsymbol{\tau}^i \cdot \boldsymbol{\varphi}^i \right)^2 + 2G_i h \left(\frac{\partial Z^i}{\partial \boldsymbol{\nu}^i} + \boldsymbol{\nu}^i \cdot \boldsymbol{\varphi}^i \right) (\boldsymbol{\nu}^i \cdot \boldsymbol{\varphi}^i) \right\} d\Sigma$$
$$= \int_{\Sigma_1} (\mathbf{m}\cdot\boldsymbol{\nu}^1)(G_1 - G_2) h \left(\frac{\partial Z^1}{\partial \boldsymbol{\tau}^1} + \boldsymbol{\tau}^1 \cdot \boldsymbol{\varphi}^1 \right)^2 d\Sigma,$$

and that

$$\sum_{i=1}^{2} \int_{\Sigma_i} \left\{ (C_i[\varepsilon(\boldsymbol{\varphi}^i)]\boldsymbol{\nu}^i) \cdot \left[(\mathbf{m}\cdot\boldsymbol{\tau}^i) \frac{\partial \boldsymbol{\varphi}^i}{\partial \boldsymbol{\tau}^i} + (1-\delta)\boldsymbol{\varphi}^i \right] d\Sigma + G_i h \left(\frac{\partial Z^i}{\partial \boldsymbol{\nu}^i} + \boldsymbol{\nu}^i \cdot \boldsymbol{\varphi}^i \right) \left[(\mathbf{m}\cdot\boldsymbol{\tau}^i) \frac{\partial Z^i}{\partial \boldsymbol{\tau}^i} + \delta Z^i \right] \right\} d\Sigma = 0.$$

□

COROLLARY 5.1. *Let* $(\boldsymbol{\Phi}^1, \boldsymbol{\Phi}^2)$ *be a sufficiently smooth solution of* (5.6)–(5.9). *Suppose that the geometric conditions* (5.12) *hold and that* $\rho_1 \le \rho_2$, $G_1 \ge G_2$. *Then*

$$\begin{aligned}
(5.19)\quad R(t) - R(0) &+ \sum_{i=1}^{2} \int_{Q_i} \{(1-\delta) m_i |\dot{Z}^i|^2 + \delta I_{\rho_i} |\dot{\boldsymbol{\varphi}}^i|^2 \\
&+ (1-\delta)(C_i[\varepsilon(\boldsymbol{\varphi}^i)], \varepsilon(\boldsymbol{\varphi}^i)) + \delta G_i h (|\nabla Z^i|^2 - |\boldsymbol{\varphi}^i|^2)\} \, dQ \\
&\le \frac{1}{2} \int_{\Sigma^D} (\mathbf{m}\cdot\boldsymbol{\nu}^2) \left\{ \rho_2 |\dot{Z}^2|^2 d + I_{\rho_2} |\dot{\boldsymbol{\varphi}}^2|^2 \right\} d\Sigma \\
&+ \frac{1}{2} \int_{\Sigma_1} (\mathbf{m}\cdot\boldsymbol{\nu}^1) \left\{ (C_1[\varepsilon(\boldsymbol{\varphi}^1)]\boldsymbol{\nu}^1) \cdot \frac{\partial \boldsymbol{\varphi}^1}{\partial \boldsymbol{\nu}^1} - (C_2[\varepsilon(\boldsymbol{\varphi}^2)]\boldsymbol{\nu}^2) \cdot \frac{\partial \boldsymbol{\varphi}^2}{\partial \boldsymbol{\nu}^2} \right. \\
&\left. - (C_1[\varepsilon(\boldsymbol{\varphi}^1)]\boldsymbol{\tau}^1) \cdot \frac{\partial \boldsymbol{\varphi}^1}{\partial \boldsymbol{\tau}^1} + (C_2[\varepsilon(\boldsymbol{\varphi}^2)]\boldsymbol{\tau}^2) \cdot \frac{\partial \boldsymbol{\varphi}^2}{\partial \boldsymbol{\tau}^2} \right\} d\Sigma.
\end{aligned}$$

PROOF. On Γ one has

$$\varepsilon_{\alpha\beta}(\boldsymbol{\varphi}^i) = \frac{1}{2} \left(\nu^i_\beta \frac{\partial \varphi^i_\alpha}{\partial \boldsymbol{\nu}^i} + \nu^i_\alpha \frac{\partial \varphi^i_\beta}{\partial \boldsymbol{\nu}^i} + \tau^i_\beta \frac{\partial \varphi^i_\alpha}{\partial \boldsymbol{\tau}^i} + \tau^i_\alpha \frac{\partial \varphi^i_\beta}{\partial \boldsymbol{\tau}^i} \right).$$

Therefore, since $C_i^{\alpha\beta} = C_i^{\beta\alpha}$, on Γ^C we have

$$(5.20)\quad \begin{aligned}(C_2[\varepsilon(\boldsymbol{\varphi}^2)]\boldsymbol{\nu}^2)\cdot\frac{\partial\boldsymbol{\varphi}^2}{\partial\boldsymbol{\nu}^2} &= \sum_{\alpha,\beta}\nu_\alpha^2 C_2^{\alpha\beta}[\varepsilon(\boldsymbol{\varphi}^2)]\frac{\partial\varphi_\beta^2}{\partial\boldsymbol{\nu}^2}\\ &= \sum_{\alpha,\beta} C_2^{\alpha\beta}[\varepsilon(\boldsymbol{\varphi}^2)]\varepsilon_{\alpha\beta}(\boldsymbol{\varphi}^2) \geq 0,\end{aligned}$$

while on Γ^D,

$$\begin{aligned}(C_2[\varepsilon(\boldsymbol{\varphi}^2)]\boldsymbol{\tau}^2)\cdot\frac{\partial\boldsymbol{\varphi}^2}{\partial\boldsymbol{\tau}^2} &= \sum_{\alpha,\beta}\tau_\alpha^2 C_2^{\alpha\beta}[\varepsilon(\boldsymbol{\varphi}^2)]\frac{\partial\varphi_\beta^2}{\partial\boldsymbol{\tau}^2}\\ &= \sum_{\alpha,\beta} C_2^{\alpha\beta}[\varepsilon(\boldsymbol{\varphi}^2)]\varepsilon_{\alpha\beta}(\boldsymbol{\varphi}^2) - \sum_{\alpha,\beta}\nu_\alpha^2 C_2^{\alpha\beta}[\varepsilon(\boldsymbol{\varphi}^2)]\frac{\partial\varphi_\beta^2}{\partial\boldsymbol{\nu}^2}\\ &= \sum_{\alpha,\beta} C_2^{\alpha\beta}[\varepsilon(\boldsymbol{\varphi}^2)]\varepsilon_{\alpha\beta}(\boldsymbol{\varphi}^2) - (C_2[\varepsilon(\boldsymbol{\varphi}^2)]\boldsymbol{\nu}^2)\cdot\frac{\partial\boldsymbol{\varphi}^2}{\partial\boldsymbol{\nu}^2}\\ &= \sum_{\alpha,\beta} C_2^{\alpha\beta}[\varepsilon(\boldsymbol{\varphi}^2)]\varepsilon_{\alpha\beta}(\boldsymbol{\varphi}^2) \geq 0.\end{aligned}$$

The estimate (5.19) therefore follows from (5.18). □

LEMMA 5.3. *Suppose that* $(\boldsymbol{\Phi}_0, \boldsymbol{\Phi}_1) \in D_A \times \mathcal{V}$ *and let* $\boldsymbol{\Phi} = (\boldsymbol{\Phi}^1, \boldsymbol{\Phi}^2)$ *be the solution of* (5.6)–(5.10). *Assume that the geometric conditions* (5.12) *hold and that* Γ^C *and* Γ^D *meet in a strictly convex angle. Suppose further that*

$$\rho_1 \leq \rho_2,\ G_1 \geq G_2,\ \mu_1 \geq \mu_2,\ \lambda_1 \geq \lambda_2.$$

Then there is a $T_0 > 0$ *such that if* $T > T_0$,

$$(5.21)\quad \|(\boldsymbol{\Phi}_0, \boldsymbol{\Phi}_1)\|^2_{\mathcal{V}\times H} \leq C_T \int_{\Sigma^D} (\mathbf{m}\cdot\boldsymbol{\nu}^2)\left(\rho_2|\dot{Z}^2|^2 + I_{\rho_2}|\dot{\boldsymbol{\varphi}}^2|^2\right) d\Sigma.$$

PROOF. Our assumptions assure that the solution $\boldsymbol{\Phi}$ has the regularity required in the two previous lemmas. We first note the elementary inequality

$$|\nabla Z^i|^2 - |\boldsymbol{\varphi}^i|^2 \geq \frac{1}{2}|\nabla Z^i + \boldsymbol{\varphi}^i|^2 - 2|\boldsymbol{\varphi}^i|^2.$$

Also, since $\Gamma^C \neq \emptyset$,

$$\sum_{i=1}^{2}\int_{\Omega_i}|\boldsymbol{\varphi}^i|^2 d\Omega \leq C_0(\Omega)\sum_{i=1}^{2}\int_{\Omega_i}(C_i[\varepsilon(\boldsymbol{\varphi}^i)], \varepsilon(\boldsymbol{\varphi}^i))\, d\Omega.$$

Therefore, if $\delta > 0$ is small enough we have the estimate

$$\sum_{i=1}^{2}\int_{Q_i}\{(1-\delta)m_i|\dot{Z}^i|^2+\delta I_{\rho_i}|\dot{\varphi}^i|^2 + (1-\delta)(C_i[\varepsilon(\varphi^i)],\varepsilon(\varphi^i))+\delta G_i h(|\nabla Z^i|^2-|\varphi^i|^2)\}\,dQ$$
$$\geq c_0\sum_{i=1}^{2}\int_{Q_i}\{m_i|\dot{Z}_i|^2+I_{\rho_i}|\dot{\varphi}_i|^2+(C_i[\varepsilon(\varphi_i)],\varepsilon(\varphi_i))+G_i h|\nabla Z_i+\varphi_i|^2\}dQ$$
$$= c_0\int_0^T(\|\dot{\Phi}(t)\|_H^2+\|\Phi(t)\|_V^2)\,dt = c_0 T\|(\Phi^0,\Phi^1)\|_{V\times H}^2$$

for some constant $c_0 > 0$. Furthermore,

$$|R(t)| \leq C_1(\Omega)\|(\Phi(t),\dot{\Phi}(t))\|_{V\times H}^2 = C_1(\Omega)\|(\Phi_0,\Phi_1)\|_{V\times H}^2.$$

It then follows from (5.19) that

$$(c_0T-T_0)\|(\Phi_0,\Phi_1)\|_{V\times H}^2 \leq \frac{1}{2}\int_{\Sigma^D}(\mathbf{m}\cdot\boldsymbol{\nu}^2)\left\{\rho_2|\dot{Z}^2|^2+I_{\rho_2}|\dot{\varphi}^2|^2\right\}d\Sigma$$
$$+\frac{1}{2}\int_{\Sigma_1}(\mathbf{m}\cdot\boldsymbol{\nu}^1)\left\{(C_1[\varepsilon(\varphi^1)]\boldsymbol{\nu}^1)\cdot\frac{\partial\varphi^1}{\partial\boldsymbol{\nu}^1}-(C_2[\varepsilon(\varphi^2)]\boldsymbol{\nu}^2)\cdot\frac{\partial\varphi^2}{\partial\boldsymbol{\nu}^2}\right.$$
$$\left.-(C_1[\varepsilon(\varphi^1)]\boldsymbol{\tau}^1)\cdot\frac{\partial\varphi^1}{\partial\boldsymbol{\tau}^1}+(C_2[\varepsilon(\varphi^2)]\boldsymbol{\tau}^2)\cdot\frac{\partial\varphi^2}{\partial\boldsymbol{\tau}^2}\right\}d\Sigma,$$

where $T_0 = 2C_1(\Omega)$. Therefore, it remains to prove that

$$\int_{\Sigma_1}(\mathbf{m}\cdot\boldsymbol{\nu}^1)\left\{(C_1[\varepsilon(\varphi^1)]\boldsymbol{\nu}^1)\cdot\frac{\partial\varphi^1}{\partial\boldsymbol{\nu}^1}-(C_2[\varepsilon(\varphi^2)]\boldsymbol{\nu}^2)\cdot\frac{\partial\varphi^2}{\partial\boldsymbol{\nu}^2}\right.$$
$$\left.-(C_1[\varepsilon(\varphi^1)]\boldsymbol{\tau}^1)\cdot\frac{\partial\varphi^1}{\partial\boldsymbol{\tau}^1}+(C_2[\varepsilon(\varphi^2)]\boldsymbol{\tau}^2)\cdot\frac{\partial\varphi^2}{\partial\boldsymbol{\tau}^2}\right\}d\Sigma\leq 0 \tag{5.22}$$

provided

$$\mu_1 \geq \mu_2, \quad \lambda_1 \geq \lambda_2. \tag{5.23}$$

The proof of (5.22) is completely analogous to that given for serially connected membranes (cf. Theorem 3.2 and Remark 3.4, Chapter VI). To see the connection, let us recall that for two serially connected membranes or, more generally, in a transmission problem for membranes, the analog of (5.22) is

$$\int_{\Sigma_1}(\mathbf{m}\cdot\boldsymbol{\nu}^1)\left\{(S_1(\varphi^1)\boldsymbol{\nu}^1)\cdot\frac{\partial\varphi^1}{\partial\boldsymbol{\nu}^1}-(S_2(\varphi^2)\boldsymbol{\nu}^2)\cdot\frac{\partial\varphi^2}{\partial\boldsymbol{\nu}^2}\right.$$
$$\left.-(S_1(\varphi^1)\boldsymbol{\tau}^1)\cdot\frac{\partial\varphi^1}{\partial\boldsymbol{\tau}^1}+(S_2(\varphi^2)\boldsymbol{\tau}^2)\cdot\frac{\partial\varphi^2}{\partial\boldsymbol{\tau}^2}\right\}d\Sigma\leq 0, \tag{5.24}$$

where $S_i(\varphi^i) = (s_i^{\alpha\beta}(\varphi^i))$ is the stress tensor and

$$s_i^{\alpha\beta}(\varphi^i) = 2\mu_i\varepsilon_{\alpha\beta}(\varphi^i) + \lambda_i\delta_{\alpha\beta}\sum_\gamma \varepsilon_{\gamma\gamma}(\varphi^i). \tag{5.25}$$

The transmission conditions for the in-plane motion of serial membranes are

$$\varphi^1 = \varphi^2, \quad S_1(\varphi^1)\boldsymbol{\nu}^1 + S_2(\varphi^2)\boldsymbol{\nu}^2 = 0 \quad \text{on } \Gamma_1, \tag{5.26}$$

and it was demonstrated in the proof of Theorem 3.2, Chapter VI (where things are written in terms of normal and tangential components) that (5.24) is valid provided $\mu_1 \geq \mu_2$ and $\lambda_1 \geq \lambda_2$.

Now recall the relation between the strains $\varepsilon_{\alpha\beta}(\varphi^i)$ and the generalized moments $C_i^{\alpha\beta}[\varepsilon(\varphi^i)]$ is

$$C_i^{\alpha\beta}[\varepsilon(\varphi^i)] = \frac{h^3}{12}[2\mu_i\varepsilon_{\alpha\beta}(\varphi^i) + \lambda_i\delta_{\alpha\beta}\sum_\gamma \varepsilon_{\gamma\gamma}(\varphi^i)], \tag{5.27}$$

and that the transmission conditions on Γ_1 are the same as (5.26) if $S_i(\varphi^i)$ is replaced by $C_i[\varepsilon(\varphi^i)]$. Thus, comparing (5.27) with (5.25), the same argument used in the case of membranes applies in the present context and we may conclude that (5.22) is valid provided (5.23) holds. □

Proposition 5.1 now follows from the last lemma by "weakening the norm" (cf. Lions [60]).

6. Systems of Linked Kirchhoff Plates

As was noted in Section 2.5, the Kirchhoff thin plate model formally corresponds to absence of transverse shearing in the Reissner thin plate model, that is, $U_\alpha = -W_{3,\alpha}$. An equivalent way of viewing the Kirchhoff model is that it corresponds to the Reissner model with an infinite shear modulus $G = \infty$. In fact, a rigorous derivation of the Kirchhoff model as the limit of the Reissner model as $G \to \infty$ may be found in [51]. Here a similar approach will be followed in order to obtain a model for a system of linked Kirchhoff plates. This program will be carried out under the assumption that all joints of the linked Reissner system are semi-rigid and non serial. This is done mainly for convenience; there is no essential difficulty in treating other types of joints by a similar method and we shall comment on these other situations in Section 6.3 below. In addition, since our main interest is in determining the form of the junction conditions for a system of linked Kirchhoff plates, it shall be assumed that all boundary forces and couples vanish and, additionally, that $\Gamma_i^C \neq \emptyset$ for at least one index i. These assumptions do not alter the form of the junction conditions but do allow for a simpler convergence analysis. Thus, the problem under consideration is the limiting behavior as $G_i \to \infty$, $i = 1, \dots, n$, of solutions of the system (4.15)–(4.20) with $\mathbf{f}_0^i = \mathbf{c}^i = 0$, $f_3^i = 0$. Let us note that the model which will be derived may also be obtained by ad hoc considerations similar to those used to derive the model of linked Reissner plates, as was done in [48], but that approach would not reveal the close relationship between the two models.

6.1. Semi-rigid Joints. We write G for the n-tuple $(G_1, \dots, G_n)$ of shear moduli, $G > 0$ to mean that every $G_i > 0$ and $G \to \infty$ to mean that $G_i \to \infty$ for every $i = 1, \dots, n$. Since $\Gamma_i^C \neq \emptyset$ for some i, according to Lemma 4.1 the seminorm $e(\cdot)$ defines a norm equivalent to the $\|\cdot\|_V$ norm on the space $\mathcal{V}$. The space $\mathcal{V}$ and the norm $e(\cdot)$ on $\mathcal{V}$ obviously depend on G and we write $\mathcal{V}_G$ in place of $\mathcal{V}$ and $\|\mathbf{R}\|_{\mathcal{V}_G}$ in place of $e_G(\mathbf{R})$ when $\mathbf{R} \in \mathcal{V}_G$.

A typical element of H is described by (4.3). We introduce the following *closed* subspace H_∞ of H:

$$H_\infty = \{\mathbf{R} \in H \,|\, W_3^i \in H^1(\Omega_i),\ \mathbf{u}^i = -\nabla W_3^i,\ W_3^i = 0 \text{ on } \Gamma_i^C\}.$$

If $\mathbf{R} \in H_\infty$, then $\mathbf{R}^i = (\mathbf{W}^i, -\nabla W_3^i)$ and

$$\|\mathbf{R}\|_{H_\infty} = \|\mathbf{R}\|_H = \left\{ \sum_{i=1}^n \int_{\Omega_i} [m_i|\mathbf{W}^i|^2 + I_{\rho_i}|\nabla W_3^i|^2]\, d\Omega \right\}^{1/2}.$$

An element $\mathbf{R}$ in the orthogonal complement $H_\infty^\perp$ of H_∞ in H is characterized by the variational equation

$$\sum_{i=1}^n \int_{\Omega_i} (m_i \mathbf{W}^i \cdot \hat{\mathbf{W}}^i - I_{\rho_i} \mathbf{u}^i \cdot \nabla \hat{W}_3^i)\, d\Omega = 0,$$

which is formally equivalent to the system

$$\mathbf{w} = 0,\ \ m_i W_3^i + I_{\rho_i} \operatorname{div} \mathbf{u}^i = 0 \ \text{ in } \Omega_i,\ \ \boldsymbol{\nu}^i \cdot \mathbf{u}^i = 0 \ \text{ on } \Gamma_i \setminus \Gamma_i^C.$$

The decomposition of an element $\mathbf{R} \in H$ into $\mathbf{R}_\infty + \mathbf{R}_\infty^\perp$, where $\mathbf{R}_\infty \in H_\infty$ and $\mathbf{R}_\infty^\perp \in H_\infty^\perp$, is given by

$$\mathbf{w}_\infty^i = \mathbf{w}^i,\ \ (\mathbf{w}_\infty^i)^\perp = 0,$$
$$m_i (W_3^i)_\infty - I_{\rho_i} \operatorname{div}(\mathbf{u}_\infty^i)^\perp = m_i W_3^i,\ \ -\nabla (W_3^i)_\infty + (\mathbf{u}_\infty^i)^\perp = \mathbf{u}^i,$$
$$(W_3^i)_\infty = 0 \ \text{ on } \Gamma_i^C,\ \ \boldsymbol{\nu}^i \cdot (\mathbf{u}_\infty^i)^\perp = 0 \ \text{ on } \Gamma_i \setminus \Gamma_i^C.$$

We similarly define a closed subspace $\mathcal{V}_\infty$ of $\mathcal{V}_G$ by

$$\mathcal{V}_\infty = \mathcal{V}_G \cap H_\infty = \{\mathbf{R} \in \mathcal{V}_G \,|\, W_3^i \in H^2(\Omega_i),\ \mathbf{u}^i = -\nabla W_3^i\}.$$

Since we are assuming that all joints are semi-rigid and non serial, functions in $\mathcal{V}_G$ satisfy the geometric conditions

$$\mathbf{R}^i \in V_i,\ \ \mathbf{R}^i = 0 \ \text{ on } \Gamma_i^C,$$

$$(6.1)\qquad \begin{cases} \mathbf{W}^i(\mathbf{p}_i^{-1}(\mathbf{p})) = \mathbf{W}^j(\mathbf{p}_j^{-1}(\mathbf{p})), \\ (\boldsymbol{\nu}^i \cdot \mathbf{u}^i)(\mathbf{p}_i^{-1}(\mathbf{p})) = \sigma_{ij}(\boldsymbol{\nu}^j \cdot \mathbf{u}^j)(\mathbf{p}_j^{-1}(\mathbf{p})), \\ \boldsymbol{\tau}^i \cdot (\mathbf{u}^i + \nabla W_3^i)(\mathbf{p}_i^{-1}(\mathbf{p})) = 0, \end{cases}$$
$$\forall \mathbf{p} \in J_k,\ \forall i,j \in \mathcal{I}(J_k),\ k = 1,\dots,m.$$

It follows that functions in $\mathcal{V}_\infty$ satisfy the geometric conditions

$$(6.2)\qquad \begin{cases} W_\alpha^i \in H^1(\Omega_i),\ \ W_3^i \in H^2(\Omega_i), \\ \mathbf{W}^i = 0,\ \ \dfrac{\partial W_3^i}{\partial \nu^i} = 0 \ \text{ on } \Gamma_i^C, \end{cases}$$

$$(6.3)\qquad \begin{cases} \mathbf{W}^i(\mathbf{p}_i^{-1}(\mathbf{p})) = \mathbf{W}^j(\mathbf{p}_j^{-1}(\mathbf{p})), \\ \dfrac{\partial W_3^i}{\partial \nu^i}(\mathbf{p}_i^{-1}(\mathbf{p})) = \sigma_{ij}\dfrac{\partial W_3^j}{\partial \nu^j}(\mathbf{p}_j^{-1}(\mathbf{p})), \end{cases}$$
$$\forall \mathbf{p} \in J_k,\ \forall i,j \in \mathcal{I}(J_k),\ k = 1,\dots,m.$$

Furthermore, for $\mathbf{R} \in \mathcal{V}_\infty$ one has

$$\|\mathbf{R}\|_{\mathcal{V}_\infty} = \|\mathbf{R}\|_{\mathcal{V}_G} = \left\{\sum_{i=1}^n \int_{\Omega_i} \{(C_i[\varepsilon(\mathbf{w}^i)], \varepsilon(\mathbf{w}^i)) + \frac{h^2}{12}(C_i[\varepsilon(\nabla W_3^i)], \varepsilon(\nabla W_3^i))\}d\Gamma\right\}^{1/2}.$$

Note that $\mathcal{V}_\infty$ is independent of G and is dense in H_∞ with continuous injection, so that if H_∞ is identified with its dual space we may write

$$\mathcal{V}_\infty \subset H_\infty \subset \mathcal{V}'_\infty,$$

where $\mathcal{V}'_\infty$ denotes the dual space of $\mathcal{V}_\infty$. Let A_G denote the Riesz isomorphism of $\mathcal{V}_G$ onto $\mathcal{V}'_G$ and A_∞ denote the Riesz isomorphism of $\mathcal{V}_\infty$ onto $\mathcal{V}'_\infty$. If $\mathbf{R} \in \mathcal{V}_\infty$, then $A_\infty\mathbf{R}$ is simply the restriction of $A_G\mathbf{R}$ to $\mathcal{V}_\infty$, i.e.,

$$\langle A_\infty \mathbf{R}, \hat{\mathbf{R}}\rangle_{\mathcal{V}_\infty} = \langle A_G \mathbf{R}, \hat{\mathbf{R}}\rangle_{\mathcal{V}_G},\quad \forall \mathbf{R}, \hat{\mathbf{R}} \in \mathcal{V}_\infty.$$

The variational formulation of the system under consideration is given by (4.25). This equation may be written in the equivalent form

$$(6.4)\qquad \ddot{\mathbf{R}}^G + A_G \mathbf{R}^G = \mathbf{G},$$

where $\dot{\mathbf{R}}$ stands for $\mathbf{R}_{,t}$, $\ddot{\mathbf{R}}$ for $\mathbf{R}_{,tt}$, and where the superscript G reflects the dependence of the solution on the shear moduli. The initial conditions are

$$(6.5)\qquad \mathbf{R}^G(0) = \mathbf{R}_0,\ \ \dot{\mathbf{R}}^G(0) = \mathbf{R}_1.$$

We assume that the data of the problem satisfy

$$\mathbf{R}_0 \in \mathcal{V}_\infty, \quad \mathbf{R}_1 \in H_\infty, \quad \mathbf{G} \in L^2(H). \tag{6.6}$$

(In (6.6) and in what follows, we shall write $L^2(H)$, $L^2(\mathcal{V}_G)$, $C(H)$, etc., to mean $L^2(0,T;H)$, $L^2(0,T;\mathcal{V}_G)$, $C([0,T];H)$, etc., where it is understood that $T > 0$ is fixed.)

REMARK 6.1. The hypotheses on the initial data mean that each plate is initially free of transverse shear and has finite (Kirchhoff) kinetic energy. Thus they are assumptions of *consistency* with the Kirchhoff plate model. □

With assumption (6.6), the problem (6.4), (6.5) has a unique solution $\mathbf{R}^G \in C(\mathcal{V}_G)$ with $\dot{\mathbf{R}}^G \in C(H)$ and $\ddot{\mathbf{R}} \in L^2(\mathcal{V}_G')$. We may decompose $\mathbf{R}^G$ and $\dot{\mathbf{R}}^G$ into

$$\mathbf{R}^G = \mathbf{R}^G_\infty + (\mathbf{R}^G_\infty)^\perp, \quad \dot{\mathbf{R}}^G = \dot{\mathbf{R}}^G_\infty + (\dot{\mathbf{R}}^G_\infty)^\perp,$$

where $\mathbf{R}^G_\infty$, $\dot{\mathbf{R}}^G_\infty \in C(H_\infty)$ and $(\mathbf{R}^G_\infty)^\perp$, $(\dot{\mathbf{R}}^G_\infty)^\perp \in C(H^\perp_\infty)$. Similarly, we write

$$\mathbf{G} = \mathbf{G}_\infty + \mathbf{G}^\perp_\infty, \quad \mathbf{G}_\infty \in L^2(H_\infty), \ \mathbf{G}^\perp_\infty \in L^2(H^\perp_\infty).$$

Let $(\ddot{\mathbf{R}}^G_\infty)(t) \in \mathcal{V}'_\infty$ denote the restriction of $\ddot{\mathbf{R}}^G(t)$ to $\mathcal{V}_\infty$. If $\hat{\mathbf{R}} \in W^1(H_\infty)$ satisfies $\hat{\mathbf{R}}(0) = \hat{\mathbf{R}}(T) = 0$, one easily verifies that

$$\int_0^T (\mathbf{R}^G_\infty, \frac{d}{dt}\hat{\mathbf{R}})_{H_\infty} dt = -\int_0^T (\dot{\mathbf{R}}^G_\infty, \hat{\mathbf{R}})_{H_\infty} dt,$$

so that

$$\dot{\mathbf{R}}^G_\infty = \frac{d}{dt}\mathbf{R}^G_\infty \text{ and, similarly, } \ddot{\mathbf{R}}^G_\infty = \frac{d}{dt}\dot{\mathbf{R}}^G_\infty.$$

THEOREM 6.1. *Assume* (6.6). *There is a function* $\mathbf{R} \in C(\mathcal{V}_\infty)$ *with* $\dot{\mathbf{R}} \in C(H_\infty)$ *and* $\ddot{\mathbf{R}} \in L^2(\mathcal{V}'_\infty)$ *such that as* $G \to \infty$,

$$\mathbf{R}^G \to \mathbf{R} \ \textit{strongly in } L^\infty(H) \textit{ and weakly* in } L^\infty(V);$$
$$\dot{\mathbf{R}}^G \to \dot{\mathbf{R}} \ \textit{weakly* in } L^\infty(H);$$
$$\dot{\mathbf{R}}^G_\infty \to \dot{\mathbf{R}} \ \textit{strongly in } L^\infty(\mathcal{V}'_\infty);$$
$$\ddot{\mathbf{R}}^G_\infty \to \ddot{\mathbf{R}} \ \textit{weakly in } L^2(\mathcal{V}'_\infty).$$

Moreover, $\mathbf{R}$ *is the unique solution of*

$$\ddot{\mathbf{R}} + A_\infty \mathbf{R} = \mathbf{G}_\infty, \tag{6.7}$$

$$\mathbf{R}(0) = \mathbf{R}_0, \quad \dot{\mathbf{R}}(0) = \mathbf{R}_1. \tag{6.8}$$

6.1.1. *Interpretation of Theorem* 6.1. Before proving Theorem 6.1, let us give its interpretation. Equation (6.7) signifies that $\mathbf{R}$ is a solution of the variational equation

$$\langle \ddot{\mathbf{R}}, \hat{\mathbf{R}} \rangle_{\mathcal{V}_\infty} + (\mathbf{R}, \hat{\mathbf{R}})_{\mathcal{V}_\infty} = (\mathbf{G}_\infty, \hat{\mathbf{R}})_{H_\infty}, \quad \forall \hat{\mathbf{R}} \in \mathcal{V}_\infty. \tag{6.9}$$

We write

$$\mathbf{G} = (\mathbf{G}^1, \dots, \mathbf{G}^n), \quad \mathbf{G}^i = (m_i^{-1}\mathbf{F}^i, I_{\rho_i}\mathbf{C}^i),$$

and $\mathbf{R} = (\mathbf{R}^1, \dots, \mathbf{R}^n)$ with

$$\mathbf{R}^i = (\mathbf{W}^i, -\nabla W_3^i), \quad \mathbf{W}^i = \sum_j W_j^i \mathbf{a}_j^i = \mathbf{w}^i + W_3^i \mathbf{a}_3^i,$$

$$\mathbf{w}^i = \sum_\alpha W_\alpha^i \mathbf{a}_\alpha^i, \quad \nabla W_3^i = \sum_\alpha W_{3,\alpha}^i \mathbf{a}_\alpha^i.$$

Then

$$(\mathbf{R}, \hat{\mathbf{R}})_{\mathcal{V}_\infty} = \sum_{i=1}^n \int_{\Omega_i} \{(C_i[\varepsilon(\mathbf{w}^i)], \varepsilon(\hat{\mathbf{w}}^i)) + \frac{h^2}{12}(C_i[\varepsilon(\nabla W_3^i)], \varepsilon(\nabla \hat{W}_3^i))\} d\Omega. \tag{6.10}$$

We have

$$\begin{aligned} \int_{\Omega_i} (C_i[\varepsilon(\mathbf{w}^i)], \varepsilon(\hat{\mathbf{w}}^i))\, d\Omega &= \frac{1}{2}\sum_{\alpha,\beta} \int_{\Omega_i} C_i^{\alpha\beta}[\varepsilon(\mathbf{w}^i)](\hat{W}_{\alpha,\beta}^i + \hat{W}_{\beta,\alpha}^i) \\ &= -\sum_{\alpha,\beta} \int_{\Omega_i} \hat{W}_\alpha^i \frac{\partial}{\partial x_\beta} C_i^{\alpha\beta}[\varepsilon(\mathbf{w}^i)]\, d\Omega \\ &\qquad + \sum_{\alpha,\beta} \int_{\Gamma_i} \hat{W}_\alpha^i \nu_\beta^i C_i^{\alpha\beta}[\varepsilon(\mathbf{w}^i)]\, d\Gamma \\ &= -\int_{\Omega_i} \hat{\mathbf{w}}^i \cdot \operatorname{div} C_i[\varepsilon(\mathbf{w}^i)]\, d\Omega + \int_{\Gamma_i} \hat{\mathbf{w}}^i \cdot (C_i[\varepsilon(\mathbf{w}^i)]\boldsymbol{\nu}^i)\, d\Gamma. \end{aligned} \tag{6.11}$$

Therefore

$$\begin{aligned} &\int_{\Omega_i} (C_i[\varepsilon(\nabla W_3^i)], \varepsilon(\nabla \hat{W}_3^i))\, d\Omega \\ &= \sum_{\alpha,\beta} \int_{\Omega_i} \hat{W}_3^i \frac{\partial^2}{\partial x_\alpha \partial x_\beta} C_i^{\alpha\beta}[\varepsilon(\nabla W_3^i)]\, d\Omega - \int_{\Gamma_i} (\nabla \hat{W}_3^i) \cdot (C_i[\varepsilon(\nabla W_3^i)]\boldsymbol{\nu}^i)\, d\Gamma \\ &= (2\mu_i + \lambda_i) h \int_{\Omega_i} \hat{W}_3^i \Delta^2 W_3^i\, d\Omega - (2\mu_i + \lambda_i) h \int_{\Gamma_i} \{\hat{W}_3^i \frac{\partial}{\partial \boldsymbol{\nu}^i}(\Delta W_3^i) \\ &\qquad - (\nabla \hat{W}_3^i) \cdot (C_i[\varepsilon(\nabla W_3^i)]\boldsymbol{\nu}^i)\}\, d\Gamma, \end{aligned} \tag{6.12}$$

where $\Delta = \sum_\alpha \partial^2/\partial x_\alpha^2$ is the harmonic operator and $\Delta^2 = \Delta\Delta$ the biharmonic operator in $\mathbb{R}^2$. In addition, at least formally we have

$$(6.13)\quad \langle \ddot{\mathbf{R}}, \hat{\mathbf{R}} \rangle_{\mathcal{V}_\infty} = \sum_{i=1}^{n} \int_{\Omega_i} (m_i \ddot{\mathbf{W}}^i \cdot \hat{\mathbf{W}}^i + I_{\rho_i} \nabla \ddot{W}_3^i \cdot \nabla \hat{W}_3^i)\, d\Omega$$

$$= \sum_{i=1}^{n} \int_{\Omega_i} [m_i \ddot{\mathbf{w}}^i \cdot \hat{\mathbf{w}}^i + (m_i \ddot{W}_3^i - I_{\rho_i} \Delta \ddot{W}_3^i) \hat{W}_3^i]\, d\Omega + \int_{\Gamma_i} I_{\rho_i} \frac{\partial \ddot{W}_3^i}{\partial \boldsymbol{\nu}^i} \hat{W}_3^i \, d\Gamma.$$

Furthermore, for every $\hat{\mathbf{R}} \in \mathcal{V}_\infty$ we have

$$(6.14)\quad (\mathbf{G}_\infty, \hat{\mathbf{R}})_{H_\infty} = (\mathbf{G}, \hat{\mathbf{R}})_H = \sum_{i=1}^{n} \int_{\Omega_i} (\mathbf{F}^i \cdot \hat{\mathbf{W}}^i - \mathbf{C}^i \cdot \nabla \hat{W}_3^i)\, d\Omega$$

$$= \sum_{i=1}^{n} \Big\{ \int_{\Omega_i} [\mathbf{F}_0^i \cdot \hat{\mathbf{w}}^i + (F_3^i + \operatorname{div} \mathbf{C}^i) \hat{W}_3^i]\, d\Omega - \int_{\Gamma_i \setminus \Gamma_i^C} \hat{W}_3^i (\boldsymbol{\nu}^i \cdot \mathbf{C}^i)\, d\Gamma,$$

where $\operatorname{div} \mathbf{C}^i = \sum_\alpha C_{\alpha,\alpha}^i$ and where we have written

$$\mathbf{F}^i = \sum_j F_j^i \mathbf{a}_j^i := \mathbf{F}_0^i + F_3^i \mathbf{a}_3^i.$$

It follows from (6.9)–(6.14) that $(\mathbf{w}^1, \dots, \mathbf{w}^n)$ and $(W_3^1, \dots, W_3^n)$ satisfy the system of partial differential equations

$$(6.15)\quad \begin{cases} m_i \ddot{\mathbf{w}}^i - \operatorname{div} C_i[\varepsilon(\mathbf{w}^i)] = \mathbf{F}_0^i, \\ m_i \ddot{W}_3^i - I_{\rho_i} \Delta \ddot{W}_3^i + D_i \Delta^2 W_3^i = F_3^i + \operatorname{div} \mathbf{C}^i, \end{cases}$$

in agreement with our formal calculation in Section 2.5, where $D_i = (2\mu_i + \lambda_i)h^3/12$. Since $\mathbf{R}$ takes is values in $\mathcal{V}_\infty$, the geometric boundary conditions (6.2) and geometric junction conditions (6.3) are also satisfied by $\mathbf{R}$. In addition, for all $\hat{\mathbf{R}} \in \mathcal{V}_\infty$ we must have

(6.16)

$$\sum_{i=1}^{n} \int_{\Gamma_i \setminus \Gamma_i^C} \Bigg\{ \hat{\mathbf{w}}^i \cdot (C_i[\varepsilon(\mathbf{w}^i)] \boldsymbol{\nu}^i) + + \hat{W}_3^i \left[I_{\rho_i} \frac{\partial \ddot{W}_3^i}{\partial \boldsymbol{\nu}^i} - D_i \frac{\partial}{\partial \boldsymbol{\nu}^i} (\Delta W_3^i) \right]$$

$$+ \frac{h^2}{12} (\nabla \hat{W}_3^i) \cdot (C_i[\varepsilon(\nabla W_3^i)] \boldsymbol{\nu}^i) \Bigg\}\, d\Gamma = -\sum_{i=1}^{n} \int_{\Gamma_i \setminus \Gamma_i^C} \hat{W}_3^i (\boldsymbol{\nu}^i \cdot \mathbf{C}^i)\, d\Gamma.$$

We may deduce the dynamic boundary and junction conditions from (6.3) and (6.16). We write $\nabla \hat{W}_3^i$ and $C_i[\varepsilon(\nabla W_3^i)] \boldsymbol{\nu}^i$ in terms of normal and tangential components:

$$\nabla \hat{W}_3^i = \frac{\partial \hat{W}_3^i}{\partial \boldsymbol{\nu}^i} \boldsymbol{\nu}^i + \frac{\partial \hat{W}_3^i}{\partial \boldsymbol{\tau}^i} \boldsymbol{\tau}^i,$$
$$C_i[\varepsilon(\nabla W_3^i)] \boldsymbol{\nu}^i = C_{\boldsymbol{\nu}^i}(\nabla W_3^i) \boldsymbol{\nu}^i + C_{\boldsymbol{\tau}^i}(\nabla W_3^i) \boldsymbol{\tau}^i,$$

where

$$C_{\nu^i}(\nabla W_3^i) = \boldsymbol{\nu}^i \cdot (C_i[\varepsilon(\nabla W_3^i)]\boldsymbol{\nu}^i),$$
$$C_{\tau^i}(\nabla W_3^i) = \boldsymbol{\tau}^i \cdot (C_i[\varepsilon(\nabla W_3^i)]\boldsymbol{\nu}^i).$$

Then (6.16) may be written

$$(6.17)\quad \sum_{i=1}^{n} \int_{\Gamma_i \backslash \Gamma_i^C} \left\{ \hat{\mathbf{w}}^i \cdot (C_i[\varepsilon(\mathbf{w}^i)]\boldsymbol{\nu}^i) + \hat{W}_3^i \left[I_{\rho_i} \frac{\partial \ddot{W}_3^i}{\partial \boldsymbol{\nu}^i} - D_i \frac{\partial}{\partial \boldsymbol{\nu}^i}(\Delta W_3^i) - \frac{h^2}{12} \frac{\partial}{\partial \boldsymbol{\tau}^i} C_{\tau^i}(\nabla W_3^i) \right] + \frac{h^2}{12} \frac{\partial \hat{W}_3^i}{\partial \boldsymbol{\nu}^i} C_{\nu^i}(\nabla W_3^i) \right\} d\Gamma = -\sum_{i=1}^{n} \int_{\Gamma_i / bs \Gamma_i^C} \hat{W}_3^i (\boldsymbol{\nu}^i \cdot \mathbf{C}^i)\, d\Gamma.$$

The dynamic boundary conditions on Γ_i^D follow immediately from (6.17); we obtain

$$(6.18)\quad \begin{cases} C_i[\varepsilon(\mathbf{w}^i)]\boldsymbol{\nu}^i = 0, \\ C_{\nu^i}(\nabla W_3^i) = 0, \\ I_{\rho_i} \dfrac{\partial \ddot{W}_3^i}{\partial \boldsymbol{\nu}^i} - D_i \dfrac{\partial}{\partial \boldsymbol{\nu}^i}(\Delta W_3^i) - \dfrac{h^2}{12} \dfrac{\partial}{\partial \boldsymbol{\tau}^i} C_{\tau^i}(\nabla W_3^i) \\ \qquad\qquad = -\boldsymbol{\nu}^i \cdot \mathbf{C}^i \quad \text{on } \Gamma_i^D. \end{cases}$$

It may be verified that these are the same boundary conditions which were formally obtained in Section 2.5.

The dynamic junction conditions along a junction J_k also follow from (6.17). They are, in variational form,

$$(6.19)\quad \sum_{i \in \mathcal{I}(J_k)} \int_{\Gamma_i(J_k)} \left\{ \left(C_i[\varepsilon(\mathbf{w}^i)]\boldsymbol{\nu}^i + \left[I_{\rho_i} \frac{\partial \ddot{W}_3^i}{\partial \boldsymbol{\nu}^i} - D_i \frac{\partial}{\partial \boldsymbol{\nu}^i}(\Delta W_3^i) - \frac{h^2}{12} \frac{\partial}{\partial \boldsymbol{\tau}^i} C_{\tau^i}(\nabla W_3^i) \right] \mathbf{a}_3^i \right) \cdot \hat{\mathbf{W}}^{i_k} + \frac{h^2}{12} \sigma_i(J_k) C_{\nu^i}(\nabla W_3^i) \frac{\partial \hat{W}_3^{i_k}}{\partial \boldsymbol{\nu}^{i_k}} \right\} d\Gamma = -\sum_{i \in \mathcal{I}(J_k)} \int_{\Gamma_i(J_k)} (\boldsymbol{\nu}^i \cdot \mathbf{C}^i)\mathbf{a}_3^i \cdot \hat{\mathbf{W}}^{i_k}\, d\Gamma.$$

We have utilized (6.3) in order to obtain (6.19) from (6.17). The dynamic junctions conditions along J_k are therefore

$$(6.20)\quad \sum_{i\in\mathcal{I}(J_k)}\left\{C_i[\varepsilon(\mathbf{w}^i)]\boldsymbol{\nu}^i+\left[I_{\rho_i}\frac{\partial\ddot{W}_3^i}{\partial\boldsymbol{\nu}^i}-D_i\frac{\partial}{\partial\boldsymbol{\nu}^i}(\Delta W_3^i)\right.\right.$$
$$\left.\left.-\frac{h^2}{12}\frac{\partial}{\partial\boldsymbol{\tau}^i}C_{\boldsymbol{\tau}^i}(\nabla W_3^i)\right]\mathbf{a}_3^i\right\}=-\sum_{i\in\mathcal{I}(J_k)}(\boldsymbol{\nu}^i\cdot\mathbf{C}^i)\mathbf{a}_3^i,\quad k=1,\dots,m,$$

and

$$(6.21)\qquad \sum_{i\in\mathcal{I}(J_k)}\sigma_i(J_k)C_{\boldsymbol{\nu}^i}(\nabla W_3^i)=0.$$

In each of the last two equations, the ith term of the sum is evaluated at $\mathbf{p}_i^{-1}(\mathbf{p})$, $\mathbf{p}\in J_k$. Equation (6.20) is a balance law for the transmission of linear momentum along J_k, while (6.21) describes how bending is transmitted there. Therefore, the dynamics of a system of linked Kirchhoff plates with semi-rigid joints are described by the equations of motion (6.15); the geometric boundary conditions (6.2); the geometric junction conditions (6.3); the dynamic boundary conditions (6.18); and the dynamic junctions conditions (6.20) and (6.21).

There is a very simple geometric interpretation of the junction condition $(\partial W_3^i/\partial\boldsymbol{\nu}^i)=\sigma_{ij}(\partial W_3^j/\partial\boldsymbol{\nu}^j)$. Suppose that the reference surfaces $\mathcal{P}_i$ and $\mathcal{P}_j$ are joined along J. The cosine of the angle between $\mathcal{P}_i$ and $\mathcal{P}_j$ is $\mathbf{a}_3^i\cdot\mathbf{a}_3^j$. On the other hand, the cosine of the angle between the deformed reference surfaces is

$$\mathbf{n}^i(\mathbf{p}_i^{-1}(\mathbf{p}))\cdot\mathbf{n}^j(\mathbf{p}_j^{-1}(\mathbf{p})),\quad \mathbf{p}\in J,$$

where $\mathbf{n}^i$, a unit vector normal to the deformed reference surface $\mathcal{P}_i$, is given by

$$\mathbf{n}^i=\frac{\mathbf{A}_1^i\times\mathbf{A}_2^i}{|\mathbf{A}_1^i\times\mathbf{A}_2^i|},\quad \mathbf{A}_\alpha^i=\mathbf{a}_\alpha^i+\mathbf{W}_{,\alpha}^i.$$

PROPOSITION 6.1. *The junction condition*

$$\frac{\partial W_3^i}{\partial\boldsymbol{\nu}^i}(\mathbf{p}_i^{-1}(\mathbf{p}))=\sigma_{ij}\frac{\partial W_3^j}{\partial\boldsymbol{\nu}^j}(\mathbf{p}_j^{-1}(\mathbf{p}))$$

is the linearization of the nonlinear constraint

$$(6.22)\qquad \mathbf{n}^i(\mathbf{p}_i^{-1}(\mathbf{p}))\cdot\mathbf{n}^j(\mathbf{p}_j^{-1}(\mathbf{p}))=\mathbf{a}_3^i\cdot\mathbf{a}_3^j.$$

Condition (6.22) obviously means that the angle between $\mathcal{P}_i$ and $\mathcal{P}_j$ is the same after deformation as before.

PROOF. The proof is by direct calculation. We have

$$\mathbf{A}_1^i \times \mathbf{A}_2^i = \mathbf{a}_3^i + \mathbf{a}_1^i \times \mathbf{W}_{,2}^i - \mathbf{a}_2^i \times \mathbf{W}_{,1}^i + \mathbf{W}_{,1}^i \times \mathbf{W}_{,2}^i,$$

so that (where $\cdots$ represent higher order terms in the displacement)

$$|\mathbf{A}_1^i \times \mathbf{A}_2^i| = 1 + \mathbf{a}_1^i \cdot \mathbf{W}_{,1}^i + \mathbf{a}_2^i \cdot \mathbf{W}_{,2}^i + \cdots,$$

$$|\mathbf{A}_1^i \times \mathbf{A}_2^i||\mathbf{A}_1^j \times \mathbf{A}_2^j| = 1 + \mathbf{a}_1^i \cdot \mathbf{W}_{,1}^i + \mathbf{a}_2^i \cdot \mathbf{W}_{,2}^i + \mathbf{a}_1^j \cdot \mathbf{W}_{,1}^j + \mathbf{a}_2^j \cdot \mathbf{W}_{,2}^j + \cdots,$$

$$(\mathbf{A}_1^i \times \mathbf{A}_2^i) \cdot (\mathbf{A}_1^j \times \mathbf{A}_2^j) = \mathbf{a}_3^i \cdot \mathbf{a}_3^j + (\mathbf{a}_3^i \times \mathbf{a}_1^j) \cdot \mathbf{W}_{,2}^j - (\mathbf{a}_3^i \times \mathbf{a}_2^j) \cdot \mathbf{W}_{,1}^j + (\mathbf{a}_3^j \times \mathbf{a}_1^i) \cdot \mathbf{W}_{,2}^i - (\mathbf{a}_3^j \times \mathbf{a}_2^i) \cdot \mathbf{W}_{,1}^i + \cdots.$$

If only linear terms are retained in (6.22) we obtain

$$0 = (\mathbf{a}_3^i \times \mathbf{a}_1^j) \cdot \mathbf{W}_{,2}^j - (\mathbf{a}_3^i \cdot \mathbf{a}_3^j)(\mathbf{a}_2^j \cdot \mathbf{W}_{,2}^j) - (\mathbf{a}_3^i \times \mathbf{a}_2^j) \cdot \mathbf{W}_{,1}^j - (\mathbf{a}_3^i \cdot \mathbf{a}_3^j)(\mathbf{a}_1^j \cdot \mathbf{W}_{,1}^j) + (\mathbf{a}_3^j \times \mathbf{a}_1^i) \cdot \mathbf{W}_{,2}^i - (\mathbf{a}_3^i \cdot \mathbf{a}_3^j)(\mathbf{a}_2^i \cdot \mathbf{W}_{,2}^i) - (\mathbf{a}_3^j \times \mathbf{a}_2^i) \cdot \mathbf{W}_{,1}^i - (\mathbf{a}_3^i \cdot \mathbf{a}_3^j)(\mathbf{a}_1^i \cdot \mathbf{W}_{,1}^i). \tag{6.23}$$

In the cross products, express $\mathbf{a}_3^i$ in terms of the $\mathbf{a}_1^j, \mathbf{a}_2^j, \mathbf{a}_3^j$ basis and $\mathbf{a}_3^j$ in terms of the $\mathbf{a}_1^i, \mathbf{a}_2^i, \mathbf{a}_3^i$ basis. Then (6.23) simplifies to

$$0 = (\mathbf{a}_3^i \cdot \mathbf{a}_2^j)(\mathbf{a}_3^j \cdot \mathbf{W}_{,2}^j) + (\mathbf{a}_3^i \cdot \mathbf{a}_1^j)(\mathbf{a}_3^j \cdot \mathbf{W}_{,1}^j) + (\mathbf{a}_3^j \cdot \mathbf{a}_2^i)(\mathbf{a}_3^i \cdot \mathbf{W}_{,2}^i) + (\mathbf{a}_3^j \cdot \mathbf{a}_1^i)(\mathbf{a}_3^i \cdot \mathbf{W}_{,1}^i),$$

that is,

$$\mathbf{a}_3^j \cdot \nabla W_3^i + \mathbf{a}_3^i \cdot \nabla W_3^j = 0, \tag{6.24}$$

where $\nabla W_3^i = \sum_\alpha W_{3,\alpha}^i \mathbf{a}_\alpha^i$. If we write

$$\mathbf{a}_3^j = (\mathbf{a}_3^j \cdot \mathbf{a}_3^i)\mathbf{a}_3^i + (\mathbf{a}_3^j \cdot \boldsymbol{\nu}^i)\boldsymbol{\nu}^i,$$

then

$$\mathbf{a}_3^j \cdot \nabla W_3^i = (\mathbf{a}_3^j \cdot \boldsymbol{\nu}^i)\frac{\partial W_3^i}{\partial \nu^i}. \tag{6.25}$$

Similarly, write

$$\begin{aligned}\mathbf{a}_3^i &= (\mathbf{a}_3^i \cdot \mathbf{a}_3^j)\mathbf{a}_3^j + (\mathbf{a}_3^i \cdot \boldsymbol{\nu}^j)\boldsymbol{\nu}^j \\ &= (\mathbf{a}_3^i \cdot \mathbf{a}_3^j)\mathbf{a}_3^j - \sigma_{ij}(\mathbf{a}_3^j \cdot \boldsymbol{\nu}^i)\boldsymbol{\nu}^j,\end{aligned}$$

so that

$$\mathbf{a}_3^i \cdot \nabla W_3^j = -\sigma_{ij}(\mathbf{a}_3^j \cdot \boldsymbol{\nu}^i)\frac{\partial W_3^j}{\partial \nu^j}. \tag{6.26}$$

It follows from (6.24)–(6.26) that

$$\mathbf{a}_3^j \cdot \boldsymbol{\nu}^i \left(\frac{\partial W_3^i}{\partial \nu^i} - \sigma_{ij} \frac{\partial W_3^j}{\partial \nu^j} \right) = 0,$$

and we note that $\mathbf{a}_3^j \cdot \boldsymbol{\nu}^i \neq 0$ when $\mathcal{P}_i$ and $\mathcal{P}_j$ are non coplanar. □

6.2. Proof of Theorem 6.1. We easily deduce from (6.4) the energy identity

$$\|\dot{\mathbf{R}}^G(t)\|_H^2 + \|\mathbf{R}^G(t)\|_{\mathcal{V}_G}^2 = \|\mathbf{R}_1\|_{H_\infty}^2 + \|\mathbf{R}_0\|_{\mathcal{V}_\infty}^2 + \int_0^t (\mathbf{G}(s), \dot{\mathbf{R}}^G(s))_H ds,$$

from which it follows that

$$(6.27)\quad \|\dot{\mathbf{R}}^G(t)\|_{L^\infty(H)}^2 + \|\mathbf{R}^G(t)\|_{L^\infty(\mathcal{V}_G)}^2 \leq C(\|\mathbf{R}_1\|_{H_\infty}^2 + \|\mathbf{R}_0\|_{\mathcal{V}_\infty}^2 + \|\mathbf{G}\|_{L^2(H)}^2)$$

for some constant C independent of G. It follows from (6.27) that as $G \to \infty$,

$$(6.28)\qquad \dot{\mathbf{R}}^G \text{ is bounded in } L^\infty(H).$$

In addition, by using (2.16) and the definition of the $\mathcal{V}_G$ norm, we may conclude from (6.27) that for $\alpha = 1, 2$ and $i = 1, \ldots, n$,

$$(6.29)\qquad W_{\alpha,\beta}^{i,G} + W_{\beta,\alpha}^{i,G} \text{ is bounded in } L^\infty(H^0(\Omega_i)),$$

$$(6.30)\qquad U_{\alpha,\beta}^{i,G} + U_{\beta,\alpha}^{i,G} \text{ is bounded in } L^\infty(H^0(\Omega_i)),$$

$$(6.31)\qquad \sqrt{G_i}|\mathbf{u}^{i,G} + \nabla W_3^{i,G}| \text{ is bounded in } L^\infty(H^0(\Omega_i)).$$

(Our notation is as follows:

$$\mathbf{R}^G = (\mathbf{R}^{1,G}, \ldots, \mathbf{R}^{n,G}), \quad \mathbf{R}^{i,G} = (\mathbf{W}^{i,G}, \mathbf{u}^{i,G}),$$
$$\mathbf{u}^{i,G} = \sum_\alpha U_\alpha^{i,G} \mathbf{a}_\alpha^i, \quad \mathbf{W}^{i,G} = \sum_j W_j^{i,G} \mathbf{a}_j^i.)$$

Since, by assumption, $\Gamma_i^C \neq \emptyset$ for at least one index i, we may assume that $\Gamma_1^C \neq \emptyset$. It then follows from (6.29)–(6.31) and Korn's lemma for two-dimensional linear elasticity that

$$(6.32)\qquad W_\alpha^{1,G},\ U_\alpha^{1,G} \text{ are bounded in } L^\infty(H^1(\Omega_1)).$$

Then, from (6.31) and (6.32) we may conclude that

$$(6.33)\qquad W_{3,\alpha}^{1,G} \text{ is bounded in } L^\infty(H^0(\Omega_1)).$$

Statement (6.28) implies that $t \mapsto \mathbf{R}^G(t) : [0, T] \mapsto H$ is equicontinuous in $G > 0$. Further (6.28) and the first assumption in (6.6) imply that

$$\mathbf{R}^G \text{ is bounded in } L^\infty(H),$$

so, in particular,

$$W_3^{1,G} \text{ is bounded in } L^\infty(H^0(\Omega_1)). \tag{6.34}$$

Taken together, (6.33) and (6.34) give

$$W_3^{1,G} \text{ is bounded in } L^\infty(H^1(\Omega_1)) \tag{6.35}$$

which, when combined with (6.32), yields

$$\mathbf{R}^{1,G} \text{ is bounded in } L^\infty(V_1). \tag{6.36}$$

Suppose that $\mathcal{P}_2$ is connected to $\mathcal{P}_1$ along a joint J. We wish to use (6.36) together with the geometric junction conditions (6.1) to deduce that

$$\mathbf{R}^{2,G} \text{ is bounded in } L^\infty(V_2). \tag{6.37}$$

Once (6.37) is established, we may conclude by iteration that

$$\mathbf{R}^{G} \text{ is bounded in } L^\infty(V). \tag{6.38}$$

From (6.36) and (6.1) with $i = 1$, $j = 2$, we obtain

$$\mathbf{W}^{2,G}|_{\Gamma_2(J)} \text{ is bounded in } L^\infty(H^{1/2}(\Gamma_2(J)),$$
$$\mathbf{u}^{2,G}|_{\Gamma_2(J)} \text{ is bounded in } L^\infty(H^{-1/2}(\Gamma_2(J)).$$

It then follows from (6.29), (6.30) and Korn's lemma (see Remark 2.2) that

$$W_\alpha^{2,G},\, U_\alpha^{2,G} \text{ are bounded in } L^\infty(H^1(\Omega_2))$$

and then, as above, that

$$W_3^{2,G} \text{ is bounded in } L^\infty(H^1(\Omega_2)).$$

This proves (6.37).

We may now conclude from (6.28), (6.38) and (6.31) that upon passing to the limit as $G \to \infty$ through an appropriate subnet of $G > 0$ we obtain, for some $\mathbf{R} \in L^\infty(V)$,

$$\mathbf{R}^G \to \mathbf{R} \text{ weak* in } L^\infty(V),$$
$$\dot{\mathbf{R}}^G \to \dot{\mathbf{R}} \text{ weak* in } L^\infty(H),$$
$$\mathbf{u}_\alpha^{i,G} + \nabla W_{3,\alpha}^{i,G} \to 0 \text{ strongly in } L^\infty(H^0(\Omega_i)).$$

Then, writing $\mathbf{R} = (\mathbf{R}^1, \ldots, \mathbf{R}^n)$, $\mathbf{R}^i = (\mathbf{W}^i, \mathbf{u}^i)$, we have

$$\mathbf{u}^i = -\nabla W_3^i.$$

Therefore $W_3^i \in L^\infty(H^2(\Omega_i))$, $\mathbf{R} \in L^\infty(\mathcal{V}_\infty)$, $\dot{\mathbf{R}} \in L^\infty(H_\infty)$. Since $\mathbf{R}^G$ is uniformly bounded and equicontinuous from $[0, T]$ into H we also have

$$\mathbf{R}^G \to \mathbf{R} \text{ in } C(H).$$

In particular, $\mathbf{R}(0) = \mathbf{R}_0$.

Let $\hat{\mathbf{R}} \in L^2(\mathcal{V}_\infty)$ and take the scalar product in H of (6.4) with $\hat{\mathbf{R}}$. One obtains

$$(6.39)\quad \int_0^T \langle \ddot{\mathbf{R}}_\infty^G, \hat{\mathbf{R}} \rangle_{\mathcal{V}_\infty} dt = \int_0^T (\mathbf{G}_\infty, \hat{\mathbf{R}})_{H_\infty} dt - \int_0^T (\mathbf{R}^G, \hat{\mathbf{R}})_{\mathcal{V}_G} dt.$$

Therefore

$$\left| \int_0^T \langle \ddot{\mathbf{R}}_\infty^G, \hat{\mathbf{R}} \rangle_{\mathcal{V}_\infty} dt \right| \le C(\|\mathbf{G}_\infty\|_{L^2(H_\infty)} + \|\mathbf{R}^G\|_{L^2(\mathcal{V}_G)}) \|\hat{\mathbf{R}}\|_{L^2(\mathcal{V}_\infty)}.$$

It follows that

$$\ddot{\mathbf{R}}_\infty^G \text{ is bounded in } L^2(\mathcal{V}'_\infty).$$

Thus

$$\ddot{\mathbf{R}}_\infty^G \to \ddot{\mathbf{R}} \text{ weakly in } L^2(\mathcal{V}'_\infty).$$

By the Arzela-Ascoli theorem,

$$\dot{\mathbf{R}}_\infty^G \to \dot{\mathbf{R}} \text{ strongly in } L^\infty(\mathcal{V}'_\infty),$$

so that $t \mapsto \dot{\mathbf{R}}(t)$ is strongly continuous into $\mathcal{V}'_\infty$ and

$$\dot{\mathbf{R}}(0) = \lim \dot{\mathbf{R}}_\infty^G(0) = \mathbf{R}_1.$$

Upon passing to the limit in (6.39) we obtain

$$\int_0^T [\langle \ddot{\mathbf{R}}, \hat{\mathbf{R}} \rangle_{\mathcal{V}_\infty} + (\mathbf{R}, \hat{\mathbf{R}})_{\mathcal{V}_\infty}] dt = \int_0^T (\mathbf{G}_\infty, \hat{\mathbf{R}})_{H_\infty} dt, \quad \forall \hat{\mathbf{R}} \in L^2(\mathcal{V}_\infty),$$

which is just the variational form of (6.7). On the other hand, the problem (6.7), (6.8) has a unique solution with $\mathbf{R} \in C(\mathcal{V}_\infty) \cap C^1(H_\infty)$. Therefore, convergence of $\mathbf{R}^G$ to this solution is through the entire net $G > 0$.

6.3. Connected, Hinged or Rigid Joints. Here we briefly consider in a formal manner the forms of the dynamic junction conditions at a connected joint, a hinged joint and at a rigid joint. In all cases, the appropriate balance laws along a joint J_k are obtained from the variational junction condition (cf. (6.19))

$$(6.40)\quad \sum_{i \in \mathcal{I}(J_k)} \int_{\Gamma_i(J_k)} \left\{ \left(C_i[\varepsilon(\mathbf{w}^i)]\boldsymbol{\nu}^i + \left[I_{\rho_i} \frac{\partial \ddot{W}_3^i}{\partial \boldsymbol{\nu}^i} - D_i \frac{\partial}{\partial \boldsymbol{\nu}^i}(\Delta W_3^i) \right.\right.\right.$$
$$\left.\left.\left. - \frac{h^2}{12} \frac{\partial}{\partial \boldsymbol{\tau}^i} C_{\boldsymbol{\tau}^i}(\nabla W_3^i) \right] \mathbf{a}_3^i \right) \cdot \hat{\mathbf{W}}^i + \frac{h^2}{12} \sigma_i(J_k) C_{\boldsymbol{\nu}^i}(\nabla W_3^i) \frac{\partial \hat{W}_3^i}{\partial \boldsymbol{\nu}^{i_k}} \right\} d\Gamma$$
$$= - \sum_{i \in \mathcal{I}(J_k)} \int_{\Gamma_i(J_k)} (\boldsymbol{\nu}^i \cdot \mathbf{C}^i) \mathbf{a}_3^i \cdot \hat{\mathbf{W}}^i \, d\Gamma,$$

in which the test functions $\hat{\mathbf{W}}^i$ must satisfy geometric constraints appropriate to the particular type of joint under consideration. At a *connected joint*, these constraints are

$$\hat{\mathbf{W}}^i(\mathbf{p}_i^{-1}(\mathbf{p})) = \hat{\mathbf{W}}^j(\mathbf{p}_j^{-1}(\mathbf{p})), \quad \forall i,j \in \mathcal{I}(J_k), \forall \mathbf{p} \in J_k; \tag{6.41}$$

at a *hinged joint* they are (6.41) and

$$\left(\frac{\partial}{\partial \tau^i}\right)^2 \hat{\mathbf{W}}^i(\mathbf{p}_i^{-1}(\mathbf{p})) = 0, \quad \forall i \in \mathcal{I}(J_k), \forall \mathbf{p} \in J_k; \tag{6.42}$$

and at a *rigid joint* they are (6.41), (6.42) and

$$\frac{\partial \hat{\mathbf{W}}_3^i}{\partial \nu^i}(\mathbf{p}_i^{-1}(\mathbf{p})) = \sigma_{ij}\frac{\partial \hat{\mathbf{W}}_3^j}{\partial \nu^j}(\mathbf{p}_j^{-1}(\mathbf{p})), \quad \forall i,j \in \mathcal{I}(J_k), \forall \mathbf{p} \in J_k. \tag{6.43}$$

We immediately deduce from (6.40) and (6.41) the dynamic junction conditions at a connected joint. They are (6.20) and

$$C_{\nu^i}(\nabla W_3^i)(\mathbf{p}_i^{-1}(\mathbf{p})) = 0, \quad \forall i \in \mathcal{I}(J_k), \forall \mathbf{p} \in J_k. \tag{6.44}$$

At a hinged joint, the dynamic conditions are (6.44) and

$$\begin{aligned} \sum_{i\in\mathcal{I}(J_k)} \int_{J_k} \Big\{ C_i[\varepsilon(\mathbf{w}^i)]\boldsymbol{\nu}^i + \Big[I_{\rho_i}\frac{\partial \ddot{W}_3^i}{\partial \nu^i} - D_i\frac{\partial}{\partial \nu^i}(\Delta W_3^i) \\ - \frac{h^2}{12}\frac{\partial}{\partial \tau^i}C_{\tau^i}(\nabla W_3^i)\Big]\mathbf{a}_3^i \Big\} \cdot \hat{\mathbf{W}}^{i_k}(\mathbf{p}_{i_k}^{-1}(\mathbf{p}))\,dJ_k \\ = - \sum_{i\in\mathcal{I}(J_k)} \int_{J_k} (\boldsymbol{\nu}^i \cdot \mathbf{C}^i)\mathbf{a}_3^i \cdot \hat{\mathbf{W}}^{i_k}(\mathbf{p}_{i_k}^{-1}(\mathbf{p}))\,dJ_k \end{aligned} \tag{6.45}$$

for all functions $\hat{\mathbf{W}}^{i_k}$ of the form (cf. Section 3.3.2)

$$\hat{\mathbf{W}}^{i_k}(\mathbf{p}_{i_k}^{-1}(\mathbf{p})) = \mathbf{C}_k^0 + s\mathbf{C}_k^1, \quad (\mathbf{p} = \mathbf{q}_k^0 + s\tau^{i_k},\ a_k \le s \le b_k),$$

where $\mathbf{C}_k^0$, $\mathbf{C}_k^1$ are fixed but arbitrary vectors in $\mathbb{R}^3$. We therefore obtain from (6.45) the nonlocal balance laws

$$\begin{aligned} \sum_{i\in\mathcal{I}(J_k)} \int_{a_k}^{b_k} \Big\{ C_i[\varepsilon(\mathbf{w}^i)]\boldsymbol{\nu}^i + \Big[I_{\rho_i}\frac{\partial \ddot{W}_3^i}{\partial \nu^i} - D_i\frac{\partial}{\partial \nu^i}(\Delta W_3^i) \\ - \frac{h^2}{12}\frac{\partial}{\partial \tau^i}C_{\tau^i}(\nabla W_3^i)\Big]\mathbf{a}_3^i \Big\}\,ds = - \sum_{i\in\mathcal{I}(J_k)} \int_{a_k}^{b_k} (\boldsymbol{\nu}^i \cdot \mathbf{C}^i)\mathbf{a}_3^i\,ds, \end{aligned} \tag{6.46}$$

$$(6.47)\quad \sum_{i\in\mathcal{I}(J_k)} \int_{a_k}^{b_k} s \left\{ C_i[\varepsilon(\mathbf{w}^i)]\boldsymbol{\nu}^i + \left[I_{\rho_i} \frac{\partial \ddot{W}_3^i}{\partial \boldsymbol{\nu}^i} - D_i \frac{\partial}{\partial \boldsymbol{\nu}^i}(\Delta W_3^i) \right.\right.$$
$$\left.\left. - \frac{h^2}{12}\frac{\partial}{\partial \boldsymbol{\tau}^i} C_{\boldsymbol{\tau}^i}(\nabla W_3^i) \right] \mathbf{a}_3^i \right\} ds = - \sum_{i\in\mathcal{I}(J_k)} \int_{a_k}^{b_k} s(\boldsymbol{\nu}^i \cdot \mathbf{C}^i)\mathbf{a}_3^i \, ds.$$

Finally, the dynamic junction conditions at a rigid joint are (6.46), (6.47) and (6.21).

CHAPTER VIII

Modeling and Controllability of Coupled Plate-Beam Systems

1. Introduction

In this chapter we shall develop a distributed parameter models for two dynamic elastic plate-beam configurations. The first consists of a thin plate to which a thin beam is rigidly and orthogonally attached to the edge of the plate in such a manner that the centerline of the beam is coplanar with the middle plane of the plate (in the equilibrium state). The second model is that of a thin plate to which a thin beam is attached orthogonally to a face of the plate (i.e., the centerline of the beam is orthogonal to a face of the plate in the equilibrium configuration). In the case of the configuration in which the beam is orthogonal to the edge of the plate, it will be proved that the dynamical system obtained is exactly controllable by means of controls applied along an appropriate portion of the edge of the plate that excludes the junction region between the plate and beam. We shall also consider a simplified model of that system obtained by passing to the limit in the original model as the shear moduli go to infinity.

Modeling of junctions between elastic elements of differing dimensions has been carried out by Gruais [32] for 2d–1d junctions and by Ciarlet, Le Dret and Nzengwa [17] for 3d–2d junctions using the Ciarlet-Dystunder method of asymptotic expansions discussed earlier. This approach requires that the "lower dimensional" body be embedded in some manner into the higher dimensional one. For the configuration in which a thin beam is attached orthogonally to the edge of a thin plate, in the limit problem (as the "thickness" parameters approach zero) the junction region appears as a slit in the middle surface of the plate. This geometry presents particular difficulties in the study of controllability of the limit system because the poor regularity properties of solutions of the uncontrolled problem in the neighborhood of the slit create problems for the multiplier methods commonly used to derive the necessary observability estimates, since such techniques can be justified only if it is known *a priori* that the solutions are sufficiently regular. (In this context see, however, the controllability results of Nicaise [79], [80] which pertain to a 3-d region with an embedded plate and to a plate with and embedded beam.) In contrast, in the framework presented below the junction region appears as a part of the *boundary* of the reference surface of the plate. Following the approach employed above for linked membranes and for linked plates, we obtain *geometric* junction conditions from the very natural requirements that the interface between the (three dimensional) plate and the (three dimensional) beam be flat and that the displacements of the plate match those of the beam along that interface, combined with certain *ad hoc* assumptions on the kinematics of the deformation. Then Hamilton's principle is employed to derive the dynamic conditions of the model. In particular, the dynamic junction conditions will turn out to be *nonlocal*, coupling averaged forces and moments induced by plate deformation along the plate-beam interface to forces and moments induced by beam deformation at that point on the centerline of the beam where the beam meets the plate.

The configuration in which the beam is attached orthogonally to the edge of the plate will be studied in Sections 2 – 5. The model equations are derived in Section 2 and, in Section 3, it is shown that this system comprises a well-posed problem (in appropriate function spaces). Exact controllability of the model equations is the subject of Section 4. There it will be established that the system is exactly controllable by means of controls acting in either geometric or mechanical boundary conditions along a certain portion of the edge of the plate that excludes the junction region. We remark that the methods employed below could also be used without essential new difficulties to treat more general configurations such as a plate with several beams attached orthogonally along its edge, or even a plate to which is attached a *network* of rigidly joined beams of the sort discussed in Chapter III. Let us also note also that exact controllability of a plate-beam distributed parameter model, but one of a completely different character than that considered below, has been established by Puel and Zuazua [84]. The reader is also referred to the paper [85], where exact controllability of a model of an elastic string connected to an elastic membrane (or, more generally, to an n-dimensional body modeled as an n-dimensional wave equation) is proved.

The geometric junction conditions which we derive in the Section 2 involve nonlinear couplings among the displacement and rotation vectors of the plate and beam. Upon linearizing these relations one finds that quantities related to motion in the reference plane of the plate are no longer coupled to the remaining quantities, which are: the transverse displacement and two transverse shear angles associated with deformation of the plate; and the transverse displacement, torsional angle and one of the shear angles associated with deformation of the beam. Thus the linearized model involving the latter quantities has six degrees of freedom. This model connects the Reissner-Mindlin plate equation to the Timoshenko beam equation and the linearized equation of torsion in a rod, with all variables coupled through the geometric and dynamic junction conditions. In order to obtain a simpler model, in Section 5 we pass to the limit in the original model as the shear modulus of the plate and the shear modulus of the beam both approach infinity. This process leads to a model in which transverse shear is not present in either the plate or beam and in which the torsional angle is spatially constant and amounts to a (time dependent) rigid rotation of the beam about its center line. The limit model is a type of *hybrid system*, involving coupled partial and ordinary differential equations. It has only three degrees of freedom and couples a Kirchhoff plate to an Euler-Bernoulli beam with the transverse displacements of each and the spatially constant torsional angle coupled through the geometric and dynamic junction conditions.

Finally, in Section 6, a distributed parameter model is derived for the plate-beam configuration in which the centerline of the beam is orthogonal to one of the faces of the plate in the equilibrium configuration. As usual, this model is obtained by imposing *ad hoc* kinematic and geometric constraints on the deformation and then applying Hamilton's principle. This model turns out to be much more complicated than the one derived in Section 2 because all variables (displacements and rotation angles of the plate and beam) are coupled. It is clear that other physically plausible constraints could be imposed instead of, or in addition to, the ones proposed below and that each different choice would lead to a different set of dynamic junction conditions. This fact emphasizes the necessity of further research devoted to simulations and model validation; for example, it would be extremely interesting to compare spectral values predicted by the model derived below with experimentally obtained spectral data.

2. Modeling of the Plate-Beam Junction: I

2.1. Geometric Conditions. Let Ω be a bounded, open, connected set in $\mathbb{R}^2$ with Lipschitz continuous boundary Γ consisting of a finite number of smooth simple curves. We consider Ω to be the reference surface of a thin plate of uniform thickness h. Points within the plate will be denoted by coordinates (x_1, x_2, x_3) with respect to the natural basis $\mathbf{a}_1, \mathbf{a}_2, \mathbf{a}_3$ for $\mathbb{R}^3$. Thus the equilibrium position of the plate is given by

$$\{x_i \mathbf{a}_i \,|\, (x_1, x_2) \in \overline{\Omega},\ |x_3| \leq h/2\}.$$

where summation convention for repeated indices has been assumed. Recall that repeated Roman indices are summed from 1 to 3, while repeated Greek indices are summed from 1 to 2.

Let $\mathbf{x}_0$ be a point of $I\!R^3$ and $\mathbf{e}_1, \mathbf{e}_2, \mathbf{e}_3$ be a right-handed orthonormal basis for $I\!R^3$. Let ω be a bounded, simply connected open set in $I\!R^2$ such that $0 \in \omega$ and which is doubly symmetric with respect to the origin. The beam, in its reference configuration, is given by

$$\{\mathbf{x}_0 + \xi_i \mathbf{e}_i \,|\, 0 \le \xi_1 \le \ell,\ (\xi_2, \xi_3) \in \overline{\omega}\}.$$

The following assumptions are made:
(A1) $\mathbf{x}_0 \in \Gamma$ and $\boldsymbol{\nu}(\mathbf{x}_0)$ exists.
(A2) $\mathbf{e}_1 = \boldsymbol{\nu}(\mathbf{x}_0)$, $\mathbf{e}_2 = \boldsymbol{\tau}(\mathbf{x}_0)$, $\mathbf{e}_3 = \mathbf{a}_3$.
(A3)

$$\omega = \{(\xi_2, \xi_3) \,|\, |\xi_2| < \frac{a}{2},\ |\xi_3| < \frac{h}{2}\}.$$

(A4) Set

$$J_a = \{\mathbf{x}_0 + \xi_2 \boldsymbol{\tau}(\mathbf{x}_0) \,|\, |\xi_2| < \frac{a}{2}\}.$$

Then $J_a \subset \Gamma$.

The region $J_a \times (-h/2, h/2)$ is the *junction region*. The above assumptions mean that the junction region is *flat*, that the beam is prismatic and its centerline is orthogonal to Γ at $\mathbf{x}_0$ and lies in the x_1x_2-plane. In terms of the $\mathbf{e}_i$ basis, the junction region is given by

$$\mathbf{x}_0 + \xi_2 \mathbf{e}_2 + \xi_3 \mathbf{e}_3, \quad |\xi_2| < \frac{a}{2},\ |\xi_3| < \frac{h}{2}.$$

The assumption (A3) that the cross sections of the beam be rectangular is unessential. At the cost of a slightly more complicated formalism, we may just as easily consider beams with doubly symmetric cross sections ω such that $\{\mathbf{x}_0 + \xi_2 \mathbf{e}_2 + \xi_2 \mathbf{e}_3 \,|\, (\xi_1, \xi_2) \in \omega\}$ is contained within the lateral boundary of the plate.

We denote by $\mathbf{R}(x_1, x_2, x_3, t)$ the *position vector* at time t to the deformed position of the material particle in the plate which is located at (x_1, x_2, x_3) when the plate is in equilibrium, and by $\mathbf{W}(x_1, x_2, t)$ the *displacement vector* of the material particle which occupies position $(x_1, x_2, 0)$ in the reference surface in equilibrium. Similarly, $\mathbf{r}(\xi_1, \xi_2, \xi_3, t)$ will denote the position vector to the deformed position of the particle originally at position $\mathbf{x}_0 + \xi_i \mathbf{e}_i$ of the beam in equilibrium, and $\mathbf{w}(\xi_1, t)$ the displacement vector of the point $\mathbf{x}_0 + \xi_1 \mathbf{e}_1$ on the centerline of the beam. The kinematics of the plate and beam are assumed to obey the following hypotheses (cf. Chapters III and VII).

Basic kinematic hypotheses.

$$\text{(2.1)} \quad \mathbf{R}(x_1, x_2, x_3, t) = x_\beta \mathbf{a}_\beta + \mathbf{W}(x_1, x_2, t) + x_3 \mathbf{A}_3(x_1, x_2, t), \qquad (x_1, x_2) \in \overline{\Omega},\ |x_3| \le h/2,$$

where $|\mathbf{A}_3| = 1$ and $\mathbf{a}_3 \cdot \mathbf{A}_3 > 0$; and

$$\text{(2.2)} \quad \mathbf{r}(\xi_1, \xi_2, \xi_3, t) = \mathbf{x}_0 + \xi_1 \mathbf{e}_1 + \mathbf{w}(\xi_1, t) + \xi_2 \mathbf{E}_2(\xi_1, t) + \xi_3 \mathbf{E}_3(\xi_1, t), \quad 0 \le \xi_1 \le \ell,\ (\xi_2, \xi_3) \in \overline{\omega},$$

where

$$\mathbf{E}_2(\xi_1, t) = O(\xi_1, t)\mathbf{e}_2, \quad \mathbf{E}_3(\xi_1, t) = O(\xi_1, t)\mathbf{e}_3$$

for some orthogonal matrix $O(\xi_1, t)$ with determinant one.

Assumption (2.2) means that cross-sections of the beam move rigidly, while (2.1) is the hypothesis of Reissner-Mindlin plate theory that the x_3 line remains straight and the extensional strain ε_{33} vanishes.

The orthogonal matrix O may be considered to be the product of three rotation matrices, the first representing rotation through an angle ϑ_1 around the ξ_1 line, the second rotation through an angle ϑ_2 around the ξ_2 line and the third rotation through an angle ϑ_3 around the ξ_3 line. It may therefore be represented with respect to the $\mathbf{e}_i$ basis as the product

$$\text{(2.3)} \quad O = \begin{pmatrix} \cos\vartheta_3 & -\sin\vartheta_3 & 0 \\ \sin\vartheta_3 & \cos\vartheta_3 & 0 \\ 0 & 0 & 1 \end{pmatrix} \begin{pmatrix} \cos\vartheta_2 & 0 & \sin\vartheta_2 \\ 0 & 1 & 0 \\ -\sin\vartheta_2 & 0 & \cos\vartheta_2 \end{pmatrix} \cdot \begin{pmatrix} 1 & 0 & 0 \\ 0 & \cos\vartheta_1 & -\sin\vartheta_1 \\ 0 & \sin\vartheta_1 & \cos\vartheta_1 \end{pmatrix},$$

where, in this representation, positive values of ϑ_i correspond to clockwise rotation around the ξ_i line. The angle ϑ_1 is a torsional rotation about the $\mathbf{e}_1$ axis and ϑ_2, ϑ_3 correspond to flexure of the beam in the $\mathbf{e}_1\mathbf{e}_3$ and $\mathbf{e}_1\mathbf{e}_2$ planes, respectively. It follows from (2.3) that (cf. (1.9), Chapter III)

$$\mathbf{E}_2 = (\sin\vartheta_1 \sin\vartheta_2 \cos\vartheta_3 - \cos\vartheta_1 \sin\vartheta_3)\mathbf{e}_1 + (\cos\vartheta_1 \cos\vartheta_3 + \sin\vartheta_1 \sin\vartheta_2 \sin\vartheta_3)\mathbf{e}_2 + (\sin\vartheta_1 \cos\vartheta_3)\mathbf{e}_3,$$

$$\mathbf{E}_3 = (\sin\vartheta_1 \sin\vartheta_3 + \cos\vartheta_1 \sin\vartheta_2 \cos\vartheta_3)\mathbf{e}_1 + (\cos\vartheta_1 \sin\vartheta_2 \sin\vartheta_3 - \sin\vartheta_1 \cos\vartheta_3)\mathbf{e}_2 + (\cos\vartheta_1 \cos\vartheta_2)\mathbf{e}_3.$$

REMARK 2.1. Assumption (2.2) is not entirely consistent with the introduction of torsion since the latter may induce cross-sectional warping, as discussed in Chapter III. However, as noted there, warping is considered negligible if the beam is thin and the torsional rotation is small, which is the situation we shall eventually consider.

We now define the *geometric junction conditions*. Consider a point $\mathbf{x}_0 + \xi_2 \boldsymbol{\tau}(\mathbf{x}_0) + \xi_3 \mathbf{a}_3$ within the junction region. We require that

$$\text{(2.4)} \quad \mathbf{R}(\mathbf{x}_0 + \xi_2 \boldsymbol{\tau} + \xi_3 \mathbf{a}_3, t) = \mathbf{r}(0, \xi_2, \xi_3, t), \quad |\xi_2| < \frac{a}{2}, \ |\xi_3| < \frac{h}{2},$$

where $\boldsymbol{\tau} = \boldsymbol{\tau}(\mathbf{x}_0)$, that is

$$(2.5) \qquad \begin{cases} \mathbf{W}(\mathbf{x}_0 + \xi_2\boldsymbol{\tau}) = \mathbf{w}(0) + \xi_2(\mathbf{E}_2(0,t) - \mathbf{e}_2), \\ \mathbf{U}(\mathbf{x}_0 + \xi_2\boldsymbol{\tau}, t) = (\mathbf{E}_3(0,t) - \mathbf{e}_3), \quad |\xi_2| < \dfrac{a}{2}, \end{cases}$$

where $\mathbf{U} = \mathbf{A}_3 - \mathbf{a}_3$ is the "rotation" of the $\mathbf{a}_3$ line. Conditions (2.5) are of *rigid type* and simply require that the displacements of the plate and beam match throughout the junction region.

Write

$$\mathbf{W} = W_i\mathbf{a}_i, \quad \mathbf{U} = U_i\mathbf{a}_i, \quad \mathbf{u} = U_\beta\mathbf{a}_\beta, \quad \mathbf{w} = w_i\mathbf{e}_i,$$

where $U_3 = -1 + (1 - |\mathbf{u}|^2)^{1/2}$. Then (2.5) may be written

$$W_3(\mathbf{x}_0 + \xi_2\boldsymbol{\tau}, t) = w_3(0,t) + \xi_2(\sin\vartheta_1 \cos\vartheta_2)(0,t),$$

$$U_3(\mathbf{x}_0 + \xi_2\boldsymbol{\tau}, t) = (\cos\vartheta_1 \cos\vartheta_2)(0,t) - 1,$$

$$W_\beta(\mathbf{x}_0 + \xi_2\boldsymbol{\tau}, t)\mathbf{a}_\beta = w_\beta(0,t)\mathbf{e}_\beta + \xi_2[(\sin\vartheta_1\sin\vartheta_2\cos\vartheta_3 - \cos\vartheta_1\sin\vartheta_3)\mathbf{e}_1 + (\cos\vartheta_1\cos\vartheta_3 + \sin\vartheta_1\sin\vartheta_2\sin\vartheta_3 - 1)\mathbf{e}_2](0,t),$$

$$\mathbf{u}(\mathbf{x}_0 + \xi_2\boldsymbol{\tau}, t) = [(\sin\vartheta_1\sin\vartheta_3 + \cos\vartheta_1\sin\vartheta_2\cos\vartheta_3)\mathbf{e}_1 + (\cos\vartheta_1\sin\vartheta_2\sin\vartheta_3 - \sin\vartheta_1\cos\vartheta_3)\mathbf{e}_2](0,t).$$

If we now linearize these last relations around the equilibrium state we obtain

$$(2.6) \qquad \begin{cases} W_3(\mathbf{x}_0 + \xi_2\boldsymbol{\tau}, t) = w_3(0,t) + \xi_2\vartheta_1(0,t), \\ \mathbf{u}(\mathbf{x}_0 + \xi_2\boldsymbol{\tau}, t) = \boldsymbol{\vartheta}(0,t), \quad |\xi_2| < \dfrac{a}{2}, \end{cases}$$

and

$$(2.7) \qquad W_\beta(\mathbf{x}_0 + \xi_2\boldsymbol{\tau}, t)\mathbf{a}_\beta = w_\beta(0,t)\mathbf{e}_\beta - \xi_2\vartheta_3(0,t)\mathbf{e}_1,$$

where

$$\boldsymbol{\vartheta} = \vartheta_2\mathbf{e}_1 - \vartheta_1\mathbf{e}_2.$$

It should be noted that in (2.6) and(2.7) there is no coupling between the variables $W_\beta, w_\beta, \vartheta_3$ related to motion in the $\mathbf{a}_1\mathbf{a}_2$-plane and the variables $W_3, \mathbf{u}, w_3, \boldsymbol{\vartheta}$.

If we recall the expression (1.52) relating the rotation angles ϕ_β of the plate to the U_β, then the second equation in (2.6) is (after linearization of (1.52))

$$\phi_2\mathbf{a}_1 - \phi_1\mathbf{a}_2 = \vartheta_1\mathbf{e}_2 - \vartheta_2\mathbf{e}_1$$

and so expresses the continuity of (vector) rotation angles across the interface.

We shall take linearized expressions (2.6) and (2.7) as the geometric junction conditions of our problem.

2.2. Dynamic conditions. The dynamic conditions will be obtained from Hamilton's Principle:

$$\delta \int_0^T [\mathcal{K}(t) - \mathcal{U}(t) + \mathcal{W}(t)]\, dt = 0, \tag{2.8}$$

where $\mathcal{K}$ and $\mathcal{U}$ represent the kinetic and strain energies, respectively, of the plate-beam system, $\mathcal{W}$ is the work done on the system by external forces and δ denotes the first variation with respect to the class of admissible displacements. The latter must obey, in particular, the geometric junction conditions (2.6) and (2.7). They must, in addition, satisfy any other geometric restrictions that are imposed on the motion. Since $W_3, \mathbf{u}, w_3, \boldsymbol{\vartheta}$ are not coupled to the quantities $W_\beta, w_\beta, \vartheta_3$ in the geometric junction conditions (nor will they be coupled in the geometric boundary conditions imposed below), in considering the dynamic conditions related to these quantities we need only take into account those portions of $\mathcal{K}$, $\mathcal{U}$ and $\mathcal{W}$ which involve $W_3, \mathbf{u}, w, \vartheta$. We shall derive the dynamic conditions only in this situation and leave the (similar) derivation of the dynamic conditions related to motion in the $\mathbf{a}_1\mathbf{a}_2$-plane to the reader.

In what follows we shall usually write W and w in place of W_3 and w_3, respectively, and ξ instead of ξ_1. We assume that $\Gamma = \overline{\Gamma}^D \cup \overline{\Gamma}^C \cup \overline{J}$, where Γ^D, Γ^C and J are relatively open in Γ and mutually disjoint. For later purposes we also assume that $\overline{\Gamma}^D \cap \overline{J} = \emptyset$, which implies in particular that $\Gamma^C \neq \emptyset$ and $\overline{\Gamma}^C \cap \overline{J} \neq \emptyset$. The part Γ^C corresponds to geometric boundary conditions

$$W = 0, \quad \mathbf{u} = 0 \quad \text{on } \Gamma^C, \tag{2.9}$$

while Γ^D corresponds to mechanical boundary conditions. We further assume that J is connected and that $\overline{J}_a \subset J$. The part $J \setminus J_a$ may be either *free*, or it will be *geometrically restricted* in the following manner. Let ζ be a $C^\infty(\overline{J})$ function such that

$$\zeta(\mathbf{x}) = \xi_2 \ \text{ if } \mathbf{x} = \mathbf{x}_0 + \xi_2 \boldsymbol{\tau}(\mathbf{x}_0) \in J_a,$$

and let σ_i, $i = 0, 1$, be $C_0^\infty(\overline{J})$ functions such that $\sigma_i = 1$ on J_a. It is required that

$$\begin{cases} W(\mathbf{x}) = \sigma_0(\mathbf{x})[w(0) + \zeta(\mathbf{x})\vartheta_1(0)], \\ \mathbf{u}(\mathbf{x}) = \sigma_1(\mathbf{x})\boldsymbol{\vartheta}(0), \quad \mathbf{x} \in J. \end{cases} \tag{2.10}$$

Conditions (2.10) require that the displacement W and rotation $\mathbf{u}$ along J change smoothly from their values on J_a to zero, and give prescribed profiles to these quantities along $J^\pm$. The quantities σ_i and ζ may be considered design parameters. We refer to Puel-Zuazua [85], where a similar boundary condition is introduced in a string-membrane problem in a neighborhood of the point where the string meets the membrane.

To summarize, the geometric conditions which the admissible displacements must obey are (2.6) and (2.9) when $J \setminus J_a$ is free, or (2.9) and (2.10) otherwise.

We write

$$\mathcal{K} = \mathcal{K}_P + \mathcal{K}_B, \quad \mathcal{U} = \mathcal{U}_P + \mathcal{U}_B, \quad \mathcal{W} = \mathcal{W}_P + \mathcal{W}_B,$$

where $\mathcal{K}_P$ and $\mathcal{K}_B$ are the kinetic energies of the plate and beam, respectively, related to the quantities $W, \mathbf{u}, w, \vartheta$, and similarly for $\mathcal{U}_P, \mathcal{U}_B$ and $\mathcal{W}_P, \mathcal{W}_B$. We assume that the plate is homogeneous and that the energy of the plate is that associated with linear Reissner-Mindlin plate theory (see (2.24), (2.26), Chapter VII). Thus we set

$$\mathcal{K}_P(t) = \frac{1}{2}\int_\Omega \left(m_1|\dot{W}|^2 + I_{\rho_1}|\dot{\mathbf{u}}|^2\right) d\Omega, \tag{2.11}$$

$$\mathcal{U}_P(t) = \frac{1}{2}\int_\Omega [(C[\varepsilon(\mathbf{u})], \varepsilon(\mathbf{u})) + G_1 h|\nabla W + \mathbf{u}|^2]\, d\Omega, \tag{2.12}$$

where m_1 is the mass of the plate per unit of reference area, $I_{\rho_1} = m_1 h^2/12$, G_1 is the shear modulus of the plate,

$$C[\varepsilon] = \frac{h^3}{12}[\lambda_1 \varepsilon_{\alpha\alpha} \mathcal{I} + 2\mu_1 \varepsilon],$$

$$\varepsilon(\mathbf{u}) = \frac{1}{2}[\nabla \mathbf{u} + (\nabla \mathbf{u})^*],$$

and $(C[\varepsilon], \varepsilon) = C^{\alpha\beta}[\varepsilon]\varepsilon_{\alpha\beta}$.

The total potential energy of the beam is given by

$$\int_0^\ell \int_\omega \Phi \, d\omega d\xi,$$

where $\Phi = (1/2)s^{ij}\varepsilon_{ij}$ is the free energy per unit of reference volume. Here the strains ε_{ij} are given by

$$\varepsilon_{ij} = \frac{1}{2}(\mathbf{r}_{,i} \cdot \mathbf{r}_{,j} - \delta_{ij}).$$

Thus the *linearized* strains are found to be

$$\begin{aligned}
\varepsilon_{11} &= w_1' - \xi_2 \vartheta_3' + \xi_3 \vartheta_2', \\
\varepsilon_{12} &= \frac{1}{2}(w_2' - \vartheta_3) - \frac{1}{2}\xi_3 \vartheta_1', \\
\varepsilon_{13} &= \frac{1}{2}(w_3' + \vartheta_2) + \frac{1}{2}\xi_2 \vartheta_1', \\
\varepsilon_{22} &= \varepsilon_{23} = \varepsilon_{33} = 0,
\end{aligned}$$

where $' = \partial/\partial\xi_1$. We assume that the beam is homogeneous and elastically isotropic, so that

$$s^{ij} = 2\mu_2\varepsilon_{ij} + \lambda_2\delta_{ij}\varepsilon_{kk},$$

where μ_2, λ_2 are the Lamé parameters of the beam. It follows that Φ is given by (cf. Chapter III)

$$\Phi = \frac{E}{2}\varepsilon_{11}^2 + 2G_2(\varepsilon_{12}^2 + \varepsilon_{13}^2),$$

where $G_2 = \mu_2$ is the shear modulus and $E = (2\mu_2 + \lambda_2)$ is the effective Young's modulus of the beam. The strain energy density is therefore

$$\int_\omega \Phi\, d\omega = \frac{E}{2}\left(A|w_1'|^2 + \frac{Aa^2}{12}|\vartheta_3'|^2 + \frac{Ah^2}{12}|\vartheta_2'|^2\right) + \frac{AG_2}{2}\left[(w_2' - \vartheta_3)^2 + (w_3' + \vartheta_2)^2 + \left(\frac{a^2}{12} + \frac{h^2}{12}\right)|\vartheta_1'|^2\right],$$

where $A = ah$ is the cross-sectional area of the beam, so that the part of the potential energy containing the quantities $w_3, \vartheta_1, \vartheta_2$ may be written

$$\text{(2.13)} \quad \mathcal{U}_B(t) = \frac{1}{2}\int_0^\ell (G_2 I|\vartheta_1'|^2 + EI_h|\vartheta_2'|^2 + G_2 A|w' + \vartheta_2|^2))\, d\xi,$$

where $w = w_3$ and

$$I_h := A\frac{h^2}{12}, \quad I_a := A\frac{a^2}{12} \quad I := I_h + I_a.$$

The kinetic energy of the beam is

$$\frac{1}{2}\int_0^\ell \int_\omega \rho_2|\dot{\mathbf{r}}|^2 d\omega d\xi,$$

where ρ_2 is the mass of the beam per unit of reference volume. It is easily found that

$$\int_0^\ell \int_\omega \rho_2|\dot{\mathbf{r}}|^2 d\omega d\xi = \int_0^\ell \rho_2 A\left[|\dot{\mathbf{w}}|^2 + \frac{a^2}{12}|\dot\vartheta_3|^2 + \frac{h^2}{12}|\dot\vartheta_2|^2 + \left(\frac{a^2}{12} + \frac{h^2}{12}\right)|\dot\vartheta_1|^2\right] d\xi,$$

hence

$$\text{(2.14)} \qquad \mathcal{K}_B(t) = \frac{1}{2}\int_0^\ell (m_2|\dot w|^2 + \rho_2 I_h|\dot\vartheta_2|^2 + \rho_2 I|\dot\vartheta_1|^2)\, d\xi,$$

where $m_2 = \rho_2 A$ is the mass of the beam per unit of reference length.

To obtain the work done on the plate-beam system the applied forces have to be specified. Let $\hat{\mathbf{F}}_P(x_1, x_2, x_3)$ be a distributed body force acting on the plate and $\hat{\mathbf{f}}_P(x_1, x_2, x_3)$ be an edge force distributed along $\Gamma^D \times (-h/2, h/2)$. The work done on the plate by these forces is approximately

$$\int_{-h/2}^{h/2} \int_{\Omega} \hat{\mathbf{F}}_P \cdot (\mathbf{W} + x_3 \mathbf{U})\, d\Omega dx_3 + \int_{-h/2}^{h/2} \int_{\Gamma^D} \hat{\mathbf{f}}_P \cdot (\mathbf{W} + x_3 \mathbf{U})\, d\Gamma dx_3.$$

Consider the resultant forces

$$\mathbf{F}_P(x_1, x_2) = \int_{-h/2}^{h/2} \hat{\mathbf{F}}_P\, dx_3, \quad \mathbf{f}_P(x_1, x_2) = \int_{-h/2}^{h/2} \hat{\mathbf{f}}_P\, dx_3,$$

and moments

$$\mathbf{M}_P(x_1, x_2) = \int_{-h/2}^{h/2} x_3 \hat{\mathbf{F}}_P\, dx_3, \quad \mathbf{m}_P(x_1, x_2) = \int_{-h/2}^{h/2} x_3 \hat{\mathbf{f}}_P\, dx_3.$$

The portion of the work done on the plate related to its transverse motion may then be written

$$\text{(2.15)} \quad \mathcal{W}_P(t) = \int_{\Omega} (F_1 W + \mathbf{M}_1 \cdot \mathbf{u})\, d\Omega + \int_{\Gamma^D} (f_1 W + \mathbf{m}_1 \cdot \mathbf{u})\, d\Gamma$$

where

$$F_1(x_1, x_2) = \mathbf{a}_3 \cdot \mathbf{F}_P(x_1, x_2), \quad f_1(x_1, x_2) = \mathbf{a}_3 \cdot \mathbf{f}_P(x_1, x_2),$$
$$\mathbf{M}_1 = (\mathbf{a}_1 \cdot \mathbf{M}_P)\mathbf{a}_1 + (\mathbf{a}_2 \cdot \mathbf{M}_P)\mathbf{a}_2, \quad \mathbf{m}_1 = (\mathbf{a}_1 \cdot \mathbf{m}_P)\mathbf{a}_1 + (\mathbf{a}_2 \cdot \mathbf{m}_P)\mathbf{a}_2.$$

Similarly, let $\hat{\mathbf{F}}_B(\xi, \xi_2, \xi_3)$ be a distributed body force acting on the beam and $\hat{\mathbf{f}}_B(\xi_2, \xi_3)$ be a force distributed over the cross-section at $\xi = \ell$. The work done on the beam by these forces is approximately

$$\int_0^{\ell} \int_{\omega} \hat{\mathbf{F}}_B \cdot (\mathbf{w}(\xi) + \xi_2(\vartheta_1(\xi)\mathbf{e}_3 - \vartheta_3(\xi)\mathbf{e}_1) + \xi_3(\vartheta_2(\xi)\mathbf{e}_1 - \vartheta_1(\xi)\mathbf{e}_2))\, d\omega d\xi$$
$$+ \int_{\omega} \hat{\mathbf{f}}_B \cdot (\mathbf{w}(\ell) + \xi_2(\vartheta_1(\ell)\mathbf{e}_3 - \vartheta_3(\ell)\mathbf{e}_1) + \xi_3(\vartheta_2(\ell)\mathbf{e}_1 - \vartheta_1(\ell)\mathbf{e}_2))\, d\omega.$$

Consider the resultant forces

$$\mathbf{F}_B(\xi) = \int_{\omega} \hat{\mathbf{F}}_B\, d\omega, \quad \mathbf{f}_B = \int_{\omega} \hat{\mathbf{f}}_B\, d\omega,$$

and set

$$M_{23}(\xi) = \int_{\omega} (\xi_3 \mathbf{e}_2 - \xi_2 \mathbf{e}_3) \cdot \hat{\mathbf{F}}_B\, d\omega, \quad M_{13}(\xi) = \int_{\omega} \xi_3 \mathbf{e}_1 \cdot \hat{\mathbf{F}}_B\, d\omega,$$

$$m_{23} = \int_{\omega} (\xi_3 \mathbf{e}_2 - \xi_2 \mathbf{e}_3) \cdot \hat{\mathbf{f}}_B\, d\omega, \quad m_{13} = \int_{\omega} \xi_3 \mathbf{e}_1 \cdot \hat{\mathbf{f}}_B\, d\omega.$$

The quantity M_{23} is a twisting moment around $\mathbf{e}_1$, M_{13} is a bending moment about $\mathbf{e}_2$ and similarly for m_{23} and m_{13}. Then the portion of the work done on the beam related its transverse motion and torsion is

$$\text{(2.16)} \quad \mathcal{W}_B(t) = \int_0^\ell [F_2 w + \mathbf{M}_2 \cdot (\vartheta_2 \mathbf{e}_1 - \vartheta_1 \mathbf{e}_2)]\, d\xi + f_2 w(\ell) + \mathbf{m}_2 \cdot (\vartheta_2 \mathbf{e}_1 - \vartheta_1 \mathbf{e}_2)(\ell),$$

where

$$F_2 = \mathbf{e}_3 \cdot \mathbf{F}_B, \quad f_2 = \mathbf{e}_3 \cdot \mathbf{f}_B,$$
$$\mathbf{M}_2 = M_{13}\mathbf{e}_1 - M_{23}\mathbf{e}_2, \quad \mathbf{m}_2 = m_{13}\mathbf{e}_1 - m_{23}\mathbf{e}_2.$$

We are now in the position to calculate (2.8), where the variation is taken with respect to displacements satisfying (2.6) and (2.9) if $J - J_a$ is free, or (2.9) and (2.10) when $J - J_a$ is geometrically restricted. We first discuss the latter situation, since this is the one which will be considered in subsequent sections. In the calculation we utilize the Green's formula (3.7), Chapter VI, in the form

$$\text{(2.17)} \quad \int_\Omega (C[\varepsilon(\mathbf{u})], \varepsilon(\mathbf{v}))\, d\Omega = -\int_\Omega \mathbf{v} \cdot (\operatorname{div} C[\varepsilon(\mathbf{u})])\, d\Omega + \int_\Gamma \mathbf{v} \cdot (C[\varepsilon(\mathbf{u})]\boldsymbol{\nu})\, d\Gamma.$$

The following dynamic equations of motion and boundary conditions are obtained.

Equations of motion of the plate:

$$\text{(2.18)} \quad \begin{cases} m_1 \ddot{W} - G_1 h \operatorname{div}(\nabla W + \mathbf{u}) = F_1, \\ I_{\rho_1} \ddot{\mathbf{u}} - \operatorname{div} C[\varepsilon(\mathbf{u})] + G_1 h(\nabla W + \mathbf{u}) = \mathbf{M}_1. \end{cases}$$

Equations of motion of the beam:

$$\text{(2.19)} \quad \begin{cases} m_2 \ddot{w} - G_2 A(w' + \vartheta_2)' = F_2, \\ \rho_2 I_h \ddot{\vartheta}_2 - E I_h \vartheta_2'' + G_2 A(w' + \vartheta_2) = M_{13}, \\ \rho_2 I \ddot{\vartheta}_1 - G_2 I \vartheta_1'' = -M_{23}. \end{cases}$$

Boundary conditions along Γ^D:

$$\text{(2.20)} \quad \begin{cases} G_1 h(\nabla W + \mathbf{u}) \cdot \boldsymbol{\nu} = f_1, \\ C[\varepsilon(\mathbf{u})]\boldsymbol{\nu} = \mathbf{m}_1. \end{cases}$$

Boundary conditions at $\xi = \ell$:

$$(2.21)\qquad \begin{cases} G_2A(w' + \vartheta_2)(\ell) = f_2, \\ EI_h\vartheta_2'(\ell) = m_{13}, \\ G_2I\vartheta_1'(\ell) = -m_{23}. \end{cases}$$

In addition, we obtain the variational junction condition

$$0 = \int_J \Big[\hat{\mathbf{u}} \cdot (C[\varepsilon(\mathbf{u})]\boldsymbol{\nu}) + G_1h\hat{W}(\nabla W + \mathbf{u}) \cdot \boldsymbol{\nu}\Big]\, d\Gamma$$
$$- G_2I\vartheta_1'(0)\hat{\vartheta}_1(0) - EI_h\vartheta_2'(0)\hat{\vartheta}_2(0) - G_2A(w' + \vartheta_2)(0)\hat{w}(0)$$

for all test functions $\hat{W}, \hat{\mathbf{u}}, \hat{w}, \hat{\boldsymbol{\vartheta}}$ which satisfy (2.10), where $\hat{\boldsymbol{\vartheta}} = \hat{\vartheta}_2\mathbf{e}_1 - \hat{\vartheta}_1\mathbf{e}_2$. The following junction conditions may be deduced from the last equation and (2.10).

Dynamic junction conditions:

$$(2.22)\qquad \begin{cases} G_2I\vartheta_1'(0) = \displaystyle\int_J \{G_1h\zeta\sigma_0\boldsymbol{\nu} \cdot (\nabla W + \mathbf{u}) \\ \qquad\qquad\qquad -\sigma_1\mathbf{e}_2 \cdot (C[\varepsilon(\mathbf{u})]\boldsymbol{\nu})\}d\Gamma, \\ EI_h\vartheta_2'(0) = \displaystyle\int_J \sigma_1\mathbf{e}_1 \cdot (C[\varepsilon(\mathbf{u})]\boldsymbol{\nu})\, d\Gamma, \\ G_2A(w' + \vartheta_2)(0) = G_1h \displaystyle\int_J \sigma_0\boldsymbol{\nu} \cdot (\nabla W + \mathbf{u})\, d\Gamma \end{cases}$$

in which it should be recalled that $\mathbf{e}_1 = \boldsymbol{\nu}(\mathbf{x}_0)$, $\mathbf{e}_2 = \boldsymbol{\tau}(\mathbf{x}_0)$. The complete description of the system when $J \setminus J_a$ is geometrically restricted therefore consists of (2.18)–(2.22) together with the geometric constraints (2.9) and (2.10).

In the case when $J \setminus J_a$ is free rather than geometrically restricted, we again obtain (2.18)–(2.22) except that the integrals are to be taken over J_a rather than all of J, and we obtain on $J \setminus J_a$ the free edge conditions

$$(2.23)\qquad G_1h(\nabla W + \mathbf{u}) \cdot \boldsymbol{\nu} = 0,\;\; C[\varepsilon(\mathbf{u})]\boldsymbol{\nu} = 0 \;\text{ on } J \setminus J_a.$$

If desired, one may in the same manner obtain the equations of motion and boundary and junction conditions for the components $\mathbf{w}_\beta, W_\beta, \vartheta_3$ related to the motion of the plate-beam system in the $\mathbf{a}_1\mathbf{a}_2$ plane.

3. Function Spaces and Well-Posedness

We shall consider below the question of exact controllability of a plate-beam system by means of controls applied along Γ^D only, which may act through either mechanical boundary conditions or geometrical boundary conditions. Only the subsystem for the components $W, \mathbf{u}, w, \vartheta_1, \vartheta_2$ will be considered; however, a similar analysis could be carried out on the subsystem satisfied by $W_\beta, w_\beta, \vartheta_3$. In preparation for this study, well-posedness of the controlled system will be considered in this section.

We suppose that all distributed forces and moments vanish, as do those at the free end of the beam. The system to be considered may then be written

$$\begin{cases} I_{\rho_1}\ddot{\mathbf{u}} - \operatorname{div} C[\varepsilon(\mathbf{u})] + G_1 h(\nabla W + \mathbf{u}) = 0, \\ m_1\ddot{W} - G_1 h \operatorname{div}(\nabla W + \mathbf{u}) = 0 \ \text{ in } \Omega; \end{cases} \tag{3.1}$$

$$\begin{cases} m_2\ddot{w} - G_2 A(w' + \vartheta_2)' = 0, \\ \rho_2 I_h \ddot{\vartheta}_2 - EI_h \vartheta_2'' + G_2 A(w' + \vartheta_2) = 0, \\ \rho_2 I \ddot{\vartheta}_1 - G_2 I \vartheta_1'' = 0, \ \ 0 < \xi < \ell; \end{cases} \tag{3.2}$$

$$G_2 A(w' + \vartheta_2)(\ell) = EI_h \vartheta_2'(\ell) = G_2 I \vartheta_1'(\ell) = 0; \tag{3.3}$$

$$W = 0, \ \ \mathbf{u} = 0 \ \text{ on } \Gamma^C; \tag{3.4}$$

$$W = \sigma_0[w(0) + \zeta\vartheta_1(0)], \ \ \mathbf{u} = \sigma_1 \boldsymbol{\vartheta}(0) \ \text{ on } J; \tag{3.5}$$

$$\begin{cases} G_2 I \vartheta_1'(0) = \displaystyle\int_J \{G_1 h \zeta \sigma_0 \boldsymbol{\nu} \cdot (\nabla W + \mathbf{u}) \\ \qquad\qquad - \sigma_1 \mathbf{e}_2 \cdot (C[\varepsilon(\mathbf{u})]\boldsymbol{\nu})\} d\Gamma, \\ EI_h \vartheta_2'(0) = \displaystyle\int_J \sigma_1 \mathbf{e}_1 \cdot (C[\varepsilon(\mathbf{u})]\boldsymbol{\nu})\, d\Gamma, \\ G_2 A(w' + \vartheta_2)(0) = G_1 h \displaystyle\int_J \sigma_0 \boldsymbol{\nu} \cdot (\nabla W + \mathbf{u})\, d\Gamma. \end{cases} \tag{3.6}$$

When the controls are in the mechanical boundary conditions the remaining boundary conditions are

$$\begin{cases} G_1 h(\nabla W + \mathbf{u}) \cdot \boldsymbol{\nu} = f, \\ C[\varepsilon(\mathbf{u})]\boldsymbol{\nu} = \mathbf{m} \ \text{ on } \Gamma^D, \end{cases} \tag{3.7}$$

where f and $\mathbf{m} = m_\beta \mathbf{a}_\beta$ are the controls. When controls are in the geometrical boundary conditions, the boundary conditions along Γ^D are

$$W = f, \ \ \mathbf{u} = \mathbf{m} \ \text{ on } \Gamma^D. \tag{3.8}$$

To complete the description of the system, initial conditions are specified:

$$(3.9)\quad \begin{cases} W|_{t=0} = W^0, \ \dot W|_{t=0} = W^1, \ \mathbf{u}|_{t=0} = \mathbf{u}^0, \ \dot{\mathbf{u}}|_{t=0} = \mathbf{u}^1, \\ w|_{t=0} = w^0, \ \dot w|_{t=0} = w^1, \ \vartheta_i|_{t=0} = \vartheta_i^0, \ \dot\vartheta_i|_{t=0} = \vartheta_i^1. \end{cases}$$

For $s \geq 0$ we set

$$\mathcal{H}^s(\Omega) := H^s(\Omega;[\mathbf{a}_1]) \bigoplus H^s(\Omega;[\mathbf{a}_2]) \bigoplus H^s(\Omega;[\mathbf{a}_3]),$$

$$\mathcal{H}^s(0,\ell) = H^s(0,\ell;[\mathbf{e}_1]) \bigoplus H^s(0,\ell;[\mathbf{e}_2]) \bigoplus H^s(0,\ell;[\mathbf{e}_3]).$$

Let

$$\boldsymbol{\Psi} = \mathbf{u} + W\mathbf{a}_3 \in \mathcal{H}^0(\Omega), \quad \mathbf{u} = U_\beta \mathbf{a}_\beta,$$

and

$$\boldsymbol{\Theta} = \boldsymbol{\vartheta} + w\mathbf{e}_3 \in \mathcal{H}^0(0,\ell), \quad \boldsymbol{\vartheta} = \vartheta_2\mathbf{e}_1 - \vartheta_1\mathbf{e}_2.$$

We define the norms

$$\|\boldsymbol{\Psi}\|_{\mathcal{H}^0(\Omega)} = \left[\int_\Omega \left(m_1|W|^2 + I_{\rho_1}|\mathbf{u}|^2\right) d\Omega\right]^{1/2},$$

$$\|\boldsymbol{\Theta}\|_{\mathcal{H}^0(0,\ell)} = \left[\int_0^\ell \left(m_2|w|^2 + \rho_2 I_h|\vartheta_2|^2 + \rho_2 I|\vartheta_1|^2\right) d\xi\right]^{1/2},$$

and we set

$$H = \mathcal{H}^0(\Omega) \times \mathcal{H}^0(0,\ell)$$

with its natural product topology. For $s > 0$ the norms on $\mathcal{H}^s(\Omega)$ and $\mathcal{H}^s(0,\ell)$ are those induced by the corresponding H^s spaces with their standard norms.

Let γ be a nonempty, relatively open subset of $\Gamma \setminus \overline{J}$, and consider the set

$$\mathcal{H}^1_\gamma(\Omega) = \{\boldsymbol{\Psi} \in \mathcal{H}^1(\Omega) \,|\, \boldsymbol{\Psi} = 0 \ \text{ on } \gamma\}.$$

For $\boldsymbol{\Psi} \in \mathcal{H}^1_\gamma(\Omega)$ we set

$$\|\boldsymbol{\Psi}\|_{\mathcal{H}^1_\gamma(\Omega)} = \left\{\int_\Omega \left[(C[\varepsilon(\mathbf{u})], \varepsilon(\mathbf{u})) + G_1 h|\nabla W + \mathbf{u}|^2\right] d\Omega\right\}^{1/2}.$$

According to Korn's Lemma, $\|\cdot\|_{\mathcal{H}^1_\gamma(\Omega)}$ defines a norm on $\mathcal{H}^1_\gamma(\Omega)$ equivalent to that induced by $\mathcal{H}^1(\Omega)$. We further introduce

$$V_\gamma = \{(\boldsymbol{\Psi}, \boldsymbol{\Theta}) \,|\, \boldsymbol{\Psi} \in \mathcal{H}^1_\gamma(\Omega), \ \ \boldsymbol{\Theta} \in \mathcal{H}^1(0,\ell) \ \text{ and } \boldsymbol{\Psi} = \sigma\boldsymbol{\Theta}(0) \text{ on } J\},$$

where σ is a 3 by 3 matrix which with respect to the $\mathbf{e}_1, \mathbf{e}_2, \mathbf{e}_3$ basis is given by

$$\sigma = \begin{pmatrix} \sigma_1 & 0 & 0 \\ 0 & \sigma_1 & 0 \\ 0 & -\sigma_0\zeta & \sigma_0 \end{pmatrix}.$$

Note that the condition $\mathbf{\Psi} = \sigma\mathbf{\Theta}(0)$ just expresses the geometric condition (3.5). We define a norm on V_γ by setting

$$\|(\mathbf{\Psi},\mathbf{\Theta})\|_{V_\gamma} = \Big\{\|\mathbf{\Psi}\|^2_{\mathcal{H}^1_\gamma(\Omega)} + \int_0^\ell (G_2 I|\vartheta_1'|^2 + EI_h|\vartheta_2'|^2 + G_2 A|w' + \vartheta_2|^2)\, d\xi\Big\}^{1/2}.$$

The Hilbert space V_γ is dense in H with compact injection, so that if H is identified with its dual space we have the compact embeddings $V_\gamma \subset H \subset V_\gamma'$, where V_γ' denotes the dual of V_γ. We denote by $\langle v', v\rangle_{V_\gamma}$ the scalar product of elements $v' \in V_\gamma'$ and $v \in V_\gamma$. Let A denote the Riesz isomorphism of V_γ onto V_γ' and set

$$D_A = \{(\mathbf{\Psi},\mathbf{\Theta}) \in V_\gamma |\, A(\mathbf{\Psi},\mathbf{\Theta}) \in H\}, \quad \|(\mathbf{\Psi},\mathbf{\Theta})\|_{D_A} = \|A(\mathbf{\Psi},\mathbf{\Theta})\|_H.$$

Then A is an isomorphism of D_A onto H, of V_γ onto V_γ' and may be extended by continuity to an isomorphism of H onto D_A' through the formula

$$\langle Au, v\rangle_{V_\gamma} = (u, Av)_H, \quad \forall u \in V,\ v \in D_A).$$

REMARK 3.1. It may be verified through integrations by parts that

$$D_A \supset D_A^0 := \{(\mathbf{\Psi},\mathbf{\Theta}) |\, (\mathbf{\Psi},\mathbf{\Theta}) \in (\mathcal{H}^2(\Omega) \times \mathcal{H}^2(0,\ell)) \cap V_\gamma,$$
$$(\mathbf{\Psi},\mathbf{\Theta}) \text{ satisfy the dynamic junction and boundary conditions}\}$$

and that, if $(\mathbf{\Psi},\mathbf{\Theta}) \in D_A^0$ and $(\hat{\mathbf{\Psi}},\hat{\mathbf{\Theta}}) \in H$,

$$\begin{aligned}(A(\mathbf{\Psi},\mathbf{\Theta}), (\hat{\mathbf{\Psi}},\hat{\mathbf{\Theta}}))_H = &-\int_\Omega \{[\operatorname{div} C[\varepsilon(\mathbf{u})] - G_1 h(\nabla W + \mathbf{u})]\cdot\hat{\mathbf{u}} \\ &+ G_1 h \operatorname{div}(\nabla W + \mathbf{u})\hat{W}\}\, d\Omega - \int_0^\ell \{G_2 I \vartheta_1''\hat{\vartheta}_1 \\ &- [EI_h\vartheta_2'' - G_2 A(w' + \vartheta_2)]\hat{\vartheta}_2 + G_2 A(w' + \vartheta_2)'\hat{w}\}\, d\Omega.\end{aligned}$$

It follows that the injection of D_A^0 (with the topology of $\mathcal{H}^2(\Omega) \times \mathcal{H}^2(0,\ell)$) into D_A is continuous. On the other hand, we know from results of Nicaise [78] that under the stated assumptions on the region Ω, an element $(\mathbf{\Psi},\mathbf{\Theta}) \in D_A$ has the following spatial regularity:

(3.10) $$\mathbf{\Psi} \in \mathcal{H}^s(\Omega),\ \forall s < s_0,\ s \le 2, \quad \text{for some } s_0 > 3/2,$$

in the case of geometrical conditions on Γ^D, and also in the case of mechanical conditions on Γ^D provided $\overline{\Gamma}^C$ and $\overline{\Gamma}^D$ either have an empty intersection or else intersect in a strictly convex corner. If $\overline{\Gamma}^C$ and $\overline{\Gamma}^D$ meet in a nonconvex corner, (3.10) holds for all $s < s_0$, $s < 3/2$, for some $s_0 > 5/4$. Since $\dim(\Omega) = 2$, it follows from a Sobolev imbedding theorem that

$$\mathbf{\Psi} \in C^\alpha(\overline{\Omega}; [\mathbf{a}_1]) \bigoplus C^\alpha(\overline{\Omega}; [\mathbf{a}_2]) \bigoplus C^\alpha(\overline{\Omega}; [\mathbf{a}_3])$$

all $\alpha \in (0, s-1]$ [30, Theorem 1.4.5.2]. With regard to $\boldsymbol{\Theta}$, we have $\boldsymbol{\Theta} \in \mathcal{H}^2(0,\ell)$. The injection of D_A into $\mathcal{H}^s(\Omega) \times \mathcal{H}^2(0,\ell)$ is continuous. □

We denote by $(\boldsymbol{\Psi}^0, \boldsymbol{\Theta}^0)$, $(\boldsymbol{\Psi}^1, \boldsymbol{\Theta}^1)$ the initial data (3.9), i.e.,

$$\boldsymbol{\Psi}^0 = \mathbf{u}^0 + W^0 \mathbf{a}_3, \quad \boldsymbol{\Psi}^1 = \mathbf{u}^1 + W^1 \mathbf{a}_3,$$
$$\boldsymbol{\Theta}^0 = \boldsymbol{\vartheta}^0 + w^0 \mathbf{e}_3, \quad \boldsymbol{\Theta}^1 = \boldsymbol{\vartheta}^1 + w^1 \mathbf{e}_3.$$

We also set

$$\mathbf{f} = m_1 \mathbf{a}_1 + m_2 \mathbf{a}_2 + f \mathbf{a}_3 = \mathbf{m} + f \mathbf{a}_3$$

and

$$U = L^2(\Gamma^D; [\mathbf{a}_1]) \bigoplus L^2(\Gamma^D; [\mathbf{a}_2]) \bigoplus L^2(\Gamma^D; [\mathbf{a}_3]).$$

THEOREM 3.1. *(Well-posedness of geometrical control problem.) Let $\gamma = \Gamma - \overline{J}$ and suppose that*

$$(\boldsymbol{\Psi}^0, \boldsymbol{\Theta}^0) \in H, \quad (\boldsymbol{\Psi}^1, \boldsymbol{\Theta}^1) \in V'_\gamma, \quad \mathbf{f} \in L^2(0,T;U).$$

Then (3.1)–(3.6), (3.8), (3.9) *has a unique solution*

$$(\boldsymbol{\Psi}, \boldsymbol{\Theta}) \in C([0,T]; H) \cap C^1([0,T]; V'_\gamma).$$

THEOREM 3.2. *(Well-posedness of mechanical control problem.) Assume that $\Gamma^C \neq \emptyset$ and set $\gamma = \Gamma^C$. Let*

$$(\boldsymbol{\Psi}^0, \boldsymbol{\Theta}^0) \in H, \quad (\boldsymbol{\Psi}^1, \boldsymbol{\Theta}^1) \in V'_\gamma, \quad \mathbf{f} \in L^2(0,T;U).$$

Then (3.1)–(3.7), (3.9) *has a unique solution*

$$(\boldsymbol{\Psi}, \boldsymbol{\Theta}) \in C([0,T]; H) \cap C^1([0,T]; V'_\gamma).$$

REMARK 3.2. The assumption in Theorem 3.2 that $\Gamma^C \neq \emptyset$ is unessential in Theorem 3.2, but if $\Gamma^C = \emptyset$ we need to replace the norm on V by $(\|(\boldsymbol{\Psi}, \boldsymbol{\Theta})\|_V^2 + \int_\Omega |\boldsymbol{\Psi}|^2 d\Omega)^{1/2}$. □

PROOF OF THEOREM 3.2. The system (3.1)–(3.7), may be written as a variational equation by forming

$$\begin{aligned}
(3.11) \quad 0 = &\int_\Omega \{[m_1 \ddot{W} - G_1 h \operatorname{div}(\nabla W + \mathbf{u})]\hat{W} \\
&+ [I_{\rho_1} \ddot{\mathbf{u}} - \operatorname{div} C[\varepsilon(\mathbf{u})] + G_1 h(\nabla W + \mathbf{u})] \cdot \hat{\mathbf{u}}\} d\Omega \\
&+ \int_0^\ell \{(m_2 \ddot{w} - G_2 A(w' + \vartheta_2)')\hat{w} + (\rho_2 I \ddot{\vartheta}_1 - G_2 I \vartheta_1'')\hat{\vartheta}_1 \\
&\qquad + (\rho_2 I_h \ddot{\vartheta}_2 - E I_h \vartheta_2'' + G_2 A(w' + \vartheta_2))\hat{\vartheta}_2\} d\xi.
\end{aligned}$$

Set

$$\hat{\boldsymbol{\Psi}} = \hat{\mathbf{u}} + \hat{W}\mathbf{a}_3, \quad \hat{\boldsymbol{\Theta}} = \hat{\vartheta}_2 \mathbf{e}_1 - \hat{\vartheta}_1 \mathbf{e}_2 + \hat{w}\mathbf{e}_3 := \hat{\boldsymbol{\vartheta}} + \hat{w}\mathbf{e}_3,$$

and suppose that $(\hat{\Psi}, \hat{\Theta}) \in V_\gamma := V$. Upon carrying out integrations by parts in (3.11) one obtains with the aid of (2.17)

$$(3.12) \quad ((\ddot{\Psi}, \ddot{\Theta}), (\hat{\Psi}, \hat{\Theta}))_H + ((\Psi, \Theta), (\hat{\Psi}, \hat{\Theta}))_V = \int_{\Gamma^D} [\mathbf{m} \cdot \hat{\mathbf{u}} + f\hat{W}]\, d\Gamma,$$

where $\Psi = \mathbf{u} + W\mathbf{a}_3$, $\Theta = \vartheta_2 \mathbf{e}_1 - \vartheta_1 \mathbf{e}_2 + w\mathbf{e}_3 = \boldsymbol{\vartheta} + w\mathbf{e}_3$. We have

$$\left| \int_{\Gamma^D} [\mathbf{m} \cdot \hat{\mathbf{u}} + f\hat{W}]\, d\Gamma \right| \leq \|\mathbf{f}\|_U \|(\hat{\Psi}, \hat{\Theta})\|_V,$$

so there is an operator $B \in \mathcal{L}(U, V')$ such that

$$(3.13) \quad \int_{\Gamma^D} [\mathbf{m} \cdot \hat{\mathbf{u}} + f\hat{W}]\, d\Gamma = \langle B\mathbf{f}, (\hat{\Psi}, \hat{\Theta}) \rangle_V.$$

Therefore (3.12) may be written

$$(\ddot{\Psi}, \ddot{\Theta}) + A(\Psi, \Theta) = B\mathbf{f} \ \text{ in } V'$$

or, alternately, as the first order system

$$(3.14) \quad \dot{X} = \mathcal{A}X + \mathcal{B}\mathbf{f},$$

where

$$X = \begin{pmatrix} (\Psi, \Theta) \\ (\dot{\Psi}, \dot{\Theta}) \end{pmatrix}, \quad \mathcal{A} = \begin{pmatrix} 0 & I \\ -A & 0 \end{pmatrix}, \quad \mathcal{B}\mathbf{f} = \begin{pmatrix} 0 \\ B\mathbf{f} \end{pmatrix}.$$

It is well known that $\mathcal{A}$, as an unbounded operator in $H \times V'$ (resp., in $V \times H$, resp., in $V' \times D'_A$) with domain $V \times H$ (resp, with domain $D_A \times V$, resp., with domain $H \times V'$), generates a C_0-unitary group on $H \times V'$ (resp., on $V \times H$, resp., on $V' \times D'_A$). The conclusion of Theorem 3.2 follows immediately, since $\mathcal{B}\mathbf{f} \in L^2(0, T; H \times V')$. □

PROOF OF THEOREM 3.1. To obtain a variational equation for the geometrical control problem (3.1)–(3.6), (3.8) we again begin with (3.11) where we now assume that $(\hat{\Psi}, \hat{\Theta}) \in D_A$. Since $(\hat{\Psi}, \hat{\Theta}) = 0$ on $\gamma = \Gamma \setminus \overline{J}$ we have

$$\begin{aligned}
&\int_\Omega \{-G_1 h \operatorname{div}(\nabla W + \mathbf{u})\hat{W} - [\operatorname{div} C[\varepsilon(\mathbf{u})] - G_1 h(\nabla W + \mathbf{u})] \cdot \hat{\mathbf{u}}\}\, d\Omega \\
&= \int_\Omega \{-G_1 h \operatorname{div}(\nabla \hat{W} + \hat{\mathbf{u}})W - [\operatorname{div} C[\varepsilon(\hat{\mathbf{u}})] - G_1 h(\nabla \hat{W} + \hat{\mathbf{u}})] \cdot \mathbf{u}\}\, d\Omega \\
&\qquad + \int_J \{-G_1 h \boldsymbol{\nu} \cdot (\nabla W + \mathbf{u})\hat{W} + G_1 h \boldsymbol{\nu} \cdot (\nabla \hat{W} + \hat{\mathbf{u}})W \\
&\qquad\qquad - \hat{\mathbf{u}} \cdot (C[\varepsilon(\mathbf{u})]\boldsymbol{\nu}) + \mathbf{u} \cdot (C[\varepsilon(\hat{\mathbf{u}})]\boldsymbol{\nu})\}\, d\Gamma \\
&\qquad\qquad + \int_{\Gamma^D} [\mathbf{m} \cdot (C[\varepsilon(\hat{\mathbf{u}})]\boldsymbol{\nu}) + G_1 h f \frac{\partial \hat{W}}{\partial \boldsymbol{\nu}}]\, d\Gamma.
\end{aligned}$$

Also

$$\int_0^\ell \{-G_2A(w'+\vartheta_2)')\hat{w} - G_2I\vartheta_1''\hat{\vartheta}_1 - [EI_h\vartheta_2'' - G_2A(w'+\vartheta_2)]\hat{\vartheta}_2\}d\xi$$
$$= \int_0^\ell \{-G_2A(\hat{w}'+\hat{\vartheta}_2)')w - G_2I\hat{\vartheta}_1''\vartheta_1 - [EI_h\hat{\vartheta}_2'' - G_2A(\hat{w}'+\hat{\vartheta}_2)]\vartheta_2\}d\xi$$
$$+ G_2A(w'+\vartheta_2)(0)\hat{w}(0) + G_2I\vartheta_1'(0)\hat{\vartheta}_1(0) + EI_h\vartheta_2'(0)\hat{\vartheta}_2(0)$$
$$- G_2A(\hat{w}'+\hat{\vartheta}_2)(0)w(0) - G_2I\hat{\vartheta}_1'(0)\vartheta_1(0) - EI_h\hat{\vartheta}_2'(0)\vartheta_2(0).$$

Since $(\mathbf{\Psi}, \mathbf{\Theta})$ and $(\hat{\mathbf{\Psi}}, \hat{\mathbf{\Theta}})$ each satisfy the geometric and dynamic junction conditions, it follows that the sum of the integrals over J with the boundary terms at zero in the last two equations vanishes; that is

$$0 = \int_J \{-G_1h\boldsymbol{\nu}\cdot(\nabla W + \mathbf{u})\hat{W} + G_1h\boldsymbol{\nu}\cdot(\nabla\hat{W} + \hat{\mathbf{u}})W$$
$$- \hat{\mathbf{u}}\cdot(C[\varepsilon(\mathbf{u})]\boldsymbol{\nu}) + \mathbf{u}\cdot(C[\varepsilon(\hat{\mathbf{u}})]\boldsymbol{\nu})\}\, d\Gamma$$
$$+ G_2A(w'+\vartheta_2)(0)\hat{w}(0) + G_2I\vartheta_1'(0)\hat{\vartheta}_1(0) + EI_h\vartheta_2'(0)\hat{\vartheta}_2(0)$$
$$- G_2A(\hat{w}'+\hat{\vartheta}_2)(0)w(0) - G_2I\hat{\vartheta}_1'(0)\vartheta_1(0) - EI_h\hat{\vartheta}_2'(0)\vartheta_2(0).$$

Therefore (3.11) may be written (see Remark 3.1)

$$\text{(3.15)}\quad ((\ddot{\mathbf{\Psi}}, \ddot{\mathbf{\Theta}}), (\hat{\mathbf{\Psi}}, \hat{\mathbf{\Theta}}))_H + ((\mathbf{\Psi}, \mathbf{\Theta}), A(\hat{\mathbf{\Psi}}, \hat{\mathbf{\Theta}}))_H = \int_{\Gamma^D} [\mathbf{m}\cdot(C[\varepsilon(\hat{\mathbf{u}})]\boldsymbol{\nu}) + G_1hf\frac{\partial\hat{W}}{\partial\boldsymbol{\nu}}]\, d\Gamma.$$

Write $(\hat{\mathbf{\Psi}}, \hat{\mathbf{\Theta}}) = A^{-1}(\tilde{\mathbf{\Psi}}, \tilde{\mathbf{\Theta}})$ where $(\tilde{\mathbf{\Psi}}, \tilde{\mathbf{\Theta}}) \in H$. Then (3.15) takes the form

$$\text{(3.16)}\quad ((\ddot{\mathbf{\Psi}}, \ddot{\mathbf{\Theta}}), A^{-1}(\tilde{\mathbf{\Psi}}, \tilde{\mathbf{\Theta}}))_H + ((\mathbf{\Psi}, \mathbf{\Theta}), (\tilde{\mathbf{\Psi}}, \tilde{\mathbf{\Theta}}))_H = \int_{\Gamma^D} [\mathbf{m}\cdot(C[\varepsilon(\hat{\mathbf{u}})]\boldsymbol{\nu}) + G_1hf\frac{\partial\hat{W}}{\partial\boldsymbol{\nu}}]\, d\Gamma.$$

One has, for some $s > 3/2$ (see Remark 3.1),

$$\left|\int_{\Gamma^D} [\mathbf{m}\cdot(C[\varepsilon(\hat{\mathbf{u}})]\boldsymbol{\nu}) + G_1hf\frac{\partial\hat{W}}{\partial\boldsymbol{\nu}}]\, d\Gamma\right| \le C\|\mathbf{f}\|_U\|\hat{\mathbf{\Psi}}\|_{\mathcal{H}^s(\Omega)}$$
$$\le C\|\mathbf{f}\|_U\|(\hat{\mathbf{\Psi}}, \hat{\mathbf{\Theta}})\|_{D_A}$$
$$= C\|\mathbf{f}\|_U\|(\tilde{\mathbf{\Psi}}, \tilde{\mathbf{\Theta}})\|_H.$$

Thus

$$\text{(3.17)}\quad \int_{\Gamma^D} [\mathbf{m}\cdot(C[\varepsilon(\hat{\mathbf{u}})]\boldsymbol{\nu}) + G_1hf\frac{\partial\hat{W}}{\partial\boldsymbol{\nu}}]\, d\Gamma = (B\mathbf{f}, (\tilde{\mathbf{\Psi}}, \tilde{\mathbf{\Theta}}))_H, \quad (\hat{\mathbf{\Psi}}, \hat{\mathbf{\Theta}}) = A^{-1}(\tilde{\mathbf{\Psi}}, \tilde{\mathbf{\Theta}}),$$

for some $B \in \mathcal{L}(U, H)$, so that (3.16) may be written

$$((\ddot{\Psi}, \ddot{\Theta}), A^{-1}(\tilde{\Psi}, \tilde{\Theta}))_H + ((\Psi, \Theta), (\tilde{\Psi}, \tilde{\Theta}))_H = (B\mathbf{f}, (\tilde{\Psi}, \tilde{\Theta}))_H, \qquad \forall (\tilde{\Psi}, \tilde{\Theta}) \in H.$$

This is the same as the equation

$$(\ddot{\Psi}, \ddot{\Theta}) + A(\Psi, \Theta) = AB\mathbf{f} \quad \text{in } D'_A, \tag{3.18}$$

where A is an isomorphism of H onto D'_A. Equation (3.18) may be written as (3.14) with

$$\mathcal{B}\mathbf{f} = \begin{pmatrix} 0 \\ AB\mathbf{f} \end{pmatrix}, \quad \mathcal{B} \in \mathcal{L}(U, V'_\gamma \times D'_A).$$

The operator $\mathcal{A}$, as an unbounded operator in $V'_\gamma \times D'_A$ with domain $H \times V'_\gamma$, generates a C_0-group of unitary operators. Therefore, for initial data

$$(\Psi^0, \Theta^0) \in V'_\gamma, \quad (\Psi^1, \Theta^1) \in D'_A,$$

the initial value problem for (3.18) has a unique solution with

$$(\Psi, \Theta) \in C([0,T], V'_\gamma) \times C^1([0,T], D'_A).$$

We need to show that if the initial data satisfy the stronger regularity assumptions of Theorem 3.1, then the solution has the regularity stated in that theorem. This result cannot be obtained from abstract semigroup theory but rather is a consequence of the following regularity estimate. Let us consider the homogeneous system

$$\begin{cases} I_{\rho_1}\ddot{\varphi} - \operatorname{div} C[\varepsilon(\varphi)] + G_1 h(\nabla Z + \varphi) = 0, \\ m_1 \ddot{Z} - G_1 h \operatorname{div}(\nabla Z + \varphi) = 0 \quad \text{in } \Omega; \end{cases} \tag{3.19}$$

$$\begin{cases} m_2 \ddot{z} - G_2 A(z' + \theta_2)' = 0, \\ \rho_2 I_h \ddot{\theta}_2 - EI_h \theta_2'' + G_2 A(z' + \theta_2) = 0, \\ \rho_2 I \ddot{\theta}_1 - G_2 I \theta_1'' = 0, \quad 0 < \xi < \ell; \end{cases} \tag{3.20}$$

$$G_2 A(z' + \theta_2)(\ell) = EI_h \theta_2'(\ell) = G_2 I \theta_1'(\ell) = 0; \tag{3.21}$$

$$Z = 0, \quad \varphi = 0 \quad \text{on } \gamma := \Gamma - \overline{J}; \tag{3.22}$$

$$Z = \sigma_0[z(0) + \zeta\theta_1(0)], \quad \varphi = \sigma_1 \boldsymbol{\theta}(0) \quad \text{on } J, \tag{3.23}$$

where $\boldsymbol{\theta} = \theta_2\mathbf{e}_1 - \theta_1\mathbf{e}_2$;

$$(3.24)\quad \begin{cases} G_2 I\theta_1'(0) = \displaystyle\int_J \{G_1 h\zeta\sigma_0\boldsymbol{\nu}\cdot(\nabla Z + \boldsymbol{\varphi}) \\ \qquad\qquad -\sigma_1\mathbf{e}_2\cdot(C[\varepsilon(\boldsymbol{\varphi})]\boldsymbol{\nu})\}\,d\Gamma, \\ EI_h\theta_2'(0) = \displaystyle\int_J \sigma_1\mathbf{e}_1\cdot(C[\varepsilon(\boldsymbol{\varphi})]\boldsymbol{\nu})\,d\Gamma, \\ G_2A(z'+\theta_2)(0) = G_1 h\displaystyle\int_J \sigma_0\boldsymbol{\nu}\cdot(\nabla Z + \boldsymbol{\varphi})\,d\Gamma \ \text{ on } J; \end{cases}$$

$$(3.25)\quad \begin{cases} Z|_{t=0} = Z^0, \ \ \dot{Z}|_{t=0} = Z^1, \ \ \boldsymbol{\varphi}|_{t=0} = \boldsymbol{\varphi}^0, \ \ \dot{\boldsymbol{\varphi}}|_{t=0} = \boldsymbol{\varphi}^1, \\ z|_{t=0} = z^0, \ \ \dot{z}|_{t=0} = z^1, \ \ \boldsymbol{\theta}|_{t=0} = \boldsymbol{\theta}^0, \dot{\boldsymbol{\theta}}|_{t=0} = \boldsymbol{\theta}^1. \end{cases}$$

As in the last section, we define

$$\boldsymbol{\Phi} = \boldsymbol{\varphi} + Z\mathbf{a}_3, \quad \boldsymbol{\Xi} = \boldsymbol{\theta} + z\mathbf{e}_3,$$

and $\boldsymbol{\Phi}^0, \boldsymbol{\Phi}^1, \boldsymbol{\Xi}^0, \boldsymbol{\Xi}^1$ with obvious connotations.

LEMMA 3.1. *Let* $(\boldsymbol{\Phi}, \boldsymbol{\Xi})$ *be the solution of* (3.19)–(3.25) *with initial data*

$$(\boldsymbol{\Phi}^0, \boldsymbol{\Xi}^0) \in V_\gamma, \quad (\boldsymbol{\Phi}^1, \boldsymbol{\Xi}^1) \in H,$$

and set $\Sigma_1 := \Gamma_1 \times (0, T)$, *where* Γ_1 *is any relatively open subset of* Γ *such that* $\overline{\Gamma}_1 \cap \overline{J} = \emptyset$. *Then*

$$(3.26)\quad \int_{\Sigma_1}\left[|C[\varepsilon(\boldsymbol{\varphi})]\boldsymbol{\nu}|^2 + G_1 h\left(\frac{\partial Z}{\partial \boldsymbol{\nu}}\right)^2\right] d\Sigma \\ \leq C_0(T+1)\|((\boldsymbol{\Phi}^0, \boldsymbol{\Xi}^0), (\boldsymbol{\Phi}^1, \boldsymbol{\Xi}^1))\|^2_{V_\gamma\times H}.$$

The proof of Lemma 3.1 is deferred until the end of this section.

To complete the proof of Theorem 3.1, we utilize the idea of transposition. Write $V := V_\gamma$. The solution of (3.18) having initial data

$$X^0 = \begin{pmatrix} (\boldsymbol{\Phi}^0, \boldsymbol{\Xi}^0) \\ (\boldsymbol{\Phi}^1, \boldsymbol{\Xi}^1) \end{pmatrix}$$

is given by

$$X(t) = \exp(t\mathcal{A})X^0 + \int_0^t \exp((t-s)\mathcal{A})\mathcal{B}\mathbf{f}(s)\,ds, \quad 0 \leq t \leq T,$$

where $\exp(t\mathcal{A})$ is the unitary group on $V' \times D_A'$ generated by $\mathcal{A}$. Suppose that $X^0 \in H \times V'$, let $Y^0 \in V \times D_A$ and $\mathcal{B}' \in \mathcal{L}(V \times D_A; U)$ be the dual of $\mathcal{B}$, defined by

$$\langle \mathcal{B}\mathbf{f}, Y^0\rangle_{V\times D_A} = (\mathbf{f}, \mathcal{B}'Y^0)_U, \quad \forall \mathbf{f} \in U, \ Y^0 \in V \times D_A.$$

Let $\tau \in (0, T]$ be fixed. We have

$$\langle X(\tau), Y^0\rangle_{V\times D_A} = \langle X^0, \exp(\tau\mathcal{A}')Y^0\rangle_{H\times V} + \int_0^\tau (\mathbf{f}(s), \mathcal{B}'\exp((\tau - s)\mathcal{A}')Y^0)_U ds. \tag{3.27}$$

Here $\mathcal{A}'$ is the dual of $\mathcal{A}$, defined by

$$\langle \mathcal{A}X^0, Y^0\rangle_{V\times D_A} = \langle X^0, \mathcal{A}'Y^0\rangle_{H\times V}, \quad \forall X^0 \in H \times V', \ Y^0 \in V \times D_A.$$

One has

$$\mathcal{A}' = \begin{pmatrix} 0 & -A \\ I & 0 \end{pmatrix}, \quad D(\mathcal{A}') = V \times D_A.$$

As is well-known, $\mathcal{A}'$ generates a unitary group $\exp(t\mathcal{A}')$ on $H \times V$ and $\exp(t\mathcal{A}')$ is the dual of the restriction of $\exp(t\mathcal{A})$ to $H \times V'$. Therefore (3.27) is the same as

$$\langle X(\tau), Y^0\rangle_{V\times D_A} = \langle X^0, Y(\tau)\rangle_{H\times V} + \int_0^\tau (\mathbf{f}(s), \mathcal{B}'Y(s))_U ds, \tag{3.28}$$

where $Y(t) := \exp((\tau - t)\mathcal{A}')Y^0$, $0 \le t \le \tau$, satisfies

$$\dot{Y} = -\mathcal{A}'Y, \quad 0 \le t \le \tau, \quad Y(\tau) = Y^0. \tag{3.29}$$

If we write

$$Y = ((\mathbf{\Phi}_1, \mathbf{\Xi}_1), (\mathbf{\Phi}, \mathbf{\Xi})), \quad Y^0 = ((\mathbf{\Phi}^1, \mathbf{\Xi}^1), (\mathbf{\Phi}^0, \mathbf{\Xi}^0)),$$

(3.29) signifies that

$$\begin{cases} (\ddot{\mathbf{\Phi}}, \ddot{\mathbf{\Xi}}) + A(\mathbf{\Phi}, \mathbf{\Xi}) = 0, \quad (\mathbf{\Phi}_1, \mathbf{\Xi}_1) = -(\dot{\mathbf{\Phi}}, \dot{\mathbf{\Xi}}), \\ (\mathbf{\Phi}(\tau), \mathbf{\Xi}(\tau)) = (\mathbf{\Phi}^0, \mathbf{\Xi}^0), \quad (\dot{\mathbf{\Phi}}(\tau), \dot{\mathbf{\Xi}}(\tau)) = -(\mathbf{\Phi}^1, \mathbf{\Xi}^1). \end{cases} \tag{3.30}$$

In addition,

$$\langle \mathcal{B}\mathbf{f}, Y\rangle_{V\times D_A} = \langle A\mathcal{B}\mathbf{f}, (\mathbf{\Phi}, \mathbf{\Xi})\rangle_{D_A} = (\mathcal{B}\mathbf{f}, A(\mathbf{\Phi}, \mathbf{\Xi}))_H.$$

From (3.17) we have

$$(\mathcal{B}\mathbf{f}, A(\mathbf{\Phi}, \mathbf{\Xi}))_H = \int_{\Gamma^D} [\mathbf{m} \cdot (C[\varepsilon(\boldsymbol{\varphi})]\boldsymbol{\nu}) + Kf\frac{\partial Z}{\partial \boldsymbol{\nu}}]\, d\Gamma,$$

hence

$$\mathcal{B}'Y = C[\varepsilon(\boldsymbol{\varphi})]\boldsymbol{\nu} + G_1 h\frac{\partial Z}{\partial \boldsymbol{\nu}}\mathbf{a}_3\Big|_{\Gamma^D}. \tag{3.31}$$

We insert this expression into (3.28) and utilize Lemma 3.1 with $\Gamma_1 = \Gamma^D$ to obtain the estimate

$$
\begin{aligned}
(3.32)\quad |\langle X(\tau), Y^0\rangle_{V\times D_A}| &\le \|X^0\|_{H\times V'}\|Y^0\|_{H\times V} \\
&\quad + \|\mathbf{f}\|_{L^2(0,T;U)}\|C[\varepsilon(\varphi)]\boldsymbol{\nu} + G_1 h\frac{\partial Z}{\partial \boldsymbol{\nu}}\mathbf{a}_3\|_{L^2(0,T;U)} \\
&\le C_0(T+1)\{\|X^0\|_{H\times V'} + \|\mathbf{f}\|_{L^2(0,T;U)}\}\|Y^0\|_{H\times V}, \quad 0\le\tau\le T.
\end{aligned}
$$

It follows that $X \in L^\infty(0,T;H\times V')$. One may pass from L^∞ to C by a standard approximation argument. □

REMARK 3.3. That $X \in C([0,T];H\times V')$ may also be obtained directly from a "lifting theorem" of Lasiecka and Triggiani [55] once Lemma 3.1 is established. □

REMARK 3.4. When $X^0 = 0$, it follows from (3.32) that

$$
(3.33)\qquad \|X(T)\|_{H\times V'} \le C_T\|\mathbf{f}\|_{L^2(0,T;U)}
$$

so that the *control-to-state* map $\mathbf{f} \mapsto X(T) : L^2(0,T;U) \mapsto H \times V'$ is continuous.

PROOF OF LEMMA 3.1. It suffices to prove (3.26) for initial data in $D_A \times V$. We first use a trick from [85]. Let $\chi \in C^\infty(\overline{\Omega})$ such that $\chi = 1$ is a neighborhood of Γ_1 and $\chi = 0$ in a neighborhood of J, and set $\tilde{\mathbf{\Phi}} = \chi\mathbf{\Phi}$. (Such a χ exists since $\overline{\Gamma}_1 \cap \overline{J} = \emptyset$.) For any $\varphi = \varphi_1\mathbf{a}_1 + \varphi_2\mathbf{a}_2$ with $\varphi_i \in H^2(\Omega)$ we have

$$
\begin{aligned}
&\operatorname{div} C[\varepsilon(\chi\varphi)] \\
&= \frac{h^3}{12}\begin{pmatrix} \frac{\partial}{\partial x_1}((\lambda_1+2\mu_1)\varepsilon_{11}(\chi\varphi) + \lambda_1\varepsilon_{22}(\chi\varphi)) + 2\mu_1\frac{\partial}{\partial x_2}\varepsilon_{12}(\chi\varphi) \\ 2\mu_1\frac{\partial}{\partial x_1}\varepsilon_{12}(\chi\varphi) + \frac{\partial}{\partial x_2}((\lambda_1+2\mu_1)\varepsilon_{22}(\chi\varphi) + \lambda_1\varepsilon_{11}(\chi\varphi)) \end{pmatrix} \\
&= \chi\operatorname{div} C[\varepsilon(\varphi)] + 2C[\varepsilon(\varphi)]\nabla\chi + \frac{h^3}{12}(\mu_1-\lambda_1)\begin{pmatrix} \varphi_{2,2} & -\varphi_{2,1} \\ -\varphi_{1,2} & \varphi_{1,1}\end{pmatrix}\nabla\chi \\
&\qquad + \frac{h^3}{12}\begin{pmatrix} \chi_{,11} + \mu_1\chi_{,22} & (\lambda_1+\mu_1)\chi_{,12} \\ (\lambda_1+\mu_1)\chi_{,12} & \chi_{,22}+\mu_1\chi_{,11}\end{pmatrix}\varphi.
\end{aligned}
$$

It follows that $\tilde{\mathbf{\Phi}}$ satisfies the equations

$$
(3.34)\qquad \begin{cases} m_1\ddot{\tilde{Z}} - G_1 h\operatorname{div}(\nabla\tilde{Z} + \tilde{\varphi}) = \tilde{F}, \\ I_{\rho_1}\ddot{\tilde{\varphi}} - \operatorname{div} C[\varepsilon(\tilde{\varphi})] + G_1 h(\nabla\tilde{Z} + \tilde{\varphi}) = \tilde{\mathbf{M}}, \end{cases}
$$

where

$$
\tilde{F} = -G_1 h\nabla\chi\cdot(\nabla Z + \varphi)
$$

and

$$\tilde{\mathbf{M}} = -2C[\varepsilon(\boldsymbol{\varphi})]\nabla\chi - \frac{h^3}{12}(\mu_1 - \lambda_1)\begin{pmatrix} \varphi_{2,2} & -\varphi_{2,1} \\ -\varphi_{1,2} & \varphi_{1,1} \end{pmatrix}\nabla\chi$$
$$- \frac{h^3}{12}\begin{pmatrix} \chi_{,11} + \mu_1\chi_{,22} & (\lambda_1 + \mu_1)\chi_{,12} \\ (\lambda_1 + \mu_1)\chi_{,12} & \chi_{,22} + \mu_1\chi_{,11} \end{pmatrix}\boldsymbol{\varphi} + G_1 h(Z\nabla\chi + \chi\boldsymbol{\varphi});$$

the boundary conditions

$$\tilde{Z} = 0, \quad \tilde{\boldsymbol{\varphi}} = 0 \text{ on } \Sigma; \tag{3.35}$$

and the initial conditions

$$\tilde{\boldsymbol{\Phi}}(0) = \chi\boldsymbol{\Phi}^0, \quad \dot{\tilde{\boldsymbol{\Phi}}}(0) = \chi\boldsymbol{\Phi}^1.$$

For a solution of (3.34), (3.35) with initial data $(\tilde{\boldsymbol{\Phi}}(0), \dot{\tilde{\boldsymbol{\Phi}}}(0)) \in \mathcal{H}^1_\Gamma(\Omega) \times \mathcal{H}^0(\Omega) := \mathcal{H}$ we have the standard energy estimate

$$\|(\tilde{\boldsymbol{\Phi}}(t), \dot{\tilde{\boldsymbol{\Phi}}}(t))\|^2_{\mathcal{H}} \le \|(\tilde{\boldsymbol{\Phi}}(0), \dot{\tilde{\boldsymbol{\Phi}}}(0))\|^2_{\mathcal{H}} + C\int_0^t \|\tilde{\mathbf{F}}(t)\|^2_{\mathcal{H}^0(\Omega)}dt,$$

where $\tilde{\mathbf{F}} = \tilde{\mathbf{M}} + \tilde{F}\mathbf{a}_3$. We shall show that

$$\int_\Sigma \left[|C[\varepsilon(\tilde{\boldsymbol{\varphi}})]\boldsymbol{\nu}|^2 + G_1 h\left(\frac{\partial \tilde{Z}}{\partial \boldsymbol{\nu}}\right)^2 \right] d\Sigma \tag{3.36}$$
$$\le C\left[\int_0^T \|\tilde{\mathbf{F}}(t)\|^2_{\mathcal{H}^0(\Omega)}dt + (T+1)\|(\tilde{\boldsymbol{\Phi}}, \dot{\tilde{\boldsymbol{\Phi}}})\|^2_{L^\infty(0,T;\mathcal{H})}\right].$$

Since

$$\|(\tilde{\boldsymbol{\Phi}}, \dot{\tilde{\boldsymbol{\Phi}}})\|^2_{L^\infty(0,T;\mathcal{H})} \le \|(\tilde{\boldsymbol{\Phi}}(0), \dot{\tilde{\boldsymbol{\Phi}}}(0))\|^2_{\mathcal{H}} + C\int_0^T \|\tilde{\mathbf{F}}(t)\|^2_{\mathcal{H}^0(\Omega)}dt$$
$$\le C\left(\|(\boldsymbol{\Phi}^0, \boldsymbol{\Phi}^1)\|^2_{\mathcal{H}^1_\gamma(\Omega)\times\mathcal{H}^0(\Omega)} + \int_0^T \|\tilde{\mathbf{F}}(t)\|^2_{\mathcal{H}^0(\Omega)}dt \right),$$

and

$$\int_0^T \|\tilde{\mathbf{F}}(t)\|^2_{\mathcal{H}^0(\Omega)}dt \le C\int_0^T \|\boldsymbol{\Phi}(t)\|^2_{\mathcal{H}^1_\gamma(\Omega)}dt$$
$$\le C\int_0^T \|((\boldsymbol{\Phi}(t), \boldsymbol{\Xi}(t)), (\dot{\boldsymbol{\Phi}}(t), \dot{\boldsymbol{\Xi}}(t)))\|^2_{V_\gamma\times H}dt$$
$$= CT\|((\boldsymbol{\Phi}^0, \boldsymbol{\Xi}^0), (\boldsymbol{\Phi}^1, \boldsymbol{\Xi}^1))\|^2_{V_\gamma\times H},$$

and since $\tilde{\boldsymbol{\Phi}} = \boldsymbol{\Phi}$ on Σ_1, Lemma 5.1 follows from (3.36).

The idea leading to (3.36) is standard (cf. [60], for example). One multiplies the two equations in (3.34) by $\mathbf{h}\cdot\nabla\tilde{Z}$ and $(\nabla\tilde{\boldsymbol{\varphi}})\mathbf{h}$ respectively, where $\mathbf{h}$ is a $W^{1,\infty}$ vector field in Ω such that $\mathbf{h} = \boldsymbol{\nu}$ on Γ, adds the products and integrates the sum over $\Omega \times (0, T)$. One thereby obtains

$$(3.37)\quad \int_0^T \int_\Omega \{[m_1 \ddot{\tilde{Z}} - G_1 h \operatorname{div}(\nabla \tilde{Z} + \tilde{\varphi})]\mathbf{h} \cdot \nabla \tilde{Z} + [I_{\rho_1} \ddot{\tilde{\varphi}} - \operatorname{div} C[\varepsilon(\tilde{\varphi})] + G_1 h(\nabla \tilde{Z} + \tilde{\varphi})] \cdot ((\nabla \tilde{\varphi})\mathbf{h})\}\, d\Omega dt = \int_0^T \int_\Omega [(\mathbf{h} \cdot \nabla \tilde{Z})\tilde{F} + ((\nabla \tilde{\varphi})\mathbf{h})\tilde{\mathbf{M}}]\, d\Omega dt.$$

One has

$$\int_0^T \int_\Omega (\mathbf{h} \cdot \nabla \tilde{Z}) \ddot{\tilde{Z}}\, d\Omega dt = \int_\Omega (\mathbf{h} \cdot \nabla \tilde{Z}) \dot{\tilde{Z}}\, d\Omega|_0^T + \frac{1}{2} \int_0^T \int_\Omega (\operatorname{div} \mathbf{h}) |\dot{\tilde{Z}}|^2 d\Omega dt,$$

$$\begin{aligned}\int_0^T \int_\Omega (\mathbf{h} \cdot \nabla \tilde{Z}) \operatorname{div}(\nabla \tilde{Z} + \tilde{\varphi})\, d\Omega dt &= \int_0^T \int_\Omega \{\operatorname{div}[(\mathbf{h} \cdot \nabla \tilde{Z})\nabla \tilde{Z}] - (\nabla \mathbf{h} \nabla \tilde{Z}) \cdot \nabla \tilde{Z} \\ &\quad - \frac{1}{2} \operatorname{div}(\mathbf{h}|\nabla \tilde{Z}|)^2 + \frac{1}{2}(\operatorname{div} \mathbf{h})|\nabla \tilde{Z}|^2 + (\operatorname{div} \tilde{\varphi})(\mathbf{h} \cdot \nabla \tilde{Z})\} d\Omega dt \\ &= \int_0^T \int_\Omega \{\frac{1}{2}(\operatorname{div} \mathbf{h})|\nabla \tilde{Z}|^2 - (\nabla h \nabla \tilde{Z}) \cdot \nabla \tilde{Z} \\ &\quad + (\operatorname{div} \tilde{\varphi})(\mathbf{h} \cdot \nabla \tilde{Z})\} d\Omega dt + \frac{1}{2} \int_\Sigma \left(\frac{\partial \tilde{Z}}{\partial \boldsymbol{\nu}}\right)^2 d\Sigma,\end{aligned}$$

$$\int_0^T \int_\Omega \ddot{\tilde{\varphi}} \cdot (\nabla \tilde{\varphi} \mathbf{h}) d\Omega dt = \int_\Omega \dot{\tilde{\varphi}} \cdot (\nabla \tilde{\varphi} \mathbf{h}) d\Omega|_0^T + \frac{1}{2} \int_0^T \int_\Omega (\operatorname{div} \mathbf{h}) |\dot{\tilde{\varphi}}|^2 d\Omega dt.$$

Also, from Lemma 3.2, Chapter VI, we have

$$\begin{aligned}\int_0^T \int_\Omega (\nabla \tilde{\varphi} \mathbf{h}) \cdot \operatorname{div} C[\varepsilon(\tilde{\varphi})]\, d\Omega dt &= -\frac{1}{2} \int_\Sigma (C[\varepsilon(\tilde{\varphi})], \varepsilon(\tilde{\varphi}))\, d\Sigma \\ &\quad + \int_\Sigma (\nabla \tilde{\varphi} \mathbf{h}) \cdot (C[\varepsilon(\tilde{\varphi})]\boldsymbol{\nu})\, d\Sigma - \int_0^T \int_\Omega Q(\nabla \tilde{\varphi})\, d\Omega dt,\end{aligned}$$

where the functional Q is defined by

$$\begin{aligned}Q(\nabla \varphi) = \frac{1}{2}(h_{1,1} - h_{2,2})[(2\mu_1 + \lambda_1)(\varphi_{1,1}^2 - \varphi_{2,2}^2) + \mu_1(\varphi_{2,1}^2 - \varphi_{1,2}^2)] \\ + (2\mu_1 + \lambda_1)(h_{1,2}\varphi_{2,1}\varphi_{2,2} + h_{2,1}\varphi_{1,1}\varphi_{1,2}) \\ + \mu_1(h_{1,2}\varphi_{1,1} + h_{2,1}\varphi_{2,2})(\varphi_{1,2} + \varphi_{2,1}) \\ + \lambda_1(h_{1,2}\varphi_{1,1}\varphi_{2,1} + h_{2,1}\varphi_{1,2}\varphi_{2,2}).\end{aligned}$$

When the last four formulas are inserted into (3.37) the result is

$$\int_\Omega \{m_1\dot{\tilde{Z}}(\mathbf{h}\cdot\nabla\tilde{Z}) + I_{\rho_1}\dot{\tilde{\boldsymbol{\varphi}}}\cdot(\nabla\tilde{\boldsymbol{\varphi}}\mathbf{h})\}d\Omega\Big|_0^T$$
$$+\frac{1}{2}\int_0^T\int_\Omega(\operatorname{div}\mathbf{h})\{m_1|\dot{\tilde{Z}}|^2 + I_{\rho_1}|\dot{\tilde{\boldsymbol{\varphi}}}|^2\}d\Omega dt$$
$$+\int_0^T\int_\Omega\{Q(\nabla\tilde{\boldsymbol{\varphi}}) + G_1h[-\frac{1}{2}(\operatorname{div}\mathbf{h})|\nabla\tilde{Z}|^2 + (\nabla\mathbf{h}\nabla\tilde{Z})\cdot\nabla\tilde{Z}$$
$$-(\operatorname{div}\tilde{\boldsymbol{\varphi}})(\mathbf{h}\cdot\nabla\tilde{Z}) + (\nabla\tilde{Z}+\tilde{\boldsymbol{\varphi}})\cdot((\nabla\tilde{\boldsymbol{\varphi}})\mathbf{h})]\}d\Omega dt$$
$$-\int_\Sigma\left\{\frac{G_1h}{2}\left(\frac{\partial\tilde{Z}}{\partial\boldsymbol{\nu}}\right)^2 + (\nabla\tilde{\boldsymbol{\varphi}}\mathbf{h})\cdot(C[\varepsilon(\tilde{\boldsymbol{\varphi}})]\boldsymbol{\nu}) - \frac{1}{2}(C[\varepsilon(\tilde{\boldsymbol{\varphi}})],\varepsilon(\tilde{\boldsymbol{\varphi}}))\right\}d\Sigma$$
$$=\int_0^T\int_\Omega[\tilde{F}(\mathbf{h}\cdot\nabla\tilde{Z}) + \tilde{\mathbf{M}}\cdot(\nabla\tilde{\boldsymbol{\varphi}}\mathbf{h})]\,d\Omega dt.$$

On Σ on has $\nabla\tilde{\boldsymbol{\varphi}}\mathbf{h} = \partial\tilde{\boldsymbol{\varphi}}/\partial\boldsymbol{\nu}$, hence

$$(\nabla\tilde{\boldsymbol{\varphi}}\mathbf{h})\cdot(C[\varepsilon(\tilde{\boldsymbol{\varphi}})]\boldsymbol{\nu}) - \frac{1}{2}(C[\varepsilon(\tilde{\boldsymbol{\varphi}})],\varepsilon(\tilde{\boldsymbol{\varphi}})) = \frac{1}{2}(C[\varepsilon(\tilde{\boldsymbol{\varphi}})]\boldsymbol{\nu})\cdot\frac{\partial\tilde{\boldsymbol{\varphi}}}{\partial\boldsymbol{\nu}} \tag{3.38}$$

(see (5.20), Chapter VII). In addition (cf. (3.34), (3.35), Chapter VI)

$$\begin{aligned}C[\varepsilon(\tilde{\boldsymbol{\varphi}})]\boldsymbol{\nu} = C^{\alpha\beta}[\varepsilon(\tilde{\boldsymbol{\varphi}})]\nu_\alpha\mathbf{a}_\beta = \frac{h^3}{12}\Bigg\{&\mu_1\left(\boldsymbol{\tau}\cdot\frac{\partial\tilde{\boldsymbol{\varphi}}}{\partial\boldsymbol{\nu}} + \boldsymbol{\nu}\cdot\frac{\partial\tilde{\boldsymbol{\varphi}}}{\partial\boldsymbol{\tau}}\right)\boldsymbol{\tau}\\ &+\left[(2\mu_1+\lambda_1)\left(\boldsymbol{\nu}\cdot\frac{\partial\tilde{\boldsymbol{\varphi}}}{\partial\boldsymbol{\nu}}\right) + \lambda_1\left(\boldsymbol{\tau}\cdot\frac{\partial\tilde{\boldsymbol{\varphi}}}{\partial\boldsymbol{\tau}}\right)\right]\boldsymbol{\nu}\Bigg\},\end{aligned} \tag{3.39}$$

$$\begin{aligned}C[\varepsilon(\tilde{\boldsymbol{\varphi}})]\boldsymbol{\tau} &= C^{\alpha\beta}[\varepsilon(\tilde{\boldsymbol{\varphi}})]\tau_\alpha\mathbf{a}_\beta\\ &= \frac{h^3}{12}\Bigg\{\left[(2\mu_1+\lambda_1)\left(\boldsymbol{\tau}\cdot\frac{\partial\tilde{\boldsymbol{\varphi}}}{\partial\boldsymbol{\tau}}\right) + \lambda_1\left(\boldsymbol{\nu}\cdot\frac{\partial\tilde{\boldsymbol{\varphi}}}{\partial\boldsymbol{\nu}}\right)\right]\boldsymbol{\tau}\\ &\qquad+\mu_1\left(\boldsymbol{\tau}\cdot\frac{\partial\tilde{\boldsymbol{\varphi}}}{\partial\boldsymbol{\nu}} + \boldsymbol{\nu}\cdot\frac{\partial\tilde{\boldsymbol{\varphi}}}{\partial\boldsymbol{\tau}}\right)\boldsymbol{\nu}\Bigg\}.\end{aligned} \tag{3.40}$$

Therefore, on Γ,

$$(C[\varepsilon(\tilde{\boldsymbol{\varphi}})]\boldsymbol{\nu})\cdot\frac{\partial\tilde{\boldsymbol{\varphi}}}{\partial\boldsymbol{\nu}} = \frac{h^3}{12}\left[\mu_1\left(\boldsymbol{\tau}\cdot\frac{\partial\tilde{\boldsymbol{\varphi}}}{\partial\boldsymbol{\nu}}\right)^2 + (2\mu_1+\lambda_1)\left(\boldsymbol{\nu}\cdot\frac{\partial\tilde{\boldsymbol{\varphi}}}{\partial\boldsymbol{\nu}}\right)^2\right], \tag{3.41}$$

and

$$|C[\varepsilon(\tilde{\boldsymbol{\varphi}})]\boldsymbol{\nu}|^2 = \left(\frac{h^3}{12}\right)^2\left[\mu_1^2\left(\boldsymbol{\tau}\cdot\frac{\partial\tilde{\boldsymbol{\varphi}}}{\partial\boldsymbol{\nu}}\right)^2 + (2\mu_1+\lambda_1)^2\left(\boldsymbol{\nu}\cdot\frac{\partial\tilde{\boldsymbol{\varphi}}}{\partial\boldsymbol{\nu}}\right)^2\right]. \tag{3.42}$$

Hence, on Γ,

$$(3.43)\qquad K^{-1}|C[\varepsilon(\tilde{\varphi})]\boldsymbol{\nu}|^2 \le (C[\varepsilon(\tilde{\varphi})]\boldsymbol{\nu})\cdot\frac{\partial\tilde{\varphi}}{\partial\boldsymbol{\nu}} \le K|C[\varepsilon(\tilde{\varphi})]\boldsymbol{\nu}|^2$$

for some positive constant K. It follows from (3.38) and (3.43) that

$$(3.44)\quad (\nabla\tilde{\varphi}\mathbf{h})\cdot(C[\varepsilon(\tilde{\varphi})]\boldsymbol{\nu}) - \frac{1}{2}(C[\varepsilon(\tilde{\varphi})],\varepsilon(\tilde{\varphi})) \ge \frac{1}{2K}|C[\varepsilon(\tilde{\varphi})]\boldsymbol{\nu}|^2.$$

We also have

$$\left|\int_0^T\int_\Omega \{\tilde{F}(\mathbf{h}\cdot\nabla\tilde{Z}) + \tilde{\mathbf{M}}\cdot(\nabla\tilde{\varphi}\mathbf{h})\}\,d\Omega dt\right| \le C\int_0^T (\|\tilde{\mathbf{F}}\|^2_{\mathcal{H}^0(\Omega)} + \|\tilde{\mathbf{\Phi}}\|^2_{\mathcal{H}^1_\Gamma(\Omega)})\,dt,$$

$$\int_\Omega \{m_1\dot{\tilde{Z}}(\mathbf{h}\cdot\nabla\tilde{Z}) + I_{\rho_1}\dot{\tilde{\varphi}}\cdot(\nabla\tilde{\varphi}\mathbf{h})\}d\Omega\Big|_0^T \le C\|(\tilde{\mathbf{\Phi}},\dot{\tilde{\mathbf{\Phi}}})\|^2_{L^\infty(0,T;\mathcal{H})},$$

$$\left|\int_0^T\int_\Omega (\operatorname{div}\mathbf{h})\{m_1|\dot{\tilde{Z}}|^2 + I_{\rho_1}|\dot{\tilde{\varphi}}|^2\}d\Omega dt\right| \le \|\mathbf{h}\|_{W^{1,\infty}(\Omega)}\int_0^T \|\dot{\tilde{\mathbf{\Phi}}}\|^2_{\mathcal{H}^0(\Omega)}dt,$$

$$\begin{aligned}\Bigg|\int_0^T\int_\Omega &\{Q(\nabla\tilde{\varphi}) + G_1h[-\frac{1}{2}(\operatorname{div}\mathbf{h})|\nabla\tilde{Z}|^2 + (\nabla\mathbf{h}\nabla\tilde{Z})\cdot\nabla\tilde{Z} \\ &- (\operatorname{div}\tilde{\varphi})(\mathbf{h}\cdot\nabla\tilde{Z}) + (\nabla\tilde{Z}+\tilde{\varphi})\cdot((\nabla\tilde{\varphi})\mathbf{h})]\}d\Omega dt\Bigg| \\ &\le C\int_0^T \|\tilde{\mathbf{\Phi}}\|^2_{\mathcal{H}^1_\Gamma(\Omega)})\,dt.\end{aligned}$$

It follows from the above estimates that

$$\begin{aligned}\int_\Sigma &\left[|C[\varepsilon(\tilde{\varphi})]\boldsymbol{\nu}|^2 + G_1h\left(\frac{\partial\tilde{Z}}{\partial\boldsymbol{\nu}}\right)^2\right]d\Sigma \\ &\le C\left[\int_0^T \left(\|\tilde{\mathbf{F}}(t)\|^2_{\mathcal{H}^0(\Omega)} + \|(\tilde{\mathbf{\Phi}},\dot{\tilde{\mathbf{\Phi}}})\|^2_{\mathcal{H}}\right)dt + \|(\tilde{\mathbf{\Phi}},\dot{\tilde{\mathbf{\Phi}}})\|^2_{L^\infty(0,T;\mathcal{H})}\right] \\ &\le C\left[\int_0^T \|\tilde{\mathbf{F}}(t)\|^2_{\mathcal{H}^0(\Omega)}dt + (T+1)\|(\tilde{\mathbf{\Phi}},\dot{\tilde{\mathbf{\Phi}}})\|^2_{L^\infty(0,T;\mathcal{H})}\right]. \qquad \square\end{aligned}$$

4. The Reachable Set

We consider the mechanical control problem (3.1)–(3.7), or the geometrical control problem (3.1)–(3.6), (3.8), with vanishing initial data: $\mathbf{\Psi}^0 = \mathbf{\Psi}^1 = 0$ and $\mathbf{\Theta}^0 = \mathbf{\Theta}^1 = 0$. The reachable set at time $T > 0$ is

$$R_T = \{((\mathbf{\Psi}(T), \mathbf{\Theta}(T)), (\dot{\mathbf{\Psi}}(T), \dot{\mathbf{\Theta}}(T)))| \, \mathbf{f} \in L^2(0,T;U)\}.$$

In order to say something useful about R_T we shall need to impose geometric restrictions on Γ. We therefore assume that there is a point $\hat{\mathbf{x}}_0 \in \mathbb{R}^2$ such that

$$(4.1) \qquad (\mathbf{x} - \hat{\mathbf{x}}_0) \cdot \boldsymbol{\nu} \le 0 \;\text{ for } \mathbf{x} \in \Gamma^C, \;\; (\mathbf{x} - \hat{\mathbf{x}}_0) \cdot \boldsymbol{\nu} < 0 \;\text{ for } \mathbf{x} \in \overline{J}.$$

Set

$$\Gamma_+^D = \{\mathbf{x} \in \Gamma^D | \, (\mathbf{x} - \hat{\mathbf{x}}_0) \cdot \boldsymbol{\nu} > 0\}, \;\; \Gamma_-^D = \{\mathbf{x} \in \Gamma^D | \, (\mathbf{x} - \hat{\mathbf{x}}_0) \cdot \boldsymbol{\nu} \le 0\}.$$

When the controls act in the geometrical boundary conditions, one may assume without loss of generality that $\Gamma_+^D = \Gamma^D$ or, to say the same thing, that the controls $\mathbf{f}$ are supported in Γ_+^D, which amounts to redefining Γ^C to be $\Gamma^C \cup \Gamma_-^D$. However, when the controls act in the mechanical boundary conditions the explicit assumption

$$(4.2) \qquad (\mathbf{x} - \hat{\mathbf{x}}_0) \cdot \boldsymbol{\nu} \ge 0 \;\text{ on } \Gamma^D$$

is needed.

THEOREM 4.1. *(Geometrical control.) Assume that* (4.1) *holds and let* $\gamma = \Gamma - \overline{J}$. *Then there is a time* $T_0 > 0$ *such that* $R_T = H \times V_\gamma'$ *for* $T > T_0$.

THEOREM 4.2. *(Mechanical control.) Assume that* (4.1) *and* (4.2) *hold, that* $\Gamma^C \neq \emptyset$, *that* $\overline{\Gamma}^C$ *and* $\overline{\Gamma}^D$ *either do not intersect or else intersect in a strictly convex corner, and set* $\gamma = \Gamma^C$. *Then there is a time* $T_0 > 0$ *such that* $V_\gamma \times H \subset R_T$ *for* $T > T_0$.

REMARK 4.1. One may eliminate the geometric condition (4.2) in Theorem 4.2 but at the expense of enlarging the control space and working in a weaker state space. For example, if one admits controls in the space

$$(4.3) \qquad (H^1(0,T;U_+))' \bigoplus L^2(0,T;\mathcal{H}^{-1}(\Gamma_-^D)),$$

then one may prove that $H \times V_\gamma' \subset R_T$ for T large enough, without assumption (4.2). In (4.3),

$$U_+ = L^2(\Gamma_+^D;[\mathbf{a}_1]) \bigoplus L^2(\Gamma_+^D;[\mathbf{a}_2]) \bigoplus L^2(\Gamma_+^D;[\mathbf{a}_2]),$$

$$\mathcal{H}^{-1}(\Gamma_-^D) = H^{-1}(\Gamma_-^D;[\mathbf{a}_1]) \bigoplus H^{-1}(\Gamma_-^D;[\mathbf{a}_2]) \bigoplus H^{-1}(\Gamma_-^D;[\mathbf{a}_3]). \quad \square$$

In the next subsection the observability estimates needed to prove Theorems 4.1 and 4.2 are presented.

4.1. Observability estimates. Let us write $\Gamma = \partial\Omega$ as $\Gamma = \overline{\Gamma}_0 \cup \overline{\Gamma}_1 \cup \overline{J}$, where Γ_0, Γ_1, J are relatively open in Γ, mutually disjoint, $\Gamma_0 \neq \emptyset$ and $\overline{\Gamma_1} \cap \overline{J} = \emptyset$. We consider the problem

$$\begin{cases} I_{\rho_1}\ddot{\varphi} - \operatorname{div} C[\varepsilon(\varphi)] + G_1 h(\nabla Z + \varphi) = 0, \\ m_1 \ddot{Z} - G_1 h \operatorname{div}(\nabla Z + \varphi) = 0 \ \text{ in } \Omega; \end{cases} \tag{4.4}$$

$$\begin{cases} m_2 \ddot{z} - G_2 A(z' + \theta_2)' = 0, \\ \rho_2 I_h \ddot{\theta}_2 - E I_h \theta_2'' + G_2 A(z' + \theta_2) = 0, \\ \rho_2 I \ddot{\theta}_1 - G_2 I \theta_1'' = 0, \ \ 0 < \xi < \ell; \end{cases} \tag{4.5}$$

$$G_2 A(z' + \theta_2)(\ell) = E I_h \theta_2'(\ell) = G_2 I \theta_1'(\ell) = 0; \tag{4.6}$$

$$Z = 0, \ \ \varphi = 0 \ \text{ on } \Gamma_0; \tag{4.7}$$

$$C[\varepsilon(\varphi)]\boldsymbol{\nu} = 0, \ \ G_1 h(\nabla Z + \varphi) \cdot \boldsymbol{\nu} = 0 \ \text{ on } \Gamma_1; \tag{4.8}$$

$$Z = \sigma_0[z(0) + \zeta\theta_1(0)], \ \ \varphi = \sigma_1 \boldsymbol{\theta}(0) \ \text{ on } J, \tag{4.9}$$

where $\boldsymbol{\theta} = \theta_2 \mathbf{e}_1 - \theta_1 \mathbf{e}_2$;

$$\begin{cases} G_2 I \theta_1'(0) = \displaystyle\int_J \{G_1 h \zeta \sigma_0 \boldsymbol{\nu} \cdot (\nabla Z + \varphi) \\ \qquad\qquad\qquad -\sigma_1 \mathbf{e}_2 \cdot (C[\varepsilon(\varphi)]\boldsymbol{\nu})\} d\Gamma, \\ E I_h \theta_2'(0) = \displaystyle\int_J \sigma_1 \mathbf{e}_1 \cdot (C[\varepsilon(\varphi)]\boldsymbol{\nu})\, d\Gamma, \\ G_2 A(z' + \theta_2)(0) = G_1 h \displaystyle\int_J \sigma_0 \boldsymbol{\nu} \cdot (\nabla Z + \varphi)\, d\Gamma; \end{cases} \tag{4.10}$$

$$\begin{cases} Z|_{t=0} = Z^0, \ \ \dot{Z}|_{t=0} = Z^1, \ \ \varphi|_{t=0} = \varphi^0, \ \ \dot{\varphi}|_{t=0} = \varphi^1, \\ z|_{t=0} = z^0, \ \ \dot{z}|_{t=0} = z^1, \ \ \boldsymbol{\theta}|_{t=0} = \boldsymbol{\theta}^0, \dot{\boldsymbol{\theta}}|_{t=0} = \boldsymbol{\theta}^1. \end{cases} \tag{4.11}$$

As in the last section, we write

$$\boldsymbol{\Phi} = \varphi + Z\mathbf{k}, \ \ \boldsymbol{\Xi} = \boldsymbol{\theta} + z\mathbf{e}_3$$

with corresponding initial values $\boldsymbol{\Phi}^0, \boldsymbol{\Phi}^1, \boldsymbol{\Xi}^0, \boldsymbol{\Xi}^1$ at $t = 0$.

Let $\hat{\mathbf{x}}_0 \in \mathbb{R}^2$ and define

$$\Gamma_\alpha^+ = \{\mathbf{x} \in \Gamma_\alpha | \, (\mathbf{x} - \hat{\mathbf{x}}_0) \cdot \boldsymbol{\nu} > 0\},$$
$$\Gamma_\alpha^- = \{\mathbf{x} \in \Gamma_\alpha | \, (\mathbf{x} - \hat{\mathbf{x}}_0) \cdot \boldsymbol{\nu} \leq 0\}, \ \ \alpha = 0, 1.$$

These sets of course depend on $\hat{\mathbf{x}}_0$. The relation between Γ_0, Γ_1 and Γ^C, Γ^D is the following: when considering the control problem with controls in the mechanical boundary conditions, we set $\Gamma_1 = \Gamma^D$ and $\Gamma_0 = \Gamma^C$. The assumptions of Theorem 4.2 then require, in particular, that $\Gamma_0^+ = \Gamma_1^- = \emptyset$. When considering the control problem with controls in the geometrical boundary conditions, we set $\Gamma_1 = \emptyset$, $\Gamma_0 = \Gamma - \overline{J}$, $\Gamma^C = \Gamma_0^-$ and $\Gamma^D = \Gamma_0^+$.

As usual, we write $\mathbf{m}(\mathbf{x}) = \mathbf{x} - \hat{\mathbf{x}}_0$ and $\Sigma_\alpha^\pm = \Gamma_\alpha^\pm \times (0,T)$. The main observability estimate is as follows.

PROPOSITION 4.1. *Let $\hat{\mathbf{x}}_0$ be such that $\mathbf{m}\cdot\boldsymbol{\nu} < 0$ on $\overline{J}$. Assume that $\overline{\Gamma}_0$ and $\overline{\Gamma}_1$ either do not intersect or else intersect in a strictly convex corner and that $\Gamma_0 \neq \emptyset$. Let*

$$(\mathbf{\Phi}^0, \Xi^0) \in D_A, \quad (\mathbf{\Phi}^1, \Xi^1) \in V_{\Gamma_0}.$$

There is a constant T_0 such that if $T > T_0$, the solution of (4.4)–(4.11) *satisfies*

$$(4.12) \quad \|((\mathbf{\Phi}^0,\Xi^0),(\mathbf{\Phi}^1,\Xi^1))\|^2_{V_{\Gamma_0}\times H} \le C_T\left[\int_{\Sigma_1^+} |\dot{\mathbf{\Phi}}|^2\,d\Sigma \right.$$
$$+\int_{\Sigma_1^-}\left(C[\varepsilon(\boldsymbol{\varphi})]\boldsymbol{\tau})\cdot\frac{\partial\boldsymbol{\varphi}}{\partial\boldsymbol{\tau}} + G_1 h|\boldsymbol{\tau}\cdot(\nabla Z+\boldsymbol{\varphi})|^2\right)d\Sigma$$
$$\left. +\int_{\Sigma_0^+}\left((C[\varepsilon(\boldsymbol{\varphi})]\boldsymbol{\nu})\cdot\frac{\partial\boldsymbol{\varphi}}{\partial\boldsymbol{\nu}} + G_1 h\left|\frac{\partial Z}{\partial\boldsymbol{\nu}}\right|^2\right)d\Sigma\right].$$

REMARK 4.2. In view of (3.43), $(C[\varepsilon(\boldsymbol{\varphi})]\boldsymbol{\nu})\cdot(\partial\boldsymbol{\varphi}/\partial\boldsymbol{\nu}) \ge 0$ on Γ_0 and this quantity be replaced by $|C[\varepsilon(\boldsymbol{\varphi})]\boldsymbol{\nu}|^2$ in the right side of (4.12). Also, since $C[\varepsilon(\boldsymbol{\varphi})]\boldsymbol{\nu} = 0$ on Γ_1, it follows from (3.39) that

$$\boldsymbol{\tau}\cdot\frac{\partial\boldsymbol{\varphi}}{\partial\boldsymbol{\nu}} + \boldsymbol{\nu}\cdot\frac{\partial\boldsymbol{\varphi}}{\partial\boldsymbol{\tau}} = 0, \quad (2\mu_1+\lambda_1)\left(\boldsymbol{\nu}\cdot\frac{\partial\boldsymbol{\varphi}}{\partial\boldsymbol{\nu}}\right) + \lambda_1\left(\boldsymbol{\tau}\cdot\frac{\partial\boldsymbol{\varphi}}{\partial\boldsymbol{\tau}}\right) = 0 \text{ on } \Gamma_1$$

and therefore from (3.40)

$$(C[\varepsilon(\boldsymbol{\varphi})]\boldsymbol{\tau})\cdot(\partial\boldsymbol{\varphi}/\partial\boldsymbol{\tau}) = \frac{h^3}{12}\frac{(2\mu_1+\lambda_1)^2-\lambda_1^2}{2\mu_1+\lambda_1}\left(\boldsymbol{\tau}\cdot\frac{\partial\boldsymbol{\varphi}}{\partial\boldsymbol{\tau}}\right)^2 \ge 0 \;\; on\; \Gamma_1. \quad \square$$

The following corollary is an immediate consequence of Proposition 4.1 and Lemma 3.1 (note that $\overline{\Gamma_0^+} \cap \overline{J} = \emptyset$) .

COROLLARY 4.1. *Let $\hat{\mathbf{x}}_0$ be such that $\mathbf{m}\cdot\boldsymbol{\nu} < 0$ on $\overline{J}$. Assume that $\Gamma_1 = \emptyset$ and that*

$$(\mathbf{\Phi}^0, \Xi^0) \in V_{\Gamma_0}, \quad (\mathbf{\Phi}^1, \Xi^1) \in H.$$

Then there is a constant T_0 such that if $T > T_0$, the solution of (4.4)–(4.11) *satisfies*

$$\text{(4.13)}\quad \|((\Phi^0, \Xi^0), (\Phi^1, \Xi^1))\|^2_{V_{\Gamma_0}\times H} \le C_T \int_{\Sigma_0^+} \left((C[\varepsilon(\varphi)]\nu) \cdot \frac{\partial \varphi}{\partial \nu} + G_1 h \left| \frac{\partial Z}{\partial \nu} \right|^2 \right) d\Sigma.$$

The corollary which follows is a consequence of Proposition 4.1 and the process of "weakening the norm."

COROLLARY 4.2. *Let $\hat{\mathbf{x}}_0$ be such that $\mathbf{m} \cdot \nu < 0$ on $\overline{J}$. Assume that $\Gamma_0^+ = \Gamma_1^- = \emptyset$, that $\Gamma_0^- \neq \emptyset$ and let*

$$(\Phi^0, \Xi^0) \in V_{\Gamma_0}, \quad (\Phi^1, \Xi^1) \in H.$$

and that $\overline{\Gamma}_0$ and $\overline{\Gamma}_1$ either do not intersect or else intersect in a strictly convex corner. Then

$$\text{(4.14)}\qquad \|((\Phi^0, \Xi^0), (\Phi^1, \Xi^1))\|^2_{H\times V'_{\Gamma_0}} \le C_T \int_{\Sigma_1} |\Phi|^2 \, d\Sigma.$$

Corollary 4.1 pertains to the problem of control in the geometric boundary conditions wherein $\Gamma^D = \Gamma_0^+$, $\Gamma^C = \Gamma_0^-$, while Corollary 4.2 pertains to the problem of control in the mechanical boundary conditions, wherein $\Gamma^D = \Gamma_1^+ = \Gamma_1$ and $\Gamma^C = \Gamma_0^- = \Gamma_0$.

PROOF OF PROPOSITION 4.1. As our starting point we apply the identity (5.17), Chapter VII, to the solution Φ of (4.4). We thus have

$$\begin{aligned}
\text{(4.15)}\quad & R(T) - R(0) + \int_Q \{(1-\delta) m_1 |\dot{Z}|^2 + \delta I_{\rho_1} |\dot{\varphi}|^2 \\
&\quad + (1-\delta)(C[\varepsilon(\varphi)], \varepsilon(\varphi)) + \delta G_1 h (|\nabla Z|^2 - |\varphi|^2)\} \, dQ \\
&= \frac{1}{2} \int_\Sigma (\mathbf{m} \cdot \nu) \{ m_1 |\dot{Z}|^2 + I_{\rho_1} |\dot{\varphi}|^2 \} \, d\Sigma \\
&\quad + \frac{1}{2} \int_\Sigma (\mathbf{m} \cdot \nu) \Big\{ (C[\varepsilon(\varphi)]\nu) \cdot \frac{\partial \varphi}{\partial \nu} - (C[\varepsilon(\varphi)]\tau) \cdot \frac{\partial \varphi}{\partial \tau} \\
&\quad + G_1 h \left(\frac{\partial Z}{\partial \nu} + \nu \cdot \varphi \right)^2 - G_1 h \left(\frac{\partial Z}{\partial \tau} + \tau \cdot \varphi \right)^2 \\
&\quad - 2 G_1 h \left(\frac{\partial Z}{\partial \nu} + \nu \cdot \varphi \right) (\nu \cdot \varphi) \Big\} \, d\Sigma \\
&\quad + \int_\Sigma \Big\{ (C[\varepsilon(\varphi)]\nu) \cdot \left[(\mathbf{m} \cdot \tau) \frac{\partial \varphi}{\partial \tau} + (1-\delta) \varphi \right] d\Sigma \\
&\quad + G_1 h \left(\frac{\partial Z}{\partial \nu} + \nu \cdot \varphi \right) \left[(\mathbf{m} \cdot \tau) \frac{\partial Z}{\partial \tau} + \delta Z \right] \Big\} \, d\Sigma,
\end{aligned}$$

where $\delta > 0$ and

$$R(t) = \int_\Omega \{m_1 \dot{Z}^i(\mathbf{m}\cdot\nabla Z + \delta Z) + I_{\rho_1}\dot{\boldsymbol{\varphi}}\cdot(\nabla\boldsymbol{\varphi}\mathbf{m} + \delta\boldsymbol{\varphi})\}\, d\Omega.$$

Since $\boldsymbol{\Phi}$ also satisfies the boundary conditions (4.7) and (4.8), the integrals on the right side of (4.15) may be written as

$$\mathcal{J}(\Sigma_0) + \mathcal{J}(\Sigma_1) + \mathcal{J}(\Sigma_J),$$

where

$$\mathcal{J}(\Sigma_0) = \frac{1}{2}\int_{\Sigma_0} (\mathbf{m}\cdot\boldsymbol{\nu})\left\{(C[\varepsilon(\boldsymbol{\varphi})]\boldsymbol{\nu})\cdot\frac{\partial\boldsymbol{\varphi}}{\partial\boldsymbol{\nu}} + G_1 h\left(\frac{\partial Z}{\partial\boldsymbol{\nu}}\right)^2\right\} d\Sigma,$$

$$\begin{aligned}\mathcal{J}(\Sigma_1) = {} & \frac{1}{2}\int_{\Sigma_1} (\mathbf{m}\cdot\boldsymbol{\nu})\left\{m_1|\dot{Z}|^2 + I_{\rho_1}|\dot{\boldsymbol{\varphi}}|^2\right\} d\Sigma \\ & - \frac{1}{2}\int_{\Sigma_1} (\mathbf{m}\cdot\boldsymbol{\nu})\left\{(C[\varepsilon(\boldsymbol{\varphi})]\boldsymbol{\tau})\cdot\frac{\partial\boldsymbol{\varphi}}{\partial\boldsymbol{\tau}} + G_1 h\left(\frac{\partial Z}{\partial\boldsymbol{\tau}} + \boldsymbol{\tau}\cdot\boldsymbol{\varphi}\right)^2\right\} d\Sigma,\end{aligned}$$

$$\begin{aligned}\mathcal{J}(\Sigma_J) = {} & \frac{1}{2}\int_{\Sigma_J} (\mathbf{m}\cdot\boldsymbol{\nu})\{m_1|\dot{Z}|^2 + I_{\rho_1}|\dot{\boldsymbol{\varphi}}|^2\}\, d\Sigma \\ & + \frac{1}{2}\int_{\Sigma_J} (\mathbf{m}\cdot\boldsymbol{\nu})\left\{(C[\varepsilon(\boldsymbol{\varphi})]\boldsymbol{\nu})\cdot\frac{\partial\boldsymbol{\varphi}}{\partial\boldsymbol{\nu}} - (C[\varepsilon(\boldsymbol{\varphi})]\boldsymbol{\tau})\cdot\frac{\partial\boldsymbol{\varphi}}{\partial\boldsymbol{\tau}}\right. \\ & \qquad + G_1 h\left(\frac{\partial Z}{\partial\boldsymbol{\nu}} + \boldsymbol{\nu}\cdot\boldsymbol{\varphi}\right)^2 - G_1 h\left(\frac{\partial Z}{\partial\boldsymbol{\tau}} + \boldsymbol{\tau}\cdot\boldsymbol{\varphi}\right)^2 \\ & \qquad \left. - 2G_1 h\left(\frac{\partial Z}{\partial\boldsymbol{\nu}} + \boldsymbol{\nu}\cdot\boldsymbol{\varphi}\right)(\boldsymbol{\nu}\cdot\boldsymbol{\varphi})\right\} d\Sigma \\ & + \int_{\Sigma_J} \left\{(C[\varepsilon(\boldsymbol{\varphi})]\boldsymbol{\nu})\cdot\left[(\mathbf{m}\cdot\boldsymbol{\tau})\frac{\partial\boldsymbol{\varphi}}{\partial\boldsymbol{\tau}} + (1-\delta)\boldsymbol{\varphi}\right] d\Sigma\right. \\ & \qquad \left. + G_1 h\left(\frac{\partial Z}{\partial\boldsymbol{\nu}} + \boldsymbol{\nu}\cdot\boldsymbol{\varphi}\right)\left[(\mathbf{m}\cdot\boldsymbol{\tau})\frac{\partial Z}{\partial\boldsymbol{\tau}} + \delta Z\right]\right\} d\Sigma.\end{aligned}$$

As in the proof of Lemma 5.3, Chapter VII, we have

$$\begin{aligned}&\int_Q \{(1-\delta)m_1|\dot{Z}|^2 + \delta I_{\rho_1}|\dot{\boldsymbol{\varphi}}|^2 + (1-\delta)(C[\varepsilon(\boldsymbol{\varphi})], \varepsilon(\boldsymbol{\varphi})) \\ &\quad + \delta G_1 h(|\nabla Z|^2 - |\boldsymbol{\varphi}|^2)\}\, dQ \geq c_0 \int_0^T \left(\|\dot{\boldsymbol{\Phi}}\|^2_{\mathcal{H}^0(\Omega)} + \|\boldsymbol{\Phi}\|^2_{\mathcal{H}^1_{\Gamma_0}(\Omega)}\right) dt\end{aligned}$$

for some constant $c_0 > 0$. Therefore, for some constant C,

(4.16)

$$R(t)|_0^T + \int_0^T \left(\|\dot{\Phi}\|^2_{\mathcal{H}^0(\Omega)} + \|\Phi\|^2_{\mathcal{H}^1_{\Gamma_0}(\Omega)}\right) dt \le C \left[\int_{\Sigma_1^+} (\mathbf{m}\cdot\boldsymbol{\nu})|\dot{\Phi}|^2 \, d\Sigma \right.$$

$$- \int_{\Sigma_1^-} (\mathbf{m}\cdot\boldsymbol{\nu}) \left((C[\varepsilon(\boldsymbol{\varphi})]\boldsymbol{\tau}) \cdot \frac{\partial \boldsymbol{\varphi}}{\partial \boldsymbol{\tau}} + G_1 h |\boldsymbol{\tau}\cdot(\nabla Z + \boldsymbol{\varphi})|^2 \right) d\Sigma$$

$$\left. + \int_{\Sigma_0^+} (\mathbf{m}\cdot\boldsymbol{\nu}) \left((C[\varepsilon(\boldsymbol{\varphi})]\boldsymbol{\nu}) \cdot \frac{\partial \boldsymbol{\varphi}}{\partial \boldsymbol{\nu}} + G_1 h \left|\frac{\partial Z}{\partial \boldsymbol{\nu}}\right|^2 \right) d\Sigma \right] + \mathcal{J}(\Sigma_J)$$

since $(C[\varepsilon(\boldsymbol{\varphi})]\boldsymbol{\tau}) \cdot \partial\boldsymbol{\varphi}/\partial\boldsymbol{\tau} \ge 0$ on Σ_1 (see Remark 4.2 above).

We proceed to estimate $\mathcal{J}(\Sigma_J)$ and are going to show that

$$\text{(4.17)} \quad \mathcal{J}(\Sigma_J) \le \frac{1}{2} \int_{\Sigma_J} (\mathbf{m}\cdot\boldsymbol{\nu}) \{ m_1 |\dot{Z}|^2 + I_{\rho_1} |\dot{\boldsymbol{\varphi}}|^2 \} \, d\Sigma$$

$$+ C_1 \int_{\Sigma_J} (\mathbf{m}\cdot\boldsymbol{\nu}) \left(\left|\frac{\partial \boldsymbol{\varphi}}{\partial \boldsymbol{\nu}}\right|^2 + \left|\frac{\partial Z}{\partial \boldsymbol{\nu}} + \boldsymbol{\nu}\cdot\boldsymbol{\varphi}\right|^2 \right) d\Sigma + C_2 \int_0^T |\Xi(0)|^2 \, dt,$$

where C_1, C_2 are positive constants independent of T. To this end, from (3.39) and (3.40) we calculate

$$(C[\varepsilon(\boldsymbol{\varphi})]\boldsymbol{\nu}) \cdot \frac{\partial \boldsymbol{\varphi}}{\partial \boldsymbol{\nu}} - (C[\varepsilon(\boldsymbol{\varphi})]\boldsymbol{\tau}) \cdot \frac{\partial \boldsymbol{\varphi}}{\partial \boldsymbol{\tau}}$$

$$= \frac{h^3}{12} \left[\mu_1 \left(\boldsymbol{\tau}\cdot\frac{\partial \boldsymbol{\varphi}}{\partial \boldsymbol{\nu}}\right)^2 + (2\mu_1 + \lambda_1) \left(\boldsymbol{\nu}\cdot\frac{\partial \boldsymbol{\varphi}}{\partial \boldsymbol{\nu}}\right)^2 \right.$$

$$\left. - (2\mu_1 + \lambda_1) \left(\boldsymbol{\tau}\cdot\frac{\partial \boldsymbol{\varphi}}{\partial \boldsymbol{\tau}}\right)^2 - \mu_1 \left(\boldsymbol{\nu}\cdot\frac{\partial \boldsymbol{\varphi}}{\partial \boldsymbol{\tau}}\right)^2 \right]$$

$$\ge \frac{h^3}{12} \left(\mu_1 \left|\frac{\partial \boldsymbol{\varphi}}{\partial \boldsymbol{\nu}}\right|^2 - (2\mu_1 + \lambda_1) \left|\frac{\partial \boldsymbol{\varphi}}{\partial \boldsymbol{\tau}}\right|^2 \right).$$

Therefore, since $\mathbf{m}\cdot\boldsymbol{\nu} < 0$ on J,

$$(4.18)\quad \int_{\Sigma_J} (\mathbf{m}\cdot\boldsymbol{\nu})\left\{ (C[\varepsilon(\boldsymbol{\varphi})]\boldsymbol{\nu})\cdot\frac{\partial\boldsymbol{\varphi}}{\partial\boldsymbol{\nu}} - (C[\varepsilon(\boldsymbol{\varphi})]\boldsymbol{\tau})\cdot\frac{\partial\boldsymbol{\varphi}}{\partial\boldsymbol{\tau}} + G_1 h\left|\frac{\partial Z}{\partial\boldsymbol{\nu}} + \boldsymbol{\nu}\cdot\boldsymbol{\varphi}\right|^2 - G_1 h\left|\frac{\partial Z}{\partial\boldsymbol{\tau}} + \boldsymbol{\tau}\cdot\boldsymbol{\varphi}\right|^2\right\} d\Sigma$$

$$\le \int_{\Sigma_J} (\mathbf{m}\cdot\boldsymbol{\nu})\left[\frac{h^3}{12}\mu_1\left|\frac{\partial\boldsymbol{\varphi}}{\partial\boldsymbol{\nu}}\right|^2 + G_1 h\left|\frac{\partial Z}{\partial\boldsymbol{\nu}} + \boldsymbol{\nu}\cdot\boldsymbol{\varphi}\right|^2 - \frac{h^3}{12}(2\mu_1+\lambda_1)\left|\frac{\partial\boldsymbol{\varphi}}{\partial\boldsymbol{\tau}}\right|^2 - G_1 h\left|\frac{\partial Z}{\partial\boldsymbol{\tau}} + \boldsymbol{\tau}\cdot\boldsymbol{\varphi}\right|^2\right] d\Sigma$$

$$\le C_1 \int_{\Sigma_J} (\mathbf{m}\cdot\boldsymbol{\nu})\left[\left|\frac{\partial\boldsymbol{\varphi}}{\partial\boldsymbol{\nu}}\right|^2 + \left|\frac{\partial Z}{\partial\boldsymbol{\nu}} + \boldsymbol{\nu}\cdot\boldsymbol{\varphi}\right|^2\right] d\Sigma + C_2\int_{\Sigma_J}\left[\left|\frac{\partial\boldsymbol{\varphi}}{\partial\boldsymbol{\tau}}\right|^2 + \left|\frac{\partial Z}{\partial\boldsymbol{\tau}} + \boldsymbol{\tau}\cdot\boldsymbol{\varphi}\right|^2\right] d\Sigma$$

for some constants C_1, C_2.

For any $\epsilon > 0$ and for some constant C_ϵ we have

$$(4.19)\quad \int_{\Sigma_J} (\mathbf{m}\cdot\boldsymbol{\nu})\left(\frac{\partial Z}{\partial\boldsymbol{\nu}} + \boldsymbol{\nu}\cdot\boldsymbol{\varphi}\right)(\boldsymbol{\nu}\cdot\boldsymbol{\varphi})\, d\Sigma \le \epsilon\int_{\Sigma_J} |\mathbf{m}\cdot\boldsymbol{\nu}|\left|\frac{\partial Z}{\partial\boldsymbol{\nu}} + \boldsymbol{\nu}\cdot\boldsymbol{\varphi}\right|^2 d\Sigma + C_\epsilon\int_{\Sigma_J} (\boldsymbol{\nu}\cdot\boldsymbol{\varphi})^2 d\Sigma.$$

Since $\mathbf{m}\cdot\boldsymbol{\nu} < 0$ on $\overline{J}$, we also have

$$(4.20)\quad \left|\int_{\Sigma_J}\left(\frac{\partial Z}{\partial\boldsymbol{\nu}} + \boldsymbol{\nu}\cdot\boldsymbol{\varphi}\right)\left[(\mathbf{m}\cdot\boldsymbol{\tau})\frac{\partial Z}{\partial\boldsymbol{\tau}} + \delta Z\right] d\Sigma\right| \le \epsilon\int_{\Sigma_J} |\mathbf{m}\cdot\boldsymbol{\nu}|\left|\frac{\partial Z}{\partial\boldsymbol{\nu}} + \boldsymbol{\nu}\cdot\boldsymbol{\varphi}\right|^2 d\Sigma + C_\epsilon\int_{\Sigma_J}\left[\left|\frac{\partial Z}{\partial\boldsymbol{\tau}}\right|^2 + |Z|^2\right] d\Sigma,$$

Further, from (3.39),

$$(4.21)\quad |C[\varepsilon(\boldsymbol{\varphi})]\boldsymbol{\nu}|^2 = 2\left(\frac{h^3}{12}\right)^2\left[(2\mu_1+\lambda_1)^2\left|\frac{\partial\boldsymbol{\varphi}}{\partial\boldsymbol{\nu}}\right|^2 + \max(\lambda_1^2,\mu_1^2)\left|\frac{\partial\boldsymbol{\varphi}}{\partial\boldsymbol{\tau}}\right|^2\right]$$

and therefore

$$\text{(4.22)}\quad \left|\int_{\Sigma_J} (C[\varepsilon(\varphi)]\nu)\cdot\left[(\mathbf{m}\cdot\boldsymbol{\tau})\frac{\partial\varphi}{\partial\boldsymbol{\tau}} + (1-\delta)\varphi\right] d\Sigma\right| \\ \leq \epsilon\int_{\Sigma_J} |\mathbf{m}\cdot\boldsymbol{\nu}|\left|\frac{\partial\varphi}{\partial\boldsymbol{\nu}}\right|^2 d\Sigma + C_\epsilon\int_{\Sigma_J}\left(\left|\frac{\partial\varphi}{\partial\boldsymbol{\tau}}\right|^2 + |\varphi|^2\right) d\Sigma.$$

It follows from (4.18)–(4.22) that for any $\epsilon > 0$,

$$\text{(4.23)}\quad \mathcal{J}(\Sigma_J) \leq \frac{1}{2}\int_{\Sigma_J} (\mathbf{m}\cdot\boldsymbol{\nu})\{m_1|\dot{Z}|^2 + I_{\rho_1}|\dot{\varphi}|^2\}\, d\Sigma \\ + (C_1-\epsilon)\int_{\Sigma_J} (\mathbf{m}\cdot\boldsymbol{\nu})\left(\left|\frac{\partial\varphi}{\partial\boldsymbol{\nu}}\right|^2 + \left|\frac{\partial Z}{\partial\boldsymbol{\nu}} + \boldsymbol{\nu}\cdot\varphi\right|^2\right) d\Sigma \\ + C_\epsilon\int_{\Sigma_J}\left(\left|\frac{\partial\varphi}{\partial\boldsymbol{\tau}}\right|^2 + |\varphi|^2 + \left|\frac{\partial Z}{\partial\boldsymbol{\tau}}\right|^2 + |Z|^2\right) d\Sigma.$$

Since $\mathbf{\Phi} = \sigma\Xi(0)$ on J, we have

$$\text{(4.24)}\quad \int_{\Sigma_J}\left(\left|\frac{\partial\varphi}{\partial\boldsymbol{\tau}}\right|^2 + |\varphi|^2 + \left|\frac{\partial Z}{\partial\boldsymbol{\tau}}\right|^2 + |Z|^2\right) d\Sigma \;\leq\; C\int_0^T |\Xi(0,t)|^2 dt,$$

where C depends on the $H^1(J)$ norms of σ_0, σ_1. The estimate (4.17) follows from (4.23) and (4.24) by choosing $\epsilon = C_1/2$, for example.

By combining (4.16) with (4.17) we obtain the estimate

$$\text{(4.25)}\\ R(t)|_0^T + \int_0^T \left(\|\dot{\mathbf{\Phi}}\|^2_{\mathcal{H}^0(\Omega)} + \|\mathbf{\Phi}\|^2_{\mathcal{H}^1_{\Gamma_0}(\Omega)}\right) dt \leq C\left[\int_{\Sigma_1^+} (\mathbf{m}\cdot\boldsymbol{\nu})|\dot{\mathbf{\Phi}}|^2\, d\Sigma\right. \\ - \int_{\Sigma_1^-} (\mathbf{m}\cdot\boldsymbol{\nu})\left((C[\varepsilon(\varphi)]\boldsymbol{\tau})\cdot\frac{\partial\varphi}{\partial\boldsymbol{\tau}} + G_1 h|\boldsymbol{\tau}\cdot(\nabla Z + \varphi)|^2\right) d\Sigma \\ \left. + \int_{\Sigma_0^+} (\mathbf{m}\cdot\boldsymbol{\nu})\left((C[\varepsilon(\varphi)]\boldsymbol{\nu})\cdot\frac{\partial\varphi}{\partial\boldsymbol{\nu}} + G_1 h\left|\frac{\partial Z}{\partial\boldsymbol{\nu}}\right|^2\right) d\Sigma\right] \\ + C_1\int_{\Sigma_J} (\mathbf{m}\cdot\boldsymbol{\nu})\left(\left|\frac{\partial\varphi}{\partial\boldsymbol{\nu}}\right|^2 + \left|\frac{\partial Z}{\partial\boldsymbol{\nu}} + \boldsymbol{\nu}\cdot\varphi\right|^2\right) d\Sigma \\ + \frac{1}{2}\int_{\Sigma_J} (\mathbf{m}\cdot\boldsymbol{\nu})\{m_1|\dot{Z}|^2 + I_{\rho_1}|\dot{\varphi}|^2\}\, d\Sigma + C_2\int_0^T |\Xi(0,t)|^2\, dt.$$

We next need to obtain an estimate analogous to (4.25) on the time integral of the total beam energy

$$\mathcal{E}_B(t) = \int_0^\ell (m_2|\dot z|^2 + \rho_2 I_h|\dot\theta_2|^2 + \rho_2 I|\dot\theta_1|^2)\,d\xi$$
$$+ \int_0^\ell (G_2 I|\theta_1'|^2 + EI_h|\theta_2'|^2 + G_2A|z' + \theta_2|^2)\,d\xi.$$

Denote by $E_B(\xi, t)$ the energy density:

$$\mathcal{E}_B(t) = \int_0^\ell E_B(\xi,t)\,d\xi.$$

It follows from [49, Proposition 3.1] that the following identity holds for Ξ:

$$r(t)|_0^T + \frac{1}{2}\int_0^T\int_0^\ell [\kappa' E_B - 2\kappa m_2 \dot z\dot\theta_2 + 2\kappa G_2 A\theta_2'(z'+\theta_2)]\,d\xi dt$$
$$= \frac{\kappa(\ell)}{2}\int_0^T E_B(\ell,t)\,dt - \frac{\kappa(0)}{2}\int_0^T E_B(0,t)\,dt,$$

where κ is an arbitrary $C^1[0,\ell]$ function and

$$r(t) = \int_0^\ell \kappa[\rho_2 I\dot\theta_1\theta_1' + m_2\dot z(z'+\theta_2) + \rho_2 I_h\dot\theta_2\theta_2']\,d\xi.$$

(This identity is proved by multiplying the three equations in (4.5) by $\kappa z'$, $\kappa\theta_2'$ and $\kappa\theta_1'$, respectively, adding the three products, integrating the sum over $[0,\ell]\times[0,T]$ and using integration by parts.)

We choose κ so that

$$\kappa \le 0,\ \ \kappa' > 0,\ \ \kappa^2 m_2 - (\kappa')^2\rho_2 I_h < 0,$$
$$\kappa^2 G_2 A - (\kappa')^2 EI_h < 0,\ \ 0 \le \xi \le \ell.$$

For example, one may choose

$$\kappa(\xi) = -\exp(-k\xi),\quad k > \max(\sqrt{m_2/\rho_2 I_h}, \sqrt{G_2A/EI_h}).$$

Then the quadratic forms

$$\kappa' m_2 q_1^2 \pm 2\kappa m_2 q_1 q_2 + \kappa'\rho_2 I_h q_2^2$$

and

$$\kappa' EI_h q_1^2 \pm 2\kappa G_2 A q_1 q_2 + \kappa' G_2 A q_2^2$$

are uniformly positive definite on $0 \le \xi \le \ell$, hence

$$\kappa' E_B - 2\kappa m_2\dot z\dot\theta_2 + 2\kappa G_2A\theta_2'(z'+\theta_2)$$
$$= \kappa'[\rho_2 I|\dot\theta_1|^2 + G_2 I|\theta_1'|^2] + [\kappa' m_2|\dot z|^2 - 2\kappa m_2\dot z\dot\theta_2 + \kappa'\rho_2 I_h|\dot\theta_2|^2]$$
$$+ [\kappa' EI_h|\theta_2'|^2 + 2\kappa G_2A\theta_2'(z'+\theta_2) + \kappa' G_2A|z'+\theta_2|^2] > c_0 E_B$$

for some $c_0 > 0$ and the following estimate holds for the solution Ξ:

$$(4.26)\qquad r(t)|_0^T + c_0 \int_0^T \mathcal{E}_B(t)\,dt \le -\frac{\kappa(0)}{2}\int_0^T E_B(0,t)\,dt.$$

We estimate in the right side of (4.26), using the junction conditions (4.10), as follows:

$$\begin{aligned}(4.27)\quad & -\frac{\kappa(0)}{2}\int_0^T E_B(0,t)\,dt \\ &= -\frac{\kappa(0)}{2}\int_0^T (m_2|\dot z(0,t)|^2 + \rho_2 I_h|\dot\theta_2(0,t)|^2 + \rho_2 I|\dot\theta_1(0,t)|^2)\,dt \\ &- \frac{\kappa(0)}{2}\int_0^T (G_2 I|\theta_1'(0,t)|^2 + EI_h|\theta_2'(0,t)|^2 + G_2A|(z'+\theta_2)(0,t)|^2)\,dt \\ &\le C\left(\int_0^T |\dot\Xi(0,t)|^2 dt + \int_J \sigma_0^2\,d\Gamma \int_{\Sigma_J} |\boldsymbol{\nu}\cdot(\nabla Z + \boldsymbol{\varphi})|^2 d\Sigma \right. \\ &\qquad \left. + \int_J \sigma_1^2\,d\Gamma \int_{\Sigma_J} |C[\varepsilon(\boldsymbol{\varphi})]\boldsymbol{\nu}|^2 d\Sigma\right).\end{aligned}$$

We then obtain from (4.26) the estimate

$$\begin{aligned}(4.28)\quad & r(t)|_0^T + c_0\left(\int_0^T \|\dot\Xi\|^2_{\mathcal{H}^0(0,\ell)} dt \right. \\ &\qquad \left. + \int_0^T\int_0^\ell (G_2I|\theta_1'|^2 + EI_h|\theta_2'|^2 + G_2A|z'+\theta_2|^2)\,d\xi dt\right) \\ &\le C\left(\int_0^T |\dot\Xi(0,t)|^2 dt + \int_{\Sigma_J} (|C[\varepsilon(\boldsymbol{\varphi})]\boldsymbol{\nu}|^2 + \boldsymbol{\nu}\cdot(\nabla Z + \boldsymbol{\varphi})|^2)\,d\Sigma\right) \\ &\le C\left[\int_0^T (|\dot\Xi(0,t)|^2 + |\Xi(0,t)|^2 dt \right. \\ &\qquad \left. + \int_{\Sigma_J} |\mathbf{m}\cdot\boldsymbol{\nu}|\left(\left|\frac{\partial\boldsymbol{\varphi}}{\partial\nu}\right|^2 + \left|\frac{\partial Z}{\partial\nu} + \boldsymbol{\nu}\cdot\boldsymbol{\varphi}\right|^2\right) d\Sigma\right],\end{aligned}$$

in view of (4.21) and the geometric junction condition, where C depends the material parameters and on the $H^1(J)$ norms of σ_0 and σ_1.

Multiply (4.28) by ϵ and add the product to (4.25) to obtain an estimate of the form

$$
\begin{aligned}
&[R(t)+\epsilon r(t)]_0^T + c_0\int_0^T \|\dot{\mathbf{\Phi}}\|^2_{\mathcal{H}^0(\Omega)}dt \\
&\quad + c_0\epsilon\int_0^T \|\dot{\Xi}\|^2_{\mathcal{H}^0(0,\ell)}dt + c_0\int_0^T \|\mathbf{\Phi}\|^2_{\mathcal{H}^1_\gamma(\Omega)}dt \\
&\quad + c_0\epsilon\int_0^T\int_0^\ell (G_2I|\theta_1'|^2 + EI_h|\theta_2'|^2 + G_2A|z'+\theta_2|^2)\,d\xi dt \\
&\quad \le C_0\Bigg[\int_{\Sigma_1^+}(\mathbf{m}\cdot\boldsymbol{\nu})|\dot{\mathbf{\Phi}}|^2\,d\Sigma + \int_0^T |\Xi(0,t)|^2\,dt \\
&\quad - \int_{\Sigma_1^-}(\mathbf{m}\cdot\boldsymbol{\nu})\left(C[\varepsilon(\boldsymbol{\varphi})]\boldsymbol{\tau})\cdot\frac{\partial\boldsymbol{\varphi}}{\partial\boldsymbol{\tau}} + G_1h|\boldsymbol{\tau}\cdot(\nabla Z+\boldsymbol{\varphi})|^2\right)d\Sigma \\
&\quad + \int_{\Sigma_0^+}(\mathbf{m}\cdot\boldsymbol{\nu})\left((C[\varepsilon(\boldsymbol{\varphi})]\boldsymbol{\nu})\cdot\frac{\partial\boldsymbol{\varphi}}{\partial\boldsymbol{\nu}} + G_1h\left|\frac{\partial Z}{\partial\boldsymbol{\nu}}\right|^2\right)d\Sigma\Bigg] \\
&\quad + (C_1-\epsilon C)\int_{\Sigma_J}(\mathbf{m}\cdot\boldsymbol{\nu})\left(\left|\frac{\partial\boldsymbol{\varphi}}{\partial\boldsymbol{\nu}}\right|^2 + \left|\frac{\partial Z}{\partial\boldsymbol{\nu}}+\boldsymbol{\nu}\cdot\boldsymbol{\varphi}\right|^2\right)d\Sigma \\
&\quad + \frac{1}{2}\int_{\Sigma_J}(\mathbf{m}\cdot\boldsymbol{\nu})\{m_1|\dot{Z}|^2 + I_{\rho_1}|\dot{\boldsymbol{\varphi}}|^2\}\,d\Sigma + \epsilon C\int_0^T |\dot{\Xi}(0,t)|^2\,dt.
\end{aligned}
$$

for some constants C_0, C_1, C independent of ϵ and T.

Choose $\epsilon > 0$ so that $C_1 - \epsilon C \ge 0$. Also, it is possible to choose ϵ so small that

$$
(4.29)\quad \frac{1}{2}\int_{\Sigma_J}(\mathbf{m}\cdot\boldsymbol{\nu})\left[m_1|\dot{Z}|^2 + I_{\rho_1}|\dot{\boldsymbol{\varphi}}|^2\right]d\Sigma + \epsilon C\int_0^T |\dot{\Xi}(0,t)|^2dt \le 0.
$$

In fact, by utilizing the geometric junction conditions one obtains

$$
\begin{aligned}
&\int_{\Sigma_J}(\mathbf{m}\cdot\boldsymbol{\nu})\left[m_1|\dot{Z}|^2 + I_{\rho_1}|\dot{\boldsymbol{\phi}}|^2\right]d\Sigma \\
&\quad = \int_{\Sigma_J}(\mathbf{m}\cdot\boldsymbol{\nu})\{m_1\sigma_0^2|\dot{z}(0,t)+\zeta\dot{\theta}_1(0,t)|^2 \\
&\qquad + I_{\rho_1}\sigma_1^2(|\dot{\theta}_1(0,t)|^2 + |\dot{\theta}_2(0,t)|^2)\}d\Sigma.
\end{aligned}
$$

Since $\mathbf{m}\cdot\boldsymbol{\nu} < 0$ on J, the region of integration on the right may be replaced by Σ_{J_a}, where $\Sigma_{J_a} = J_a \times (0,T)$. On J_a we have $\sigma_i(\mathbf{x}) = 1$, $\zeta(\mathbf{x}) = \xi_2$, where $\mathbf{x} = \mathbf{x}_0 + \xi_2\boldsymbol{\tau}(\mathbf{x}_0)$, and

$$
\begin{aligned}
(\mathbf{m}\cdot\boldsymbol{\nu})(\mathbf{x}) &= (\mathbf{x}_0 + \xi_2\boldsymbol{\tau}(\mathbf{x}_0) - \hat{\mathbf{x}}_0)\cdot\boldsymbol{\nu}(\mathbf{x}_0) \\
&= (\mathbf{x}_0 - \hat{\mathbf{x}}_0)\cdot\boldsymbol{\nu}(\mathbf{x}_0) = \text{constant} < 0.
\end{aligned}
$$

Since

$$\int_{J_a} d\Gamma = a, \quad \int_{J_a} \zeta d\Gamma = 0, \quad \int_{J_a} \zeta^2 d\Gamma = \frac{a^3}{12},$$

we have

$$\begin{aligned}\int_{\Sigma_J} (\mathbf{m}\cdot\boldsymbol{\nu}) \left[m_1|\dot{Z}|^2 + I_{\rho_1}|\dot{\boldsymbol{\phi}}|^2\right] d\Sigma & \\ \leq (\mathbf{x}_0 - \hat{\mathbf{x}}_0)\cdot\boldsymbol{\nu}(\mathbf{x}_0) \int_0^T & \{m_1|\dot{z}(0,t)|^2 \\ & + a\left(m_1\frac{a^2}{12} + I_{\rho_1}\right)|\dot{\theta}_1(0,t)|^2 + aI_{\rho_1}|\dot{\theta}_2(0,t)|^2\Big\}\, dt,\end{aligned}$$

from which (4.29) immediately follows.

We have therefore proved the inequality

$$\begin{aligned}(4.30)\quad [R(t) + \epsilon r(t)]_0^T &+ c_0\int_0^T \|((\mathbf{\Phi},\Xi),(\dot{\mathbf{\Phi}},\dot{\Xi}))\|^2_{V_{\Gamma_0}\times H} dt \\ &\leq C_0\Bigg[\int_{\Sigma_1^+}(\mathbf{m}\cdot\boldsymbol{\nu})|\dot{\mathbf{\Phi}}|^2\, d\Sigma + \int_0^T |\Xi(0,t)|^2\, dt \\ &- \int_{\Sigma_1^-}(\mathbf{m}\cdot\boldsymbol{\nu})\left(C[\varepsilon(\boldsymbol{\varphi})]\boldsymbol{\tau})\cdot\frac{\partial\boldsymbol{\varphi}}{\partial\boldsymbol{\tau}} + G_1h|\boldsymbol{\tau}\cdot(\nabla Z + \boldsymbol{\varphi})|^2\right) d\Sigma \\ &+ \int_{\Sigma_0^+}(\mathbf{m}\cdot\boldsymbol{\nu})\left((C[\varepsilon(\boldsymbol{\varphi})]\boldsymbol{\nu})\cdot\frac{\partial\boldsymbol{\varphi}}{\partial\boldsymbol{\nu}} + G_1h\left|\frac{\partial Z}{\partial\boldsymbol{\nu}}\right|^2\right) d\Sigma\Bigg]\end{aligned}$$

for some positive constants c_0, C_0. Since the solution of (4.4)–(4.11) is given by a unitary group on $V_{\Gamma_0}\times H$, we have

$$\int_0^T \|((\mathbf{\Phi},\Xi),(\dot{\mathbf{\Phi}},\dot{\Xi}))\|^2_{V_\gamma\times H} dt = T\|((\mathbf{\Phi}^0,\Xi^0),(\mathbf{\Phi}^1,\Xi^1))\|^2_{V_{\Gamma_0}\times H}.$$

Moreover, it is easy to see that

$$\begin{aligned}|R(t) + \epsilon r(t)| &\leq C\|((\mathbf{\Phi},\Xi),(\dot{\mathbf{\Phi}},\dot{\Xi}))\|^2_{V_{\Gamma_0}\times H} \\ &= C\|((\mathbf{\Phi}^0,\Xi^0),(\mathbf{\Phi}^1,\Xi^1))\|^2_{V_{\Gamma_0}\times H}.\end{aligned}$$

Therefore, (4.30) may be rewritten

$$
\begin{aligned}
(c_0T - 2C)\|((\Phi^0, \Xi^0), (\Phi^1, \Xi^1))\|^2_{V_{\Gamma_0}\times H} \\
\leq C_0 \Bigg[\int_{\Sigma_1^+} (\mathbf{m}\cdot\boldsymbol{\nu})|\dot{\Phi}|^2\, d\Sigma + \int_0^T |\Xi(0,t)|^2\, dt \\
- \int_{\Sigma_1^-} (\mathbf{m}\cdot\boldsymbol{\nu}) \left(C[\varepsilon(\boldsymbol{\varphi})]\boldsymbol{\tau}) \cdot \frac{\partial \boldsymbol{\varphi}}{\partial \boldsymbol{\tau}} + G_1 h|\boldsymbol{\tau}\cdot(\nabla Z + \boldsymbol{\varphi})|^2 \right) d\Sigma \\
+ \int_{\Sigma_0^+} (\mathbf{m}\cdot\boldsymbol{\nu}) \left((C[\varepsilon(\boldsymbol{\varphi})]\boldsymbol{\nu}) \cdot \frac{\partial \boldsymbol{\varphi}}{\partial \boldsymbol{\nu}} + G_1 h \left|\frac{\partial Z}{\partial \boldsymbol{\nu}}\right|^2 \right) d\Sigma \Bigg],
\end{aligned}
$$

hence for $T > T_0 := 2C/c_0$,

$$
\begin{aligned}
(4.31) \quad \|((\Phi^0, \Xi^0), (\Phi^1, \Xi^1))\|^2_{V_{\Gamma_0}\times H} \leq C_T \Bigg[\int_{\Sigma_1^+} |\dot{\Phi}|^2\, d\Sigma \\
+ \int_{\Sigma_1^-} \left(C[\varepsilon(\boldsymbol{\varphi})]\boldsymbol{\tau}) \cdot \frac{\partial \boldsymbol{\varphi}}{\partial \boldsymbol{\tau}} + G_1 h|\boldsymbol{\tau}\cdot(\nabla Z + \boldsymbol{\varphi})|^2 \right) d\Sigma \\
+ \int_{\Sigma_0^+} \left((C[\varepsilon(\boldsymbol{\varphi})]\boldsymbol{\nu}) \cdot \frac{\partial \boldsymbol{\varphi}}{\partial \boldsymbol{\nu}} + G_1 h \left|\frac{\partial Z}{\partial \boldsymbol{\nu}}\right|^2 \right) d\Sigma + \int_0^T |\Xi(0,t)|^2\, dt \Bigg].
\end{aligned}
$$

where

$$
C_T = \frac{C_0}{c_0T - 2C} \max_{\Gamma} |\mathbf{m}\cdot\boldsymbol{\nu}|.
$$

Except for the last term on the right side, (4.31) is the estimate of Proposition 4.1. Therefore, to complete the proof, it suffices to establish the following result.

LEMMA 4.1. *Under the assumptions of Proposition* 4.1 *we have*

$$
\begin{aligned}
(4.32) \quad \int_0^T |\Xi(0,t)|^2 dt \leq C \Bigg[\int_{\Sigma_1^+} |\dot{\Phi}|^2\, d\Sigma \\
+ \int_{\Sigma_1^-} \left(C[\varepsilon(\boldsymbol{\varphi})]\boldsymbol{\tau}) \cdot \frac{\partial \boldsymbol{\varphi}}{\partial \boldsymbol{\tau}} + G_1 h|\boldsymbol{\tau}\cdot(\nabla Z + \boldsymbol{\varphi})|^2 \right) d\Sigma \\
+ \int_{\Sigma_0^+} \left((C[\varepsilon(\boldsymbol{\varphi})]\boldsymbol{\nu}) \cdot \frac{\partial \boldsymbol{\varphi}}{\partial \boldsymbol{\nu}} + G_1 h \left|\frac{\partial Z}{\partial \boldsymbol{\nu}}\right|^2 \right) d\Sigma \Bigg].
\end{aligned}
$$

PROOF. The proof uses a standard compactness-uniqueness argument.

One first proves that

$$\|((\mathbf{\Phi}^0, \Xi^0), (\mathbf{\Phi}^1, \Xi^1))\|^2_{V_{\Gamma_0} \times H} \le C_T \Bigg[\int_{\Sigma_1^+} |\dot{\mathbf{\Phi}}|^2 \, d\Sigma + \int_{\Sigma_1^-} \left(C[\varepsilon(\boldsymbol{\varphi})]\boldsymbol{\tau}) \cdot \frac{\partial \boldsymbol{\varphi}}{\partial \boldsymbol{\tau}} + G_1 h |\boldsymbol{\tau} \cdot (\nabla Z + \boldsymbol{\varphi})|^2 \right) d\Sigma + \int_{\Sigma_0^+} \left((C[\varepsilon(\boldsymbol{\varphi})]\boldsymbol{\nu}) \cdot \frac{\partial \boldsymbol{\varphi}}{\partial \boldsymbol{\nu}} + G_1 h \left| \frac{\partial Z}{\partial \boldsymbol{\nu}} \right|^2 \right) d\Sigma + \|\mathbf{\Phi}\|^2_{L^\infty(0,T;\mathcal{H}^0(\Omega))} \Bigg]. \tag{4.33}$$

for $T > T_0$ by showing that

$$\int_0^T |\Xi(0,t)|^2 dt \le C_T \Bigg[\int_{\Sigma_1^+} |\dot{\mathbf{\Phi}}|^2 \, d\Sigma + \|\mathbf{\Phi}\|^2_{L^\infty(0,T;\mathcal{H}^0(\Omega))} + \int_{\Sigma_1^-} \left(C[\varepsilon(\boldsymbol{\varphi})]\boldsymbol{\tau}) \cdot \frac{\partial \boldsymbol{\varphi}}{\partial \boldsymbol{\tau}} + G_1 h |\boldsymbol{\tau} \cdot (\nabla Z + \boldsymbol{\varphi})|^2 \right) d\Sigma + \int_{\Sigma_0^+} \left((C[\varepsilon(\boldsymbol{\varphi})]\boldsymbol{\nu}) \cdot \frac{\partial \boldsymbol{\varphi}}{\partial \boldsymbol{\nu}} + G_1 h \left| \frac{\partial Z}{\partial \boldsymbol{\nu}} \right|^2 \right) d\Sigma \Bigg]. \tag{4.34}$$

Estimate (4.34) may be proved by contradiction, using (4.31) and the compactness of the injection (see Simon [102, Section 8])

$$(\mathbf{\Phi}, \Xi) \in L^\infty(0,T; V_\gamma) \cap W^{1,\infty}(0,T;H) \mapsto L^\infty(0,T;[V_\gamma, H]_\alpha), \quad 0 < \alpha \le 1,$$

where $[\cdot,\cdot]_\alpha$ denotes the interpolation space of order α. In fact, if (4.34) is false, there is a sequence of initial data $((\mathbf{\Phi}_n^0, \Xi_n^0), (\mathbf{\Phi}_n^1, \Xi_n^1)) \in D_A \times V$ (where $V := V_{\Gamma_0}$) such that the corresponding solution $(\mathbf{\Phi}_n, \Xi_n)$ of (4.4)–(4.11) satisfies

$$\int_0^T |\Xi_n(0,t)|^2 dt = 1, \tag{4.35}$$

$$\|\mathbf{\Phi}_n\|_{L^\infty(0,T;\mathcal{H}^0(\Omega))} \to 0, \tag{4.36}$$

$$\int_{\Sigma_1^+} |\dot{\mathbf{\Phi}}_n|^2 \, d\Sigma + \int_{\Sigma_0^+} \left((C[\varepsilon(\boldsymbol{\varphi}_n)]\boldsymbol{\nu}) \cdot \frac{\partial \boldsymbol{\varphi}_n}{\partial \boldsymbol{\nu}} + G_1 h \left| \frac{\partial Z_n}{\partial \boldsymbol{\nu}} \right|^2 \right) d\Sigma + \int_{\Sigma_1^-} \left(C[\varepsilon(\boldsymbol{\varphi}_n)]\boldsymbol{\tau}) \cdot \frac{\partial \boldsymbol{\varphi}_n}{\partial \boldsymbol{\tau}} + G_1 h |\boldsymbol{\tau} \cdot (\nabla Z_n + \boldsymbol{\varphi}_n)|^2 \right) d\Sigma \to 0. \tag{4.37}$$

It follows from (4.31), (4.35) and (4.37) that $((\mathbf{\Phi}_n^0, \Xi_n^0), (\mathbf{\Phi}_n^1, \Xi_n^1))$ is bounded in $V \times H$ and, therefore

$$(\mathbf{\Phi}_n, \Xi_n) \text{ is bounded in } L^\infty(0,T;V) \cap W^{1,\infty}(0,T;H).$$

We may then conclude that, as $n \to \infty$ through some subset of positive integers,

$$(\mathbf{\Phi}_n, \Xi_n) \text{ converges } \textit{strongly} \text{ in } L^\infty(0,T;[V,H]_\alpha), \quad 0 < \alpha \le 1.$$

In view of (4.35) and (4.36) we may then conclude that

$$\mathbf{\Phi}_n \to 0 \text{ strongly in } L^\infty(0,T;[\mathcal{H}^1_{\Gamma_0}(\Omega), \mathcal{H}^0(\Omega)]_\alpha) \tag{4.38}$$

and

$$\int_0^T |\Xi_n(0,t)|^2 dt \to \int_0^T |\Xi(0,t)|^2 dt = 1. \tag{4.39}$$

From (4.38),

$$\mathbf{\Phi}_n|_{\Sigma_J} \to 0 \text{ strongly in } L^\infty(0,T;L^2(J)).$$

Since the geometric junction condition $\mathbf{\Phi}_n = \sigma \Xi_n(0)$ holds along J,

$$\Xi_n(0,\cdot) \to 0 \text{ strongly in } L^\infty(0,T),$$

which contradicts (4.39). This proves (4.34) and hence (4.33).

Let $\mathcal{X}$ be the linear space of all $(\mathbf{\Phi}, \Xi)$ which satisfy the conditions

$$\begin{cases} (\mathbf{\Phi}, \Xi) \in L^\infty(0,T;V) \cap W^{1,\infty}(0,T;H), \\ (\mathbf{\Phi}, \Xi) \text{ satisfy (4.4)–(4.10)}, \\ C[\varepsilon(\boldsymbol{\varphi})]\boldsymbol{\nu} = 0, \quad \dfrac{\partial Z}{\partial \boldsymbol{\nu}} = 0 \text{ on } \Sigma_0^+, \\ \dot{\boldsymbol{\varphi}} = 0, \quad \dot{Z} = 0 \text{ on } \Sigma_1^+, \\ \boldsymbol{\tau} \cdot \dfrac{\partial \boldsymbol{\varphi}}{\partial \boldsymbol{\tau}} = 0, \quad \dfrac{\partial Z}{\partial \boldsymbol{\tau}} + \boldsymbol{\tau} \cdot \boldsymbol{\varphi} = 0 \text{ on } \Sigma_1^- \end{cases} \tag{4.40}$$

with norm

$$\|(\mathbf{\Phi}, \Xi)\|_{\mathcal{X}} = \|((\mathbf{\Phi}, \Xi), (\dot{\mathbf{\Phi}}, \dot{\Xi}))\|_{L^\infty(0,T;V \times H)}.$$

We wish to show that $\mathcal{X} = \{0\}$. To this end, we set $X = (\mathbf{\Phi}, \Xi)$ and suppose that $X \in \mathcal{X}$. Then the difference quotient $(X(\cdot + h) - X(\cdot))/h \in \mathcal{X}$ and from (4.33) we have

$$\begin{aligned} &\|(X(h) - X(0))/h, (\dot{X}(h) - \dot{X}(0))/h)\|_{V \times H} \\ &\qquad \le C_T \|(\mathbf{\Phi}(\cdot + h) - \mathbf{\Phi}(\cdot))/h\|_{L^\infty(0,T;\mathcal{H}^0(\Omega))}. \end{aligned} \tag{4.41}$$

Since $X \in \mathcal{X}$, the right side of (4.41) converges to $\|\dot{\mathbf{\Phi}}\|_{L^\infty(0,T;\mathcal{H}^0(\Omega))}$ as $h \to 0$, hence $(X(h) - X(0))/h \to \dot{X}(0)$ weakly in V, $(\dot{X}(h) - \dot{X}(0))/h \to \ddot{X}(0)$

weakly in H and

$$\begin{aligned}\|(\dot{X},\ddot{X})\|_{L^\infty(0,T;V\times H)} &\le C_T\|\dot{\Phi}\|_{L^\infty(0,T;\mathcal{H}^0(\Omega))}\\ &\le C_T\|(X,\dot{X})\|_{L^\infty(0,T;V\times H)} = C_T\|X\|_{\mathcal{X}}.\end{aligned}$$

Hence $\dot{X}\in\mathcal{X}$ and the linear map

$$(\mathbf{\Phi},\Xi)\mapsto(\dot{\mathbf{\Phi}},\dot{\Xi}):\mathcal{X}\mapsto\mathcal{X} \tag{4.42}$$

is continuous. It then follows by the Arzela-Ascoli Theorem that the injection

$$\{(\mathbf{\Phi},\Xi)\in\mathcal{X}\,|\,(\dot{\mathbf{\Phi}},\dot{\Xi})\in\mathcal{X}\}\mapsto\mathcal{X}$$

is compact, hence $\mathcal{X}$ itself is compact and, therefore, finite dimensional.

Since the map (4.42) is continuous, it has an eigenvalue η (here we need to work in the complexification of $\mathcal{X}$). In particular, $\dot{\mathbf{\Phi}}=\eta\mathbf{\Phi}$ so that for each $t\in(0,T)$, $\mathbf{\Phi}=\varphi+Z\mathbf{k}$ satisfies

$$\begin{cases} m_1\eta^2 Z - G_1 h\,\mathrm{div}(\nabla Z+\varphi)=0,\\ I_{\rho_1}\eta^2\varphi-\mathrm{div}(C[\varepsilon(\varphi)]+G_1h(\nabla Z+\varphi)=0 \ \text{ in } \Omega,\end{cases} \tag{4.43}$$

$$\begin{cases} \varphi=0,\ \ Z=0,\ \ C[\varepsilon(\varphi)]\boldsymbol{\nu}=0,\ \ \dfrac{\partial Z}{\partial\boldsymbol{\nu}}=0 \ \text{ on } \Gamma_0^+,\\ \dot{\varphi}=0,\ \ \dot{Z}=0,\ \ C[\varepsilon(\varphi)]\boldsymbol{\nu}=0,\ \ \dfrac{\partial Z}{\partial\boldsymbol{\nu}}+\boldsymbol{\nu}\cdot\varphi=0 \ \text{ on } \Gamma_1^+.\end{cases} \tag{4.44}$$

Since (4.43) is a second order elliptic system with constant coefficients whose solution has vanishing Cauchy data on Γ_0^+, if $\Gamma_0^+\neq\emptyset$ we may conclude that $\varphi=0$ and $Z=0$ in $\Omega\times(0,T)$. The geometric and dynamic junction conditions then give $\Xi(0,\cdot)=0$ and $\Xi'(0,\cdot)=0$, hence $\Xi=0$ in $(0,\ell)\times(0,T)$ since $\Xi=\boldsymbol{\theta}+z\mathbf{e}_3$ satisfies the system of ordinary differential equations

$$\begin{cases} m_2\eta^2 z - G_2A(z'+\theta_2)'=0,\\ \rho_2 I_h\eta^2\theta_2 - EI_h\theta_2''+G_2A(z'+\theta_2)=0,\\ \rho_2 I\eta^2\theta_1 - G_2I\theta_1''=0,\ \ 0<\xi<\ell.\end{cases}$$

Therefore $\mathcal{X}=\{0\}$ if $\Gamma_0^+\neq\emptyset$. On the other hand, if $\Gamma_0^+=\emptyset$, then necessarily $\Gamma_1^+\neq\emptyset$. The pair $\dot{\varphi},\dot{Z}$ also satisfies (4.43) and the Cauchy conditions

$$\dot{\varphi}=0,\ \ \dot{Z}=0,\ \ C[\varepsilon(\dot{\varphi})]\boldsymbol{\nu}=0,\ \ \frac{\partial\dot{Z}}{\partial\boldsymbol{\nu}}=0 \ \text{ on } \Gamma_1^+.$$

Thus $\dot{\varphi} = 0$ and $\dot{Z} = 0$ in $\Omega \times (0,T)$ and then, as above, we obtain $\dot{\Xi} = 0$ $(0,\ell) \times (0,T)$. It then follows from (4.4)–(4.10) that $(\mathbf{\Phi}, \Xi)$ satisfies the homogeneous, stationary problem

$$\begin{cases} \operatorname{div} C[\varepsilon(\varphi)] - G_1 h(\nabla Z + \varphi) = 0, \\ G_1 h \operatorname{div}(\nabla Z + \varphi) = 0 \ \text{ in } \Omega; \end{cases}$$

$$\begin{cases} G_2 A(z' + \theta_2)' = 0, \\ EI_h \theta_2'' - G_2 A(z' + \theta_2) = 0, \\ G_2 I \theta_1'' = 0, \ \ 0 < \xi < \ell; \end{cases}$$

$$G_2 A(z' + \theta_2)(\ell) = EI_h \theta_2'(\ell) = G_2 I \theta_1'(\ell) = 0;$$

$$Z = 0, \ \ \varphi = 0 \ \text{ on } \Gamma_0;$$

$$C[\varepsilon(\varphi)]\boldsymbol{\nu} = 0, \ \ G_1 h(\nabla Z + \varphi) \cdot \boldsymbol{\nu} = 0 \ \text{ on } \Gamma_1;$$

$$Z = \sigma_0[z(0) + \zeta\theta_1(0)], \ \ \varphi = \sigma_1 \boldsymbol{\theta}(0) \ \text{ on } J;$$

$$\begin{cases} G_2 I \theta_1'(0) = \displaystyle\int_J \{G_1 h \zeta \sigma_0 \boldsymbol{\nu} \cdot (\nabla Z + \varphi) \\ \qquad\qquad\qquad - \sigma_1 \mathbf{e}_2 \cdot (C[\varepsilon(\varphi)]\boldsymbol{\nu})\} d\Gamma, \\ EI_h \theta_2'(0) = \displaystyle\int_J \sigma_1 \mathbf{e}_1 \cdot (C[\varepsilon(\varphi)]\boldsymbol{\nu})\, d\Gamma, \\ G_2 A(z' + \theta_2)(0) = G_1 h \displaystyle\int_J \sigma_0 \boldsymbol{\nu} \cdot (\nabla Z + \varphi)\, d\Gamma \ \text{ on } J. \end{cases}$$

If $\Gamma_0 \neq \emptyset$, the only solution of this problem is $\mathbf{\Phi} = 0$, $\Xi = 0$ and once again we conclude that $\mathcal{X} = \{0\}$.

To complete the proof of the lemma, let us suppose that (4.32) is false. Then there is a sequence of initial data $((\mathbf{\Phi}_n^0, \Xi_n^0), (\mathbf{\Phi}_n^1, \Xi_n^1)) \in D_A \times V$ such that the corresponding solution $(\mathbf{\Phi}_n, \Xi_n)$ of (4.4)–(4.11) satisfies

$$\int_0^T |\Xi_n(0,t)|^2 dt = 1, \tag{4.45}$$

$$\int_{\Sigma_1^+} |\dot{\mathbf{\Phi}}_n|^2 d\Sigma + \int_{\Sigma_0^+} \left(|C[\varepsilon(\varphi_n)]\boldsymbol{\nu}|^2 + \left|\frac{\partial Z_n}{\partial \boldsymbol{\nu}}\right|^2 \right) d\Sigma \tag{4.46}$$
$$+ \int_{\Sigma_1^-} \left(C[\varepsilon(\varphi_n)]\boldsymbol{\tau}) \cdot \frac{\partial \varphi_n}{\partial \boldsymbol{\tau}} + G_1 h |\boldsymbol{\tau} \cdot (\nabla Z_n + \varphi_n)|^2 \right) d\Sigma \to 0.$$

It follows from (4.31) that $((\mathbf{\Phi}_n^0, \Xi_n^0), (\mathbf{\Phi}_n^1, \Xi_n^1))$ is bounded in $V \times H$ and, therefore

$$(\mathbf{\Phi}_n, \Xi_n) \ \text{ is bounded in } L^\infty(0,T;V) \cap W^{1,\infty}(0,T;H).$$

We may then conclude that, as $n \to \infty$ through some subset of positive integers,

$$(4.47)\quad (\mathbf{\Phi}_n, \Xi_n) \text{ converges weak* in } L^\infty(0,T;V) \cap W^{1,\infty}(0,T;H) \text{ and strongly in } L^\infty(0,T;[V,H]_\alpha),\ \ 0 < \alpha \le 1.$$

If $(\mathbf{\Phi}, \Xi)$ denotes the limit function, then $(\mathbf{\Phi}, \Xi)$ satisfies, in view of (4.45)–(4.47),

$$(4.48)\qquad \int_0^T |\Xi(0,t)|^2 dt = 1,$$

$$\begin{cases} (\mathbf{\Phi}, \Xi) \in L^\infty(0,T;V) \cap W^{1,\infty}(0,T;H), \\ (\mathbf{\Phi}, \Xi) \ \text{ satisfy (4.4)–(4.10)}, \\ C[\varepsilon(\boldsymbol{\varphi})]\boldsymbol{\nu} = 0, \ \dfrac{\partial Z}{\partial \boldsymbol{\nu}} = 0 \ \text{ on } \Sigma_0^+, \\ \dot{\boldsymbol{\varphi}} = 0, \ \ \dot{Z} = 0 \ \text{ on } \Sigma_1^+, \\ \boldsymbol{\tau} \cdot \dfrac{\partial \boldsymbol{\varphi}}{\partial \boldsymbol{\tau}} = 0, \ \ \dfrac{\partial Z}{\partial \boldsymbol{\tau}} + \boldsymbol{\tau} \cdot \boldsymbol{\varphi} = 0 \ \text{ on } \Sigma_1^-, \end{cases}$$

that is, $(\mathbf{\Phi}, \Xi) \in \mathcal{X}$. Therefore $(\mathbf{\Phi}, \Xi) = 0$, contradicting (4.48). □

4.2. Proofs of Theorems 4.1 and 4.2. We begin with the proof of Theorem 4.1. Consider the control-to-state map

$$S_T\mathbf{f} = \int_0^T \exp((T-s)\mathcal{A})\mathcal{B}\mathbf{f}(s)\, ds.$$

We have already proved that $S_T \in \mathcal{L}(L^2(0,T;U), H \times V_\gamma')$ for each $T > 0$ (see (3.33)). Since

$$(4.49)\qquad \langle S_T\mathbf{f}, Y^0\rangle_{H\times V_\gamma} = \int_0^T (\mathbf{f}(s), \mathcal{B}'Y(s))_U ds,$$

the dual map $S_T' : H \times V_\gamma \mapsto L^2(0,T;U)$ is given by $S_T'Y^0 = \mathcal{B}'Y(\cdot)$, where $Y = ((\mathbf{\Phi}_1, \Xi_1), (\mathbf{\Phi}, \Xi))$ satisfies (3.30) with $\tau = T$. Therefore Range$(S_T) = H \times V_\gamma'$ is equivalent to

$$(4.50)\qquad \|\mathcal{B}'Y(\cdot)\|_{L^2(0,TU)} \ge c_0\|Y^0\|_{H\times V_\gamma}, \ \ \forall Y^0 \in H \times V_\gamma,$$

for some $c_0 > 0$. From (3.31) it is seen that (4.50) is equivalent to showing that

$$\int_{\Sigma^D} \{|C[\varepsilon(\boldsymbol{\varphi})]\boldsymbol{\nu}|^2 + K_1 \left|\frac{\partial Z}{\partial \boldsymbol{\nu}}\right|^2\} d\Sigma \ge c_0\|((\mathbf{\Phi}^0, \Xi^0), (\mathbf{\Phi}^1, \Xi^1))\|^2_{V_\gamma \times H}.$$

For T large enough the last estimate follows from Corollary 4.1.

The proof of Theorem 4.2 is similar. In this case one again has (4.49), where now $\mathcal{B}'$ is given by (see (3.13))

$$\mathcal{B}'Y = \mathbf{\Phi}|_{\Sigma_1} = \boldsymbol{\varphi} + Z\mathbf{k}|_{\Sigma_1}.$$

However, (4.50) will not hold in general, so to prove Theorem 4.2 we use the Hilbert Uniqueness Method. We define a semi-norm

$$\|Y^0\|_{\mathcal{F}} = \|\mathcal{B}'Y(\cdot)\|_{L^2(0,T;U)}, \quad Y^0 \in H \times V_\gamma$$

which, according to Corollary 4.2, is a *norm* if $T > T_0$, and we define a space $\mathcal{F}$ which is the completion of $H \times V_\gamma$ with respect to $\|\cdot\|_{\mathcal{F}}$. Then S_T' is an isometry of $\mathcal{F}$ onto $L^2(0,T;U)$. Since

$$\langle S_T S_T' Z^0, Y^0 \rangle_{\mathcal{F}} = (Z^0, Y^0)_{\mathcal{F}},$$

$S_T S_T'$ is the Riesz isomorphism of $\mathcal{F}$ onto $\mathcal{F}'$, the dual space of $\mathcal{F}$. Therefore $R_T = \mathcal{F}'$ for $T > 0$. On the other hand, according to Corollary 4.2 one has $\mathcal{F} \subset V_\gamma' \times H$ for $T > T_0$ and, therefore, $V_\gamma \times H \subset \mathcal{F}'$ for $T > T_0$. If $X^0 \in R_T$ and $Y^0 = (S_T S_T')^{-1} X^0$, then $\mathbf{f} = S_T' Y^0$ is the control of minimum norm in $L^2(0,T;U)$ for which $S_T \mathbf{f} = X^0$.

5. Limit Model as the Shear Moduli Approach Infinity

In this section we consider the limit as $G_1 \to \infty$ and $G_2 \to \infty$ in the model

$$\begin{cases} I_{\rho_1} \ddot{\mathbf{u}} - \operatorname{div} C[\varepsilon(\mathbf{u})] + G_1 h(\nabla W + \mathbf{u}) = \mathbf{M}_1, \\ m_1 \ddot{W} - G_1 h \operatorname{div}(\nabla W + \mathbf{u}) = F_1 \quad \text{in } \Omega; \end{cases} \tag{5.1}$$

$$\begin{cases} m_2 \ddot{w} - G_2 A(w' + \vartheta_2)' = F_2, \\ \rho_2 I_h \ddot{\vartheta}_2 - E I_h \vartheta_2'' + G_2 A(w' + \vartheta_2) = M_{22}, \\ \rho_2 I \ddot{\vartheta}_1 - G_2 I \vartheta_1'' = M_{21}, \quad 0 < \xi < \ell; \end{cases} \tag{5.2}$$

$$G_2 A(w' + \vartheta_2)(\ell) = E I_h \vartheta_2'(\ell) = G_2 I \vartheta_1'(\ell) = 0; \tag{5.3}$$

$$W = 0, \quad \mathbf{u} = 0 \quad \text{on } \Gamma_0; \tag{5.4}$$

$$\begin{cases} G_1 h(\nabla W + \mathbf{u}) \cdot \boldsymbol{\nu} = 0, \\ C[\varepsilon(\mathbf{u})]\boldsymbol{\nu} = 0 \quad \text{on } \Gamma_1; \end{cases} \tag{5.5}$$

$$W = w(0) + \xi_2 \vartheta_1(0), \quad \mathbf{u} = \boldsymbol{\vartheta}(0) \quad \text{on } J_a, \tag{5.6}$$

where $\boldsymbol{\vartheta} = \vartheta_2 \mathbf{e}_1 - \vartheta_1 \mathbf{e}_2$;

$$(5.7)\qquad \begin{cases} G_2 I \vartheta_1'(0) = \displaystyle\int_{J_a} \{G_1 h \xi_2 \boldsymbol{\nu} \cdot (\nabla W + \mathbf{u}) \\ \qquad\qquad\qquad -\mathbf{e}_2 \cdot (C[\varepsilon(\mathbf{u})]\boldsymbol{\nu})\} d\Gamma, \\ E I_h \vartheta_2'(0) = \displaystyle\int_{J_a} \mathbf{e}_1 \cdot (C[\varepsilon(\mathbf{u})]\boldsymbol{\nu})\, d\Gamma, \\ G_2 A(w' + \vartheta_2)(0) = G_1 h \displaystyle\int_{J_a} \boldsymbol{\nu} \cdot (\nabla W + \mathbf{u})\, d\Gamma. \end{cases}$$

It is assumed that Ω is bounded, open and connected with a Lipschitz boundary consisting of a finite number of smooth, simple curves, that $\partial\Omega := \Gamma = \overline{\Gamma}_0 \cup \overline{\Gamma}_1 \cup \overline{J}_a$ where Γ_0, Γ_1, J_a are relatively open in Γ, mutually disjoint and $J_a \neq \emptyset$. For simplicity we assume that $\Gamma_0 \neq \emptyset$. This model corresponds to the situation, discussed in Section 2, where $J - \overline{J}_a$ is free, where there are no external forces or moments acting on Γ^D and where we set $\Gamma_0 = \Gamma^C$, $\Gamma_1 = \Gamma^D \cup (J - \overline{J}_a)$.

We write G for the pair (G_1, G_2) of shear moduli, $G > 0$ and $G \to \infty$ to mean that $G_i > 0$ and $G_i \to \infty$ respectively, for $i = 1, 2$. As above, we set

$$\boldsymbol{\Psi} = \mathbf{u} + W\mathbf{a}_3, \quad \boldsymbol{\Theta} = \boldsymbol{\vartheta} + w\mathbf{e}_3,$$

with corresponding initial functions $\boldsymbol{\Psi}^0, \boldsymbol{\Psi}^1, \boldsymbol{\Theta}^0, \boldsymbol{\Theta}^1$. We further denote the pair $(\boldsymbol{\Psi}, \boldsymbol{\Theta})$ by $\mathbf{R}$ and the pairs $(\boldsymbol{\Psi}^0, \boldsymbol{\Theta}^0)$, $(\boldsymbol{\Psi}^1, \boldsymbol{\Theta}^1)$ by $\mathbf{R}_0$ and $\mathbf{R}_1$, respectively. As discussed earlier, the appropriate function space setting for the above problem is $V \times H$, where $V := V_{\Gamma_0}$, and its variational formulation is

$$(5.8)\quad \langle \ddot{\mathbf{R}}, \hat{\mathbf{R}} \rangle_V + (\mathbf{R}, \hat{\mathbf{R}})_V = \int_\Omega \mathbf{F}_1 \cdot \hat{\boldsymbol{\Psi}}\, d\Omega + \int_0^\ell \mathbf{F}_2 \cdot \hat{\boldsymbol{\Theta}}\, d\xi, \quad \forall \hat{\mathbf{R}} = (\hat{\boldsymbol{\Psi}}, \hat{\boldsymbol{\Theta}}) \in V,$$

$$(5.9)\qquad \mathbf{R}(0) = \mathbf{R}_0, \quad \dot{\mathbf{R}}(0) = \mathbf{R}_1,$$

where

$$\mathbf{F}_1 = \mathbf{M}_1 + F_1 \mathbf{a}_3, \quad \mathbf{F}_2 = M_{22}\mathbf{e}_1 - M_{21}\mathbf{e}_2 + F_2 \mathbf{e}_3.$$

It follows immediately that if

$$\mathbf{R}_0 \in V, \quad \mathbf{R}_1 \in H, \quad (\mathbf{F}_1, \mathbf{F}_2) \in L^1(H),$$

(5.8), (5.9) has a unique solution with regularity

$$\mathbf{R} \in C(V) \cap C^1(H) \cap L^1(V').$$

We have written, in order to simplify the notation, $L^1(H)$, $C(V)$, etc., in place of $L^1(0, T; H)$, $C([0, T]; V)$, etc.

The space V and solution $\mathbf{R}$ both depend on G. We indicate this dependence explicitly by writing V_G and $\mathbf{R}^G$ in their places. Thus we write (5.8), (5.9) as

$$(5.10)\quad \langle \ddot{\mathbf{R}}^G, \hat{\mathbf{R}}\rangle_{V_G} + (\mathbf{R}^G, \hat{\mathbf{R}})_{V_G} = (\mathbf{G}, \hat{\mathbf{R}})_H, \quad \forall \hat{\mathbf{R}} = (\hat{\mathbf{\Psi}}, \hat{\mathbf{\Theta}}) \in V_G,$$

$$(5.11)\qquad \mathbf{R}^G(0) = \mathbf{R}_0, \quad \dot{\mathbf{R}}^G(0) = \mathbf{R}_1,$$

where $\mathbf{G} = (\mathbf{G}_1, \mathbf{G}_2)$ with

$$\mathbf{G}_1 = I_{\rho_1}^{-1}\mathbf{M}_1 + m_1^{-1}F_1\mathbf{a}_3, \quad \mathbf{G}_2 = \rho_2 I_h^{-1}M_{22}\mathbf{e}_1 - I^{-1}M_{21}\mathbf{e}_2 + m_2^{-1}F_2\mathbf{e}_3.$$

Our goal is to pass to the limit as $G \to \infty$ in (5.10) and (5.11). To this end, we introduce a *closed* subspace H_∞ of H as follows:

$$H_\infty = \{(\mathbf{\Psi}, \mathbf{\Theta}) \in H \,|\, W \in H^1(\Omega),\ w \in H^1(0,\ell),\ \vartheta_1 \in H^1(0,\ell),$$
$$\mathbf{u} = -\nabla W,\ \vartheta_2 = -w',\ \vartheta_1' = 0\}.$$

If $\mathbf{R} = (\mathbf{\Phi}, \mathbf{\Theta}) \in H_\infty$, then

$$\mathbf{\Phi} = -\nabla W + W\mathbf{a}_3, \quad \mathbf{\Theta} = -w'\mathbf{e}_1 - \vartheta_1\mathbf{e}_2 + w\mathbf{e}_3, \quad \vartheta_1 \equiv \text{const},$$

and

$$(5.12)\quad \|\mathbf{R}\|_{H_\infty} = \|\mathbf{R}\|_H = \Big[\int_\Omega (m_1|W|^2 + I_{\rho_1}|\nabla W|^2)\,d\Omega$$
$$+ \int_0^\ell (m_2|w|^2 + \rho_2 I_h|w'|^2)\,d\xi + I\ell|\vartheta_1|^2\Big]^{1/2}.$$

An element $\mathbf{R}$ in the orthogonal complement $H_\infty^\perp$ of H_∞ in H is characterized by the variational equation

$$\int_\Omega (m_1 W\hat{W} - I_{\rho_1}\mathbf{u}\cdot\nabla\hat{W})\,d\Omega$$
$$+ \int_0^\ell (m_2 w\hat{w} - \rho_2 I_h\vartheta_2\hat{w}' + I\vartheta_1\hat{\vartheta}_1)\,d\xi = 0, \quad \forall \hat{\mathbf{R}} \in H_\infty,$$

which is formally equivalent to the system

$$m_1 W + I_{\rho_1}\operatorname{div}\mathbf{u} = 0 \ \text{ in } \Omega, \quad \boldsymbol{\nu}\cdot\mathbf{u}|_{\Gamma_1} = 0;$$
$$m_2 w + I_{\rho_1}\vartheta_2' = 0 \ \text{ in } (0,\ell), \quad \vartheta_2(\ell) = 0;$$
$$I\int_0^\ell \vartheta_1 d\xi = I_{\rho_1}\int_{J_a} \xi_2(\boldsymbol{\nu}\cdot\mathbf{u})\,d\Gamma, \quad \rho_2 I_h\vartheta_2(0) = -I_{\rho_1}\int_{J_a}\boldsymbol{\nu}\cdot\mathbf{u}\,d\Gamma.$$

We further introduce

$$V_\infty = V_G \cap H_\infty = \{(\mathbf{\Psi}, \mathbf{\Theta}) \in V_G \,|\, W \in H^2(\Omega),\ w \in H^2(0,\ell),$$
$$\mathbf{u} = -\nabla W,\ \vartheta_2 = -w',\ \vartheta_1' = 0\}.$$

Functions in V_∞ satisfy the geometric boundary conditions

$$W = \frac{\partial W}{\partial \boldsymbol{\nu}} = 0 \ \text{ on } \Gamma_0 \tag{5.13}$$

and the geometric junction conditions

$$W = w(0) + \xi_2 \vartheta_1, \quad -\nabla W = -w'(0)\boldsymbol{\nu}(\mathbf{x}_0) - \vartheta_1 \boldsymbol{\tau}(\mathbf{x}_0) \ \text{ on } J_a,$$

where $\vartheta_1 \equiv$const.; equivalently

$$W = w(0) + \xi_2 \vartheta_1, \quad \frac{\partial W}{\partial \boldsymbol{\nu}} = w'(0) \ \text{ on } J_a. \tag{5.14}$$

If $\mathbf{R} \in V_\infty$, then

(5.15)

$$\|\mathbf{R}\|_{V_\infty} = \|\mathbf{R}\|_{V_G} = \left\{ \int_\Omega (C[\varepsilon(\nabla W)], \varepsilon(\nabla W))\, d\Omega + EI_h \int_0^\ell |w''|^2 d\xi \right\}^{1/2}$$

The space $(V_\infty, \|\cdot\|_{V_\infty})$ is dense in $(H_\infty, \|\cdot\|_{H_\infty})$ with continuous injection so that we may write, as usual,

$$V_\infty \subset H_\infty \subset V'_\infty.$$

Let A_G denote the Riesz isomorphism of V_G onto V'_G and A_∞ the Riesz isomorphism of V_∞ onto V'_∞. If $\mathbf{R} \in V_\infty$ then $A_\infty \mathbf{R}$ is the restriction of $A_G \mathbf{R}$ to V_∞, that is

$$\langle A_\infty \mathbf{R}, \hat{\mathbf{R}} \rangle_{V_\infty} = \langle A_G \mathbf{R}, \hat{\mathbf{R}} \rangle_{V_G}, \quad \forall \mathbf{R}, \hat{\mathbf{R}} \in V_\infty.$$

In order to pass to the limit in (5.10), (5.11), we assume that the data of the problem satisfy the following conditions:

$$\mathbf{R}_0 \in V_\infty, \quad \mathbf{R}_1 \in H_\infty, \quad \mathbf{G} \in L^2(H). \tag{5.16}$$

As in Section 6, Chapter VII, the restrictions on the initial data are conditions of consistency of the data with the limit model.

The solution $\mathbf{R}^G$ of (5.10), (5.11) may be decomposed into

$$\mathbf{R}^G = \mathbf{R}^G_\infty + (\mathbf{R}^G_\infty)^\perp, \quad \dot{\mathbf{R}}^G = \dot{\mathbf{R}}^G_\infty + (\dot{\mathbf{R}}^G_\infty)^\perp,$$

where $\mathbf{R}^G_\infty$, $\dot{\mathbf{R}}^G_\infty \in C(H_\infty)$ and $(\mathbf{R}^G_\infty)^\perp$, $(\dot{\mathbf{R}}^G_\infty)^\perp \in C(H^\perp_\infty)$. Similarly, we write

$$\mathbf{G} = \mathbf{G}_\infty + \mathbf{G}^\perp_\infty, \quad \mathbf{G}_\infty \in L^2(H_\infty), \ \mathbf{G}^\perp_\infty \in L^2(H^\perp_\infty).$$

Let $(\ddot{\mathbf{R}}^G_\infty)(t) \in V'_\infty$ denote the restriction of $\ddot{\mathbf{R}}^G(t)$ to V_∞. Then one has

$$\dot{\mathbf{R}}^G_\infty = \frac{d}{dt}\mathbf{R}^G_\infty \ \text{ and, similarly, } \ddot{\mathbf{R}}^G_\infty = \frac{d}{dt}\dot{\mathbf{R}}^G_\infty.$$

In analogy with Theorem 6.1 of Chapter VII, we have

THEOREM 5.1. *Assume* (5.16). *There is a function* $\mathbf{R} \in C(V_\infty)$ *with* $\dot{\mathbf{R}} \in C(H_\infty)$ *and* $\ddot{\mathbf{R}} \in L^2(V'_\infty)$ *such that as* $G \to \infty$,

$$\begin{aligned} \mathbf{R}^G &\to \mathbf{R} \ \textit{strongly in } L^\infty(H); \\ \mathbf{R}^G &\to \mathbf{R} \ \textit{weakly* in } L^\infty(\mathcal{H}^1_{\Gamma_0}(\Omega) \times \mathcal{H}^1(0,\ell)); \\ \dot{\mathbf{R}}^G_\infty &\to \dot{\mathbf{R}} \ \textit{strongly in } L^\infty(V'_\infty); \\ \dot{\mathbf{R}}^G &\to \dot{\mathbf{R}} \ \textit{weakly* in } L^\infty(H); \\ \ddot{\mathbf{R}}^G_\infty &\to \ddot{\mathbf{R}} \ \textit{weakly in } L^2(V'_\infty). \end{aligned}$$

Moreover, $\mathbf{R}$ *is the unique solution of*

$$\ddot{\mathbf{R}} + A_\infty \mathbf{R} = \mathbf{G}_\infty, \tag{5.17}$$

$$\mathbf{R}(0) = \mathbf{R}_0, \ \ \dot{\mathbf{R}}(0) = \mathbf{R}_1. \tag{5.18}$$

PROOF. From (5.10), (5.11) and assumption (5.16) we have

$$\begin{aligned} \|\dot{\mathbf{R}}^G(t)\|^2_H + \|\mathbf{R}^G(t)\|^2_{V_G} &= \|\mathbf{R}_1\|^2_{H_\infty} + \|\mathbf{R}_0\|^2_{V_\infty} + \int_0^T (G(s), \dot{\mathbf{R}}^G(s))_H ds \\ &\le \|\mathbf{R}_1\|^2_{H_\infty} + \|\mathbf{R}_0\|^2_{V_\infty} + \|\dot{\mathbf{R}}^G\|_{L^\infty(H)} \|\mathbf{G}\|_{L^1(H)}, \end{aligned}$$

hence

$$\|\dot{\mathbf{R}}^G\|^2_{L^\infty(H)} + \|\mathbf{R}^G\|^2_{L^\infty(V_G)} \le C\left(\|\mathbf{R}_1\|^2_{H_\infty} + \|\mathbf{R}_0\|^2_{V_\infty} + \|\mathbf{G}\|^2_{L^1(H)}\right).$$

Therefore

$$\dot{\mathbf{R}}^G \ \text{ is bounded in } L^\infty(H), \tag{5.19}$$

$$\|\mathbf{R}^G\|_{L^\infty(V_G)} \ \text{ is bounded},. \tag{5.20}$$

From (5.19) and assumption (5.16) we may conclude that

$$\mathbf{R}^G \ \text{ is bounded in } W^{1,\infty}(H).$$

Therefore, upon passing to the limit as $G \to \infty$ through some subnet of $G > 0$ we obtain, for some $\mathbf{R} \in W^{1,\infty}(H)$,

$$\begin{cases} \mathbf{R}^G \to \mathbf{R} \ \text{ strongly in } L^\infty(H), \\ \dot{\mathbf{R}}^G \to \dot{\mathbf{R}} \ \text{ weakly* in } L^\infty(H). \end{cases} \tag{5.21}$$

In particular, the map $t \mapsto \mathbf{R}(t) : [0,T] \mapsto H$ is continuous and $\mathbf{R}(0) = \mathbf{R}_0$.

Let us write

$$\begin{aligned} \mathbf{R}^G = (\mathbf{\Psi}^G, \mathbf{\Theta}^G), \ \ \mathbf{\Psi}^G = \mathbf{u}^G + W^G \mathbf{a}_3, \ \ \mathbf{\Theta}^G = \vartheta^G_2 \mathbf{e}_1 - \vartheta^G_1 \mathbf{e}_2 + w^G \mathbf{e}_3, \\ \mathbf{R} = (\mathbf{\Psi}, \mathbf{\Theta}), \ \ \mathbf{\Psi} = \mathbf{u} + W \mathbf{a}_3, \ \ \mathbf{\Theta} = \vartheta_2 \mathbf{e}_1 - \vartheta_1 \mathbf{e}_2 + w \mathbf{e}_3. \end{aligned}$$

From (5.20) and the definition of the V_G norm we have

$$\varepsilon_{\alpha\beta}(\mathbf{u}^G) \text{ is bounded in } L^\infty(L^2(\Omega)), \tag{5.22}$$

$$\sqrt{G_1}|\nabla W^G + \mathbf{u}^G| \text{ is bounded in } L^\infty(L^2(\Omega)), \tag{5.23}$$

$$(\vartheta_2^G)' \text{ is bounded in } L^\infty(L^2(0,\ell)), \tag{5.24}$$

$$\sqrt{G_2}|(w^G)' + \vartheta_2^G| \text{ is bounded in } L^\infty(L^2(0,\ell)), \tag{5.25}$$

$$\sqrt{G_2}(\vartheta_1^G)' \text{ is bounded in } L^\infty(L^2(0,\ell)). \tag{5.26}$$

Since, by assumption, $\Gamma_0 \neq \emptyset$, (5.22) and Korn's Lemma imply that

$$U_\alpha^G \text{ is bounded in } L^\infty(H^1(\Omega)) \tag{5.27}$$

and then, by (5.23),

$$W_{,\alpha}^G \text{ is bounded in } L^\infty(L^2(\Omega)). \tag{5.28}$$

Thus by (5.21), (5.23), (5.27) and (5.28),

$$\begin{gathered} \mathbf{\Psi}^G \text{ is bounded in } L^\infty(\mathcal{H}^1_{\Gamma_0}(\Omega)), \\ |\nabla W^G + \mathbf{u}^G| \to 0 \text{ strongly in } L^\infty(L^2(\Omega)) \end{gathered}$$

and therefore

$$\begin{cases} \mathbf{\Psi}^G \to \mathbf{\Psi} \text{ weakly* in } L^\infty(\mathcal{H}^1_{\Gamma_0}(\Omega)), \\ \mathbf{u} = -\nabla W. \end{cases} \tag{5.29}$$

Similarly, from (5.21), (5.24)–(5.26) we conclude that

$$\begin{gathered} \vartheta_2^G \text{ is bounded in } L^\infty(H^1(0,\ell)), \\ w^G \text{ is bounded in } L^\infty(H^1(0,\ell)), \\ (\vartheta_1^G)' \to 0 \text{ strongly in } L^\infty(L^2(0,\ell)), \\ (w^G)' + \vartheta_2^G \to 0 \text{ strongly in } L^\infty(L^2(0,\ell)), \end{gathered}$$

hence

$$\begin{cases} \mathbf{\Theta}^G \to \mathbf{\Theta} \text{ weakly* in } L^\infty(\mathcal{H}^1(0,\ell)), \\ \vartheta_2 = -w', \quad \vartheta_1' = 0. \end{cases} \tag{5.30}$$

It follows from (5.21), (5.29) and (5.30) that $\mathbf{R} \in L^\infty(V_\infty) \cap C(H_\infty)$ and $\dot{\mathbf{R}} \in L^\infty(H_\infty)$.

The remainder of the proof is *identical* to the last part of the proof of Theorem 6.1, Chapter VII, beginning with (6.39), and is omitted. □

5.1. Interpretation of the Limit Model. It remains to *interpret* (5.17), (5.18). Equation (5.17) may be written

$$(5.31)\qquad \langle \ddot{\mathbf{R}}, \hat{\mathbf{R}} \rangle_{V_\infty} + (\mathbf{R}, \hat{\mathbf{R}})_{V_\infty} = (\mathbf{G}_\infty, \hat{\mathbf{R}})_{H_\infty}, \quad \forall \hat{\mathbf{R}} \in V_\infty.$$

Let us calculate each of the three terms in (5.31). One has, at least formally,

$$(5.32)\quad \begin{aligned} \langle \ddot{\mathbf{R}}, \hat{\mathbf{R}} \rangle_{V_\infty} &= \int_\Omega (m_1 \ddot{W}\hat{W} + I_{\rho_1} \nabla \ddot{W} \cdot \nabla \hat{W})\, d\Omega \\ &\quad + \int_0^\ell (m_2 \ddot{w}\hat{w} + \rho_2 I_h \ddot{w}'\hat{w}')\, d\xi + I\ell \ddot{\vartheta}_1 \hat{\vartheta}_1 \\ &= \int_\Omega (m_1 \ddot{W} - I_{\rho_1} \Delta \ddot{W})\hat{W}\, d\Omega + \int_{\Gamma\setminus\Gamma_0} I_{\rho_1} \hat{W} \frac{\partial \ddot{W}}{\partial \boldsymbol{\nu}}\, d\Gamma \\ &\quad + \int_0^\ell (m_2 \ddot{w} - \rho_2 I_h \ddot{w}'')\hat{w}\, d\xi + \rho_2 I_h [\ddot{w}'\hat{w}]_0^\ell + I\ell \ddot{\vartheta}_1 \hat{\vartheta}. \end{aligned}$$

Next, we calculate

$$(\mathbf{R}, \hat{\mathbf{R}})_{V_\infty} = \int_\Omega (C[\varepsilon(\nabla W)], \varepsilon(\nabla \hat{W}))\, d\Omega + EI_h \int_0^\ell w''\hat{w}''d\xi.$$

The first term on the right was evaluated in Chapter VI (see (6.12)) with the result

$$\begin{aligned} \int_\Omega (C[\varepsilon(\nabla W)], \varepsilon(\nabla \hat{W}))\, d\Omega &= D_1 \bigg\{ \int_\Omega \hat{W} \Delta^2 W\, d\Omega \\ &\quad - \int_{\Gamma\setminus\Gamma_0} \Big[\hat{W} \frac{\partial \Delta W}{\partial \boldsymbol{\nu}} - \nabla \hat{W} \cdot (C[\varepsilon(\nabla W)]\boldsymbol{\nu}) \Big]\, d\Gamma \bigg\} \\ &= D_1 \bigg\{ \int_\Omega \hat{W} \Delta^2 W\, d\Omega + \int_{\Gamma\setminus\Gamma_0} \frac{\partial \hat{W}}{\partial \boldsymbol{\nu}} C_\nu(\nabla W)\, d\Gamma \\ &\quad - \int_{\Gamma\setminus\Gamma_0} \hat{W} \Big[\frac{\partial \Delta W}{\partial \boldsymbol{\nu}} + \frac{\partial}{\partial \boldsymbol{\tau}} C_\tau(\nabla W) \Big]\, d\Gamma \bigg\}, \end{aligned}$$

where $D_1 = (2\mu_1 + \lambda_1)h^3/12$ and

$$C_\tau(\nabla W) = \boldsymbol{\tau} \cdot (C[\varepsilon(\nabla W)]\boldsymbol{\nu}), \quad C_\nu(\nabla W) = \boldsymbol{\nu} \cdot (C[\varepsilon(\nabla W)]\boldsymbol{\nu}).$$

Thus

$$(5.33)\quad (\mathbf{R},\hat{\mathbf{R}})_{V_\infty} = D_1\left\{\int_\Omega \hat{W}\Delta^2 W\,d\Omega + \int_{\Gamma\setminus\Gamma_0} \frac{\partial\hat{W}}{\partial\boldsymbol{\nu}} C_\nu(\nabla W)\,d\Gamma - \int_{\Gamma\setminus\Gamma_0} \hat{W}\left[\frac{\partial\Delta W}{\partial\boldsymbol{\nu}} + \frac{\partial}{\partial\boldsymbol{\tau}}C_\tau(\nabla W)\right]d\Gamma\right\} + EI_h\left\{\int_0^\ell w''''\hat{w}\,d\xi + [w''\hat{w}' - w'''\hat{w}]_0^\ell\right\}.$$

Finally, let us calculate, for $\hat{\mathbf{R}} \in H_\infty$,

$$(5.34)\quad (\mathbf{G}_\infty,\hat{\mathbf{R}})_{H_\infty} = (\mathbf{G},\hat{\mathbf{R}})_H = \int_\Omega (F_1\hat{W} - \mathbf{M}_1\cdot\nabla\hat{W})\,d\Omega + \int_0^\ell (F_2\hat{w} - M_{22}\hat{w}' + M_{21}\hat{\vartheta}_1)\,d\xi$$
$$= \int_\Omega (F_1 + \operatorname{div}\mathbf{M}_1)\hat{W}\,d\Omega + \int_0^\ell (F_2 + M_{22}')\hat{w}\,d\xi - \int_{\Gamma\setminus\Gamma_0} (\boldsymbol{\nu}\cdot\mathbf{M}_1)\hat{W}\,d\Gamma + \hat{\vartheta}_1\int_0^\ell M_{21}\,d\xi - [M_{22}\hat{w}]_0^\ell.$$

It follows from (5.31)–(5.34) that W, w, ϑ_1 satisfy the following system of equations, boundary and junction conditions.

$$(5.35)\quad m_1\ddot{W} - I_{\rho_1}\Delta\ddot{W} + D_1\Delta^2 W = F_1 - \operatorname{div}\mathbf{M}_1 \text{ in } \Omega;$$

$$(5.36)\quad W = \frac{\partial W}{\partial\boldsymbol{\nu}} = 0 \text{ on } \Gamma_0;$$

$$(5.37)\quad \begin{cases} I_{\rho_1}\dfrac{\partial\ddot{W}}{\partial\boldsymbol{\nu}} - D_1\left[\dfrac{\partial\Delta W}{\partial\boldsymbol{\nu}} + \dfrac{\partial}{\partial\boldsymbol{\tau}}C_\tau(\nabla W)\right] = -\boldsymbol{\nu}\cdot\mathbf{M}_1, \\ C_\nu(\nabla W) = 0 \text{ on } \Gamma_1; \end{cases}$$

$$(5.38)\quad m_2\ddot{w} - \rho_2 I_h\ddot{w}'' + EI_h w'''' = F_2 + M_{22}' \text{ in } (0,\ell);$$

$$(5.39)\quad \begin{cases} w''(\ell,\cdot) = 0, \\ (\rho_2 I_h\ddot{w}' - EI_h w''')(\ell,\cdot) = -M_{22}(\ell,\cdot); \end{cases}$$

$$(5.40)\quad W = w(0,\cdot) + \xi_2\vartheta_1(\cdot),\quad \frac{\partial W}{\partial\boldsymbol{\nu}} = w'(0,\cdot) \text{ on } J_a.$$

There is, in addition, the variational junction condition

$$\int_{J_a} I_{\rho_1} \hat{W} \frac{\partial \ddot{W}}{\partial \nu}\, d\Gamma - D_1 \int_{J_a} \hat{W} \left[\frac{\partial \Delta W}{\partial \nu} + \frac{\partial}{\partial \tau} C_\tau(\nabla W) \right] d\Gamma$$
$$+ D_1 \int_{J_a} \frac{\partial \hat{W}}{\partial \nu} C_\nu(\nabla W)\, d\Gamma - \rho_2 I_h \ddot{w}'(0)\hat{w}(0) + I\ell \ddot{\vartheta}_1 \hat{\vartheta}$$
$$+ EI_h[-w''(0)\hat{w}'(0) + w'''(0)\hat{w}(0)]$$
$$= -\int_{J_a} (\boldsymbol{\nu} \cdot \mathbf{M}_1)\hat{W}\, d\Gamma + \hat{\vartheta}_1 \int_0^\ell M_{21}\, d\xi + M_{22}(0)\hat{w}(0), \quad \forall \hat{\mathbf{R}} \in V_\infty.$$

The test functions $\hat{\mathbf{R}}$ satisfy

$$\hat{W} = \hat{w}(0) + \xi_2 \hat{\vartheta}_1, \quad \frac{\partial \hat{W}}{\partial \nu} = \hat{w}'(0) \text{ on } J_a,$$

where $\hat{w}(0), \hat{w}'(0)$ and $\hat{\vartheta}_1$ are arbitrary real numbers. We may therefore deduce from the last relation the following dynamic junction conditions.

$$\rho_2 I_h \ddot{w}'(0,\cdot) - EI_h w''(0,\cdot) + M_{22}(0,\cdot)$$
$$= \int_{J_a} \left\{ I_{\rho_1} \frac{\partial \ddot{W}}{\partial \nu} - D_1 \left[\frac{\partial \Delta W}{\partial \nu} + \frac{\partial}{\partial \tau} C_\tau(\nabla W) \right] \right\} d\Gamma; \tag{5.41}$$

$$EI_h w''(0,\cdot) = D_1 \int_{J_a} C_\nu(\nabla W)\, d\Gamma; \tag{5.42}$$

$$I\ell \ddot{\vartheta}_1(\cdot) - \int_0^\ell M_{21}(\xi,\cdot)\, d\xi$$
$$= \int_{J_a} \xi_2 \left\{ I_{\rho_1} \frac{\partial \ddot{W}}{\partial \nu} - D_1 \left[\frac{\partial \Delta W}{\partial \nu} + \frac{\partial}{\partial \tau} C_\tau(\nabla W) \right] \right\} d\Gamma. \tag{5.43}$$

The complete system of equations, boundary and junction conditions is (5.35)–(5.43). It is a type of *hybrid* system in that it couples partial differential equations for w, W with an ordinary differential equation for ϑ_1.

Under assumption (5.16), the initial conditions (5.18) reduce to

$$W|_{t=0} = W^0, \quad \dot{W}|_{t=0} = W^1 \text{ in } \Omega,$$
$$w|_{t=0} = w^0, \quad \dot{w}|_{t=0} = w^1 \text{ in } (0,\ell),$$
$$\vartheta_1|_{t=0} = \vartheta_1^0 \equiv \text{const.}, \quad \dot{\vartheta}_1|_{t=0} = \vartheta_1^1 \equiv \text{const.},$$

where $W^0 \in H^2_{\Gamma_0}(\Omega)$, $W^1 \in H^1_{\Gamma_0}(\Omega)$, $w^0 \in H^2(0,\ell)$, $w^1 \in H^1(0,\ell)$ and

$$W^0 = w^0(0) + \xi_2 \vartheta_1, \quad \frac{\partial W^0}{\partial \nu} = (w^0)'(0) \text{ on } J_a.$$

Since ϑ_1^0 and ϑ_1^1 are constants, it follows that ϑ_1 is independent of ξ and amounts to a (time dependent) rigid rotation of the beam around its axis (caused by the deflection of the plate). The relation (5.41) may be recognized as a balance law for linear momentum in the $\mathbf{a}_3$ direction along J_a, (5.42) as a balance law for angular momentum about $\boldsymbol{\tau}(\mathbf{x}_0)$ and (5.43) as a balance law for angular momentum about the axis of the beam. Of course, ϑ_1 may be eliminated from the system by integrating (5.43) and substituting for ϑ_1 in (5.40), but this would make the physical interpretation of the junction condition (5.40) less transparent.

6. Modeling of a Plate-Beam Junction: II

6.1. Geometric Conditions. Let Ω and ω be bounded, open, connected sets in $\mathbb{R}^2$ with Lipschitz continuous boundaries consisting of a finite number of smooth simple curves. We further assume that ω is simply connected, that $\mathbf{0} \in \omega$ and that ω is doubly symmetric with respect to $\mathbf{0}$. We consider Ω to be the reference surface of a thin plate of uniform thickness h. Points within the plate will be denoted by coordinates (x_1, x_2, x_3) with respect to the natural basis $\mathbf{a}_1, \mathbf{a}_2, \mathbf{a}_3$ for $\mathbb{R}^3$. Thus the equilibrium position of the plate is given by

$$\{x_i \mathbf{a}_i \,|\, (x_1, x_2) \in \overline{\Omega},\ |x_3| \le h/2\}.$$

where summation convention for repeated indices has been assumed as in Section 2.

Let $\mathbf{x}_0$ be a point of $\mathbb{R}^3$ and $\mathbf{e}_1, \mathbf{e}_2, \mathbf{e}_3$ be a right-handed orthonormal basis for $\mathbb{R}^3$. The beam, in its reference configuration, is given by

$$\{\mathbf{x}_0 + \xi_i \mathbf{e}_i \,|\, 0 \le \xi_1 \le \ell,\ (\xi_2, \xi_3) \in \overline{\omega}\}.$$

The following assumptions are made:

(A1) $\overline{\omega} \subset \Omega$.
(A2) $\mathbf{x}_0 = (h/2)\mathbf{a}_3$, $\mathbf{e}_1 = \mathbf{a}_3$, $\mathbf{e}_2 = \mathbf{a}_1$, $\mathbf{e}_3 = \mathbf{a}_2$.

Thus the beam is attached to and orthogonal to the upper face of the plate and ω is the junction region between the plate and the beam.

As in Section 2, we denote by $\mathbf{R}(x_1, x_2, x_3, t)$ the *position vector* at time t to the deformed position of the material particle in the plate which is located at (x_1, x_2, x_3) when the plate is in equilibrium, by $\mathbf{W}(x_1, x_2, t)$ the *displacement vector* of the material particle which occupies position $(x_1, x_2, 0)$ in the reference surface in equilibrium, by $\mathbf{r}(\xi_1, \xi_2, \xi_3, t)$ the position vector to the deformed position of the particle originally at position $\mathbf{x}_0 + \xi_i \mathbf{e}_i$ of the beam in equilibrium, and $\mathbf{w}(\xi_1, t)$ the displacement vector of the point $\mathbf{x}_0 + \xi_1 \mathbf{e}_1$ on the centerline of the beam. We further retain the kinematic hypotheses (2.1) and (2.2) of Section 2.

As a *geometric junction condition* we impose the continuity condition

$$\mathbf{R}(x_1, x_2, h/2, t) = \mathbf{r}(0, x_1, x_2, t), \quad (x_1, x_2) \in \omega,\ t > 0. \tag{6.1}$$

Since $\mathbf{a}_1 = \mathbf{e}_2$ and $\mathbf{a}_2 = \mathbf{e}_3$, (6.1) is the same as

$$\text{(6.2)}\quad \mathbf{W}(x_1, x_2, t) + \frac{h}{2}\mathbf{U}(x_1, x_2, t) = \mathbf{w}(0,t) \\ + x_1(\mathbf{E}_2(0,t) - \mathbf{e}_2) + x_2(\mathbf{E}_3(0,t) - \mathbf{e}_3), \quad (x_1, x_2) \in \omega,\ t > 0.$$

In terms of components, the last relation is

$$W_1(x_1, x_2, t) + \frac{h}{2}U_1(x_1, x_2, t) = w_2(0,t) \\ + x_1(\cos\vartheta_1 \cos\vartheta_3 + \sin\vartheta_1 \sin\vartheta_2 \sin\vartheta_3)(0,t) \\ + x_2(\cos\vartheta_1 \sin\vartheta_2 \sin\vartheta_3 - \sin\vartheta_1 \cos\vartheta_3)(0,t),$$

$$W_2(x_1, x_2, t) + \frac{h}{2}U_2(x_1, x_2, t) = w_3(0,t) \\ + x_1(\sin\vartheta_1 \cos\vartheta_3)(0,t) + x_2(\cos\vartheta_1 \cos\vartheta_2 - 1)(0,t),$$

and

$$W_3(x_1, x_2, t) + \frac{h}{2}U_3(x_1, x_2, t) = w_1(0,t) \\ + x_1(\sin\vartheta_1 \sin\vartheta_2 \cos\vartheta_3 - \cos\vartheta_1 \sin\vartheta_3)(0,t) \\ + x_2(\sin\vartheta_1 \sin\vartheta_3 + \cos\vartheta_1 \sin\vartheta_2 \cos\vartheta_3)(0,t),$$

where

$$\mathbf{W} = W_i \mathbf{a}_i, \quad \mathbf{U} = U_i \mathbf{a}_i, \quad , \mathbf{w} = w_i \mathbf{e}_i.$$

When linearized, the above scalar relations reduce to

$$\text{(6.3)}\quad \begin{cases} W_1(x_1, x_2, t) + \dfrac{h}{2}U_1(x_1, x_2, t) = w_2(0,t) - x_2\vartheta_1(0,t), \\ W_2(x_1, x_2, t) + \dfrac{h}{2}U_2(x_1, x_2, t) = w_3(0,t) + x_1\vartheta_1(0,t), \\ W_3(x_1, x_2, t) = w_1(0,t) - x_1\vartheta_3(0,t) + x_2\vartheta_2(0,t). \end{cases}$$

We note that in (6.3) the quantities $W_3, w_1, \vartheta_2, \vartheta_3$ are not coupled to $W_\alpha, U_\alpha, w_2, w_3, \vartheta_1$. However, all quantities will turn out to be coupled once the dynamic conditions are taken into account. We shall take the linearized expressions (6.3) as the geometric junction conditions of our problem.

6.2. Dynamic Conditions. As usual, the dynamic conditions will be obtained from Hamilton's Principle:

$$\text{(6.4)}\quad \delta \int_0^T [\mathcal{K}(t) - \mathcal{U}(t) + \mathcal{W}(t)]\, dt = 0,$$

where $\mathcal{K}$ and $\mathcal{U}$ represent the kinetic and strain energies, respectively, of the plate-beam system, $\mathcal{W}$ is the work done on the system by external forces and δ denotes the first variation with respect to the class of admissible displacements. The latter must obey, in particular, the geometric junction conditions (6.3).

We assume that $\Gamma = \overline{\Gamma}^D \cup \overline{\Gamma}^C$ where Γ^D and Γ^C are relatively open in Γ and mutually disjoint. The part Γ^C corresponds to geometric boundary conditions

$$W = 0, \quad \mathbf{v} = 0, \quad \mathbf{u} = 0 \quad \text{on } \Gamma^C, \tag{6.5}$$

while Γ^D corresponds to mechanical boundary conditions.

We write

$$\mathcal{K} = \mathcal{K}_P + \mathcal{K}_B, \quad \mathcal{U} = \mathcal{U}_P + \mathcal{U}_B, \quad \mathcal{W} = \mathcal{W}_P + \mathcal{W}_B,$$

where $\mathcal{K}_P$ and $\mathcal{K}_B$ are the kinetic energies of the plate and beam, respectively, and similarly for $\mathcal{U}_P, \mathcal{U}_B$ and $\mathcal{W}_P, \mathcal{W}_B$. We assume that the beam is homogeneous and elastically isotropic, that the plate is homogeneous and that the energy of the plate is that associated with linear Reissner-Mindlin plate theory. Thus we set

$$\mathcal{K}_P(t) = \frac{1}{2}\int_\Omega \left(m_1|\dot{\mathbf{W}}|^2 + I_{\rho_1}|\dot{\mathbf{u}}|^2\right) d\Omega, \tag{6.6}$$

$$\mathcal{U}_P(t) = \frac{1}{2}\int_\Omega [(C[\varepsilon(\mathbf{v})], \varepsilon(\mathbf{v})) + \frac{h^2}{12}(C[\varepsilon(\mathbf{u})], \varepsilon(\mathbf{u})) + G_1 h|\nabla W_3 + \mathbf{u}|^2]\, d\Omega, \tag{6.7}$$

where we have set $\mathbf{v} = W_\alpha \mathbf{a}_\alpha$ and where m_1, $I_{\rho_1} = m_1 h^2/12$ and G_1 are as in Section 2,

$$C[\varepsilon] = h[\lambda_1 \varepsilon_{\alpha\alpha}\mathcal{I} + 2\mu_1\varepsilon],$$

$$\varepsilon(\mathbf{u}) = \frac{1}{2}[\nabla\mathbf{u} + (\nabla\mathbf{u})^*],$$

and $(C[\varepsilon], \varepsilon) = C^{\alpha\beta}[\varepsilon]\varepsilon_{\alpha\beta}$. We further have

$$\mathcal{W}_P(t) = \int_\Omega (\mathbf{F}_P \cdot \mathbf{W} + \mathbf{M}_P \cdot \mathbf{U})\, d\Omega + \int_{\Gamma^D} (\mathbf{f}_P \cdot \mathbf{W} + \mathbf{m}_P \cdot \mathbf{U})\, d\Gamma, \tag{6.8}$$

where $\mathbf{F}_P, \mathbf{M}_P, \mathbf{f}_P, \mathbf{m}_P$ are various resultant (in x_3) forces and moments as discussed in Section 2.

The total kinetic and potential energies of the beam are calculated as in Section 2 (cf. Chapter III). One finds that

$$\mathcal{K}_B(t) = \frac{1}{2}\int_0^\ell (m_2|\dot{\mathbf{w}}|^2 + \rho_2 I_2|\dot{\vartheta}_3|^2 + \rho_2 I_3|\dot{\vartheta}_2|^2 + \rho_2 I|\dot{\vartheta}_1|^2)\, d\xi_1, \tag{6.9}$$

where ρ_2 is the mass of the beam per unit of reference volume, $m_2 = \rho_2 A$ is mass of the beam per unit of reference length, A denoting the area of ω, and

$$I_2 = \int_\omega \xi_2^2 \, d\omega, \quad I_3 = \int_\omega \xi_3^2 \, d\omega.$$

Also

$$(6.10)\quad \mathcal{U}_B(t) = \frac{1}{2}\int_0^\ell \{EA|w_1'|^2 + EI_2|\vartheta_3'|^2 + EI_3|\vartheta_2'|^2 + G_2 I|\vartheta_1'|^2 + G_2A|w_2' - \vartheta_3|^2 + G_2A|w_3' + \vartheta_2|^2\}\, d\xi_1,$$

where E and G_2 have the same meaning as in Section 2, and

$$(6.11)\quad \mathcal{W}_B(t) = \int_0^\ell (\mathbf{F}_B \cdot \mathbf{w} - M_{23}\vartheta_1 + M_{13}\vartheta_2 - M_{12}\vartheta_3)(\xi, t)\, d\xi + \mathbf{f}_B \cdot \mathbf{w}(\ell, t) - m_{23}\vartheta_1(\ell, t) + m_{13}\vartheta_2(\ell, t) - m_{12}\vartheta_3(\ell, t),$$

where $\mathbf{F}_B, \mathbf{f}_B, M_{23}, M_{13}, m_{23}, m_{13}$ are various resultant (over ω) forces and moments as defined in Section 2 and

$$M_{12} = \int_0^\ell \xi_2 \mathbf{F}_B \cdot \mathbf{e}_1 \, d\omega, \quad m_{12} = \int_0^\ell \xi_2 \mathbf{f}_B \cdot \mathbf{e}_1 \, d\omega.$$

We may now calculate (6.4), where the variation is taken with respect to displacements satisfying (6.3) and (6.5). In the calculation we utilize the Green's formula (2.17). Standard computations lead to the following dynamic equations.

Equations of motion of the plate: For $(x_1, x_2) \in \Omega \setminus \omega$ and $t > 0$,

$$(6.12)\quad \begin{cases} m_1 \ddot{\mathbf{v}} - \operatorname{div} C[\varepsilon(\mathbf{v})] = (\mathbf{F}_P \cdot \mathbf{a}_\beta)\mathbf{a}_\beta, \\ m_1 \ddot{W}_3 - G_1 h \operatorname{div}(\nabla W_3 + \mathbf{u}) = \mathbf{F}_P \cdot \mathbf{a}_3, \\ I_{\rho_1} \ddot{\mathbf{u}} - \dfrac{h^2}{12} \operatorname{div} C[\varepsilon(\mathbf{u})] + G_1 h(\nabla W_3 + \mathbf{u}) = (\mathbf{M}_P \cdot \mathbf{a}_\beta)\mathbf{a}_\beta. \end{cases}$$

Equations of motion of the beam: For $\xi_1 \in (0, \ell)$ and $t > 0$,

$$(6.13)\quad \begin{cases} m_2 \ddot{w}_1 - EAw_1'' = \mathbf{f}_B \cdot \mathbf{e}_1, \\ m_2 \ddot{w}_2 - G_2A(w_2' - \vartheta_3)' = \mathbf{F}_B \cdot \mathbf{e}_2, \\ m_2 \ddot{w}_3 - G_2A(w_3' + \vartheta_2)' = \mathbf{F}_B \cdot \mathbf{e}_3, \\ \rho_2 I \ddot{\vartheta}_1 - G_2 I \vartheta_1'' = -M_{23}, \\ \rho_2 I_3 \ddot{\vartheta}_2 - EI_3 \vartheta_2'' + G_2A(w_3' + \vartheta_2) = M_{13}, \\ \rho_2 I_2 \ddot{\vartheta}_3 - EI_2 \vartheta_2'' - G_2A(w_2' - \vartheta_3) = -M_{12}. \end{cases}$$

Boundary conditions along Γ^D, $t > 0$:

$$\begin{cases} C[\varepsilon(\mathbf{v})]\boldsymbol{\nu} = (\mathbf{f}_P \cdot \mathbf{a}_\beta)\mathbf{a}_\beta, \\ G_1 h(\nabla W_3 + \mathbf{u}) \cdot \boldsymbol{\nu} = \mathbf{f}_P \cdot \mathbf{a}_3, \\ \dfrac{h^2}{12} C[\varepsilon(\mathbf{u})]\boldsymbol{\nu} = (\mathbf{m}_P \cdot \mathbf{a}_\beta)\mathbf{a}_\beta. \end{cases} \tag{6.14}$$

Boundary conditions at $\xi = \ell$, $t > 0$:

$$\begin{cases} EAw_1' = \mathbf{f}_B \cdot \mathbf{e}_1, \\ G_2 A(w_2' - \vartheta_3) = \mathbf{f}_B \cdot \mathbf{e}_2, \\ G_2 A(w_3' + \vartheta_2) = \mathbf{f}_B \cdot \mathbf{e}_3, \\ G_2 I \vartheta_1' = -m_{23}, \\ EI_3 \vartheta_2' = m_{13}, \\ EI_2 \vartheta_3' = -m_{12}. \end{cases} \tag{6.15}$$

There is, in addition, the variational junction condition

$$\begin{aligned} 0 = EAw_1'(0)\hat{w}_1(0) &+ G_2 A(w_2' - \vartheta_3)(0)\hat{w}_2(0) + G_2 A(w_3' + \vartheta_2)(0)\hat{w}_3(0) \\ &+ G_2 I \vartheta_1'(0)\hat{\vartheta}_1(0) + EI_3 \vartheta_2'(0)\hat{\vartheta}_2(0) + EI_2 \vartheta_3'(0)\hat{\vartheta}_3(0) \\ &+ \int_\omega \{\hat{\mathbf{v}} \cdot \operatorname{div} C[\varepsilon(\mathbf{v})] + \frac{h^2}{12}\hat{\mathbf{u}} \cdot \operatorname{div} C[\varepsilon(\mathbf{u})]\}\, d\omega \\ &+ \int_\omega \{G_1 h \operatorname{div}(\nabla W_3 + \mathbf{u})\hat{W}_3 - G_1 h(\nabla W_3 + \mathbf{u}) \cdot \hat{\mathbf{u}}\}\, d\omega \\ &- \int_\omega (m_1 \ddot{\mathbf{W}} \cdot \hat{\mathbf{W}} + I_{\rho_1} \ddot{\mathbf{u}} \cdot \hat{\mathbf{u}})\, d\omega \end{aligned}$$

for all test functions $\hat{\mathbf{W}}, \hat{\mathbf{u}}, \hat{\mathbf{w}}, \hat{\vartheta}_i$ which satisfy (6.3). Because of the decoupling of $\hat{W}_3, \hat{w}_1, \hat{\vartheta}_2, \hat{\vartheta}_3$ from the remaining variables in (6.3), the last equation is equivalent to the two variational equations

$$\begin{aligned} 0 = &\int_\omega \{G_1 h \operatorname{div}(\nabla W_3 + \mathbf{u}) - m_1 \ddot{W}_3\}\hat{W}_3\, d\omega \\ &+ EAw_1'(0)\hat{w}_1(0) + EI_3 \vartheta_2'(0)\hat{\vartheta}_2(0) + EI_2 \vartheta_3'(0)\hat{\vartheta}_3(0) \end{aligned} \tag{6.16}$$

and

$$
\begin{aligned}
(6.17)\quad 0 = {} & G_2A(w_2' - \vartheta_3)(0)\hat{w}_2(0) \\
& + G_2A(w_3' + \vartheta_2)(0)\hat{w}_3(0) + G_2I\vartheta_1'(0)\hat{\vartheta}_1(0) \\
& + \int_\omega \{\frac{h^2}{12} \operatorname{div} C[\varepsilon(\mathbf{u})] - G_1h(\nabla W_3 + \mathbf{u}) - I_{\rho_1}\ddot{u}\} \cdot \hat{\mathbf{u}}\, d\omega \\
& + \int_\omega \{\operatorname{div} C[\varepsilon(\mathbf{v})] - m_1\ddot{u}\} \cdot \hat{\mathbf{v}}\, d\omega.
\end{aligned}
$$

By utilizing the last of the junction conditions in (6.3) in (6.16) we obtain the dynamic junction conditions

$$
(6.18)\quad
\begin{cases}
\int_\omega \{m_1\ddot{W}_3 - G_1h \operatorname{div}(\nabla W_3 + \mathbf{u})\}\, d\omega = EAw_1'(0), \\
\int_\omega x_2\{m_1\ddot{W}_3 - G_1h \operatorname{div}(\nabla W_3 + \mathbf{u})\}\, d\omega = EI_3\vartheta_2'(0), \\
\int_\omega x_1\{m_1\ddot{W}_3 - G_1h \operatorname{div}(\nabla W_3 + \mathbf{u})\}\, d\omega = EI_2\vartheta_3'(0).
\end{cases}
$$

These three equations may be recognized as a balance law for linear momentum in the $\mathbf{a}_3$ direction, a balance law for angular momentum around the $\mathbf{a}_2$ axis and a balance law for angular momentum around the $\mathbf{a}_1$ axis, respectively.

In the variational equation (6.17), we may set

$$
\hat{\mathbf{v}} = -\frac{h}{2}\hat{\mathbf{u}} + \hat{w}_2(0)\mathbf{a}_1 + \hat{w}_3(0)\mathbf{a}_2 + (x_1\mathbf{a}_2 - x_2\mathbf{a}_1)\hat{\vartheta}_1(0)
$$

in view of (6.3). We thereby obtain

$$
\begin{aligned}
0 = {} & G_2A(w_2' - \vartheta_3)(0)\hat{w}_2(0) \\
& + G_2A(w_3' + \vartheta_2)(0)\hat{w}_3(0) + G_2I\vartheta_1'(0)\hat{\vartheta}_1(0) \\
& + \frac{h}{2}\int_\omega \{m_1\ddot{\mathbf{v}} - \operatorname{div} C[\varepsilon(\mathbf{v})]\} \cdot \hat{\mathbf{u}}\, d\omega \\
& - \int_\omega \{I_{\rho_1}\ddot{\mathbf{u}} - \frac{h^2}{12} \operatorname{div} C[\varepsilon(\mathbf{u})] + G_1h(\nabla W_3 + \mathbf{u})\} \cdot \hat{\mathbf{u}}\, d\omega \\
& - \int_\omega \{m_1\ddot{\mathbf{v}} - \operatorname{div} C[\varepsilon(\mathbf{v})]\} \cdot \{\hat{w}_2(0)\mathbf{a}_1 + \hat{w}_3(0)\mathbf{a}_2 + (x_1\mathbf{a}_2 - x_2\mathbf{a}_1)\hat{\vartheta}_1(0)\}\, d\omega,
\end{aligned}
$$

where $\hat{\mathbf{u}}, \hat{w}_2(0), \hat{w}_3(0), \hat{\vartheta}_1(0)$ are arbitrary. Therefore

$$(6.19)\quad \begin{cases} \displaystyle\int_\omega (m_1\ddot{W}_1 - \mathbf{a}_1 \cdot \operatorname{div} C[\varepsilon(\mathbf{v})])\, d\omega = G_2 A(w_2' - \vartheta_3)(0), \\ \displaystyle\int_\omega (m_1\ddot{W}_2 - \mathbf{a}_2 \cdot \operatorname{div} C[\varepsilon(\mathbf{v})])\, d\omega = G_2 A(w_3' + \vartheta_2)(0), \\ \displaystyle\int_\omega (x_1\mathbf{a}_2 - x_2\mathbf{a}_1) \cdot (m_1\ddot{\mathbf{v}} - \cdot \operatorname{div} C[\varepsilon(\mathbf{v})])\, d\omega = G_2 I\vartheta_1'(0), \end{cases}$$

and

$$(6.20)\quad \int_\omega \{I_{\rho_1}\ddot{\mathbf{u}} - \frac{h^2}{12}\operatorname{div} C[\varepsilon(\mathbf{u})] + G_1 h(\nabla W_3 + \mathbf{u})\}\, d\omega$$
$$= \frac{h}{2}\int_\omega \{m_1\ddot{\mathbf{v}} - \operatorname{div} C[\varepsilon(\mathbf{v})]\}\, d\omega$$
$$= \frac{h}{2} G_2 A[(w_2' - \vartheta_3)\mathbf{a}_1 + (w_3' + \vartheta_2)\mathbf{a}_2](0).$$

The first two relations in (6.19) are balance laws of linear momentum in the $\mathbf{a}_1$ and $\mathbf{a}_2$ directions, respectively, and the third is a balance law for angular momentum around $\mathbf{a}_3$. The right side of (6.20) is a torque applied to the reference surface of the plate (a force in the $\mathbf{a}_1\mathbf{a}_2$ plane at $\xi_1 = 0$ acting through the distance $h/2$) caused by the rotation of the cross-section of the beam at $\xi_1 = 0$. Equation (6.20) matches this torque with the one applied at the base of the beam by the rotation of the plate filaments $\{x_3\mathbf{a}_3 \,|\, |x_3| \le h/2,\ (x_1, x_2) \in \omega\}$.

The complete model of the plate-beam configuration under consideration therefore consists of (6.3), (6.5), (6.12)–(6.15) and (6.18)–(6.20).

Remark 6.1. Although we shall not provide details, since the above model is derived from Hamilton's principle it is fairly obvious that it has a natural variational formulation (the Principle of Virtual Work) which, under reasonable assumptions on the external loads, comprises a well-posed problem in an appropriate "energy" space (a product of H^1 spaces such that the geometric junction and boundary conditions are respected). We have not investigated the reachability problem or the stabilizability problem for this system but it is to be expected that the system may be exactly controlled, or uniformly stabilized, through the action of appropriate controls acting on Γ^D alone, provided the geometry of the triple $(\Omega, \Gamma^C, \Gamma^D)$ is suitably restricted.

Remark 6.2. The above model has eleven degrees of freedom. As in section 5, a model with fewer degrees of freedom may be obtained by passing to the limit as G_1, G_2 go to infinity. The limit model will have seven degrees of freedom: w_i, W_i and ϑ_1, where ϑ_1 is spatially constant.

Bibliography

[1] S. S. Antman, *The theory of rods*, Handbuch der Physik Vol. 6/2a (Berlin) (Flügge, ed.), Springer-Verlag, Berlin, 1972, pp. 643 –703.

[2] ———, *The equations for large vibrations of strings*, Math Monthly **87** (1980), 359–370.

[3] S. S. Antman and Arje Nachman, *Large buckled states of rotating rods*, Nonlinear Analysis: Theory, Methods and Appl. **4** (1980), 303–327.

[4] S. S. Antman and W. H. Warner, *Dynamical theory of hyperelastic rods*, Arch. Rational Mech. Anal. **40** (1966), 135–162.

[5] J. Bailleul and M. Levi, *Rotational elastic dynamics*, Pfysica D **27** (1987), 43–62.

[6] ———, *Constrained relative motions in rotational mechanics*, Arch. Rational Mech. Anal. **115** (1991), 101–135.

[7] A. V. Balakrishnan, *On the Wittrick–Williams algorithm for modes of flexible strucutres*, to appear.

[8] V. Brasey and E. Hairer, *Half-explicit Runge-Kutta methods for differential-algebraic systems of index 2*, SIAM J. Numer. Anal. **30** (1993), no. 2, 538–552.

[9] H. Bremer, *Dynamik und Regelung mechanischer Systeme*, Teubner Studienbücher Mechanik, vol. 67, B.G. Teubner, Stuttgart, 1988.

[10] J. A. C. Bresse, *Cours de mechanique applique*, Mallet Bachellier, 1859.

[11] H. Brezis, *Operateurs maximaux monotones*, North-Holland/American Elsevier, Amsterdam, London, New York, 1973.

[12] H. Cabannes, *Mouvement d'une corde vibrante soumise à un fottement solide*, C. R. Acad. Sc. Paris **287** (1978), 671–673.

[13] ———, *Propagation des discontinuités dans les cordes vibrante soumises à un frottement solide*, C. R. Acad. Sc. Paris **289** (1979), 127–130.

[14] G. F. Carrier, *On the non-linear vibration problem of the elastic string*, Quart. Appl. Math. **2** (1945), 157–165.

[15] G. Chen, M. Coleman, and H.H. West, *Pointwise stabilization in the middle of the span for second order systems, nonuniform and uniform exponential decay of solutions*, SIAM J. Control Optim. **47** (1987), no. 4, 751–779.

[16] G. Chen and D. L. Russell, *A mathematical model for linear elastic systems with structural damping*, Quartely Appl. Math **39** (1982), 433–455.

[17] P. G. Ciarlet, H. Le Dret, and R. Nzengwa, *Junctions between three-dimensional and two-dimensional linearly elastic structures*, J. Math. Pures Appl. **68** (1989), 261–295.

[18] Ph. G. Ciarlet, *Mathematical Elasticity, Vol.I: Three-Dimensional Elasticity*, North–Holland Elsevier Science Publ., 1988.

[19] ———, *Plates and junctions in elastic multi–structures: An asymptotic analysis*, Masson and Springer– Verlag, 1990.

[20] F. Conrad and M. Pierre, *Stabilization of second order evolution equations by unbounded nonlinear feedback*, Tech. Report RR 1679, INRIA, 1992.

[21] R. F Curtain and A. J. Pritchard, *Infinite dimensional system theory*, Lect. Notes in Control and Inf. Sc., vol. 8, Springer-Verlag, 1978.

[22] D. A. Danielson and D. H. Hodges, *A beam theory for large global rotation, moderate local rotation and small strain*, J. Appl. Mech. **55** (1988), 179–185.

[23] R. Datko, *Not all feedback stabilized hyperbolic systems aer robust with respect to small time delays in their feedbacks*, SIAM J. Control Optim. **26** (1988), 697–713.

[24] ———, *A rank-one perturbation result on the spectra of certain operators*, Proceedings of the Royal Society of Edinburgh **111A** (1989), 337–346.

[25] R. Datko, J. E. Lagnese, and M. P. Polis, *An example of the effect of time delays in boundary feedback stabilization of wave equations*, SIAM J. Control Optim. **24** (1986), 152–156.

[26] M. C. Delfour, M. Kern, L. Passeron, and B. Sevenne, *Modelling of a rotating flexible beam*, Control of distributed parameter systems, 4th IFAC Symposium, Los Angeles 1986 (H. E. Rauch, ed.), Pergamon Press, 1986, pp. 383–387.

[27] Ph. Dystunder, *Remarks on substructuring*, Eur. J. Mech. **8** (1989), 201–218.

[28] I. Ekeland and T. Temam, *Convex analysis and variational problems*, North-Holland/American Elsevier, Amsterdam, Oxford, New York, 1976.

[29] A. E. Green, *The equilibrium of rods*, Arch. Rational Mech. Anal. **3** (1959), 417–421.

[30] P. Grisvard, *Elliptic problems in nonsmooth domains*, Pitman, London, 1985.

[31] ———, *Singularités en élasticité*, J. Math. Pures Appl. **107** (1989), 157–180.

[32] I. Gruais, *Modélisation de la jonction entre une poutre et une plaque*, Ph.D. thesis, Université Pierre et Marie Curie, 1990.

[33] E. Hairer and G. Wanner, *Solving ordinary differential equations ii. stiff and differential algebraic systems*, Springer-Verlag, Berlin, 1987.

[34] S. Hanson and E. Zuazua, *Exact controllability and stabilization of a vibrating string with an interior point mass*, Tech. Report 1140, Institute for Mathematics and its Applications, University of Minnesota, 1993.

[35] A. Haraux, *Comportement à l'infini pour une equation d'ondes non linéaire dissipative*, C. R. Acad. Sc. Paris **287** (1978), 507–509.

[36] ———, *Nonlinear evolution equations - global behavior of solutions*, Lect. Notes in Mathematics, vol. 841, Springer-Verlag, Berlin, 1981.

[37] Lop Fat Ho, *Exact controllability of second order hyperbolic systems with control in the dirichlet boundary conditions*, J. Math. Pures Appl. **66** (1987), 363–368.

[38] S. M. Howard, *Transient stress waves in trusses and frames*, Ph.D. thesis, Cornell University, 1990.

[39] T. J. R. Hughes, *The finite element method. Linear static and dynamic finite element analysis*, Prentice-Hall, New Yersey, 1987.

[40] G.Leugering J.E. Lagnese and E.J.P.G. Schmidt, *On the analysis and control of hyperbolic systems associated with vibrating networks*, Proc. Royal Soc. Edinburgh, to appear.

[41] T. R. Kane, R.R. Ryan, and A. K. Barnerjee, *Dynamics of a beam attached to a moving base*, AIAA J.Guidance, Control and Dynamics **10** (1987), 139–151.

[42] F. Khorrami, *Analysis of multi–link flexible manipulators via asymptotic expansions*, Proceedings of the 28th Conference on Decision and Control, Tampa,Fl, 1989 (L. Shaw, ed.), IEEE, 1989, pp. 2089–2094.

[43] V. Komornik, *A mathematics paper*, J. Math. Pures Appl. **115** (1987), 1–18.

[44] J. E. Lagnese, *The Hilbert uniqueness method: A retrospective*, Optimal Control of Partial Differential Equations (Berlin) (K. H. Hoffmann and W.Krabs, eds.), Springer-Verlag, Berlin, 1991, pp. 158–181.

[45] ———, *Uniform asymptotic energy estimates for solutions of the equations of dynamic plane elasticity with nonlinear dissipation at the boundary*, Nonlinear Analysis: Theory, Methods and Appl. **16** (1991), 35–54.

[46] ———, *Boundary controllability in transmission problems for thin plates*, Differential Equations, Dynamical Systems and Control Science: A Festschrift in Honor of Lawrence Markus (New York) (K. D. Elworthy, W. N. Everitt, and E. B. Lee, eds.), Marcel Dekker, New York, 1993.

[47] J. E. Lagnese and G. Leugering, *Uniform stabilization of a nonlinear beam by nonlinear boundary feedback*, J. Diff. Eqns. **91** (1991), 355–388.

[48] ———, *Modelling of dynamic networks of thin elastic plates*, Math. Methods Appl. Sci. **16** (1993), 379–407.

[49] J. E. Lagnese, G. Leugering, and E. J. P. G. Schmidt, *Controllability of planar networks of Timoshenko beams*, SIAM J. Control Optim. **31** (1993), 780– 811.

[50] ———, *Modelling of dynamic networks of thin thermoelastic beams*, Math. Methods. Appl. Sci. **16** (1993), 327–358.

[51] J. E. Lagnese and J. L. Lions, *Modelling, analysis and control of thin plates*, Collection RMA, vol. 6, Masson, Paris, 1988.

[52] J.E. Lagnese, *Boundary stabilization of thin plates*, Studies in Applied Mathematics, vol. 10, SIAM, Philadelphia, 1989.

[53] I. Lasiecka, *Stabilization of wave and plate-like equations with nonlinear dissipation on the boundary*, J.Diff..Equs. **79** (1989), 340–381.

[54] I. Lasiecka and D. Tataru, *Uniform boundary stabilization of semilinear wave equations with nonlinear boundary damping*, J.Diff.Int.Equ. (19), –.

[55] I. Lasiecka and R. Triggiani, *A lifting theorem for the time regularity of solutions to abstract equations with unbounded operators and applications to hyperbolic equations*, Proc. Amer. Math. Soc. **104** (1988), 745–755.

[56] H. LeDret, *Modelling of the junction between two rods*, J. Math. Pures et Appl. **68** (1989), 365–397.

[57] ———, *Problèmes variationnels dans les multi-domains: Modélisation des jonctions et applications*, Masson, Paris, 1991.

[58] G. Leugering and E. J. P. G. Schmidt, *On the control of networks of vibrating strings and beams*, Proceedings of the 28th IEEE Conference on Decision and Control, vol. 3, IEEE, 1989, pp. 2287–2290.

[59] L.Halpern, *Absorbing boundary conditions for the discretization of the one-dimensional wave equation*, Mathematics of Computation **38** (1982), no. 158, 415–429.

[60] J.-L. Lions, *Contrôlabilité exacte, perturbations et stabilisation de systèmes distribués, Tome I: Contrôlabilté exacte*, Collection RMA, vol. 8, Masson, Paris, 1988.

[61] ———, *Exact controllability, stabilization and perturbations for distributed systems*, SIAM Review **30** (1988), 1–68.

[62] J.-L. Lions and E. Magenes, *Non-homogeneous boundary value problems and applications*, vol. I, Springer-Verlag, Berlin, 1972.

[63] A. E. H. Love, *A treatise on the mathematical theory of elasticity*, Cambridge Univ. Press, 1927.

[64] Ch. Lubich, *Extrapolation integrators for constrained multibody systems*, Impact of Computing in Science and Engineering (1991), 213–234.

[65] Ch. Lubich, U. Nowak, U. Pöhle, and Ch. Engstler, *MEXX-Numerical software for the integration of constrained multibody systems*, Preprint SC 92-12, 1992.

[66] G. Lumer, *Espaces ramifiés et diffusions sur des résaux topologiques*, Comptes Rendus Acad. Science Paris **201** (1980), 627–630.

[67] F. Ali Mehmeti, *Regular solutions of transmission and interaction problems for wave equations*, Math. Meth. in the Appl. Sciences (1989), 656–685.

[68] ———, *Existence and regularity of solutions of cauchy problems for inhomogenous wave equations with interaction*, Oper. Theory Adv. and Appl. **50** (1991), 23–34.

[69] ———, *Nonlinear interaction problems*, Nonlinear Analysis: Theory, Methods and Appl. **20** (1993), 281–295.

[70] F. Ali Mehmeti and S. Nicaise, *Nonlinear interaction problems*, Nonlinear Analysis: Theory, Methods and Appl., to appear.

[71] ———, *Some realizations of interaction problems*, Semigroup theory and evolution equations (New York) (E. Mitidieri Ph. Clément and B. de Pagter, eds.), Lect. Notes Pure and Appl. Math., vol. 135, Marcel Dekker, New York, 1991, pp. 15–27.

[72] Wei Ming, *Controllability of networks of strings and masses*, Ph.D. thesis, McGill University, December 1993.

[73] S. Nicaise, *General interface problems I*, Math. Methods Appl. Sci., to appear.

[74] ———, *General interface problems II*, Math. Methods Appl. Sci., to appear.

[75] ———, *Some results on spectral theory over networks applied to nerve impulse transmission*, Lecture Notes in Mathematics (Brezinski, ed.), Springer-Verlag, 1985, pp. 533–541.

[76] ———, *Diffusion sur les espaces ramifiés*, Ph.D. thesis, Thése du Doctorat, Mons, 1986.

[77] ———, *Spectre des ésaux topologique fines*, Bull. Sci. Math. **111** (1987), 401–413.

[78] ———, *About the Lamé system in a polygonal or a polyhedral domain and a coupled problem between the Lamé system and the plate equation. I: regularity of the solutions*, Ann. Scuola Normale Sup. Pisa **XIX** (1992), 327–361.

[79] ———, *Exact controllability of a pluridimensional coupled problem*, Revista Matematica **5** (1992), 91–135.

[80] ———, *About the Lamé system in a polygonal or a polyhedral domain and a coupled problem between the Lamé system and the plate equation. II: exact controllability*, Ann. Scuola Normale Sup. Pisa **XX** (1993), 163–191.

[81] ———, *Polygonal interface problems*, Peter Lang, Frankfurt am Main, 1993.

[82] O. A. Oleinik, A. S. Shamaev, and G. A. Yosifian, *Mathematical problems in elasticity and homogenization*, North Holland, Amsterdam, 1992.

[83] A. Pazy, *Semigroups of linear operators and applications to partial differential equations*, Springer Verlag, New York, Berlin, Heidelberg, Tokyo, 1983.

[84] J. P. Puel and E. Zuazua, *Controllability of a multi-dimensional system of Schrödinger equations: application to a system of plate and beam equations*, To appear, 1992.

[85] ———, *Exact controllability for a model of multidimensional flexible structure*, Proc. Royal Soc. Edinburgh (1993), 323–344.

[86] Bopeng Rao, *Stabilisation du modéle SCOLE par un contrôle a priori borné*, C.R. Acad. Sci. Paris **316** (1993), 1061–1066.

[87] J. Rauch and M. Taylor, *Exponential decay of solutions to hyperbolic equations in bounded domains*, Indiana Univ. Math. Jour. **24** (1974), 79–86.

[88] E. Reissner, *On bending of elastic plates*, Quart. Appl. Math. **V** (1947), 55– 68.

[89] ———, *On one dimensional large-displacement finite-strain beam theory*, Studies in Applied Mathematics **52** (1973), 87–95.

[90] R. C. Rogers, *Derivation of linear beam equations using nonlinear continuum mechanics*, VPI preprint 1989, 1989.

[91] A. Y. Le Roux, *Stability of numerical schemes solving quasi-linear wave equations*, Mathematics of Computation **36** (1981), no. 153, 93–105.

[92] D. L. Russell, *Controllability and stabilizability theory for linear partial differential equations: recent progress and open questions*, SIAM Review **20** (1978), 639–739.

[93] E. J. P. G. Schmidt, *On the modelling and exact controllability of networks of vibrating strings*, SIAM J. of Control and Optimization **30** (1992), 229–245.

[94] ———, *On the modelling and exact controllability of networks of vibrating strings*, SIAM J. Control Optim. **30** (1992), 229–245.

[95] E. J. P. G. Schmidt and Wei Ming, *On the modelling and analysis of networks of vibrating strings and masses*, Tech. Report 91-13, Department of Mathematics and Statistics, McGill University, 1991.

[96] E.J.P.G. Schmidt and Wei Ming, *On the modelling, analysis and control of vibrating networks of strings and masses*, submitted manuscript, 1993.

[97] J. Schmidt, *Entwurf von Reglern zur aktiven Schwingungsdämpfung an flexiblen mechanischen Strukturen*, Ph.D. thesis, Technische Hochschule Darmstadt, Fachbereich Mechanik, Darmstadt Germany, 1987.

[98] A. Schneider, *On spectral theory for the linear sel fadjoint equation* $fy = \lambda gy$, Ordinary and Partial Differential equations, Lecture Notes in Mathematics, vol. 846, Springer-Verlag, 1981, pp. 306–332.

[99] A. A. Shabana, *Dynamics of multibody systems*, John Wiley, New York, 1989.

[100] J. C. Simo, *A finite strain beam formulation, Part 1: the three-dimensional dynamic problem*, Comp. Meth. in Appl. Mech. and Engng. **28** (1986), 79–116.

[101] J. C. Simo, J.E. Marsden, and P. S. Krishnaprasad, *The hamiltonian structure of nonlinear elasticity: the material and convective representation of solids, rods and plates*, Arch. Rat. Mech. and Anal. **104** (1988), 125–185.

[102] J. Simon, *Compact sets in* $L^p(0,T;B)$, Ann. Mat. Pura et Applicate **CXLVI** (1987), 65–96.

[103] A. Simpson and B. Tabarrok, *On Korn's eigenvalue procedure and related methods of frequency analysis*, Quart. Journal Mech. and Applied Math. **21** (1968), 1–39.

[104] I. S. Sokolnikoff, *Mathematical theory of elasticity*, R. E. Krieger, Malabar,FL, 1983.

[105] J. von Below, *A characteristic equation associated to an eigenvalue problem on* c^2*-networks*, Lin. Alg. and its Appl. **71** (1985), 309–325.

[106] ———, *Classical solvability of linear equations on networks*, J. Diff. Eqns. **72** (1988), 316– 337.

[107] ———, *Parabolic network equations*, Ph.D. thesis, Habilitationsschrift, Universität Tübingen, 1993.

[108] G. Wempner, *Mechanics of solids with applications to thin bodies*, Sijthoff and Noordhoff, Rockville, Maryland, 1981.

[109] J. Wittenburg, *Nonlinear equations of motion for arbitrary systems of interconnected rigid bodies*, Dynamics of Multibody Systems (K. Magnus, ed.), Springer–Verlag, 1978, pp. 357–369.

[110] W. H. Wittrick and F. W. Williams, *A general algortihm for computing natural frequencies of elastic structures*, Quart. Journal Mech. and Applied Math. **XXIV/3** (1971), 263–284.

[111] ———, *Natural frequencies of repetitive structures*, Quart. Journal Mech. and Applied Math. **XXIV/3** (1971), 285–308.

Index

Systems & Control: Foundations & Applications

Series Editor
Christopher I. Byrnes
School of Engineering and Applied Science
Washington University
Campus P.O. 1040
One Brookings Drive
St. Louis, MO 63130-4899
U.S.A.

Systems & Control: Foundations & Applications publishes research monographs and advanced graduate texts dealing with areas of current research in all areas of systems and control theory and its applications to a wide variety of scientific disciplines.

We encourage the preparation of manuscripts in TeX, preferably in Plain or AMS TeX—LaTeX is also acceptable—for delivery as camera-ready hard copy which leads to rapid publication, or on a diskette that can interface with laser printers or typesetters.

Proposals should be sent directly to the editor or to: Birkhäuser Boston, 675 Massachusetts Avenue, Cambridge, MA 02139, U.S.A.

Estimation Techniques for Distributed Parameter Systems
H.T. Banks and K. Kunisch

Set-Valued Analysis
Jean-Pierre Aubin and Hélène Frankowska

Weak Convergence Methods and Singularly Perturbed Stochastic Control and Filtering Problems
Harold J. Kushner

Methods of Algebraic Geometry in Control Theory: Part I
Scalar Linear Systems and Affine Algebraic Geometry
Peter Falb

H - Optimal Control and Related Minimax Design Problems
Tamer Başar and Pierre Bernhard

Identification and Stochastic Adaptive Control
Han-Fu Chen and Lei Guo

Viability Theory
Jean-Pierre Aubin

Representation and Control of Infinite Dimensional Systems, Vol. I
A. Bensoussan, G. Da Prato, M.C. Delfour and S.K. Mitter

Disease Dynamics
Alexander Asachenkov, Guri Marchuk, Ronald Mohler, Serge Zuev

Theory of Chattering Control with Applications to Astronautics, Robotics, Economics, and Engineering
Michail I. Zelikin and Vladimir F. Borisov

Modeling, Analysis and Control of Dynamic Elastic Multi-Link Structures
J. E. Lagnese, Günter Leugering, E. J. P. G. Schmidt

The manufacturer's authorised representative in the EU is Springer Nature Customer Service Centre GmbH, Europaplatz 3, 69115 Heidelberg, Germany. If you have any concerns regarding our products, please contact ProductSafety@springernature.com

Printed and bound by CPI Group (UK) Ltd, Croydon, CR0 4YY
29/04/2026
02099472-0009